Earth and Environmental Sciences Library

Earth and Environmental Sciences Library (EESL) is a multidisciplinary book series focusing on innovative approaches and solid reviews to strengthen the role of the Earth and Environmental Sciences communities, while also providing sound guidance for stakeholders, decision-makers, policymakers, international organizations, and NGOs.

Topics of interest include oceanography, the marine environment, atmospheric sciences, hydrology and soil sciences, geophysics and geology, agriculture, environmental pollution, remote sensing, climate change, water resources, and natural resources management. In pursuit of these topics, the Earth Sciences and Environmental Sciences communities are invited to share their knowledge and expertise in the form of edited books, monographs, and conference proceedings.

More information about this series at https://link.springer.com/bookseries/16637

Neloy Khare
Editor

Assessing the Antarctic Environment from a Climate Change Perspective

An Integrated Approach

Springer

Editor
Neloy Khare
Government of India
Ministry of Earth Sciences
New Delhi, India

ISSN 2730-6674 ISSN 2730-6682 (electronic)
Earth and Environmental Sciences Library
ISBN 978-3-030-87077-5 ISBN 978-3-030-87078-2 (eBook)
https://doi.org/10.1007/978-3-030-87078-2

This Springer imprint is published by the registered company Springer Nature Switzerland AG
The registered company address is: Gewerbestrasse 11, 6330 Cham, Switzerland

Foreword

Antarctica is the last great untouched wilderness. Antarctica's frozen continent is an incredible continent of stunning and alien beauty with a rich history of adventure, exploration, and discovery. It is known for diversified uniqueness and is a key to understanding how anthropogenic activities adversely impact the world's climate and carry an associated impact on it. Indubitably, Antarctica is also essential for science because of its profound effect on the Earth's climate and ocean systems which has also revealed much about the impact of human activity on the natural world.

The looming danger of global warming on Antarctica is not confined but spreading fast across the continent, long thought to be untouched by warming. But now, the glaciers and ice shelves in this frigid region are showing signs of melting. Such unprecedented development portends dramatic rises in sea levels in this century and beyond. The collapse of the Larsen C ice shelf warns us against the Antarctic's fragile environment. The subtle climatic changes may primarily pose dire global consequences because collapsing ice shelves prompt the glaciers behind them to retreat more quickly, causing further sea-level rise, thus increasing peril, especially for island countries.

Some of the changes Antarctica is facing are already irreversible, and the situation may turn out to be devastating if the issues are not timely managed through appropriate strategies. To tackle the potential impact of climate, it becomes imperative to prepare futuristic climate change trends to prepare humankind in a larger perspective. It necessitates an in-depth assessment of the Antarctic Environment through an integrated approach.

Over four decades, India has been actively pursuing Antarctic research commensurate with its scientific strength and global visibility. A particular focus has been paid towards climate change. The present book, *Assessing the Antarctic Environment from Climate Change perspective: An Integrated Approach,* provides a comprehensive overview of Antarctic Environmental changes in space and time and assesses climate change scenarios in the present context global warming. It is aptly brought out with eighteen dedicated chapters, where each chapter has its specific significance.

The book begins with *Dhanasree Jayaram's* detailed account on current geopolitical issues arising out of ongoing environmental shift due to unfavourable activities elsewhere, causing damage to the icy continent's pristine nature requiring a firm committed and transparent Antarctic Governance. *Dastidar and Khare* used the data obtained from the web of science and analysed various trends and patterns from the scientific literature on Antarctic Climate Change science. Such analyses significantly impact the direction of the present research to help understand climate change and variability. Subsequently, a detailed assessment is made by *Choudhary and Khare* on how climate change over the Antarctic and the Southern Ocean impacts the global climate system. Gleaning clues drawn from the marine sedimentary records. *Singh et al.* illustriously elaborated Cenozoic Evolution of Antarctic Ice sheet, Circum Antarctic Circulation and Antarctic climate.

To understand the Antarctic region's climate scenarios, a firm understanding of the past climatic evolution is exciting and a key factor. *Baba et al.* studied the variations in the cosmogenic radionuclides. They reconstructed the climatic conditions and glacial history over the DronningMaudland region. Whereas' *Shrivastava et al.* utilised yet another proxy (Terrestrial Diamicts and Lacustrine Sediments) to illuminate Late Quaternary Climate Change in Schirmacher Region, East Antarctica. It is well corroborated with *Govil and Mazumder's* focused review on lacustrine signatures of the palaeoclimatic conditions. Glacial-interglacial paleoenvironmental records have been retrieved from lake sediments of Schirmacher Oasis, East Antarctica, by *Warier et al.*

On the contrary, Nutrient cycling and productivity in Antarctic lakes have been detailed by *Choudhary et al.* In contrast, the Chemical and isotopic characterisation of lakes in the Larsemann Hills, East Antarctica, has been addressed by *Reshmi et al.*

Gwal et al. studied the effect of Ionospheric scintillation and observed the loss of lock-in GPS signals. Further the effect of Ionospheric Scintillation on the positional error and loss of lock of GPS Signal have also been investied in details by *Gwal et al.* Towards understanding biological response to ongoing climate changes over the Antarctic region, *Pande* and *Kuppusamy* highlighted that the rapid changes in the physical environment of the Antarctic and the Southern Ocean affect marine life at all trophic levels, from the primary prey species (zooplankton including Antarctic Krill) to mesopredators (like squids) to top predators such as marine mammals and seabirds. They also postulated that the Seabird populations across the globe are threatened with human-induced changes. Long-term monitoring programs have highlighted exciting trends, including foreseen threats and the declining status of seabirds worldwide.

Similarly, *Nayaka and Rai* have examined the response of Antarctic lichen to climate change. Their evaluation was primarily based on the evidence from natural gradients and temperature enchantment. Simultaneously, *Singh et al.* found a higher Pigment Synthesis rate in Antarctic Plants as an adaptive survival strategy under U.V. radiation.

Catherine et al. have provided an overview of Antarctica's Geoscience studies. In contrast, the Antarctic region's seismogenesis and seismic potential have been assessed for the future comprehensive study by *Mishra.* On the contrary, *Sunil et al.* demonstrated the Antarctic plate's new kinematics using GPS and GRACE data.

Understanding is a continuous process, so the Scientific advancements in Antarctic Science may pose a more significant database to attend to challenging scientific questions. It requires enhanced monitoring, long-ranged time series climate data, efficient models and strong international collaborations to help understand Antarctic climates' evolution.

This book aptly consolidates recent scientific findings and insights related to the ongoing climate change in and around the Antarctic region through an integrated approach. This book will act as a ready reference to all avid researchers and students.

This book will be a good source of information about the Antarctic climate and act as a reference for students, professionals and researchers.

April 2021

A. E. Muthunayagam
Former Secretary, Department
of Ocean Development
Government of India
Thiruvananthapuram, India

Preface

Over 100 million years ago, Antarctica was part of the supercontinent Gondwana. Gondwana gradually broke apart with passing time, and Antarctica in its present situation was formed around 25 million years ago, owing to the opening of the Drake Passage between it and South America. The vast frozen landmass at the southernmost part of the planet is more than just spectacular icing worldwide. The Antarctic ice deflects some of the sun's rays away from the Earth, keeping temperatures liveable. It could be vital for our survival too.

Historically, the first confirmed sighting of mainland Antarctica was recorded on January 27, 1820. Antarctica's discovery is attributed to the Russian expedition led by Fabian Gottlieb von Bellingshausen and Mikhail Lazarev. They discovered an ice shelf at Princess Martha Coast, subsequently known as the Fimbul Ice Shelf. This continent carries many superlatives like it holds most of the world's freshwater but remains a desert. Antarctica used to be as warm as any other tropical place. The Antarctic, which has active volcanoes and several subglacial lakes, has no time zone.

Owing to ongoing global warming, the Antarctic Peninsula has become one of Earth's most rapidly warming areas. The high ice sheet and the polar location make Antarctica a powerful heat sink that strongly affects the climate of the whole Earth. The Antarctic ice sheet contains sufficient ice to raise worldwide sea level by more than 60 meters if melted completely. Through Antarctica, we can understand the Earth's past, present, and future. It also exhibits a platform to understand and value our planet. The ice sheets over the Antarctic region also holds over half-million-years old climatic change signatures. The major threats to this pristine region are climate change which is the greatest long-term threat to the area, increased fishing pressure and illegal fishing, marine pollution, persistent organic pollutants (POPs), and invasive species. It is now a fact that Antarctica and its surrounding waters are under pressure from a variety of forces that are already transforming the area. The most immediate threats are regional warming, ocean acidification, and sea ice loss, all linked to global levels of carbon dioxide. Environmental impacts in Antarctica occur at a range of scales. Global warming, ozone depletion, and global contamination have planet-wide consequences. These affect Antarctica at the most significant scale. Fishing and hunting have more localized impacts but still, have the potential to cause region-wide effects. Indubitably, if all the ice covering Antarctica, Greenland, and mountain glaciers worldwide were to melt, the sea level would rise about 70 meters. The ocean would cover all the coastal cities, and the land area would shrink significantly. However, all the ice is not going to melt. Altogether, Greenland and Antarctica have lost 6.4 trillion tons of ice since the 1990s. The resulting meltwater boosted global sea levels by 0.7 inches. Therefore, it is essential but vital to understand the environmental conditions vis-a-vis the impact of global climate change on this icy continent.

The present book *Assessing the Antarctic Environment from Climate Change perspective: An Integrated Approach* attempts to address various facets of the climate change being witnessed over the Antarctic region. The book begins with the Geopolitics, Environmental Change and Antarctic Governance ably highlighted by *Dhanasree Jayaram*. Although the Antarctic Treaty (AT) is considered a successful example of science diplomacy, as countries have set aside their territorial claims and the continent is a nuclear-free zone by shifting focus to scientific cooperation, its future remains uncertain with these developments. Science diplomacy always goes hand in hand with geopolitics. The AT that reflects Cold War geopolitics needs to be modified to represent present-day geopolitical realities for it to be enduring. A transformative approach to Antarctic governance (including the Southern Ocean), especially in terms of its resources, needs to be adopted. This chapter is followed by detailed data analyses obtained from the web of science dealing with the climate change-related research over Antarctica by world's researchers by *Dastidar and Khare*. Their efforts observe peculiar trends and patterns in the climate change research suggesting priority for climate change research since the 1970s. *Choudhary and Khare* have addressed the climate change over the Antarctic and the Southern Ocean and its impact and bearing on the global climate system. They advocated for a thorough understanding and knowledge of the causes and impacts of climate change and the duration and rates of change, requiring the integration of observational and modelling knowledge from all Earth system-based scientific disciplines.

In a significant manner, *Singh et al.* put forth the evolution of the Antarctic Ice sheet, Circum Antarctic Circulation, and Antarctic climate during Cenozoic by gleaning clues from marine sedimentary records. This chapter ably covers the geological evidence for the origin and evolution of the Antarctic Ice sheet, which primarily includes marine sediments deposited from southern to lower latitudes and summarizes crucial research regarding the origin and development of the Antarctic Ice Sheet (AIS) and offers some future directions for research.

While *Baba et al.* utilized cosmogenic radionuclides to reconstruct the glacial history of the Dronning Maudland region of East Antarctica, this chapter deliberates on the comprehensive outline of DML, basics of cosmogenic radionuclide and its application, and major glacial events from DML. Further, meltwater pulse due to deglaciation of EAIS and evidence related to the marine isotope stages are discussed to understand the impact of deglaciation on the global ocean. This region shows sparse or no evidence of ice thickening during the last glacial maximum (LGM). Field observations and ice core models show that the ice sheet's interior parts, the ice dome, were possibly 100 m lower during LGM than the present. On the other hand, *Shrivastava et al.* discussed the Late Quaternary climate change in the Schirmacher region, based on the terrestrial diamicts and lacustrine sediments. The multi-proxy data, generated from the moraines and sediment cores of a variety of lakes from the Schirmacher region, cDML, has provided better insight into the Late Pleistocene to Holocene paleoclimatic evolution of the region during the Late Quaternary. This chapter highlights the glacial signatures, which are very well preserved in all kinds of sediments of this region. The clay minerals indicate a gradual shift in the weathering regime and that in climate from strongly glacial to fluvio-glacial during Late Quaternary. The results of surface textures of quartz grains have been discussed depth wise and in the same samples. In general, it shows dominant glacial and glaciofluvial actions. The OSL chronology on moraines has provided information on different events of deglaciation in Schirmacher region, East Antarctica. An overview of the paleoclimate changes archived in the lacustrine sediments has been provided by *Govil and Mazumder*. Major palaeoclimatic/palaeoenvironmental studies from this region reveal the presence of several episodes of alternating warm–cold events during Holocene and even beyond, based on a large number of proxies, mainly biological, geochemical, and sedimentological parameters. On the basis of these data, the morphological evolution of the lakes and palaeoclimatic reconstruction of the Schirmacher Oasis have been deciphered. The lacustrine signatures have further been explored to synthesize a glacial–interglacial paleoenvironmental conditions of Schirmacher Oasis, East Antarctica. Paleoclimatic studies using lake sediments drew scientific attention due to its efficiency to record long, high-resolution climate records. Recent studies have employed multiple proxies like environmental magnetism, isotope geochemistry, petrography, sedimentology, and geochronology on lake sediments of Schirmacher Oasis to decipher the past climate and the prevailing environmental conditions. The existing studies poorly record climatic events such as the Mid-Holocene Warm period, Hypsithermal and neoglacial cooling. Despite better chronometric control in these studies, coarse temporal resolution and sparsely documented finer scale climatic variations place the need for future high-resolution works in the East Antarctic region. To fill up this gap, *Warier et al.* have synthesized glacial–interglacial paleoenvironmental records from lake sediments of Schirmacher Oasis, East Antarctica. The nutrient cycling and productivity in Antarctic lakes have been addressed by *Choudhary et al.,* who used sedimentary organic matter from the Antarctic lakes as the source of various proxies to study productivity changes. A total of three sediment cores have been analyzed for TOC, total nitrogen (TN), total phosphorus (TP), biogenic silica (BSi), and their ratios were computed to understand the nutrient cycling and productivity in Antarctic lakes. On the contrary, *Reshmi et al.* studied chemical and isotopic characterization of lakes in the Larsemann Hills, East Antarctica. The ionic and isotopic ratios of some lakes kinetic controlled ice-water fractionation and evaporation processes are found to affect the isotopic evolution of lake water.

In the field of Antarctic ionospheric research, the effect of ionosphere scintillations on the loss of lock in GPS signals has been detailed by *Gwal et al.* during the five most disturbed days of the December 2006. During all the disturbed days weak, moderate or intense geomagnetic storms were observed. The amplitude and phase scintillation become frequently, the visible PRNSs does not remain stable for longer periods. The loss of lock occurs frequently whenever the GPS signals scintillate. Similarly, *Gwal et al.* also attempted to understand the effect of Ionospheric Scintillation on the Positional Error and Loss of Lock of GPS Signal in Antarctic Region. They reported that during all the five geomagnetically disturbed days, the positional error increases significantly.

In Antarctic Biology, *Pande and Kuppusamy* provided an insight into the impacts of ongoing climate changes on Antarctic birds. They have assessed the population status, distribution, and genetic structure of key seabird species (Adelie penguin, snow petrel, south polar skua, Wilson's storm petrel) breeding around Indian research stations, whereas *Nayaka and Rai* have detailed the response of lichens to temperature rise. The extreme climatic conditions such as temperature, precipitation, and smaller ice-free regions allow only cryptogams like bryophytes and lichens to grow dominantly. Lichens are well-known biomonitors and bioindicators of climate change, environmental pollution, and anthropogenic perturbations, they have explored their potential. The study points out that climate warming will cause the extinction of sensitive species. Simultaneously, some will increase their geographical extension due to the increased water availability and nutrients in changed ecosystems. On the contrary, *Singh et al.* assessed the Strategy of Antarctic plants for survival under UV radiations. Antarctic cryptogams are growing in the photosynthetically active radiation (PAR), and ultraviolet radiations (UV-R), closely associated with the synthesis of photosynthetic and photo protective pigments. Antarctic cryptogams cope with high UV radiation stress by synthesizing UV-absorbing compounds; UV-B absorbs pigments and other compounds; the pigment synthesis provides protection to cryptogamic flora.

An exhaustive collation of significant geoscientific studies in Antarctica has been made by *Catherine et al.*, with special mention of CSIR-NGRI which has been participating in these expeditions and has established seismological and GPS observatories for monitoring seismicity and to understand the tectonics of the Antarctica plate and has carried out some geological and geophysical studies too. They summarize some of the important contributions of CSIR-NGRI. While seismogenesis and seismic potential of Antarctic region has not yet been well understood because of unique and complex tectonic settings of the region, besides several causative factors associated with natural and anthropogenic, which are still enigmatic in sense to unravel the fact what and how the genesis of earthquakes is related to the glacial dynamics. *Mishra* have ably provided a comprehensive overview on these aspects to suggest future course of action to undertake detailed study on the glacial mass change induced earthquakes (GMCIE) for the Antarctic regions. The book ends with a dedicated chapter dealing with the Gravity Recovery and Climate Experiment (GRACE) satellites data by *Sunil et al.* which indicates global sea level rise by 0.35 millimetres per year. Antarctic GPS and GRACE data were reanalyzed and discussed to understand in this chapter to understand and estimate the kinematics of Antarctic plate.

The present book shall act as a ready reference to all researchers/academicians who have curiosity to know about various dimensions of the Antarctic climate. The latest insight and data shall be of immense use to climate scientists and policy makers.

New Delhi, India
April 2021

Neloy Khare

Acknowledgements

Generous support and help received from all contributing authors are sincerely acknowledged. Without their valuable inputs on various facets of Climate change over the Antarctic region, the book would not have been possible. Various learned experts who have reviewed different Chapters are sincerely acknowledged for their timely, essential, and critical reviews.

Professor Goverdhan Mehta, FRS, has always been the source of inspiration for my scientific pursuit, and I owe a lot for his kind encouragement, moral support, and guidance. I express my sincere thanks to Dr. M. N. Rajeevan, Secretary to the Government of India, at the Ministry of Earth Sciences, New Delhi (India), for his various valuable inputs and suggestions. Dr. Prem Chand Pandey, founder Director, National Centre for Polar and Ocean Research, Goa (India) and Dr. K. J. Ramesh, former Director-General, India Meteorological Department (IMD), New Delhi (India), have always been supportive as a true well-wisher.

Professor Anil Kumar Gupta, Indian Institute of Technology, Kharagpur (India) and Dr. Rajiv Nigam, former Adviser at the National Institute of Oceanography, Goa (India), are deeply acknowledged for providing many valuable suggestions to this book. This book has received significant support from Akshat Khare and Ashmit Khare, who helped me during the book preparation. Dr. Rajni Khare has unconditionally supported enormously during various stages of this book. Dr. Shabnam Choudhary, Shri Hemant Kumar, and Shri Hari Dass Sharma from the Ministry of Earth Sciences, New Delhi (India), have helped immensely formatting the text and figures of this book and bring the book to its present form. Publishers (Springer Nature) have done a commendable job and are sincerely acknowledged.

New Delhi, India
April 2021

Neloy Khare

Dedication

Late Professor Surendra Kumar
(14.8.1942 – 20.4.2021)

Professor Surendra Kumar was born in Lucknow. His schooling and college education are from Jubilee Inter College, Lucknow, and thereafter shifted to Lucknow University for graduation, postgraduation, and Ph. D. He worked on Almora Crystallines, Lesser Himalaya. He joined the Department of Geology as a faculty in September 1967.

He was a noted sedimentologist, however, also contributed significantly on Precambrian stromatolites and microbiota. He has worked on Lesser Himalaya, Tethys Himalaya, Vindhyan Supergroup, Marwar Supergroup, and Ganga River basin. His contributions on Vindhyan stromatolites are noteworthy in Indian geology as established and reported many new forms including *Maiharia maiharensis*. Similarly, reassessment and refinement of the Marwar Supergroup stratigraphy, based on organo-sedimentary structures, is again quite significant. Closely associated people are well aware of his systematic approach, planning, and skill of minute observations during the field work. Either as a leader or member of expedition, he visited the high-altitude, hostile, and glacial terrain of Malla Johar area, Tethys Himalaya for 3–4 timed and scaled more than 20,000 ft for sample and data collections. He was also concerned with the recent problems of river pollution and environmental hazards, therefore, concentrated on various aspects of the Ganga basin, helpful in planning and management for both State and Centre governments.

He is also the recipient of prestigious Alaxender von Humboldt Fellowship at Heidelberg. He has also been awarded the National Mineral Award by the Government of India in the year 1999. Served as an active member of Palaeontological Society of India in various capacities including the responsibility of the Editor of the Journal for many years. He has more than 100 research articles of national and international reputes in his credit besides three books/field guides on Lameta, Vindhyan Basin, and Marwar Supergroup. Superannuated in the year 2005, but committed to research activity for more than a decade. His sudden demise on April 20, 2021, has created a void in the field of Geology As his student, I pay my sincere tribute to his noble soul and humbly dedicate this book to him.

Neloy Khare

Contents

Geopolitics, Environmental Change and Antarctic Governance: A Region in Need of a Transformative Approach to Science Diplomacy 1
Dhanasree Jayaram

Impact of Antarctic Science on Climate Change Research: Global Research Landscape 19
Prabir G. Dastidar and Neloy Khare

Climate Change Over the Antarctic and the Southern Ocean and Its Impact and Bearing on the Global Climate System 37
Shabnam Choudhary and Neloy Khare

Cenozoic Evolution of Antarctic Ice Sheet, Circum Antarctic Circulation and Antarctic Climate: Evidence from Marine Sedimentary Records 47
Ashutosh K. Singh, Devesh K. Sinha, Vikram Pratap Singh, Kirtiranjan Mallick, Ankush Shrivastava, and Tushar Kaushik

Dronning Maud Land (Antarctica) and Reconstruction of Its Glacial History with Cosmogenic Radionuclides 73
Waseem Ahmad Baba, Pankaj Kumar, Jitendra Kumar Pattanaik, and Neloy Khare

Late Quaternary Climate Change in Schirmacher Region, East Antarctica: As Revealed from Terrestrial Diamicts and Lacustrine Sediments 97
Prakash Kumar Shrivastava, Rajesh Asthana, and Sandip Roy

A Review of the paleoclimatic Studies from Lake Sediments of Schirmacher Oasis, East Antarctica 107
Pawan Govil and Abhijit Mazumder

A Synthesis of Glacial-Interglacial Paleoenvironmental Records from Lake Sediments of Schirmacher Oasis, East Antarctica 127
Anish Kumar Warrier, B. S. Mahesh, Joju George Sebastian, A. S. Yamuna Sali, and Rahul Mohan

Nutrient Cycling and Productivity in Antarctic Lakes 141
Shabnam Choudhary, G. N. Nayak, and Neloy Khare

Chemical and Isotopic Characterization of Lakes in the Larsemann Hills, East Antarctica 153
T. R. Resmi, Girish Gopinath, P. S. Sunil, M. Praveenbabu, P. Arjun, and Rahul Rawat

Effect of Ionosphere Scintillations on the Loss of Lock-In GPS Signals at Antarctica Region 167
A. K. Gwal, Suryanshu Choudhary, and Ritesh Yadav

Study of Positional Error on Ionospheric Scintillation Over Antarctic Region and Loss due to Locking of GPS signal 189
A. K. Gwal, Suryanshu Choudhary, and Ritesh Yadav

Climate Change and Seabirds: Insights from Ecological Monitoring of Snow Petrels in the Indian Antarctic Program 207
Anant Pande and Kuppusamy Sivakumar

Antarctic Lichen Response to Climate Change: Evidence from Natural Gradients and Temperature Enchantment Experiments 235
Sanjeeva Nayaka and Himanshu Rai

Higher Rate of Pigment Synthesis in Antarctic Plants: A Strategy of Survival Under UV Radiations 255
Jaswant Singh, Rudra P. Singh, and Rajni Khare

Geoscience Studies in Antarctica by CSIR-National Geophysical Research Institute, Hyderabad 277
Joshi K. Catherine, Ch Mohana Lakshmi, Amit Kumar, Saroj K. Mondal, V Rajesheswar Rao, and Vineet K. Gahalaut

Seismo-Geophysical Studies in the Antarctic Region: Geodynamical Implications 287
O. P. Mishra

Revealing the Contemporary Kinematics of Antarctic Plate Using GPS and GRACE Data 343
P. S. Sunil, Ajish P. Saji, K. Vijay Kumar, M. Ponraj, S. Amirtharaj, and Ajay Dhar

Editors and Contributors

About the Editor

Dr. Neloy Khare presently Adviser/Scientist G to the Government of India at MoES. has a very distinctive acumen not only of administration but also of quality science and research in his areas of expertise covering a large spectrum of geographically distinct locations like Antarctic, Arctic, Southern Ocean, Bay of Bengal, Arabian Sea, Indian Ocean, etc. He has almost 30 years of experience in the field of paleoclimate research using paleobiology Paleontology)/teaching/science management/administration/coordination for scientific programmes (including Indian Polar Programme), etc. Having completed his doctorate (Ph.D.) on tropical marine region and Doctor of Science (D.Sc.) on Southern High latitude marine regions towards environmental/climatic implications using various proxies including foraminifera (micro-fossil), have made significant contributions in the field of palaeoclimatology of Southern high latitude regions (Antarctic and Southern Ocean) using Micropaleontology as a tool. These studies coupled with his palaeoclimatic reconstructions from tropical regions helped understand causal linkages and teleconnections between the processes taking place in Southern high latitudes with that of climate variability occurring in tropical regions. He has been conferred Honorary Professor and Adjunct Professor by many Indian universities. He has a very impressive list of publications to his credit (125 research articles in National and International Scientific journals; 3 special Issues of National Scientific Journals as Guest Editor; Edited Special Issue of Polar Science (Elsevier) as its Managing Editor, Quaternary International and Journal of Asian Earth Sciences as Guest Editor Authored/edited many books, 130 abstracts have been contributed to various seminars; 23 popular Science Articles; 5 technicalreports). Government of India and many professional bodies have bestowed him with many prestigious awards for his humble scientific contributions to Past climate changes/Oceanography/Polar Science and Southern Oceanography The most coveted award is Rajiv Gandhi National Award—2013 conferred by Honourable President of India. Other include ISCA Young Scientist Award, Boyscast Fellowship, CIES French Fellowship, Krishnan Gold Medal, Best Scientist Award, Eminent Scientist

Award, ISCA Platinum Jubilee Lecture, IGU Fellowship, besides many. He has made tremendous efforts to popularize ocean science and polar science across the country by the way of delivering many invited lectures, radio talks, and publishing popular science articles.

He has sailed in the Arctic Ocean as a part of "Science PUB" in 2008 during the International Polar Year campaign for scientific and became the first Indian to sail in the Arctic Ocean.

Contributors

Waseem Ahmad Baba Department of Geology, Central University of Punjab, Punjab, India

S. Amirtharaj Indian Institute of Geomagnetism, Mumbai, India

P. Arjun Centre for Water Resources Development and Management, Kozhikode, Kerala, India

Rajesh Asthana Geological Survey of India, Kolkata, India

Joshi K. Catherine CSIR-National Geophysical Research Institute, Uppal Raod, Hyderabad, India

Shabnam Choudhary National Centre for Polar and Ocean Research, Vasco da Gama, Goa-403802, India

Suryanshu Choudhary Department of Physics, Rabindranath Tagore University, Raisen, India;
Department of Physics, Dr A P J Abdul Kalam University, Indore, India

Prabir G. Dastidar Ministry of Earth Sciences, Prithvi BhawanNew Delhi, India

Ajay Dhar Indian Institute of Geomagnetism, Mumbai, India

Vineet K. Gahalaut CSIR-National Geophysical Research Institute, Uppal Raod, Hyderabad, India

Girish Gopinath Centre for Water Resources Development and Management, Kozhikode, Kerala, India;
Kerala University of Fisheries and Ocean Studies, Kochi, Kerala, India

Pawan Govil Birbal Sahni Institute of Palaeosciences, Lucknow, UP, India

A. K. Gwal Department of Physics, Rabindranath Tagore University, Raisen, India;
Department of Physics, Dr A P J Abdul Kalam University, Indore, India

Dhanasree Jayaram Department of Geopolitics and International Relations, Old TAPMI Building, Manipal Academy of Higher Education, Manipal, Karnataka, India

Tushar Kaushik Biodiversity and Palaeobiology, Agharkar Research Institute, Pune, India

Neloy Khare Ministry of Earth Sciences, Prithvi Bhawan, New Delhi, India

Rajni Khare Delhi Public School, Sector Gama-2, Greater Noida, Gautam Buddh Nagar, Uttar Pradesh, India

Jitendra Kumar Pattanaik Department of Geology, Central University of Punjab, Punjab, India

Prakash Kumar Shrivastava Geological Survey of India, Kolkata, India

Amit Kumar CSIR-National Geophysical Research Institute, Uppal Raod, Hyderabad, India

K. Vijay Kumar Indian Institute of Geomagnetism, Mumbai, India

Pankaj Kumar AMS Division & Geochronology Group, Inter-University Accelerator Centre, New Delhi, India

Sivakumar Kuppusamy Department of Endangered Species Management, Wildlife Institute of India, Dehradun, Uttarakhand, India

B. S. Mahesh National Centre for Polar and Ocean Research, Ministry of Earth Sciences, Vasco-da-Gama, Goa, India

Kirtiranjan Mallick Department of Geology, Utkal University, Bhubaneswar, Odisha, India

Abhijit Mazumder Birbal Sahni Institute of Palaeosciences, Lucknow, UP, India

O. P. Mishra National Centre for Seismology, Ministry of Earth Sciences, New Delhi, India

Rahul Mohan National Centre for Polar and Ocean Research, Ministry of Earth Sciences, Vasco-da-Gama, Goa, India

Ch Mohana Lakshmi CSIR-National Geophysical Research Institute, Uppal Raod, Hyderabad, India

Saroj K. Mondal CSIR-National Geophysical Research Institute, Uppal Raod, Hyderabad, India

G. N. Nayak Marine Sciences, School of Earth, Ocean and Atmospheric Sciences, Goa University, Taleigao, Goa, India

Sanjeeva Nayaka Lichenology Laboratory, Plant Diversity, Systematics and Herbarium Division, CSIR-National Botanical Research Institute, Lucknow, Uttar Pradesh, India

Anant Pande Department of Endangered Species Management, Wildlife Institute of India, Dehradun, Uttarakhand, India

M. Ponraj Indian Institute of Geomagnetism, Mumbai, India

M. Praveenbabu Centre for Water Resources Development and Management, Kozhikode, Kerala, India

Vikram Pratap Singh Department of Geology, Indira Gandhi National Tribal University, Amarkantak, India

Himanshu Rai Lichenology Laboratory, Plant Diversity, Systematics and Herbarium Division, CSIR-National Botanical Research Institute, Lucknow, Uttar Pradesh, India

V Rajesheswar Rao CSIR-National Geophysical Research Institute, Uppal Raod, Hyderabad, India

Rahul Rawat Indian Institute of Geomagnetism, Mumbai, Maharashtra, India

T. R. Resmi Centre for Water Resources Development and Management, Kozhikode, Kerala, India

Sandip Roy Geological Survey of India, Kolkata, India

Ajish P. Saji Indian Institute of Geomagnetism, Mumbai, India

A. S. Yamuna Sali Centre for Climate Studies / Department of Civil Engineering (Manipal Institute of Technology), Manipal Academy of Higher Education, Manipal, Karnataka, India

Joju George Sebastian Centre for Climate Studies / Department of Civil Engineering (Manipal Institute of Technology), Manipal Academy of Higher Education, Manipal, Karnataka, India

Ankush Shrivastava Department of Geology, Mohanlal Sukhadia University, Udaipur, Rajasthan, India

Jaswant Singh Department of Environmental Sciences, Dr R M. L. Avadh University, Ayodhya, Uttar Pradesh, India

Ashutosh K. Singh Department of Geology, Centre of Advanced Studies, University of Delhi, Delhi, India

Rudra P. Singh Department of Environmental Sciences, Dr R M. L. Avadh University, Ayodhya, Uttar Pradesh, India

Devesh K. Sinha Department of Geology, Centre of Advanced Studies, University of Delhi, Delhi, India

P. S. Sunil Department of Marine Geology and Geophysics, CUSAT-NCPOR Centre for Polar Sciences, School of Marine Sciences, Cochin University of Science and Technology, Kochi, Kerala, India

Ritesh Yadav Department of Physics, Rabindranath Tagore University, Raisen, India;
Department of Physics, Dr A P J Abdul Kalam University, Indore, India

Anish Kumar Warrier Centre for Climate Studies / Department of Civil Engineering (Manipal Institute of Technology), Manipal Academy of Higher Education, Manipal, Karnataka, India

Geopolitics, Environmental Change and Antarctic Governance: A Region in Need of a Transformative Approach to Science Diplomacy

Dhanasree Jayaram

Abstract Antarctica, a continent that has been dedicated to scientific cooperation for decades, is increasingly coming under the pressure of several environmental, climatic, geopolitical (including the rise of Asian powers such as China) and geoeconomic changes (including fishing and bioprospecting). Although the Antarctic Treaty (AT) is considered a successful example of science diplomacy, as countries have set aside their territorial claims and the continent is a nuclear-free zone by shifting focus to scientific cooperation, its future remains uncertain with these developments. Science diplomacy always goes hand in hand with geopolitics. The AT that reflects Cold War geopolitics needs to be modified to represent present-day geopolitical realities for it to be enduring. It is also critical for the Antarctic Treaty System to continue maintaining the continent as a peace zone, environmental conservation and protection, and scientific collaboration. In this context, this chapter analyses the recent geopolitical trends associated with the Antarctic (against the backdrop of climatic and environmental change) and argues that the Antarctic Treaty System (and specific agreements under it) need to be reviewed. A transformative approach to Antarctic governance (including the Southern Ocean), especially in terms of its resources, needs to be adopted.

Keywords Antarctic treaty system · Science diplomacy · Environmental change · Antarctic governance · Transformative approach · Polar geopolitics

1 Introduction

Antarctica is one of the world's most significant regions regarding climate change and geopolitics are concerned. Being one of the global commons on the one hand and subjected to global climate change with grave repercussions for the entire world on the other, Antarctica could potentially emerge as a hotspot in terms of governance

---

D. Jayaram (✉)
Department of Geopolitics and International Relations, Old TAPMI Building, Manipal Academy of Higher Education, Manipal 576104, Karnataka, India
e-mail: dhanasree.j@manipal.edu

N. Khare (ed.), *Assessing the Antarctic Environment from a Climate Change Perspective*, Earth and Environmental Sciences Library,
https://doi.org/10.1007/978-3-030-87078-2_1

and security. While the focus of the larger international community seems to be on the Arctic, Antarctica's geopolitical dimensions have principally been directed by the evolution of the Antarctic Treaty System (ATS).[1] Very often, due to the region's scientific importance, remoteness (territorial) or even the absence of the native human population, Antarctica has not been at the centre of geopolitical discussions. However, with climate change, more recently discovered geoeconomic and geostrategic potential, growing scientific opportunities as well as longstanding territorial ambitions, the need for a transformative approach beyond the ATS has assumed significance more than ever before. This is critical to preserving the region's environmental and scientific integrity and maintaining peace and stability. This becomes even more relevant in the criticality of this region's ecological stability for the rest of the world.

Antarctic governance has, therefore, been complicated by various factors, with the future being uncertain in many ways. Even though the likelihood of an open conflict in or over the region may be negligible, confrontations and disagreements may arise that could even dilute the AT and related agreements. Already there are instances in which fissures have appeared between the original 12 signatories that were active during the International Geophysical Year of 1957–58 and the late entrants such as China. In this context, it becomes pertinent to analyse the various ways geopolitics affect Antarctic governance and how, in particular, recent geopolitical developments affect the future of the AT and related agreements. Many agreements about the Antarctic concern governance of environmental and resource issues. Governance essentially connotes the act of making "collective decisions", choosing "collective goals", and taking "action to achieve those goals." As an extension, environmental governance "addresses issues of access, use, protection, and management of common-pool natural resources" (Chaffin et al. 2016). Hence, this chapter deals with the interplay between Antarctic geopolitics and governance with the primary focus on environmental and climate change. Since science diplomacy has been used as a vital tool to bring together various countries to cooperate and collaborate in the Antarctic, the chapter looks into this concept and its multiple dimensions. It has been argued that science diplomacy is not devoid of geopolitical underpinnings and considers today's geopolitical realities. The Antarctic requires a transformative approach to science diplomacy that could help reinvigorate Antarctic governance, achieve sustainability and maintain peace in the region.

2 Science Diplomacy in the Antarctic

Science diplomacy has been in play since time immemorial. The ATS and many other subsequent decisions and agreements reached by the international community in Antarctica and the Southern Ocean are attributed to science diplomacy. The use

[1] The Antarctic Treaty and related agreements can be found at: https://www.ats.aq/e/antarctictreaty.html.

of science as a tool for diplomacy is not the only element of science diplomacy. Beyond this, the science-policy interface, which manifests in science to design institutions and implement policies, or for international collaboration, or management of natural resources and environmental concerns, is critical to the concept and practice of science diplomacy. Therefore, it is imperative to reflect on the various aspects of science diplomacy and their relevance to Antarctic governance. At the same time, can science diplomacy be disassociated from geopolitics? As many examples, including from the Antarctic, would suggest, they are interrelated, and science diplomacy is often practised by the existing geopolitical scenario. Hence, in this section, the interrelationship between science diplomacy and geopolitics would be explained to provide a context to the future of governance in the Antarctic.

The United Nations Educational, Scientific and Cultural Organization (UNESCO) describes science diplomacy[2] as:

> Science, due to its international and universal nature, has the power to cross borders and connect different peoples, communities, and societies…a tool to achieve foreign policy objectives where, not only the research outcomes but also science itself as a process and way of communicating, may serve to promote peace and sustainable development.

This definition tends to delink science from politics, which is debatable in an age where the politics or the politicisation of science is often used as a tool to influence decision-making (Gibbons 1995). By treating science as 'international' and 'universal', it is also rendered 'rational', 'objective', and so on (Sabbagh 2017). This is where geopolitics also assumes significance as countries engage in scientific exploration, research and related diplomatic initiatives only in tune with the prevailing geopolitics and not separate from it. However, this definition brings out certain critical elements of science diplomacy—facilitation of communication and connection, the achievement of and alignment with foreign policy objectives, and promotion of peace and sustainable development. These are essentially the need of the hour, more so when the credibility of science (such as climate science) is being discarded by several sections of the political class.

Science diplomacy can be understood in many ways. First, "science in diplomacy", wherein the focus is on how science could inform foreign policy. Science has become integral to several foreign policy options and goals adopted by nation-states, whether related to transboundary resource management, trade, security-related issues or other sectors. Second, "diplomacy for science", wherein formal diplomacy becomes the means for achieving scientific goals. The Montreal Protocol on Substances that Deplete the Ozone Layer is a classic example of a treaty that was reached through successful diplomatic efforts to address the issue of 'ozone hole' or depletion of the ozone layer in the stratosphere. All the countries have committed to phasing out or have already phased out ozone-depleting substances (ODS) such as chlorofluorocarbons (CFCs), based on a scientific agenda. Third, "science for diplomacy", wherein science becomes a tool for collaboration, cooperation, engagement

---

[2] More on UNESCO's interpretation of science diplomacy and its practice can be found at: http://www.unesco.org/new/en/natural-sciences/science-technology/science-policy-and-society/science-diplomacy/.

and peacebuilding (Walport 2014). Initiatives such as "Science for Peace" in Cyprus have primarily been engendered by scientific communities to promote engagement and peace between North and South Cyprus that continue to be politically divided.[3] With science diplomacy, scientists' role in foreign policy development and international relations has also been reinforced more directly. The contribution of the Intergovernmental Panel on Climate Change (IPCC) to the global climate change negotiations can be cited as a case in point. On foreign policy developments in sectors such as nuclear and space, too, the scientific community has at times had a direct role to play.

If one takes Antarctica's case, one could argue that all three versions are applicable and have come into use at various points in time. Even at the peak of the Cold War, this region witnessed international collaboration and efforts to bring peace between countries. Perhaps this is the only global commons with so-called 'claimant states'—Argentina, Australia, Chile, France, New Zealand, Norway and the United Kingdom (UK). Besides, the United States (US) and Russia (erstwhile Soviet Union) have consistently opposed these claims. Concerns about nuclear tests and dumping had also gathered momentum. India was among the countries that first demanded that Antarctica be used only for peaceful purposes. It was under these circumstances that the ATS was signed with the active involvement of both the US and the Soviet Union, who were in agreement that Antarctica should not become the centre of East–West conflict. Territorial sovereignty was set aside by most nation-states. The movement for establishing peace gained momentum during the 1950s, which led to the promotion of scientific research and cooperation in Antarctica that later were integrated into the ATS. Science diplomacy has become a cornerstone of several other measures taken in the region, such as the ones aimed at biodiversity protection. One such initiative is the science-based Convention on the Conservation of Antarctic Marine Living Resources (CCAMLR) that is considered to be the ocean arm of the ATS. Science diplomacy in the Antarctic context has been used to attain scientific goals, achieve sustainable development, establish peace, promote international collaboration, and inform foreign policy objectives, among others.

3 The Antarctic Treaty System and Its Future

The Antarctic Treaty (AT) is considered a success story, primarily because it was arrived during the peak of the Cold War and survived many decades of rivalries and conflicts. The reason why the AT could be reached is at times attributed to the deliberate attempt by the chief negotiators to keep many provisions largely ambiguous, with mechanisms for compliance, enforcement and governance relatively weak. One could argue that science trumped geopolitics in this case, albeit as discussed, the two

[3] More on Science for Peace Initiative Cyprus can be found at: https://medium.com/naturewords/the-science-of-building-peace-with-nature- 838b36cd5bfb

go hand in hand. The AT needs to be seen through the prism of science diplomacy—how science played an important role in reaching intergovernmental agreements, including on marine living and mineral resources. With support from groups such as the Scientific Committee on Antarctic Research (SCAR), much before the CCAMLR came into force in 1982, a few more agreements were reached by the international community. These include Agreed Measures for the Conservation of Antarctic Fauna and Flora (1964) and Conservation for Antarctic Seals (1972). Science contributed to the adoption of precautionary and ecosystem approaches within these agreements, whether it is to "maintain populations that are the target of harvesting at healthy levels" or to "prevent irreversible damage in the Antarctic marine ecosystem" (Scully 2011, p. 3). The most recent step undertaken under the ATS is that of entry into force of the Ross Sea Marine Protected Area in 2017. This is the world's largest marine protected area (MPA) and will be in power for 35 years (Dodds and Brooks 2018).

The AT also needs to be seen against the backdrop of "scientific internationalism" that has dominated much of the debates on Antarctic governance, especially in the twentieth century. Scientific internationalism is based on three principles. First, the epistemological principle pertains to knowledge and asserts that scientific knowledge is borderless and universal (belongs to all). Second, the organisational focus concerns cooperation in scientific research, standardisation of research methods, discussion and interpretation (such as peer review), and information exchange, among others that could reduce costs and duplication of efforts. Third, the welfare principle "involves solidarity and the application of the fruits of science for the benefit of all humankind, including the distribution of its goods." As argued by Elzinga (2011, p. 59), the AT has been successful only on the epistemological principle, while on the other two, there are still gaps. Calls for declaring Antarctica as a "heritage of humankind" (to be put under the United Nations) or as a "World Natural Park" (by several environmental organisations) have not been endorsed even at the conceptual level, thereby highlighting the continuing role of territoriality and exclusivity attached to this continent. Nevertheless, these principles have reflected to some extent in the International Polar Year and International Geophysical Year.

The First International Polar Year (1882–83) was steered by the quest for knowledge, scientific exploration and cooperation. The second International Polar Year (1932–33) or IPY2 came more than a decade after the First World War, during which scientific internationalism suffered a setback. IPY2 also happened amid territorial claims by several nations—starting from the UK (along with Australia and New Zealand) during the early 1900s to France and Norway in the 1920s and 1930s. After IPY2, Argentina and Chile also entered the fray in the 1940s and their territorial claims overlapped with that of the UK. The International Geophysical Year (1957–58) or IGY laid the AT's foundation as the US, under former President Dwight D. Eisenhower, spearheaded preparatory talks and negotiations. The IGY saw 67 nation-states collaborating for scientific endeavours, out of which 12 were mainly involved in the Antarctic (Joyner 2011).

To contend that the IGY and the subsequently inked treaty were exceptions to the bitter Cold War rivalry and devoid of geopolitical motives would be naïve. The IGY had security, strategic and foreign policy objectives. The US saw "international scientific cooperation and data exchange" as a potentially "powerful, yet neglected, vehicle for promoting American interests and values as well as for the collection of intelligence of use to the American state" (Naylor et al. 2008). By merely suspending all the territorial claims and allowing all countries to enter the continent freely and fully (as a global commons), the US and the erstwhile Soviet Union could establish their research bases/stations all over the mainland. Simultaneously, the claimant states (traditionally) restricted construction and scientific activities to their claimed territories only. Thus, the US built one of its research stations—Amundsen-Scott South Pole Station—at the much-coveted geographic South Pole.[4] In any case, the future of the Antarctic region as a nuclear-free zone for 'scientific' purposes was concretised.

Despite these success stories, questions are still raised over the future of governance in the region due to its relevance for the fishing industry (particularly krill and toothfish), the whaling industry, access to fresh water (since the area is home to 70 per cent of the world's freshwater), bioprospecting, mineral exploration (and potential exploitation), tourism and other human activities (Joyner 2011). Science may be the binding factor fostering international collaboration; it is also visibly being used as a geopolitical tool to expand research on climate change through physical presence on the continent that may even metamorphose into an opportunity to fulfil military or security or strategic objectives. The Antarctic is abundantly rich in resources (including marine life), with many species and resources still undiscovered. Both human activities and environmental change are, however, putting pressure on the region's ecosystems.

Whether or not the ATS has been able to check such activities and whether or not it will be able to do so in the future are questions that require further analysis. The Protocol on Environmental Protection to the Antarctic Treaty or the Madrid Protocol (concluded in 1991 and entered into force in 1998) has banned any activity related to mineral resources other than scientific research. However, there are no comprehensive frameworks and policy regimes to regulate other activities such as bioprospecting[5] in the region (except two resolutions adopted by the Antarctic Treaty Consultative Parties or ATCPs). There has already been the collection of microorganisms in Antarctica for various purposes (such as pharmaceutical and healthcare), and the interest in bioprospecting activities is growing further (SCAR 2009). In such a scenario, there could be conflicts on issues related to patenting, information exchange and benefit-sharing (Australian Antarctic Division 2004). Even the progress assessments made against the Strategic Plan for Biodiversity (includes a goal on "Benefits from biodiversity and ecosystem services"), adopted under the purview of the Convention on Biological Diversity (for until 2020), do not include Antarctica and the Southern Ocean (Chown et al. 2017).

---

[4] More information regarding the US' research station at the South Pole can be found at: https://www.nsf.gov/geo/opp/support/southp.jsp.

[5] Bioprospecting refers to "exploration of naturally occurring microorganisms, plants and animals for commercially valuable genetic and biochemical resources." The definition and other details can be found at: https://documents.ats.aq/ATCM25/wp/ATCM25_wp043_e.pdf.

The CCAMLR, which currently has 26 Members and 10 Acceding States (as of 2019),[6] successfully established the Ross Sea MPA. Being one of the most pristine and most productive (in terms of healthy marine ecosystems) stretches of the Southern Ocean, the demand for an MPA remained fervent from several parties, particularly New Zealand and the US (that also put forth a joint proposal). Scientifically too, it is one of the best-studied Antarctic continental shelf systems. The need for Ross Sea MPA was mainly driven by the rapidly intensifying fishing activities in the region. Setting a limit on fishing through the MPA became a priority for several CCAMLR parties. However, the journey was not smooth. In the initial years when the US and New Zealand came up with independent proposals, there were many technical differences and competing values between them and other member states. While the US hosts the most extensive scientific base in the Ross Sea, New Zealand has a historical claim over the area. The US is not involved in the toothfish industry and, therefore, recommended an MPA that would take away a more prominent place off the limits of toothfish fishery. New Zealand, on the other hand, sought to develop the Ross Sea fishery in the past and even claimed exclusive access to it, which was eventually denied as other CCAMLR members objected to it (Dodds and Brooks 2018).

Even in the discussions that led up to the establishment of Ross Sea MPA, there were a few parties that held out until the last minute. For example, during the discussions, China opposed no-fishing zones without enough scientific evidence of threat (based on the precautionary approach). Russia endorsed this view, too (Brooks et al. 2018). To accommodate the fishing interests of many parties, the area that was initially proposed for the MPA had to be trimmed by almost 40 per cent in 2013. This was the only way the initiative could have gained more support, and since the CCAMLR works on consensus, it was essential to have the affirmative vote of all parties. How geopolitics influences these discussions can be best exemplified by how the Ukraine crisis and tensions between the US and Russia influenced the discussions. In the end, the efforts of former Secretary of State John Kerry played a crucial role in bringing on board China and Russia. As remarked by Brooks, "It was not only an environmental win for Antarctica and the whole world but also a diplomatic win. It felt like a peace agreement, especially in heightened geopolitical tensions between the US and Russia. It made me realise that we still do have exceptional governance despite tensions and contested resource frontiers" (Jayaram 2019).

The Antarctic krill are increasingly coming under pressure from overfishing (illegal fishing too) and climate change. They are mainly used in aquaculture, omega-fatty-acid supplements and livestock foods. Although they are present in humungous numbers in the Southern Ocean, unregulated fishing could disrupt the overall ecosystems as other mammals (such as whales), and birds feed on the krill directly or indirectly (as a part of the food chain). The pressure on Antarctic species is immense, with several of them being almost driven to extinction, and only a few can recover. These include elephant seals, blue whales, marbled rock cod and king penguins. If one takes just the case of krill, around 300,000 tonnes are caught annually (Brooks

[6] More information regarding the CCAMLR can be found at: https://www.ccamlr.org/en/organisation/who-involved-ccamlr.

et al. 2018). While fishing has been regulated in the Ross Sea, other areas such as the Weddell Sea continue to be threatened by overfishing. A 2018 decision by the krill fishing companies that account for 85 per cent of krill fishing decided to stop fishing in areas of the Antarctic Peninsula just ahead of the CCAMLR meeting is a welcome step. However, as scientists observe, that may not resolve the problem, as other industries such as toothfish and icefish ones are still highly active in many areas. This is why the support for an EU proposal on Weddell Sea MPA is growing among the scientific and environmental conservation communities (Marshall 2018).

4 Climate Change and Changing Geopolitics in the Antarctic

Much before climate change became the centre-point of global discussions, Antarctica grabbed the spotlight when the stratospheric ozone hole was unexpectedly discovered over it (first reported in 1985). This revolutionised both science and environmental policymaking as it laid the foundation for the Montreal Protocol (Solomon 2019). Recent reports in 2019 suggest that the ozone hole is the smallest on record since it was discovered and that it could recover entirely by 2040 if the Montreal Protocol is adhered to by all the parties (Convey and Peck 2019). Despite being geographically remote and still inaccessible to many, Antarctica's scientific research opened up possibilities for further collaboration and corrective measures in the political domain. Today the region is no longer considered inaccessible as thousands of tourists flock to this continent. From around 5,000 every year in the early 1990s to about 45,000 in 2016–17, the number of tourists visiting Antarctica has risen and continues to increase. To a large extent, these numbers are partly explained by the phenomenon of "last-chance tourism", which implies that people are increasingly keen on visiting places/attractions that are threatened by natural or human factors such as climate change or over-tourism and may not be accessible or available for visits in the future (Abedi 2018).

The extensive impacts of climate change on the earth's cryosphere have been highlighted by several scientific groups, including the IPCC. These impacts are, however, not uniform in the Antarctic region due to which certain parts of the area are warming up more severely than others (such as the Antarctic Peninsula). At the same time, in some other places, the sea ice extent is increasing. According to the 2019 IPCC Special Report on the Ocean and Cryosphere in a Changing Climate, "Antarctic ice loss is dominated by acceleration, retreat and rapid thinning of major West Antarctic Ice Sheet (WAIS) outlet glaciers (very high confidence), driven by melting of ice shelves by warm ocean waters (high confidence)" (Meredith et al. 2019). Not only is the Antarctic adversely affected by climate change—in the form of increased temperature, ocean acidification and thinning of the ice shelf (also contributing to retreating ice sheet)—it also poses the potential risk of rising global sea levels as the grounded ice enters the oceans (Gudmundsson et al. 2019). Another recent study reveals that the Antarctic Ice Sheet "lost 2,720 ± 1,390 billion tonnes of ice between 1992 and 2017, which corresponds to an increase in mean sea level of 7.6 ± 3.9 mm" (Shepherd 2018, p. 219). The west coast of the Antarctic Peninsula is reportedly one of the most rapidly warming regions of the world, with an increase of 3 °C recorded in air temperature. These warming trends have adverse impacts on the region's ecosystems, including the penguin and Antarctic krill. The prevailing measures, including under the CCAMLR, do not take into consideration long-term climate change scenarios. Because climate change has adverse impacts on these species, specific decisions such as fishing quotas or catches shares should account for it.

Climate change research in the Antarctic is critical to predicting the future of the earth's climate system. The Southern Ocean is crucial for the global climate and ecological systems as it serves as the link between the Atlantic, Pacific and Indian Oceans (Sallée 2018). Owing to the centrality of this region to the global climate system and ongoing scientific research activities, the Antarctic's scientific profile has grown over the years, leading to the arrival of many more scientific expeditions and the establishment of more scientific stations in the continent. The number of research stations and camps in Antarctica on the rise (Kotecki 2018). Yet unlike the Arctic that has been geopolitically volatile more than ever before in the event of increasing climate change effects, Antarctica has not been previously considered a geopolitical hotspot, mainly due to the existence and continuance of the ATS. However, recent developments show that with varying attitudes of different countries towards the region and its resources, contestations are expected to grow. The rise of Asian countries and their increasing involvement in the area, in particular, are being seen as trigger-points for the emergence of new rivalries—just as in the Arctic.

China's ambition to be a "polar great power" (Brady 2017) has been sufficiently explicated in its white paper on "Antarctic activities," in which it vows to abide by the principles of the ATS, including non-militarisation research (on issues such as climate change) and environmental protection that would entail a commitment to the existing ban on commercial resource extraction in the region (Tiantian 2017). At the same time, in its 13th Five-Year Plan for fishing industry technology (released in 2017), its intent "to increase its krill fishing and processing capacity and improve its fishing technology and competitiveness to support the growth of a krill industrial chain" is emphasised (Chun and Damin 2018). China currently has three research stations and one camp in Antarctica—Changcheng, Zhongshan, Kunlun and Taishan—with a fifth one built on the Ross Sea Ice Shelf (2022). Kunlun and Taishan lie within Australia's Antarctic claim. China is a signatory to the ATS and ratified the 1991 Madrid Protocol (in 1998) that prohibits any activity concerning exploitation of mineral resources in Antarctica. However, this will be open for review in 2048, and as pointed out by some analysts, the consensus reached by many countries on imposing a ban on mining may not hold water later in the current century (Liu 2019). In any case, commercial mining in the region is not viable at this stage (or shortly). Therefore, it might be too early to be speculating about the mining interests of countries such as China in the region. This does not mean that strategic competition over resources (including fossil fuels and rare earths) in the area is impossible in the future as they become scarce in other parts of the world.

Nevertheless, China has been more overt in showing its interest in fisheries, bioprospecting, scientific research, tourism and shipping. For instance, China National Fisheries Corporation (CNFC) is one of the leading krill fishing companies in the region. Krill fishing was once dominated by the Soviet Union, and after its collapse, Japan took over. Currently, Norway, South Korea and China are ahead of the rest, with Russia also making a comeback gradually (Stark et al. 2019, pp. 37–38). China's krill fishing has increased over some time, touching 65,000 tonnes in 2016. The depleted fisheries in its backyard due to overfishing and marine pollution is pushing China to explore resources outside its territorial waters and exclusive economic zone (EEZ). According to the China Fisheries Bureau, under CCAMLR's umbrella, there is scope for further exploration and exploitation of krill resources in the Antarctic. Hence, China should invest more in such explorations, enhancing fishing activities, developing polar fishing technology and building krill fishing vessels for commercialisation (Liu and Brooks 2018). According to CCAMLR data, "China has more krill fishing vessels and fishes over a wider area than any other nation." Since the 2016/17 CCAMLR fishing season, China has also started to venture into East Antarctica to fish for krill. It is among the countries opposed to establishing the East Antarctic Marine Protected Area, proposed by Australia and the European Union (EU) (Liu and Brooks 2018).

It is rather apparent that besides the commercial and scientific relevance of the region, the Antarctic represents geopolitical and strategic space too in international politics. Countries would be interested in vying for influence in the region through increasing presence (bases, research stations, cultural edifices and so on). Russia, for instance, has built the continent's first Orthodox Church. Since the collapse of the Soviet Union, Russia has found it difficult to bounce back into the pinnacle of international politics. However, it is re-emerging as a world power with military, strategic, scientific and resource interests in Antarctica and the Southern Ocean using its "smart power strategy" (use of both hard and soft power) (Carter, Brady and Pavlov 2016). Even while actively participating in the multilateral arrangements concerning the Antarctic, it has gone ahead and deployed measuring stations for the GLONASS global satellite navigation system (akin to the US-led Global Positioning System or GPS) and plans to build additional ones (Kezina 2015). It plans to reopen its Russkaya station in 2021 as Roscosmos (state space corporation) has proposed to "install equipment for GLONASS" at the station (Xinhua 2019a).

Although the ATS forbids Antarctica's use for military purposes, it does not prevent military personnel or equipment in the region for scientific and peaceful purposes. However, it must be noted here that the ATS was negotiated and signed in the Cold War era in which one of its most important objectives was to keep Antarctica out of the conflict between the two superpowers (and their allies) as well as to suspend all territorial claims over the continent. The focus now was on preventing nation-states from using this continent for "military manoeuvres" or building "military bases and fortifications" and, most notably, nuclear weapons testing and radioactive waste disposal. The present-age technologies in the space and cyber domains, without doubt, dual-use, are allowed under the ATS as long as they are used to meet scientific objectives. However, the global navigation satellite systems such as US-led GPS, Russia's GLONASS, China's BeiDou and Norway's Troll Satellite Station are further developed in Antarctica. These could be used during wartime for "missile targeting and timing, as well as access to fleet-based broadband for unclassified and classified systems, and environmental, situational awareness" (Brady 2018).

In response to these recent developments, Australia is rethinking its positions to regain its influence in the Antarctic—primarily by deploying its military equipment and defence technology (Gothe-Snape 2019a). There were also disagreements between Australia and China over the latter's proposal for a "code of conduct" at "Dome A" (considered to be the "best location for space observation on Earth due to its high elevation and outstanding visibility") that again falls within the former's Antarctic claim (Gothe-Snape 2019b). Although these disagreements may not be based on Australia's territorial claims, there are concerns regarding the use of such codes to be used in the future by nation-states to claim territory (especially since even under the ATS, the US and Russia do reserve the right to make territorial claims in the future). Since the ATS has frozen all territorial claims, Australia's claim over 42 per cent of the territory is not recognised. Yet, Australia has time and again stood by the need for preserving the integrity of the ATS to maintain its sovereignty over a part of Antarctica's territory (strategic interests) and at the same time achieve environmental/scientific goals (Bray 2016).

Similarly, in its 2018 Strategic Defense Policy Statement, New Zealand has expressed its apprehension over the "difficulty in distinguishing between allowed and prohibited activities" under the ATS, which could effectively be exploited by some parties to use the continent for "military or security-related activities." It has reinvigorated its commitment towards Antarctica and the Southern Ocean, especially to fight climate change, under the pretext of heightened presence and other countries' activities, including China, South Korea and Italy in the region. Even on mining, it acknowledges the fact that although mining activities are currently banned under the ATS, in the future, the treaty may not be able to prevent them (Ministry of Defence 2018). While resource exploitation might be on the cards in the long term, it could be deemed a long shot at this stage as the region's rough topography, the scarcity of adequate infrastructure, economic non-viability, and other factors are likely to derail any such efforts. However, this does not imply that advances in technology cannot overcome these bottlenecks in the future.

5 The Road Ahead: Science Diplomacy and a Transformative Approach

The current geopolitical realities call for a reinvigorated approach towards Antarctic governance. While geopolitics may not blatantly threaten the future of the Antarctic and its longstanding stature as a zone of peace and scientific collaboration, in light of the above-discussed developments, there would likely be disagreements over a host of new activities (such as tourism, bioprospecting, fisheries and potential research exploration) that are now central to Antarctica. Impacts of climate change are further complicating these developments. On the one hand, climate change could be seen as a uniting factor that could bring countries together for further research that could more precisely predict the changing patterns. On the other hand, it could also become a dividing factor that pits nations against each other in resource and benefit-sharing. One must also not forget that nation-states could use climate change to gain more and more access to the continent. As more human population enters the Antarctic, its environmental integrity and sustainability could come under more pressure. The use of scientific facilities for military purposes is another concern, despite inspections under Article VII of the treaty.

Nation-states are vying for influence in the region through more scientific bases and personnel on the ground. However, it is not just the quantity that matters here; quality is equally important, as there is a push for all-year bases rather than just summer-only ones. The location, type and purpose of the grounds are also integral to the strategy adopted by countries. Overlapping territorial claims, competing prerogatives and values over resource and benefit-sharing, and bases for global navigation satellite systems, among others, dictate the nature and extent of influence that a nation-state could exert in the region. These efforts are being strengthened by increased spending on Antarctic-related activities in many countries –operational costs and research funding, and capacity building (including icebreakers). For example, in terms of the budget's size in Antarctica, China leads the world (Brady 2014). China's first indigenously built polar icebreaker—Xuelong 2—completed its maiden voyage for its 36th Antarctic expedition in 2019 (Xinhua 2019b). While many strategic analysts see China's rise and increasing influence as a threat or menace, the reality is that its ambitions and interests are not very different from that of the US and Russia, if not the claimant nations. China's quest for superpower status invariably entails a greater emphasis on Polar Regions. Therefore, although the possibility of conflict or confrontation cannot be ruled out entirely, there is ample scope for collaboration, cooperation and partnership that has, in any case, prevailed in the Antarctic for decades.

As far as India is concerned, Antarctica's presence is relatively marginal (compared to the US, Russia, China, Australia and others). It has so far set up three stations—Dakshin Gangotri (that has now become a supply base and transit camp), Maitri and Bharati (established in 2013) (Press Information Bureau 2013a). With the establishment of Bharati, India has joined the elite club of countries that have multiple research stations in Antarctica. However, in the long term, India has interests in the region for both scientific purposes as well as meeting diplomatic objectives (mainly to establish its place at the international level as a scientific power, as it has been able to achieve in other domains such as space, as this is indispensable to India's recognition as a world power). This is reflected in some of the recent decisions and steps were taken by the Government of India, including a Memorandum of Understanding (MoU) with Argentina on Antarctic cooperation—in earth sciences and environmental/marine conservation and protection (Press Information

Bureau 2013b). Being a party to the AT and a 15th Consultative Member of the Antarctic Treaty, India is increasingly becoming aware of climate change's geopolitical and security implications in the region. Thereby, there is a need for reviewing the effectiveness of the ATS and possible alternative governance solutions, which Chaturvedi (2016) labels the "climate security dilemma". At the same time, India's longstanding engagement with the Antarctic has not yet translated into a coherent Antarctic policy, partly attributed to its lack of "strategic culture" and "institutionalisation of the country's foreign policymaking" (Chaturvedi 2012). This would have to change significantly if India sees itself engaging in the region in a more strategic manner—an endeavour in which its growing emphasis on science diplomacy could play a critical role.

One needs to ask how to address climate change, resource, and environmental imperatives in the region by accommodating various countries' differing geopolitical interests. Science diplomacy could provide answers to this question. However, as we live in a world where science is continuously challenged and/or is politicised, the dangers of science diplomacy backfiring are also valid. The AT and related agreements could be reviewed. Climate change presents a window of opportunity to open up new research avenues into its impacts on the region and the entire world. Science diplomacy could also pave the way for comprehensive regulations on tourism and bioprospecting that are gathering momentum. There are various ways of bolstering science diplomacy efforts—from promoting science education and research through increased funding domestically to encouraging collaborative and joint use of scientific infrastructure (Antarctica). One of the purposes of science diplomacy is to overcome political and other differences. Hence, if there are mutual benefits to be gained, science diplomacy could foster scientific collaboration. Science diplomacy should ideally involve all communities, including indigenous peoples, that have been left out of the governance and decision-making mechanisms for a long time, even in the case of polar governance (in the Arctic). This is why the most recent International Polar Year (2007–09)[7] is distinct in many ways as it involved the indigenous peoples. Still, their integration with the governance regimes in a more meaningful way is yet to happen. In all these matters, the need of the hour is to strengthen the science-policy interface through co-creation and co-production. This would be possible by a more significant engagement level between different actors—both state and non-state—and from varying backgrounds (disciplinary, professional and bureaucratic). Even though this has been facilitated mainly under the ATS, non-governmental scientific and environmental organisations' participation needs to be boosted further as their outreach to the public and other stakeholders are immensely required to address various concerns about climate change, tourism and bioprospecting.

Science diplomacy could become more effective if a transformative approach to environmental governance is adopted in the Antarctic. By 'transformative approach', the goal would be to revise and redesign the existing structures, models and processes

[7] More information regarding the IPY 2007–08 can be found at: http://library.arcticportal.org/1211/1/IPY_Summary.pdf.

that characterise a particular governance regime, architecture or system. Environmental governance in this context becomes the binding factor as the Antarctic has emerged as a centre of geopolitical contestation on account of its rich resources. A transformative approach to environmental governance in the Antarctic, based on the changing climate and geopolitical realities in the region, needs to align scientific, resource and geostrategic objectives. In this endeavour, the transformation could be achieved through informal and non-governmental initiatives' more prominent involvement. As has been seen in the case so far, science has played a vital role in Antarctic governance and to maintain the collaborative nature of engagement between nation-states in the region. New governance models must be invented in place of the new geopolitical scenario marked by the emergence of new powers such as China. Transformative approaches to environmental governance in the Antarctic need to be precautionary as well. Different actors, including scientists, need to come together and collaborate on co-producing knowledge using various methods such as modelling, scenario building (involving geopolitical variables), experimentation and so on, and using it to frame a governance regime that is sustainable and cooperative (irrespective of the overall competing values and interests).

Instead of circumventing geopolitical realities, a transformative approach to science diplomacy in Antarctica needs to transform the concerned stakeholders' values, interests, and beliefs. This would entail systemic shifts in how diplomacy is practised; science is communicated, and internalised collaborative governance attitudes. Both framings of issues and agenda-setting within the governance architecture are crucial to adopting a transformative approach. Therefore, it is high time that the Cold War-based values embedded in the ATS are discarded, including the consensus-based system that has could stand in the way of adopting strong measures to check illegal and suspicious activities in the region. One such instance is unlawful fishing carried out by a South Korean fishing vessel in 2011, after which it could not be blacklisted (as per the Report of the Thirtieth Meeting of CCAMLR) (Commission for the Conservation of Antarctic Marine Living Resources 2011). South Korea has recently (in 2019) been warned by the US for illegal fishing once again (Yeon-soo 2019).

A multi-level governance regime with a nested leadership that does not promote hegemony or reinforce geopolitical rivalries may be better placed to maintain the 'global commons' status of the Antarctic. The new geopolitical realities reflect multipolar characteristics—multiple power centres with actors at different strategic capabilities, influencing capacity and geopolitical reach. Even if the US continues to be the preeminent power after the end of the Cold War, other power centres (such as China and Russia) are becoming geoeconomically and geopolitically more vital day by day. Claimant states such as Australia and New Zealand are becoming exceedingly wary of various geopolitical developments and the enhanced presence of a host of Antarctica countries, driving them to reassert their positions and claims. These developments are also adding to the environmental pressures of the continent. On climate change specifically, there is a need for concerted efforts to align scientific research in Antarctica and climate policy related to the region into intergovernmental mechanisms such as the United Nations Framework Convention on Climate Change (UNFCCC), which is yet to materialise in a meaningful manner. The ATS cannot be the sole architecture that governs this remote, harsh and fragile region. Science diplomacy is premised on the idea that global problems such as climate change require global solutions. The Antarctic, central to several critical international science diplomacy initiatives, needs a transformative agenda that promotes scientific cooperation, sustainability and peace.

References

Abedi M (2018) Antarctica keeps attracting visitors—and it may be 'last-chance tourism'. Global News. https://globalnews.ca/news/4072700/antarctica-tourism-climate-change/. Accessed 26 Dec 2019

Australian Antarctic Division (2004) Antarctic bioprospecting, benefit sharing and cooperation in Antarctic science. Australian Antarctic Magazine. http://www.antarctica.gov.au/magazine/2001-2005/issue-6-autumn-2004/feature/antarctic-bioprospecting-benefit-sharing-and-cooperation-in-antarctic-science. Accessed 26 Dec 2019

Brady A (2014) Evaluating China as an Antarctic state. Kissinger Institute on China and the United States (Wilson Center). https://www.wilsoncenter.org/event/china-and-antarctica. Accessed 26 Dec 2019

Brady A (2017) China as a polar great power. Cambridge University Press, Cambridge

Brady A (2018) China, Russia push GPS rival into Antarctica. The Australian. https://www.theaustralian.com.au/nation/inquirer/china-russia-push-gps-rival-into-antarctica/news-story/1faeb3222806f61110c016ff00390357. Accessed 26 Dec 2019

Bray D (2016) The geopolitics of Antarctic governance: sovereignty and strategic denial in Australia's Antarctic policy. Aust J Int Aff 70(3):256–274. https://doi.org/10.1080/10357718.2015.1135871

Brooks CM, Ainley DG, Abrams PA, Dayton PK, Hofman RJ, Jacquet J, Siniff DB (2018) Antarctic fisheries: factor climate change into their management. Nature 558:177–180. https://doi.org/10.1038/d41586-018-05372-x

Carter P, Brady A, Pavlov E (2016) Russia's "smart power" foreign policy and Antarctica. Polar J 6(2):259–272. https://doi.org/10.1080/2154896x.2016.1257102

Chaffin BC, Garmestani AS, Gunderson LH, Benson MH, Angeler DG, Arnold CA, Cosens B, Craig RK, Ruhl JB, Allen CR (2016) Transformative environmental governance. Annu Rev Environ Resour 41:399–423. https://doi.org/10.1146/annurev-environ-110615-085817

Chaturvedi S (2012) India and Antarctica: towards post-colonial engagement? In: Brady A (ed) The emerging politics of Antarctica. Routledge, London, pp 50–74

Chaturvedi S (2016) Indian contributions to Antarctic social sciences. Proc Indian Natn Sci Acad 83 (2):505–512. https://doi.org/10.16943/ptinsa/2017/48965

Chown SL, Brooks CM, Terauds A, Le Bohec C, van Klaveren-Impagliazzo C et al (2017) Antarctica and the strategic plan for biodiversity. PLoS Biol 15(3). https://doi.org/10.1371/journal.pbio.2001656

Chun Z, Damin T (2018) Krill fishers agree to partial ban in the Antarctic. China Dialogue. https://chinadialogueocean.net/3891-krill-fishers-agree-to-partial-ban-in-antarctic/. Accessed 26 Dec 2019

Commission for the Conservation of Antarctic Marine Living Resources (2011) Report of the Commission's Thirtieth meeting. https://www.ccamlr.org/en/system/files/e-cc-xxx.pdf. Accessed 26 Dec 2019

Convey P, Peck LS (2019) Antarctic environmental change and biological responses. Sci Adv 5(11):eaaz0888. https://doi.org/10.1126/sciadv.aaz0888

Dodds K, Brooks CM (2018) Antarctic geopolitics and the Ross Sea Marine protected area. E-International Relations. https://www.e-ir.info/2018/02/20/antarctic-geopolitics-and-the-ross-sea-marine-protected-area/. Accessed 26 Dec 2019

Elzinga A (2011) Origin and limitations of the Antarctic Treaty. In: Berkman PA, Lang MA, Walton DWH, Young OR (eds) Science diplomacy: Antarctica, science, and the governance of international spaces. Smithsonian Institution Scholarly Press, Washington, DC, pp 59–68

Gibbons JH (1995) The politics of science. Science 269(5221):143. https://doi.org/10.1126/science.269.5221.143

Gothe-Snape J (2019a) Defence wants to roll out military tech in Antarctica despite the Treaty ban on military activity. ABC News. https://www.abc.net.au/news/2019-08-19/australia-antarctica-military-dual-use-technology/11427226. Accessed 26 Dec 2019

Gothe-Snape J (2019b) Australia declares China's Antarctic conduct plan has 'no formal standing'. ABC News. https://www.abc.net.au/news/2019-07-30/antarctica-china-code-of-conduct-dome-a/11318646. Accessed 26 Dec 2019

Gudmundsson GH, Paolo FS, Adusumilli S, Fricker HA (2019) Instantaneous Antarctic ice- sheet mass loss was driven by thinning ice shelves. Geophys Res Lett 46(13):13903–13909. https://doi.org/10.1029/2019GL085027

Jayaram D (2019) Webinar series: science diplomacy and environmental peacebuilding. Future Earth. https://futureearth.org/2019/06/14/webinar-series-science-diplomacy-and-environmental-peace-building/. Accessed 26 Dec 2019

Joyner CC (2011) United States foreign policy interests in the Antarctic. Polar J 1(1):17–35. https://doi.org/10.1080/2154896X.2011.569384

Kezina D (2015) New GLONASS stations to appear in Antarctica. Russia Beyond. https://www.rbth.com/science_and_tech/2015/04/24/new_glonass_stations_to_appear_in_antarctica_45513.html. Accessed 23 Dec 2019

Kotecki P (2018) Stunning photos show what daily life in Antarctic research stations is really like. Business Insider. https://www.businessinsider.in/stunning-photos-show-what-daily-life-in-antarctic-research-stations-is-really-like/articleshow/66696208.cms. Accessed 23 Dec 2019

Liu N (2019) What Are China's Intentions in Antarctica? https://thediplomat.com/2019/06/what-are-chinas-intentions-in-antarctica/. Accessed 24 Dec 2019

Liu N, Brooks CM (2018) China's changing position towards marine protected areas in the Southern Ocean: implications for future Antarctic governance. Mar Policy 94:189–195. https://doi.org/10.1016/j.marpol.2018.05.011

Marshal C (2018) Krill companies limit Antarctic fishing. BBC. https://www.bbc.com/news/science-environment-44771943. Accessed 24 Dec 2019

Meredith M, Sommerkorn M, Cassotta S, Derksen C, Ekaykin A, Hollowed A, Kofinas G, Mackintosh A, Melbourne-Thomas J, Muelbert MMC, Ottersen G, Pritchard H, Schuur EAG (2019) Polar regions. In: Pörtner H-O, Roberts, DC, Masson-Delmotte V, Zhai P, Tignor M, Poloczanska E, Mintenbeck K, Alegría A, Nicolai M, Okem A, Petzold J, Rama B, Weyer NM (eds) IPCC special report on the ocean and cryosphere in a changing climate. In press

Ministry of Defence, New Zealand Government (2018) Strategic Defence Policy Statement 2018. https://defence.govt.nz/assets/Uploads/8958486b29/Strategic-Defence-Policy-Statement-2018.pdf. Accessed 24 Dec 2019

Naylor S, Siegert M, Dean K, Turchetti S (2008) Science, geopolitics, and Antarctica's governance. Nat Geosci 1:143–145. https://doi.org/10.1038/ngeo138

Press Information Bureau, Government of India, Ministry of Earth Science (2013a) Research stations in Arctic and Antarctica. https://pib.gov.in/newsite/PrintRelease.aspx?relid=98006. Accessed 24 Dec 2019

Press Information Bureau, Government of India, Cabinet (2013b) Cabinet apprised about MoU on Antarctic cooperation between India and Argentina. https://pib.gov.in/newsite/PrintRelease.aspx?relid=189561. Accessed 24 Dec 2019

Sabbagh U (2017) Science has always been inseparable from politics. Scientific American. https://blogs.scientificamerican.com/guest-blog/science-has-always-been-inseparable-from-politics/. Accessed 24 Dec 2019

Sallée J-B (2018) Southern ocean warming. Oceanography 31(2):52–62. https://doi.org/10.5670/oceanog.2018.215

SCAR (2009) Biological prospecting in the Antarctic: an update on the review by SCAR. https://www.scar.org/antarctic-treaty/actm-papers/atcm-xxxii-and-cep-xii-2009/2885-atcm32-ip065/file/. Accessed 26 Dec 2019

Scully T (2011) The development of the Antarctic treaty system. In: Berkman PA, Lang MA, Walton DWH, Young OR (eds) Science diplomacy: Antarctica, science, and the governance of international spaces. Smithsonian Institution Scholarly Press, Washington, DC, pp 29–38

Shepherd A, Ivins E, Rignot E et al (2018) Mass balance of the Antarctic ice sheet from 1992 to 2017. Nature 558:219–222. https://doi.org/10.1038/s41586-018-0179-y

Solomon S (2019) The discovery of the Antarctic ozone hole. Nature 575:46–47. https://doi.org/10.1038/d41586-019-02837-5

Stark JS, Raymond T, Deppeler SL, Morrison AK (2019) Antarctic seas. In: Sheppard C (ed) World seas: an environmental evaluation, 2nd edn. Elsevier, London, pp 1–38

Tiantian B (2017) China releases 1st Antarctic paper. Global Times. http://www.globaltimes.cn/content/1048187.shtml. Accessed 26 Dec 2019

Walport M (2014) Science diplomacy: a big country perspective. Paper presented at the Science Advice to Governments conference held in Auckland. https://www.slideshare.net/bis_foresight/science-diplomacy-a-big-country-perspective-auckland-conference?next_slideshow=1. Accessed 26 Dec 2019

Xinhua (2019a) Russia to reopen mothballed Antarctic station. Xinhuanet. http://www.xinhuanet.com/english/2019-05/24/c_138083964.htm. Accessed 26 Dec 2019

Xinhua (2019b) China's icebreakers to conduct researches in the Southern Ocean. China Daily. https://www.chinadailyhk.com/articles/235/141/2/1577257353310.html. Accessed 29 Dec 2019

Yeon-soo K (2019) Korea warned for illegal fishing activities by the US. The Korea Times. https://www.koreatimes.co.kr/www/biz/2019/09/367_275879.html. Accessed 29 Dec 2019

Impact of Antarctic Science on Climate Change Research: Global Research Landscape

Prabir G. Dastidar and Neloy Khare

Abstract Data obtained from the web of science has been subjected to detailed analyses to understand various trend and pattern emerged from the massive scientific literature available on Antarctic Climate Change science. Such results and the impact of previous studies have been exhaustively explained, discussed and collated in this chapter.

Keywords Web of science · Antarctica · Bibliometrics indicators · Intellectual structure · Research landscape

1 Introduction

Antarctica is the fifth-largest continent on earth. It is a cold desert and surrounded by the southern ocean. Gondwana was a supercontinent that existed 550 million years ago and started breaking during the Jurassic about 180 million years ago (Fig. 1). For millions of years, the fragmentation and drifting of the landmass (tectonic plates) have given rise to the present-day continents. Because of the continents' unique location, surrounded by turbulent ocean currents, the Antarctica plays a pivotal role in stabilizing the earth's climate system.

The ice sheet covers 98% of the Antarctic continent. Around 30 million years ago, separating South America and surrounding Antarctica Tectonic plate by Antarctic Circumpolar current made Antarctica more polar and cold. The vast ice sheet and southern ocean together act as a buffering zone and protects Antarctica from direct anthropogenic influences.

Rising temperatures and increasing snowmelt and ice loss are the reflections of the climate change in Antarctica. According to a recent reporting (https://phys.org/news/2020-02-antarctica-broken.html) of the summary study in 2018 incorporating calculations and data from many other studies estimated that total ice loss in Antarctica due to climate change was 43 gigatons per year on average during the period

P. G. Dastidar (✉) · N. Khare
Ministry of Earth Sciences, Lodhi Road, Prithvi Bhawan, New Delhi 110 003, India

N. Khare (ed.), *Assessing the Antarctic Environment from a Climate Change Perspective*, Earth and Environmental Sciences Library,
https://doi.org/10.1007/978-3-030-87078-2_2

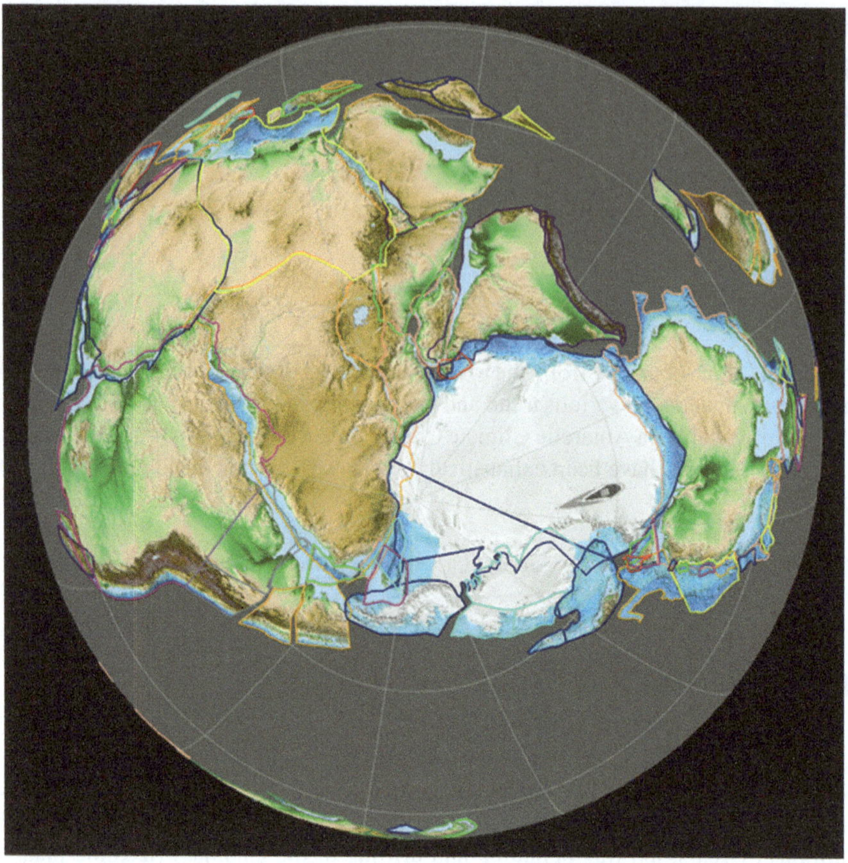

Fig. 1 Gondwanaland-existed 550 million years the Antarctic was the part of this supercontinent then. (*Source* https://commons.wikimedia.org/wiki/File:Gondwana_420_Ma.png)

from 1992 to 2002 but has accelerated to an average of 220 gigatons per year during the five years from 2012 to 2017 (Clem et al. 2020).

Some of Antarctica has been warming up with the extreme warming on the Antarctic Peninsula. On the contrary, it has also been observed that Antarctica's continent-wide average surface temperature trend was slightly positive from 1957 to 2006 (Dastidar and Ramchandran 2008). Over the second half of the twentieth century, the Antarctic Peninsula was the fastest-warming place on earth. West Antarctica followed it.

In the early 21st-century (https://discoveringantarctica.org.uk/challenges/sustainability/impacts-of-climate-change/), these trends were weakened. Interestingly, the South Pole in East Antarctica barely warmed last century. However, conversely, in the previous three decades, the temperature increase has been more than three times greater than the global average (Gillett et al. 2008). In February 2020, the continent recorded its highest temperature of 18.3 °C, a degree higher than the previous record of 17.5 °C in March 2015 (Glasser 2008).

Some evidence suggests that surface warming in Antarctica is owing to human greenhouse gas emissions (https://www.cnn.com), but this is difficult to determine due to internal variability (https://www.nytimes.com/2009/04/05/science/earth/05antarctica.html). Antarctica's main component of climate variability is the Southern Annular Mode, which showed strengthened winds around Antarctica in the summer of the later decades of the twentieth century, associated with cooler temperatures over the continent. The trend was at a scale unprecedented over the last 600 years; the most dominant driver of this mode of variability is likely the depletion of ozone above the continent (http://news.bbc.co.uk/2/hi/7984054.stm).

In 2002 the Antarctic Peninsula's Larsen-B ice shelf collapsed (https://www.cnn.com). Between 28 February and 8 March 2008, about 570 km^2 of ice from the Wilkins Ice Shelf on the southwest part of the peninsula collapsed, putting the remaining 15,000 km^2 of the ice shelf at risk. The ice was being held back by a "thread" of ice about 6 km (4 mi) wide (Meredith et al. 2019; Shepherd et al. 2018) before its collapse on 5 April 2009 (Stammerjohn and Scambos 2008; Steig et al. 2009).

Antarctica air temperature, particularly in the peninsular region, has experienced a rise of 3 °C. It is now established that Antarctic Circumpolar Current is warming more rapidly than the global ocean as a whole.

The warming of the Antarctic Peninsula is manifested in the changes to Antarctica's physical and living environment. The distribution of penguin colonies has changed as the sea ice conditions alter. Melting of perennial snow and ice covers has resulted in increased colonization by plants. The animals that inhabit the Antarctic will be affected by changing temperatures, sea ice levels, food availability, among many other adverse factors. In the past several years, ice loss in Antarctica has at least tripled, resulting in sea level rises worldwide. Undoubtedly, Antarctica's air and ocean are warming up. In the last 50 years, the peninsula warmed almost 3 degrees C, significantly higher than the global average of 0.9 degrees C. It is the highest temperature ever recorded in continental Antarctica.

It may be understood that significant threats to the icy and pristine continent 'Antarctica' have been climate change, Increased fishing pressure and illegal fishing, marine pollution, persistent organic pollutants (POPs) among others.

Studies on Climate change in Antarctica will throw light on climate change dynamics and their role in the global climate system. It also has an enormous impact on the Antarctic Food Web. The studies will also help politicians and policymakers to formulate policies to manage this global issue (Steig et al. 2013).

The importance of climate changes over Antarctica was perceived long back and documented in research publications in the early seventies. Gradually the importance of such detailed investigations of climate science over Antarctica and the surrounding Southern ocean started reflecting in the science mandate of every Antarctic Treaty Nation over this paper attempt have been made to visualize the global international research initiatives to address the issues related to the role of Antarctica in climate change and global warming.

Recognizing Antarctica's role in the global climate system, Antarctica Treaty Consultative Meeting (ATCM) formulated the Environmental Protocol, stating the position of research and activities in Antarctica's pristine environment, a continent for science and peace as declared by the United Nations.

There are three distinct primary objectives for the environmental Protocol:

(1) to protect the scientific value of the Antarctic.
(2) to help in the continuous improvement of Antarctic ecological management, and.
(3) to meet the legal requirements of the Protocol and national legislation.

Enormous data has been generated and is available in a plethora of research publications worldwide which ought to be collated to arrive at some informed conclusions on the prioritization of the climate science domain, recent trends and pattern of such climate-related research activities and also the ongoing climate issues having a direct or indirect impact over the Antarctic region. Such an analysis will help generate scenarios of climate change research over the Antarctic area. Thus, it may help policy decisions on the futuristic climate change issue over Antarctica and the entire world.

Bibliometrics indicators are routinely used to assess the growth of knowledge development in space and time. Reconstructing Antarctic science's intellectual structure was used to model the growth of the subject (https://www.webofscience.com/wos/woscc/basic-search). Considering the importance of Antarctic Science and its relevance in climate change research, it was felt that analyzing the research area will be of great help to understand the degree of research initiatives and their impact on the global climate system. Prolific organizations are involved in research on this subject of international importance. Most researched subject categories indicate global concerns addressed by scientists across the countries.

In this chapter, we have attempted to visualize global research initiatives on this topic. Considering the importance of the subject, there is a substantial rise in research activities (Tables 1 and 3).

Table 1 Number of research articles, highly cited and hot papers published in various journals by various authors from different Countries and Organization

Total papers	3467	Out of 171 million research articles published in impact factor journals between 1900 to present was used for the study
No. of highly cited papers	63	
No of hot papers	6	
Open access papers	1516	

Table 2 Parameters used to visualize the structure of research and innovation in Antarctic climate change research

Sl. No	Parameters	Remarks
1	Number of research articles	The number of research articles indicates the volume of research going on in a particular subject of interest. It may vary with time and space. Depending on the social requirement and economic and intellectual relevance research, this number may change from time to time
2	Number and quality of journals	Journals contain a wealth of information about the research articles—on research topics, the research's impact, its temporal and spatial distribution
3	Most cited research papers	The more a paper is cited, the more will be the impact of the paper. It may be arranged as per rank order, or may be categorized as: **Highly Cited Research papers:** Papers received top 1% citations in an academic field in a publication year **Hot papers:** Papers received top 0.1% citation in last two years in a particular academic field
4	Research area	Research articles indexed in Web of Science grouped under 250 research areas

2 Materials and Methods

Data was downloaded from the Web of Science Database (https://en.wikipedia.org/wiki/Climate_change_in_Antarctica). Following the search, the strategy was used for downloading data.

TITLE: antarctic* and ((global <near/2> warming) or (climate <near/2> change)).
OR

Abstract: antarctic* and ((global <near/2> warming) or (climate <near/2> change)).
OR

Author Keyword: antarctic* and ((global <near/2> warming) or (climate <near/2> change)).

We got the below-mentioned result for the above search query: (As of 27 January 2021).

The research landscape can be constructed using various bibliographic parameters, grouped under publication and citation. The number of research articles published in peer-reviewed journals, proceedings by countries, organizations, authors in different time frames indicates the importance, growth, evolution and focus of the subject to solve national, international and intellectual challenges. While citations reflect a degree of impact, a piece of research work is.

The following parameters were studied to visualize the intellectual structure of research and innovation in the realm of global warming and climate change from the perspective of Antarctica (Table 2).

The total number of research publications focused on climate change aspects in different scientific journals is reflected in Table 3. Interestingly Table 4 reflects the top 25 research publications in the field of Antarctic climate science.

The in-depth analyses suggest that since the early seventies, a burgeoning interest has been generated among polar scientists to pursue their research in the domain of climate change over the Antarctic region. Figure 2 indicates an exponential growth in climate science research, thus suggesting a worldwide consensus to underpin the importance of global climate changes and their profound impact on the polar regions.

Similarly, Fig. 3a indicates Top 10 subject categories involved in research, Fig. 3b Indicates Top 10 out of 105 countries involved in research in climate change vis-a-vis Antarctica.

Figure 3c indicates Top ten journals out of 923 journals publishing articles on Climate Change vis-a-vis Antarctica research.

Figure 3d indicates Top 15 top organizations out of 2544 involved in Antarctica and climate change research.

3 Conclusions

It may be concluded that phenomenal growth has been observed in Antarctic Climate science's research domain. The global interest and significance have been observed in the following aspects of the Antarctic Climate Change research.

Global analyses of sea surface temperature, sea ice, and night marine air temperature since the late nineteenth century; Climate and atmospheric history of the past 420,000 years from the Vostok ice core, Antarctica; Climate is forcing by anthropogenic aerosols; The phanerozoic record of global sea-level change; The Last Glacial Maximum; Representing twentieth-century space–time climate variability. Part II: Development of 1901–96 monthly grids of terrestrial surface climate; Representing twentieth-century space–time climate variability. Part I: Development of a 1961–90 mean monthly terrestrial climatology; Volcanic eruptions and climate; Interpretation of recent Southern Hemisphere climate change Sea-Level Rise and Its Impact on Coastal Zones; Hydrological changes in the African tropics since the Last Glacial Maximum; Climate variations and the physiological basis of temperature-dependent biogeography: systemic to molecular hierarchy of thermal tolerance in animals Abrupt Deep-Sea Warming, Palaeoceanographic Changes And Benthic Extinctions At The End of The Paleocene; The physiology of climate change: how potentials for acclimatization and genetic adaptation will determine 'winners' and 'losers'; Hemispheric and large-scale surface air temperature variations: An extensive revision and an update to 2001; One-to-one coupling of glacial climate variability in Greenland and Antarctica; The anthropogenic greenhouse era began thousands of years ago; Atmospheric CO_2 concentrations over the last glacial termination; Attributing physical and biological impacts to anthropogenic climate change; Timing of millennial-scale climate change in Antarctica and Greenland during the last glacial period; Oxygen- and capacity-limitation of thermal tolerance: a matrix for integrating

Table 3 Number of research publications on climate science between 1972 and 2020

As of 27 January 2021	
Publication years	No of the research articles
2020	431
2019	416
2018	440
2017	391
2016	340
2015	315
2014	318
2013	357
2012	301
2011	276
2010	236
2009	191
2008	213
2007	181
2006	170
2005	135
2004	126
2003	101
2002	90
2001	77
2000	69
1999	50
1998	57
1997	49
1996	22
1995	23
1994	20
1993	26
1992	28
1991	18
1990	3
1989	2
1988	1
1987	1
1986	1
1983	1
1977	1
1972	1

Table 4 The top 25 research publications in the field of Antarctic climate science

S. No	Title	Authors	Journal name	Research area	Year	Times cited	Journal normalized citation impact	Category normalized citation impact	Journal impact factor	Highly cited (Yes/No)
1	Global analyses of sea surface temperature, sea ice, and night marine air temperature since the late nineteenth century	Rayner, NA; Parker, DE; Horton, EB; Folland, CK; Alexander, LV; Rowell, DP; Kent, EC; Kaplan, A	Journal Of Geophysical Research-Atmospheres	Meteorology & Atmospheric Sciences	2003	5930	86.38	122.52	3.82	No
2	Climate and atmospheric history of the past 420,000 years from the Vostok ice core, Antarctica	Petit, JR; Jouzel, J; Raynaud, D; Barkov, NI; Barnola, JM; Basile, I; Bender, M; Chappellaz, J; Davis, M; Delaygue, G; Delmotte, M; Kotlyakov, VM; Legrand, M; Lipenkov, VY; Lorius, C; Pepin, L; Ritz, C; Saltzman, E; Stievenard, M	Nature	Multidisciplinary Sciences	1999	3650	9.32	79.31	42.78	No
3	CLIMATE FORCING BY ANTHROPOGENIC AEROSOLS	Charlson, RJ; Schwartz, SE; Hales, JM; Cess, RD; Coakley, JA; Hansen, JE; Hofmann, DJ	Science	Meteorology & Atmospheric Sciences	1992	2547	8.52	58.23	41.85	No
4	The phanerozoic record of global sea-level change	Miller, KG; Kominz, MA; Browning, JV; Wright, JD; Mountain, GS; Katz, ME; Sugarman, PJ; Cramer, BS; Christie-Blick, N; Pekar, SF	Science	GEOSCIENCES, MULTI-DISCIPLINARY	2005	1858	1.87	15.63	41.85	No
5	The Last Glacial Maximum	Clark, Peter U.; Dyke, Arthur S.; Shakun, Jeremy D.; Carlson, Anders E.; Clark, Jorie; Wohlfarth, Barbara; Mitrovica, Jerry X.; Hostetler, Steven W.; McCabe, A. Marshall	Science	GEOSCIENCES, MULTI-DISCIPLINARY	2009	1573	4.61	53.55	41.85	No
6	Representing twentieth-century space–time climate variability. Part II: Development of 1901–96 monthly grids of terrestrial surface climate	New, M; Hulme, M; Jones, P	Journal Of Climate	METEOROLOGY & ATMOSPHERIC SCIENCES	2000	1558	14.03	32.75	5.71	No

(continued)

Table 4 (continued)

S. No	Title	Authors	Journal name	Research area	Year	Times cited	Journal normalized citation impact	Category normalized citation impact	Journal impact factor	Highly cited (Yes/No)
7	Representing twentieth-century space–time climate variability. Part I: Development of a 1961–90 mean monthly terrestrial climatology	New, M; Hulme, M; Jones, P	Journal Of Climate	METEOROLOGY & ATMOSPHERIC SCIENCES	1999	1398	12.79	30.58	5.71	No
8	Volcanic eruptions and climate	Robock, A	Reviews Of Geophysics	GEOCHEMISTRY & GEOPHYSICS	2000	1313	5.08	10.05	21.45	No
9	Interpretation of recent Southern Hemisphere climate change	Thompson, DWJ; Solomon, S	Science	METEOROLOGY & ATMOSPHERIC SCIENCES	2002	1192	2.61	25.67	41.85	No
10	Sea-Level Rise and Its Impact on Coastal Zones	Nicholls, Robert J.; Cazenave, Anny	Science	GEOSCIENCES, MULTI-DISCIPLINARY	2010	1105	1.67	10.15	41.85	Yes
11	Hydrological changes in the African tropics since the Last Glacial Maximum	Gasse, F	Quaternary Science Reviews	GEOSCIENCES, MULTI-DISCIPLINARY; GEOGRAPHY, PHYSICAL	2000	1039	7.94	24.10	3.80	No
12	Climate variations and the physiological basis of temperature-dependent biogeography: systemic to molecular hierarchy of thermal tolerance in animals	Portner, HO	Comparative Biochemistry And Physiology A-Molecular & Integrative Physiology	PHYSIOLOGY; ZOOLOGY; BIOCHEMISTRY & MOLECULAR BIOLOGY	2002	886	21.1	22.59	1.97	No

(continued)

Table 4 (continued)

S. No	Title	Authors	Journal name	Research area	Year	Times cited	Journal normalized citation impact	Category normalized citation impact	Journal impact factor	Highly cited (Yes/No)
13	Abrupt Deep-Sea Warming, Palaeoceanographic Changes And Benthic Extinctions At The End Of The Paleocene	Kennett, JP; Stott, LD	*Nature*	MULTI-DISCIPLINARY SCIENCES	1991	886	2.63	26.67	42.78	No
14	The physiology of climate change: how potentials for acclimatization and genetic adaptation will determine 'winners' and 'losers.'	Somero, G. N	Journal Of Experimental Biology	BIOLOGY	2010	876	23.42	31.99	3.01	Yes
15	Hemispheric and large-scale surface air temperature variations: An extensive revision and an update to 2001	Jones, PD; Moberg, A	Journal Of Climate	Meteorology & Atmospheric Sciences	2003	864	9.03	17.85	5.71	No

(continued)

Table 4 (continued)

S. No	Title	Authors	Journal name	Research area	Year	Times cited	Journal normalized citation impact	Category normalized citation impact	Journal impact factor	Highly cited (Yes/No)
16	One-to-one coupling of glacial climate variability in Greenland and Antarctica	Barbante, C.; Barnola, J. -M.; Becagli, S.; Beer, J.; Bigler, M.; Boutron, C.; Blunier, T.; Castellano, E.; Cattani, O.; Chappellaz, J.; Dahl-Jensen, D.; Debret, M.; Delmonte, B.; Dick, D.; Falourd, S.; Faria, S.; Federer, U.; Fischer, H.; Freitag, J.; Frenzel, A.; Fritzsche, D.; Fundel, F.; Gabrielli, P.; Gaspari, V.; Gersonde, R.; Graf, W.; Grigoriev, D.; Hamann, I.; Hansson, M.; Hoffmann, G.; Hutterli, M. A.; Huybrechts, P.; Isaksson, E.; Johnsen, S.; Jouzel, J.; Kaczmarska, M.; Karlin, T.; Kaufmann, P.; Kipfstuhl, S.; Kohno, M.; Lambert, F.; Lambrecht, Anja; Lambrecht, Astrid; Landais, A.; Lawer, G.; Leuenberger, M.; Littot, G.; Loulergue, L.; Luethi, D.; Maggi, V.; Marino, F.; Masson-Delmotte, V.; Meyer, H.; Miller, H.; Mulvaney, R.; Narcisi, B.; Oerlemans, J.; Oerter, H.; Parrenin, F.; Petit, J. -R.; Raisbeck, G.; Raynaud, D.; Roethlisberger, R.; Ruth, U.; Rybak, O.; Severi, M.; Schmitt, J.; Schwander, J.; Siegenthaler, U.; Siggaard-Andersen, M. -L.; Spahni, R.; Steffensen, J. P.; Stenni, B.; Stocker, T. F.; Tison, J. -L.; Traversi, R.; Udisti, R.; Valero-Delgado, F.; van den Broeke, M. R.; van De Wal, R. S. W.; Wagenbach, D.; Wegner, A.; Weiler, K.; Wilhelms, F.; Winther, J. -G.; Wolff, E	Nature	Multidisciplinary Sciences	2006	797	1.95	12.31	42.78	No

(continued)

Table 4 (continued)

S. No	Title	Authors	Journal name	Research area	Year	Times cited	Journal normalized citation impact	Category normalized citation impact	Journal impact factor	Highly cited (Yes/No)
17	The anthropogenic greenhouse era began thousands of years ago	Ruddiman, WF	Climatic Change	Environmental Sciences; Meteorology & Atmospheric Sciences	2003	782	10.16	16.92	4.13	No
18	Atmospheric CO2 concentrations over the last glacial termination	Monnin, E; Indermuhle, A; Dallenbach, A; Fluckiger, J; Stauffer, B; Stocker, TF; Raynaud, D; Barnola, JM	Science	Multidisciplinary Sciences	2001	776	1.72	7.43	41.85	No
19	Attributing physical and biological impacts to anthropogenic climate change	Rosenzweig, Cynthia; Karoly, David; Vicarelli, Marta; Neofotis, Peter; Wu, Qigang; Casassa, Gino; Menzel, Annette; Root, Terry L.; Estrella, Nicole; Seguin, Bernard; Tryjanowski, Piotr; Liu, Chunzhen; Rawlins, Samuel; Imeson, Anton	Nature	Meteorology & Atmospheric Sciences	2008	771	1.89	21.77	42.78	No
20	Timing of millennial-scale climate change in Antarctica and Greenland during the last glacial period	Blunier, T; Brook, EJ	Science	Geosciences, Multidisciplinary	2001	748	1.66	20.60	41.85	No
21	Oxygen- and capacity-limitation of thermal tolerance: a matrix for integrating climate-related stressor effects in marine ecosystems	Poertner, H.-O	Journal Of Experimental Biology	Biology	2010	738	19.73	26.95	3.01	Yes

(continued)

Table 4 (continued)

S. No	Title	Authors	Journal name	Research area	Year	Times cited	Journal normalized citation impact	Category normalized citation impact	Journal impact factor	Highly cited (Yes/No)
22	Recent rapid regional climate warming on the Antarctic Peninsula	Vaughan, DG; Marshall, GJ; Connolley, WM; Parkinson, C; Mulvaney, R; Hodgson, DA; King, JC; Pudsey, CJ; Turner, J	Climatic Change	Meteorology & Atmospheric Sciences; Environmental Sciences	2003	709	2.03	4.46	4.13	No
23	Climate change and temperature-dependent biogeography: oxygen limitation of thermal tolerance in animals	Portner, HO	Naturwissenschaften ???	Physiology	2001	688	9.42	3.36	n/a	No
24	Antarctic climate change during the last 50 years	Turner, J; Colwell, SR; Marshall, GJ; Lachlan-Cope, TA; Carleton, AM; Jones, PD; Lagun, V; Reid, PA; Iagovkina, S	International Journal Of Climatology	Meteorology & Atmospheric Sciences	2005	678	3.6	15.03	3.93	No
25	Antarctic ice-sheet loss driven by basal melting of ice shelves	Pritchard, H. D.; Ligtenberg, S. R. M.; Fricker, H. A.; Vaughan, D. G.; van den Broeke, M. R.; Padman, L	Nature	Geosciences, Multidisciplinary	2012	659	1.81	26.81	42.78	YEs

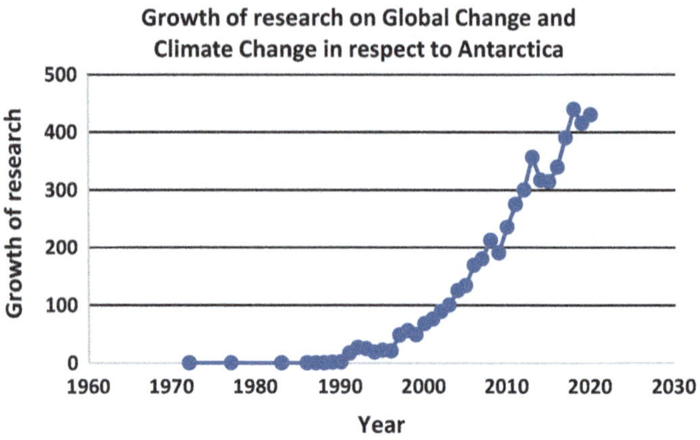

Fig. 2 Substantial increase in research initiatives in recent past

climate-related stressor effects in marine ecosystems; Recent rapid regional climate warming on the Antarctic Peninsula; Climate change and temperature-dependent biogeography: oxygen limitation of thermal tolerance in animals; Antarctic climate change during the last 50 years and Antarctic ice-sheet loss driven by basal melting of ice shelves.

Based on the research, it has now been indubitably agreed that the Antarctic is warming up and showing the symptoms of global climate changes by way of Antarctica's air and ocean are heating up; Ice is retreating and melting rapidly; Some penguin populations are shrinking; Snow is turning red while the land is getting greener.

The anthropogenic climate change known as global warming is now a reality. The concentrations of 'greenhouse gases' in the atmosphere are above their pre-industrial levels. It has happened mainly due to the overuse of fossil fuels (coal, oil, and natural gas) coupled with deforestation and agricultural practices. As greenhouse gases accumulate in the atmosphere, more heat energy from the earth's surface is intercepted before escaping to space, thereby causing the atmosphere to warm.

Nevertheless, the entire blame for ongoing climate change worldwide cannot be attributed to human interferences; instead, natural variations in climate occur on a range of different timescales. Therefore, it is made sense to understand that there are variations, both natural and human-caused, in how the greenhouse gases themselves cycle between atmosphere, land, and oceans. Understanding and predicting climate change are far more complex than perceived earlier.

Studies of Antarctica, particularly analyses of Antarctic ice cores, have played a crucial role in this endeavour. These ice cores have helped reconstruct past climate changes spanning hundreds of thousands of years. Past variations in climate due to natural causes are now far better understood due to multi-disciplinary integrated proxy records of past climate changes. Such baseline paleoclimate assessment has provided vital context against which to assess more recent changes caused by human

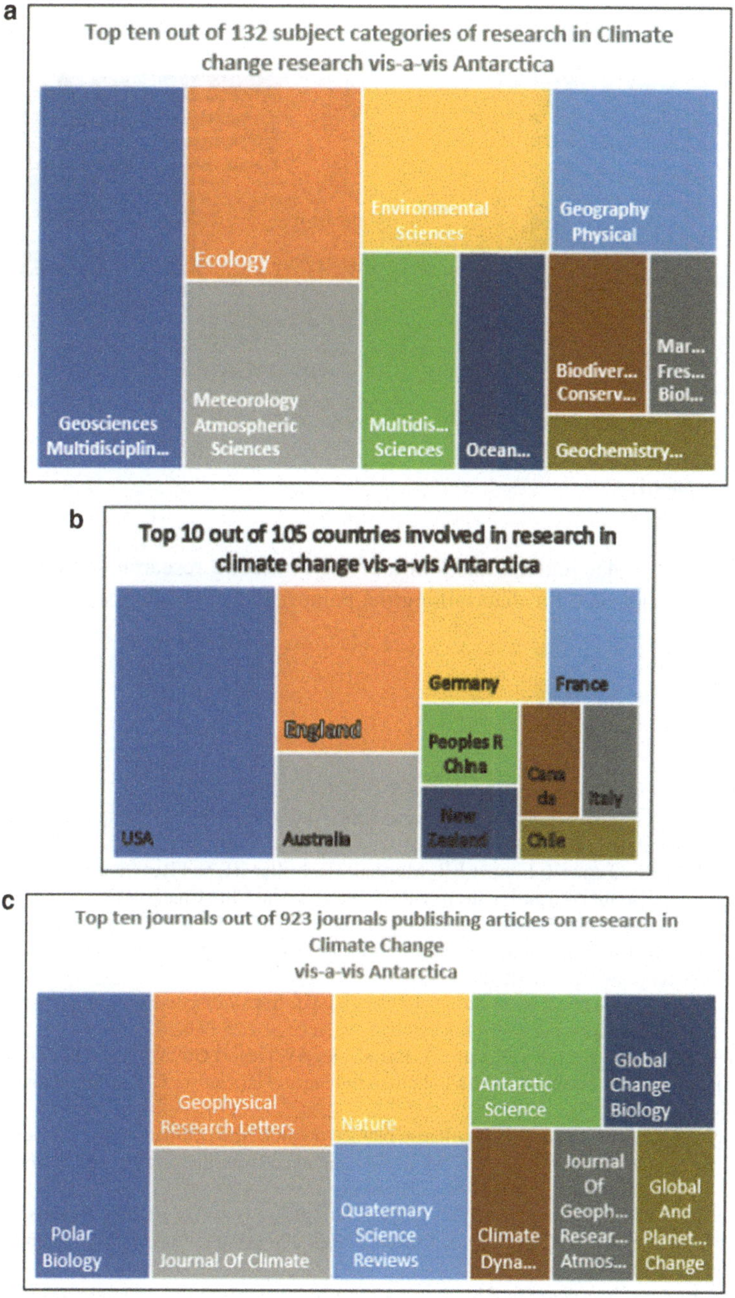

Fig. 3 **a** Top 10 subject categories involved in research **b** Top 10 out of 105 countries involved in research in climate change vis-a-vis Antarctica **c** Top ten journals out of 923 journals publishing articles on research in Climate Change vis-a-vis Antarctica **d** Top 15 top organizations out 2544 involved in research in Antarctica and climate change research

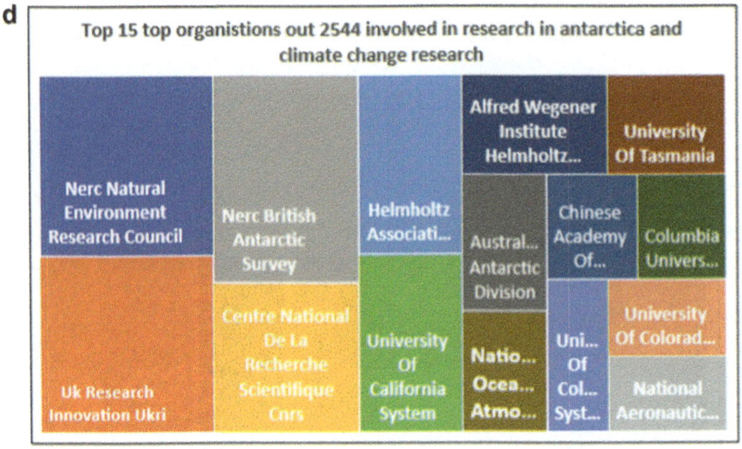

Fig. 3 (continued)

activities. The polar regions and some of the Antarctic research stations provide essential weather data for determining the polar response to the rise in greenhouse gases.

References

Antarctica appears to have broken a heat record. Retrieved 9 February 2020. https://phys.org/news/2020-02-antarctica-broken.html

Clem KR, Fogt RL, Turner J, Lintner BR, Marshall GJ, Miller JR, and Renwick JA (2020) Record warming at the South Pole during the past three decades. Nat Clim Chang 10(8):762–770. https://doi.org/10.1038/s41558-020-0815-z.ISSN1758-6798.S2CID220261150

Dastidar PG, and Ramachandran S (2008) Intellectual structure of antarctic science: a 25-years analysis. Scientometrics 77(3):3890–4414

Discovering antarctica (2021) Impact of climate change. https://discoveringantarctica.org.uk/challenges/sustainability/impacts-of-climate-change/. Accessed 25 Feb 2021

Gillett NP, Stone DIA, Stott PA, Nozawa T, Karpechko AY, Hegerl GC, Wehner MF, and Jones PD (2008) Attribution of polar warming to human influence. Nat Geosci 1(11):750. https://doi.org/10.1038/ngeo338

Glasser N (2008) Antarctic ice shelf collapse blamed on more than climate change. ScienceDaily

Huge Antarctic ice chunk collapses. CNN (2008) Archived from the original on 29 March 2008. Retrieved 25 March 2008 (https://www.cnn.com)

Ice bridge holding Antarctic shelf in place shatters (2009) The New York Times. Reuters. Archived from the original on 16 April 2009. Retrieved 5 April 2009 (https://www.nytimes.com/2009/04/05/science/earth/05antarctica.html)

Ice bridge ruptures in Antarctic. BBC News (2009) Archived from the original on 6 April 2009. Retrieved 5 April 2009 (http://news.bbc.co.uk/2/hi/7984054.stm)

Massive ice shelf on verge of breakup. CNN (2008) Archived from the original on 29 March 2008. Retrieved 26 March 2008 (https://www.cnn.com)

Meredith M et al (2019) Chapter 3: Polar regions (PDF). IPCC special report on the ocean and cryosphere in a changing climate. p 212

Shepherd A et al (2018) Mass balance of the Antarctic Ice Sheet from 1992 to 2017 (PDF). Nature 558(7709):219–222. https://doi.org/10.1038/s41586-018-0179-y

Stammerjohn SE, and Scambos TA (2020) Warming reaches the South Pole. Nat Clim Chang 10(8):710–711. https://doi.org/10.1038/s41558-020-0827-8.ISSN1758-6798.S2CID220260051

Steig EJ, Schneider DP, Rutherford SD, Mann ME, Comiso JC, and Shindell DT (2009) Warming of the Antarctic ice-sheet surface since the 1957 International Geophysical Year. Nature 457(7228):459–462. https://doi.org/10.1038/nature07669

Steig EJ, Ding Q, White JWC, Küttel M, Rupper SB, Neumann TA, Neff PD, Gallant AJE, Mayewski PA, Taylor KC, Hoffmann G, Dixon DA, Schoenemann SW, Markle BR, Fudge TJ, Schneider DP, Schauer AJ, Teel RP, Vaughn BH, Burgener L, Williams J, and Korotkikh E (2013) Recent climate and ice-sheet changes in West Antarctica compared with the past 2,000 years. Nat Geosci 6(5):372. https://doi.org/10.1038/ngeo1778

Web of Science. https://www.webofscience.com/wos/woscc/basic-search

https://en.wikipedia.org/wiki/Climate_change_in_Antarctica

Climate Change Over the Antarctic and the Southern Ocean and Its Impact and Bearing on the Global Climate System

Shabnam Choudhary and Neloy Khare

Abstract Antarctica and surrounding regions will have global consequences as a result of climate change. It is relentless to understand how the anthropogenic climate change will affect the vast continent and surrounding ocean, but for the impacts such as sea-level rise accurate projections must be prepared. A thorough understanding of the causes and effects of climate change and the duration and rates of change is required involving the integration of observational and modelling knowledge from Earth systems.

Keywords Antarctica · Climate · Southern Annular Mode (SAM) · Ozone · Stratosphere

1 Introduction

Climate change and the impacts over the Arctic are well studied and discussed often, however, in the Southern Hemisphere, Antarctic region climate change is comparatively neglected or reported misleadingly. Antarctica is divided into two geologically distinct regions, East Antarctica and West Antarctica. These two regions are separated by Trans-Antarctic mountains but are joined together by a massive ice sheet (Fig. 1). Antarctica acts as a powerful heat sink affecting strongly the Earth's climate because of the massive ice sheet and its polar location. The annual sea ice cover around the continent modulates the exchange of heat, moisture, and gases between the atmosphere and the ocean. It forms the cold oceanic bottom waters through salt rejection when it freezes, which spread out under the world's oceans. Any changes or perturbations to this system will influence the climate throughout the globe.

S. Choudhary (✉)
National Centre for Polar and Ocean Research, Vasco da Gama, Goa-403802, India

Present Address:
N. Khare
Ministry of Earth Sciences, Lodhi Road, Prithvi Bhawan, New Delhi 110 003, India

N. Khare (ed.), *Assessing the Antarctic Environment from a Climate Change Perspective*, Earth and Environmental Sciences Library,
https://doi.org/10.1007/978-3-030-87078-2_3

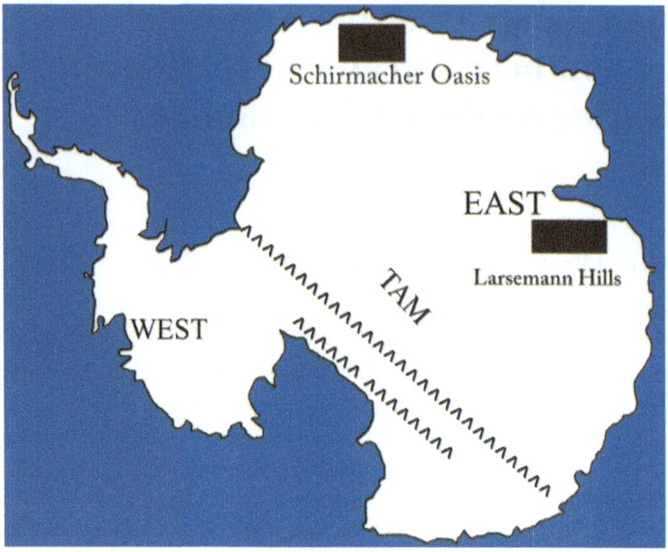

Fig. 1 East Antarctica and West Antarctica separated by Trans-Antarctic Mountains (TAM)

Overall sea ice in the Antarctic region had a slight increase from 1981 to 2010, while areas, such as the western part of the Antarctic Peninsula, showed a decreasing trend (Fig. 2). Short-term trends such as those observed, in the Southern Ocean, occur readily from the natural variability of the sea ice, ocean and atmosphere. The change in sea ice pattern in the Antarctic is attributed to the changes in surface wind patterns around the continent and freshwater that is cold coming from melting ice shelves have also played an important role. However, the Antarctic ice extent began to decline after 2014, reducing a record low in 2017 and further low in the next two years.

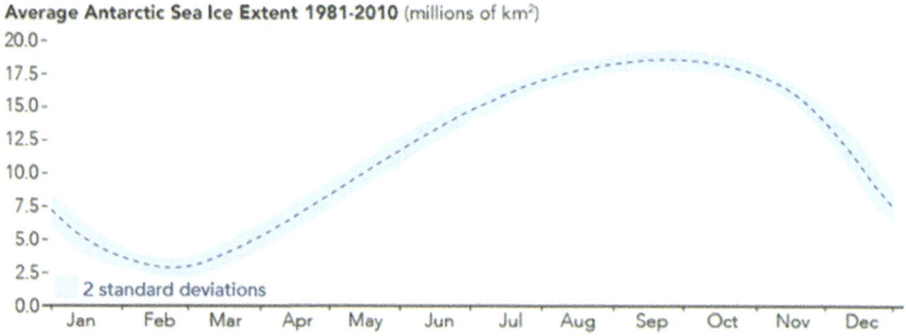

Fig. 2 Sea ice around Antarctica peaks in september and reaches a minimum in february (National Snow and Ice Data Center (NSIDC)

During the annual cycle, approximately 15 million sq. Km of ice melt and freeze (National Snow and Ice Data Center). The ice sheet and the sea ice are subject to climate change potentially. The West Antarctic ice sheet (WAIS) possess a significant threat to the inhabited world as it rests on a bed far below sea level and has the potential for shrinking rapidly. The Antarctic is so vast, remote, and challenging to monitor. The physical behaviour of the ice sheet is so complex that it does not provide any proof of definite changes. Even though in the northern part of the continent pronounced climatic warming is going on. The Antarctic ice sheet consists of a huge amount of ice if melted completely can raise worldwide sea level by more than 60 m. The annual amount of snow deposited on the ice sheet is equal to about 5 mm of global sea level. It is the mean annual discharge of ice back into the ocean. Thus, a significant contributor to the present-day rise in sea level (1.5–2 mm per year) along with the uncertainty is an imbalance between the input and output of ice.

The west coast of the Antarctic Peninsula is one of the rapidly warming portions of the planet over the past 50 years. This warming can also be observed in the Southern Ocean and is not only limited to the land. Towards the west of the Antarctic Peninsula, Upper-ocean temperatures have increased over 1 °C since 1955. The Antarctic Circumpolar Current is warming more rapidly than the global ocean as a whole it has been very well established.

In Antarctica studying climate change is necessary because it enables scientists to predict future climate change more accurately and serve as an important piece of information to politicians and policymakers. It is believed by the Antarctic Southern Ocean and Coalition (ASOC) that in Antarctica understanding climate change impacts are a matter of critical importance for the continent itself and the whole world.

2 Changes in Southern Hemisphere Climate

The Southern Ocean is surrounded by the Antarctic circumpolar current (ACC) which regulates the mid-to high-latitude southern hemisphere climate and its variability (Olbers et al. 2004). The Antarctic Circumpolar Current is an eastward geostrophic current around the Antarctic continent displaying zonal circulation in the form of series of fronts and turbulent jets (Mayewski et al. 2009). In this region, the southern annular mode (SAM), also called the Antarctic Oscillation, is the important mode of atmospheric variability controlling the intraseasonal-to-interannual variability in wind, temperature and precipitation. These variables display important and widespread changes that follow the SAM (van den Broeke and van Leipzig 2004). The atmospheric pressure gradient between the Antarctic low and the sub-Antarctic high regulates the westerlies' strength. The SAM index is used to quantify the strength variation of the SAM. It is defined as the main component of the 700 hPa atmospheric geopotential height anomalies in the Antarctic region. It is known that eastward wind stress over the Southern Ocean regulates the Antarctic Circumpolar Current (Thompson and Wallace 2000). In the southern part of the world, the Southern

Ocean and Westerly Winds are a critically coupled system regulating the climate. This system is related to the position of the Intertropical Convergence Zone and releasing of Carbon dioxide from the deep ocean.

The Antarctic Circumpolar Current (ACC) is driven by the westerly winds in the southern hemisphere and encircles Antarctica. It can flow unobstructed around the continent of Antarctica through the Drake Passage and becoming more stronger as the Westerly Winds are contracting towards the pole in the Southern Annular Mode positive phase as in the current scenario. The northern boundary of the Antarctic Circumpolar Current is the Subantarctic Front (SAF), and Polar Front (PF) forms the Southern Boundary. Circumpolar Deep Water (CDW) is a warmer layer of water beneath the calmer surface centred at a depth of ~500 m. It is dense and warm, with high salinity. The depth of the Antarctic continental shelf is around 600 m and as it heads inland it deepens. The height of the CDW in the water column and the local bathymetry determines the Circumpolar Deep Water's ability to access the continental shelf i.e. more the height of Circumpolar Deepwater there is a higher possibility that it will reach the continental shelf. The westerly winds in the southern region have become stronger leading to the increased circulation of circumpolar deep water onto the continental shelf, where it can reach the grounding lines percolating through the cavities in the ice shelf which ultimately can lead to ice sheet instability and then collapse.

2.1 The Southern Annular Mode

The Southern Annular Mode is the movement of a south-westerly wind belt in a north–south direction over a decadal to century timescale. It is a vital component of climate affecting the glaciers in the Southern Hemisphere and their response to changing climate. Factors driving the glaciation in the southern hemisphere and the bihemispheric approach comparing both the hemispheres, whether there is phase lag in Antarctica or arctic is explained by the Southern Annular Mode (Antarctic Oscillation; AO). The difference in the zonal mean sea level pressure at 40°S (mid-latitudes) and 65°S (Antarctica) is defined as the Southern Annular Mode. The strength and position of the westerly winds are driven by the changes in air pressure distribution. Southern Westerly Wind strength is usually measured by the SAM index (Holland et al. 2020).

2.2 Positive Southern Annular Mode

In the Positive phase of the Southern Annular Mode, Antarctica experience lower air pressure which is abnormal while mid-latitudes experiences higher air pressure (Lee et al. 2019. This is a present scenario where the westerly wind belt strengthens and contracts towards Antarctica and becomes weak in the mid-latitudes. This creates drier conditions over Patagonia resulting in receding glaciers. Upwelling of Circumpolar Deepwater onto the continental shelf in Antarctica will help in receding glaciers and ice sheets (Rignot et al. 2019) (Fig. 3a).

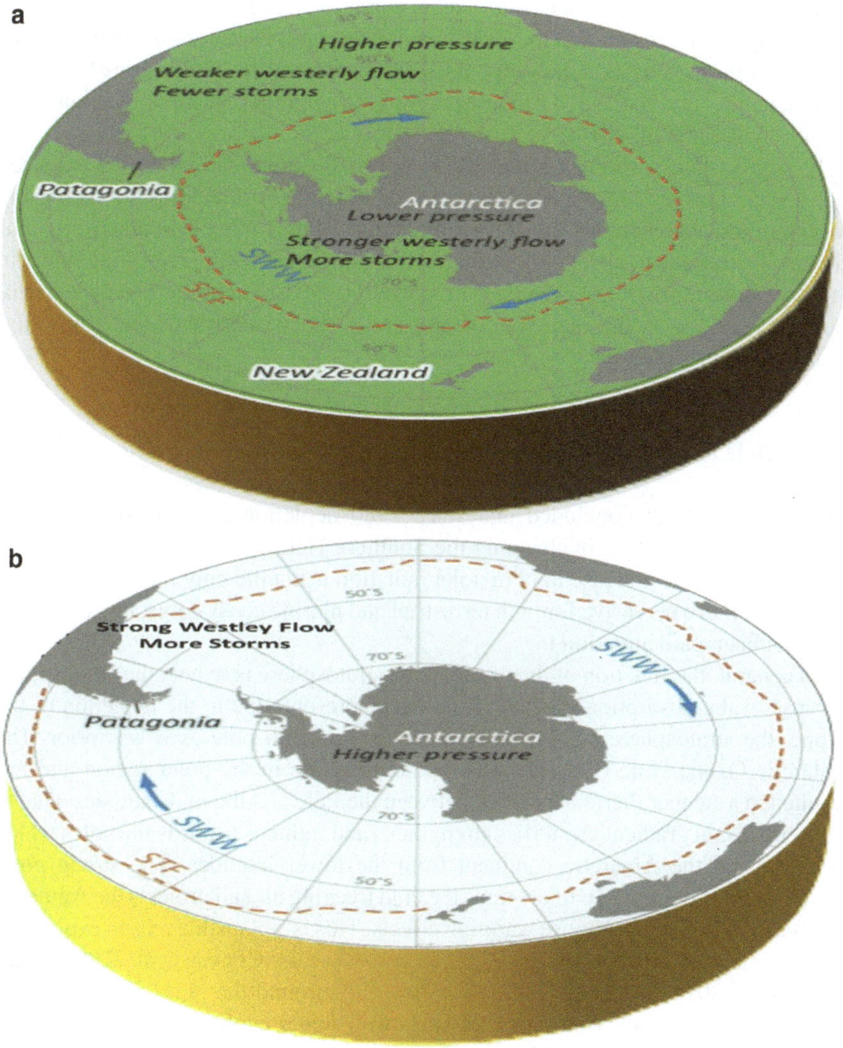

Fig. 3 **a** Positive (+) Southern Annular Mode (*Source* Antarctic glaciers.org) **b** Negative (−) Phase of Southern Annular Mode (*Source* Antarctic Glaciers.org)

2.3 Negative Southern Annular Mode

In a Negative phase of Southern Annular Mode, the southern belt of Westerly Winds moves towards the north near the equatorial region. This leads to glacier advances in Patagonia as a result of wet and cold weather conditions. Consequently, the upwelling of circumpolar Deepwater on the shelf region of Antarctica decreases. In this phase, winds are relatively weaker. In Patagonia, these conditions prevailed during Holocene neoglaciations and the Last Glacial Maximum (Sagredo et al. 2018; Kaplan et al. 2020; Reynhout et al. 2019).

Presently, the Southern Annular Mode is in a Positive phase and it is expected that it will continue to be in a positive phase which is attributed to the increasing concentration of greenhouse gas emissions. Increasing Positivity in Southern Annular mode will continue to change the strength of southwesterly winds resulting in warm and dry conditions over Patagonia. The upwelling of Circumpolar Deepwater will increase resulting in the recession of glaciers in the Western part and peninsular regions of Antarctica (Fig. 3b).

2.3.1 Role of Ozone Hole in Southern Hemisphere Climate

Climatologists have concluded that since 1980 depletion of stratospheric ozone played a significant role in changing the Southern Hemisphere climate. Due to the depletion of ozone increase in ultraviolet radiation is not the only impact, the implications are more pervasive for both terrestrial and marine ecosystems, however, they have not been paid attention to.

In general, the function of the ozone in the stratosphere is to heat the stratosphere because of the absorption of solar radiation. Therefore, due to the depletion of the ozone, the stratosphere has become cooler in the south pole as it was prior. The Antarctic Ozone Hole (Fig. 4) cools the stratosphere in the polar region and has resulted in a steeper thermal gradient between the Pole and the mid-latitudes. Due to steeper thermal gradient westerlies strengthens and tightens towards the poles which have isolated the Antarctic continent from the lower latitudes. The strong polar vortex on the Antarctic landmass traps the cold freezing air and protects the Antarctic landmass from the greenhouse warming effects. This is a plausible reason explaining the cooling trend in East Antarctica for the past 30 years (Convey et al. 2009; Turner et al. 2009, 2014; Wu et al. 2013). The heat flux around the poles is reduced due to the loss in stratospheric ozone which showed subtle cooling in East Antarctica. Ozone depletion has protected Antarctica largely from the warming in the southern hemisphere radiating across the globe. Effectively, ozone depletion has helped shield a large part of Antarctica from the Southern Hemisphere warming with consequences that radiate across the globe. The ecosystem of the Antarctica and rest of the globe is affected by maintaining very low temperatures over the Antarctic region. (Turner et al. 2014). It is expected by the scientific community that recovery of ozone concentrations by the next century then Southern Annular mode will be relatively

Fig. 4 Ozone Hole over Antarctica

weaker. Due to which warming in the southern hemisphere can affect Antarctica which can be studied by understanding the impact of Greenhouse gases on the Southern Annular Mode index (Abram et al. 2014; Dixon et al. 2012; Perlwitz 2011).

The sea and the atmosphere closely couple the climate, physical and biological properties of the Antarctic continent and the surrounding oceans with other parts of the globe. For example, the most important scientific discoveries of the last thousand years are the Antarctic ozone hole. It has protected Antarctica from greenhouse warming for the last 30 years and the Southern Ocean continued warming. Across the Antarctic Peninsula, plant communities have been expanded rapidly. Paleoclimatic studies in the Antarctic region exhibited that the current changes in climate are unusual and they are not natural as they are taking place at an unprecedented rate. Antarctica is expected to get warmer by approx. 3 °C if the concentration of the greenhouse gas is doubled by the next century. Still, there are significant knowledge gaps although the data are collected, analysed and modelled daily. Change in Global and Antarctic climate will always be a significant component of SCAR and an area of interest for researchers. Researchers have brought the dynamics of the polar climate system at its elementary level however, still, there are some limitations in predicting climate change. There is still a lot to be done to improve the predictions of climate change and the role of Antarctica in that. The Antarctic region is experiencing climate changes that have a global effect. Shreds of evidence of anthropogenic contributions of changing climate in the Antarctic atmosphere and the Southern Ocean have increased. According to Lenaerts et al. (2018) decrease in springtime stratospheric ozone (the 'ozone hole') had led to an increase in snowfall across the continent. It is affected by the Antarctic ice sheet mass balance and ultimately the rise in sea level. The observed and modelled warming of the Southern Ocean by Swart et al. (2018) showed that it is not consistent with the natural variabilities. The primary source of these variabilities is considered to be greenhouse gas concentrations induced by anthropogenic activities. From 1979 to 2017, the Antarctica ice sheet has lost ice relatively six-fold annually. Between 1992 and 2017, the Antarctic Ice Sheet has lost 2,720 ± 1,390 gigatons (Gt) of ice which is attributed to 7.6 ± 3.9 mm of increase in mean sea level. During this period, West Antarctica has lost the ice roughly triple the amount driven by the oceans and ice shelf collapse to almost five folds (The IMBIE team, 2018). A rise in sea level averaged 3.6 ± 0.5 mm per decade with a cumulative 14.0 ± 2.0 mm since 1979 is contributed by the Antarctic ice sheet. In the Western part of Antarctica, largely the ice loss occurs because of the melting of the ice shelves from below due to the warm ocean water entrainment as compared to East Antarctica and the Peninsular region (Rintoul 2018). Ice sheet stability is decided by the grounding line of glaciers it is the point where the ice starts to float. Between 2010 and 2016 it has been observed by Konrad et al. (2018) that the grounding lines are retreating at 25 m per year or more than that. It is concluded by satellite observations of changing elevations of ice. The retreat rate of the grounding lines was different for West Antarctica, East Antarctica and Antarctic Peninsula which is 22%, 3% and 10% respectively. Since 2014 extent of Arctic Sea ice declined to a greater extent. There is a significant increase in Antarctic Sea ice from 1970 to 2014 and in 2014 it has reached a maximum annual mean extent of 17.41 million sq. Km. In the satellite era, this increase in the southern hemisphere is followed by an unprecedented decrease of 16 million sq. Km in 2017 (Turner et al. 2017). Tropical convection in the Western Pacific and Indian oceans along with the polar vortex's of the stratosphere has created very low summer temperature in the Antarctic sea ice observed by Wang et al. (2019) in 2016. In the ocean–atmosphere system, unusual natural variability is exhibited by these processes. In the Indian Ocean, it might have been influenced by anthropogenic forcings. The gradual recovery of the ice extent started in 2018 which was much lower than the annual mean. Ice shelves can become prone to breakage in the absence of sea ice. In recent decades, the collapse of Larsen A and B and Wilkins ice shelves are important glaciological events. Decreasing sea ice extent has exposed the ice shelves for coastal erosion with the help of ocean swells resulting in the weakening of ice shelves by calving (Massom et al. 2018). Antarctic bottom water' (AABW). is cold and dense, its composition is changing continuously. Antarctic bottom water regulates the climate globally as it is an important component of the overturning circulation. Since 1944 in the Indian sector of the Southern Ocean Bathymetric surveys have been carried out repeatedly and revealed that Antarctic Bottom Water is becoming fresh, warmer and less dense and spreading climate change signals all over the globe (Menezes et al. 2017). Long term climate changes are explained by the ice cores in West Antarctica. Ross sea is experiencing Dipole effects known as "Ross Sea Dipole" showing the reverse pattern of temperature between the Eastern and western parts of the Ross Sea. This has been revealed by the ice core dataset of the last 2700 years from the Ross Ice shelf, West Antarctic Ice sheet and Roosevelt Island. Southern Annular mode and tropical forcing appear to be a response of Dipole. During this period the west Antarctic was cooling and the eastern ross sea became warmer with an increased accumulation of snow while the western part of the Ross Sea does not show any significant trend. According to the Rice community (2017) all the regions are showing signs of warming from 17th century. Different climatological models have predicted that with increasing global warming as a result of an increase in greenhouse concentrations sea ice extent in Antarctica will decrease approximately 1/3rd by the end of this century. Since 2014, we have observed the decrease but we are not sure whether it has started a decline for the long term it is just an indication of the variabilities to be natural.

3 Conclusions

To understand the variability in climate and the forces that control future changes and responses to change, a detailed understanding of past climate is essential. Antarctica is ideal for studying local-to-global scale climate change as the continent is remote from direct human influence. No other approach or experiment can provide perspectives across a range of time scales different from deciphering past climate change through proxies archived in ice and sedimentary records. To fill the knowledge gap in past climate, retrieval of ice and sedimentary records is a prerequisite. Obtaining geological records of past Antarctic ice sheet dynamics and integrate this knowledge into coupled ice sheet-climate models are the primary objectives of the geoscience community. Improved models are critical to constrain and improve predictions of future changes in global and regional temperatures, ocean acidification, and sea level. To decipher Paleoclimate records and to improve integrated Earth system models, much remains to be accomplished.

References

ATCM XLII and CEP XXII (2019) Prague, Czech Republic IP136: Antarctic climate change and the environment

Abram NJ, Mulvaney R, Vimeux F, Phipps SJ, Turner J, England MH (2014) Evolution of the southern annular mode during the past millennium. Nat Clim Chang 4:564–569

Van den Broeke MR, Van Lipzig NPM (2004) Changes in Antarctic temperature, wind, and precipitation respond to the Antarctic oscillation. Ann Glaciol 39(1):119–126

Convey PC, Bindschadler R, Di Prisco G, Fahrbach E, Gutt J, Hodgson DA, Mayewski PA, Summerhayes CP, Turner J, Robinson SA (2009) Antarctic climate change and the environment. Antarct Sci 21:541–563

Dixon DA, Mayewski PA, Goodwin ID, Marshall GJ, Freeman R, Maasch KA, Sneed SB (2012) An ice-core proxy for northerly air mass incursions into West Antarctica. Int J Climatol 32:1455–1465

Holland DM, Nicholls KW, Basinski A (2020) The southern ocean and its interaction with the Antarctic ice sheet science 367(6484):1326 LP–1330

Kaplan MR, Strelin JA, Schaefer JM, Peltier C, Martini MA, Flores E, Winckler G, Schwartz R (2020) Holocene glacier behaviour worldwide Antarctic Peninsula and possible causes. Earth Planet Sci Lett 534:116077

Konrad H, Shepherd A, Gilbert L, Hogg AE, Mcmillan M, Muir A, Slater T (2018) Net retreat of Antarctic glacier grounding lines. Natures Geosci 11:258–262

Lee DY, Petersen MR, Lin W (2019) The southern annular mode and southern ocean surface westerly winds in E3SM. Earth Space Sci 6:2624–2643

Lenaerts J, Fyke J, Medley B (2018) The signature of ozone depletion in recent Antarctic precipitation change: a study with the community earth system model. Geophys Res Lett 45
Massom R, Scambos T, Bennets LG, Reid P, Squire VA, Stammerjohn S (2018) Antarctic ice shelf disintegration triggered by sea ice loss and ocean swell. Nature 558:383–389
Mayewski PA, Meredith MP, Summerhayes CP, Turner J, Worby A, Barrett PJ, Casassa G, Bertler NA, Bracegirdle T, Naveira Garabato AC, Bromwich D (2009) State of the Antarctic and Southern Ocean climate system. Rev Geophys 47(1)
Mayewski PA, Maasch KA, Dixon D (2013) West Antarctica's sensitivity to natural and human-forced climate change over the Holocene. J Quat Sci 28:40–48
Menezes V, Macdonald AM, Schatzman C (2017) Accelerated freshening of Antarctic bottom water over the last decade in the southern Indian ocean. Sci Adv 3:e-1601426
Nowlin WD Jr, Klinck JM (1986) The physics of the Antarctic circumpolar current. Rev Geophys 24(3):469–491
Olbers D, Borowski D, Völker C, Wölf JO (2004) The dynamical balance, transport and circulation of the Antarctic circumpolar current. Antarct Sci 16(4):439–470
Perlwitz J (2011) Tug of war on the Jet stream. Nat Clim Chang 1:29–31
RICE Community (2017) The Ross sea dipole—temperature, snow accumulation and sea ice variability in the Ross sea region, Antarctica, over the past 2,700 years. The climate of the past discussion
Reynhout SA, Sagredo EA, Kaplan MR, Aravena JC, Martini MA, Moreno PI, Rojas M, Schwartz R, Schaefer JM (2019) Holocene glacier fluctuations in Patagonia are modulated by summer insolation intensity and paced by southern Annular mode-like variability. Quatern Sci Rev 220:178–187
Rignot E, Mouginot J, Scheuchl B, Van Den Broeke M, Van Wessem MJ, Morlighem M (2019) Four decades of Antarctic Ice Sheet mass balance from 1979–2017. Proc Nat Acad Sci 116(4):1095–1103
Rintoul SR (2018) The global influence of localized dynamics in the southern ocean. Nature 558:209–218
Rintoul SR, Hughes C, Olbers D (2001) The Antarctic circumpolar current system. In: Siedler G, Church J, Gould J (eds) Ocean circulation and climate. Academic Press, London, pp 271–302
Sagredo EA, Kaplan MR, Araya PS, Lowell TV, Aravena JC, Moreno PI, Kelly MA, Schaefer JM (2018) Trans-pacific glacial response to the Antarctic cold reversal in the southern mid-latitudes. Quatern Sci Rev 188:160–166
Swart NC, Gille JC, Fyfe JC, Gillette NP (2018) Recent southern ocean warming and freshening driven by greenhouse gas emissions and ozone depletion. Nat Geosci 11
Thompson DWJ, Wallace JM (2000) Annular modes in the extratropical circulation. Part I: month-to-month variability. J Climatol 13:1000–1016
Turner J, Barrand NE, Bracegirdle TJ, Convey P, Hodgson DA, Jarvis M, Jenkins A, Marshall G, Meredith MP, Roscoe H, Shanklin J (2014) Antarctic climate change and the environment a update polar. Record 50:237–259
Turner J, Bindschadler RA, Convey P, Di Prisco G, Fahrbach E, Gutt J, Hodgson DA, Summerhayes CP, Mayewski PA (2009) Antarctic climate change and the environment. Scientific Committee on Antarctic Research, Cambridge
Turner J, Phillips T, Marshall GJ, Hosking JS, Pope JO, Bracegirdle TJ, Deb P (2017) Unprecedented springtime retreat of Antarctic Sea ice in 2016. Geophys Res Lett 44(13):6868–6875
Wang G, Hendon HH, Arblaster JM, Lim EP, Abhik S, Van Rensch P (2019) Compounding tropical and stratospheric forcing of the record low Antarctic sea-ice in 2016. Nat Commun 10(1):1–9
Wu Y, Polvani LM, Seager R (2013) The Montreal Protocol's importance in protecting Earth's hydroclimate. J Clim 26:4049–4068

Cenozoic Evolution of Antarctic Ice Sheet, Circum Antarctic Circulation and Antarctic Climate: Evidence from Marine Sedimentary Records

Ashutosh K. Singh, Devesh K. Sinha, Vikram Pratap Singh, Kirtiranjan Mallick, Ankush Shrivastava, and Tushar Kaushik

Abstract The last few decades have witnessed intensive studies regarding the origin and development of permanent ice sheets in Antarctica. The studies were primarily focused on determining the paleopositions of Antarctica in various plate assemblies, journey of Antarctica to the southernmost position, atmospheric carbon dioxide concentrations connected to Antarctic cooling, earliest ice-rafted debris deposition in the southern ocean, oxygen isotope composition of the marine sediments to understand ice volume, the opening of the southern ocean gateways and development of the Circum Antarctic Circulation and thermal isolation of Antarctica, meridional heat transport to Antarctica, and the response of Antarctic ice sheet to orbital forcing. All these studies point towards the fact that the Antarctic ice sheet was triggered by more than one mechanism. However, most of the studies and debates revolve around the causes for lowering of temperature rather than the causative factors that contributed to moisture for the massive ice sheets in Antarctica. Interestingly, the geological evidence for the origin and evolution of the Antarctic Ice sheet has mainly been reported from places outside Antarctica, including marine sediments deposited from southern to lower latitudes. Erosion by the probable ice sheets on Antarctica in the form of ice-rafted debris gives one of the earliest clues to ice near continental shelves during the Early Oligocene. Studies on the Carbon dioxide concentration point towards a lowering of CO_2 levels from 3000 to 800 ppm in the Early Oligocene. A synthesis of the oxygen isotope record of foraminifera also indicates that the

A. K. Singh · D. K. Sinha (✉)
Department of Geology, Centre of Advanced Studies, University of Delhi, Delhi 110007, India
e-mail: dsinha@geology.du.ac.in

V. Pratap Singh
Department of Geology, Indira Gandhi National Tribal University, Amarkantak 484886, India

K. Mallick
Department of Geology, Utkal University, Vani Vihar, Bhubaneswar, Odisha 751004, India

A. Shrivastava
Department of Geology, Mohanlal Sukhadia University, Udaipur, Rajasthan 313001, India

T. Kaushik
Biodiversity and Palaeobiology, Agharkar Research Institute, Pune 411004, India

N. Khare (ed.), *Assessing the Antarctic Environment from a Climate Change Perspective*, Earth and Environmental Sciences Library,
https://doi.org/10.1007/978-3-030-87078-2_4

cooling started in the Early Oligocene. Other causative factors proposed include the opening of the southern ocean gateways and the development of circum Antarctic Current leading to the thermal isolation of Antarctica. Recent studies have revealed that the ice sheets are not inert to the global carbon cycle and have played an active role. Thus under the climate change scenario, it is essential to understand the history of the Antarctic Ice sheet. The chapter summarizes the crucial researches regarding the origin and development of the Antarctic Ice Sheet (AIS) and offers some future directions for research.

Keywords Antarctic ice sheet (AIS) · Circum Antarctic Circulation · Marine sedimentary records · Cenozoic · Climate

1 Introduction

The icy continent Antarctica, occupying approximately 14 million km^2, has travelled through various climatic regimes of the Earth before reaching its present position as the southernmost continent. This southernmost position was the primary factor for the development and growth of the Antarctic Ice sheet during later times. This fascinating, convincing, and long journey through the ages has primarily been revealed by the rock records of different periods present on Antarctica and their similarities in many ways with contemporary rocks present on other continents. Webb (1962) presented one of the earliest compilations of absolute dates obtained from Antarctic rocks in a tabular form. This work compiled ninety radiometric dates obtained from Antarctic rocks ranging from Precambrian to Miocene and helped to a great extent in the understanding geological evolution of Antarctica. The compilation by Webb (1962), Sobotovich et al. (1976), Stuiver and Braziunas (1985) also provides testimony that Antarctica had batchmates of all the ages who might have been attached to her and later drifted far as a result of plate motions. After the hypothesis of continental drift proposed by Alfred Wegener in 1912, there have been diverse views concerning splitting, assembly, and reorganization of the Earth's continents, including Antarctica, through geological time. While some believed that there were few very big supercontinents in different time slices, the others envisaged many dispersed smaller continents separated by the oceans. The criteria used for continent assemblies in Precambrian differ from those used for Phanerozoic. For the paleogeographic reconstructions during Precambrian, the primary measures employed include similarity of rocks, the grade of metamorphism, palaeomagnetism, economic mineral deposits, and dyke swarms. However, for the Phanerozoic paleogeographic reconstructions, the palaeontological evidence, including similarity of fauna and flora, diachronism in appearance and extinction of species and related migratory routs, paleobiogeographic evidence for opening and closing of ocean gateways are considered (Jenkins 1974; Kennett 1977; Srinivasan and Sinha 1998). Paleogeographic reconstructions are generally accepted when the evidence comes from multi-proxy data. The debates arise when interpretations drawn from different proxies fail to corroborate with each other.

Paleogeographic reconstructions involving Antarctica and her possible attached neighbours in various geological times have also been a subject matter of intense debate, particularly for the Precambrian Eon. Several studies based on geophysical and palaeontological criteria have been made on such reconstructions. The modern distribution of rocks in Antarctica and other continents reveals that the remnants of all the envisioned supercontinents can be found in some or all of them depending upon the absolute ages assigned to the rocks. Though Antarctica's older rocks history is mainly obscured by thick ice sheet cover, some idea about Antarctica's sub-ice geology has been revealed by GOCE (Gravity Field and Steady-State Ocean Circulation Explorer) studies (Ebbing et al. 2018). The study found that interior East Antarctica appears to include at least three major heterogeneous lithospheric domains (Ebbing et al. 2018). Direct evidence points towards the fact that Antarctica contains rocks as old as 3.5 billion years (Elliot 1975; Tingey 1991) to young rocks belonging to the Cenozoic (Webb 1962). The Transantarctic Mountains divide East and West Antarctica, which lie to the Indian and Pacific Ocean sides. Whereas East Antarctica mainly comprises Precambrian rocks (Phillips et al. 2006), the West Antarctica and Transantarctic mountains contain Phanerozoic rocks (Vaughan and Livermore 2005).

Antarctica now occupies the southernmost position on Earth. The Antarctic plate, which hosts Antarctica and surrounding oceans, shares boundaries with the African, Australian, Pacific, and South American Plate. Ghavri et al. (2017), based on the GPS measurements, concluded that the Antarctic plate's net motion is 8 mm/year towards the north, and also it spins because divergent margins surround it. It is worthwhile to take a cursory look at the journey of Antarctica as a continent during Earth's history before reaching its present location to enable us to answer such exciting questions as

a. How has plate tectonics driven Antarctica through contrasting climatic conditions as a result of inferred latitudinal shifting?
b. What role did plate tectonics play in contributing to Antarctic ice sheet formation?

Though the scope of this chapter is limited to Cenozoic, it is worthwhile to look at the journey of Antarctica through geological time based on published literature.

2 Pre-Cenozoic Paleopositions of Antarctica

The chronicles of Precambrian crustal evolution revealed from the world's cratons, including Antarctica, have many geological similarities. Based on the similarities and absolute ages of the rocks, Antarctica's paleo-positions during various Precambrian segments have been inferred. Since the advent of plate tectonics in the Archean, the continents have separated, rejoined, and again separated, as explained in the Wilson

Cycle. The Antarctic continent has been a member of various supercontinent assemblies at different times in the history of the Earth. Though there have been debates regarding its position in pre-Pangaea supercontinents, its position in Gondwanaland is more or less accepted. Although the continent exposes only two percent of the area as ice-free, the rocks reveal a lot about its membership to various supercontinent assemblies during Earth's history. After Earth's origin and its early differentiation into the core, mantle, and crust, the major cratons of the world, including those in Antarctica, stabilized by the end of the Archean. A critical review of the early crustal evolution of major cratons of the world provides clues to Antarctica's paleo-positions in Precambrian supercontinent assemblies. Antarctica's paleo-positions are revealed from the rocks of different ages preserved on the Antarctic continent and their affinity and similarity from rocks of similar ages from other cratons of the world. Also, the geometrical fits tried by several workers helped to ascertain Antarctica's paleo-positions. It thus seems likely that Antarctica hosts remnants of most, if not all, of Earth's supercontinents, and Antarctic research continues to provide insights into their palaeogeographical reconstructions and geological evolution (Harley et al. 2013).

Yakubchuk (2019) proposed three stages in the tectonic evolution of the Earth: (1) nucleation, from the origin of protocratons to their assembly into the Kenorland supercontinent (2.7–2.5 Ga); (2) cratonization, from the breakup of Kenorland (2.45 Ga) to the assembly of Columbia (1.85 Ga) and its reorganization into Rodinia (1.0–0.72 Ga); and (3) modern plate tectonics, from the breakup of Rodinia 720 Ma until the present.

Vaalbara was probably the first supercontinent after the early differentiation of Earth into its almost concentric layers (~3.6–2.803 Ga), followed by Ur (~3.0 Ga), Kenorland (2.72–2.1), Columbia (1.82–1.35), Rodinia (1.071– ~0.75 Ga), Pannotia (650–540 Ma) and Pangaea (335–173 Ma). Though there are debates about accepting all of these supercontinents and their ages, it is generally accepted that the Precambrian crustal evolution has similarities in different Cratons now separated by oceans and orogenic belts. Antarctica has probably been a member of all these supercontinents, as revealed by its rocks. The amalgamation of continents and their breakup and re-amalgamation profoundly affected the continental climate due to two factors. On one hand the merger and separation both resulted in continents traversing various climatic regimes of the Earth, while on the other hand the reorganization of continents resulted in different land-sea distribution and changing ocean circulation, which had significant control over the climate of adjacent regions. In a later section, it is explained how the icy continent of Antarctica owes its present ice sheet to some extent to continental breakup and ocean circulation changes. Despite being primarily covered beneath a widespread and thick ice sheet, Antarctica contains rocks spanning some 3.5 billion years of history, and it is likely to have fragments of all Earth's recognized supercontinents (Harley et al. 2013).

2.1 Antarctica in Vaalbara and Ur: (~3.6–2.803 Ga)

There are debates about the earliest supercontinent, Vaalbara consisting of Kappavaal (South Africa) and Pilbara (Australia) cratons, and the supercontinent Ur consisting of India, Madagascar, and Australia. Jones et al. (2003) suggested that the Grunehogna Craton of eastern Antarctica can be included in the Vaalbara supercontinent configuration. Antarctica, India, Australia, and Kalahari should be clubbed as Ur due to the similarity between the Archean granite-greenstone terranes (Rogers 1993). Marschall et al. (2010), based on the dating of Granites, concluded that Kaapvaal (South Africa) -Grunehogna Craton (Antarctica) played a significant role in the mechanical stabilization of the continental crust during the establishment of the cratons in the Mesoarchaean. Cheney (1996) and Bleeker (2003) were also of the view that the Grunehogna Craton in western Dronning Maud Land, Antarctica, appears to be related to the Archaean rocks of southernmost Africa and is interpreted to have been part of the Vaalbara Supercraton.

2.2 Antarctica in Kenorland (2.72–2.1 Ga).

The Napier Complex consisting of high-grade metamorphic rocks in Antarctica, has an age range between 2.6–2.5 Ga attributed to the assembly of "Kenorland" or "Sclavia." The Napier Complex and Vestfold Hills portions of Antarctica were attached to the Dharwar and Singhbhum cratons of India (Veevers 2012). Uranium-lead isotopic data indicated that granulitic facies rocks of Napier Complex of Enderby Land, Antarctica, were cut by charnokitic pegmatites 2.5 billion years ago and that the Napier complex may be an extension of Archean granulitic terrene of Southern India (Grew and Manton 1979).

2.3 Antarctica in Columbia (Nuna) and Rodinia ((1.071–~0.75 Ga)

The famous SWEAT (Southwest USA and East Antarctica) configuration envisaged connections between Antarctica and North America during Proterozoic. It was part of the early Neoproterozoic supercontinent known as Rodinia (Harley et al. 2013). Remnants of older rocks of Paleoproterozoic and Mesoproterozoic present in Antarctica suggest that the continent was part of the supercontinent known as Columbia or Nuna. These rocks are present along with scattered fragments of Archaean rocks that might have been part of one or more 'supercratons's during the Neoarchaean. Several workers, including Zhao et al. (2004), Reddy and Evans (2009), Evans and Mitchell (2011), and Payne et al. (2009), proposed a "Mawson Continent" (the Gawler–Ade´lie Craton in southern Australia and Antarctica, and crust of the Miller Range, Transantarctic Mountains) for reconstructions of Nuna or Columbia–which is believed to have amalgamated along Palaeoproterozoic orogens during 1.9–1.8 Ga.

2.4 Antarctica in Pangaea

Pangaea, consisting of Laurasia and Gondwanaland, was perhaps the last supercontinent that broke up into the above two of its major divisions separated by intervening Tethys, part of the superocean Panthalassa. Antarctica was part of the previous supercontinent Gondwanaland around 150 Ma, with the neighbours of East Antarctica as Australia, India, and Africa.

3 Reaching the South Pole

Antarctica's journey to its present southernmost position after the Gondwanaland's breakup continued with plate motion rates of 24 mm/year to 9 mm/year during the Cretaceous (Lawrence et al. 1992; Kent and Gradstein 1986). The journey started with the breakup of Gondwanaland in Late Triassic –Early Cretaceous times with the separation of East Antarctica from Australia. This was followed by the rifting of India from East Antarctica. By this time, significant pieces of West Antarctica were nearly in their present-day positions with respect to East Antarctica, as inferred from the paleomagnetic data of Grunow et al. (1991). By the Late Cretaceous, Antarctica reached its present place surrounded closely by very shallow seas. The free circum Antarctic circulation wasn't possible as the drake passage and Tasmanian seaway didn't open. All its neighbours, Australia, India, South America, and Africa, gradually drifted north, and Antarctica was ready for a big future climatic event – the formation of the permanent ice cap. Paleomagnetic evidence shows that the Antarctic continent has essentially been in a polar position since at least the Late Cretaceous (Kennett 1977; McElhinny 1973; Lowrie and Hays 1975).

4 Initiation of Antarctic Ice Sheet—Understanding in the Huttonian Way

Before discussing how and when did the Antarctic Ice sheet first form, it is worthwhile to apply the Hutonian Principle of "the present is the key to the past." A discussion on the growth of some recent ice sheets and their proposed mechanisms would help us to understand the factors responsible for the origin and development of the permanent ice sheet in Antarctica. We have several examples from the near past when the ice sheets have been initiated and grown, and the most conspicuous of them is Northern Hemisphere Glaciations. The terms "glaciations" and "cooling" have often been used in literature as substitutes while describing the Earth's climate history. This usage is not appropriate. Mere cooling may not result in an ice sheet if there is no abundant supply of moisture. Based on such concepts, Haug and Tiedemann (1998) proposed that the increased atmospheric moisture content was a necessary precondition for Northern Hemisphere ice-sheet growth. This was caused by the closure of the Central American Seaway and the resultant intensification of the Gulf Stream. Logically, this hypothesis is appealing as the cold temperatures alone would not be sufficient for the ice sheet initiation, and enough moisture supply is necessary. Though in the Earth's climatic history, often, the glacial ages, ice-sheet expansion, and global cooling events are considered the same and occurring together. However, this notion needs to be carefully evaluated. The essential material which the thick ice sheets need is enough moisture, and that can usually be supplied if the climate is warm and humid. Thus the conditions necessarily comprise a situation where moist air is provided to a frigid land. In such a case, ocean circulation's role becomes vital as the ocean currents are great transporters of heat and moisture to the Polar regions, even in the modern world. Applying the same logic to Antarctic glaciations, we can reach some preconditions for the initiation of the ice sheet.

a. Antarctica should have been situated in a sufficiently cold climate.
b. Antarctica should have been thermally isolated from warm ocean currents / meridional heat flow.
c. There should be enough supply of moisture to Antarctica for the initiation of ice sheets.

The stability of an ice sheet depends on a tussle between precipitation and ablation. Ice core data from the Antarctic and Greenland does give us some clue to the process by which ice sheets have accumulated. The studies by Cuffey and Clow (1997), van der Veen (2002), Alley et al. (1993), and Lourious et al. (1979) show a dramatic increase in snow accumulation rates during warmer periods (Gildor 2003). Thus one needs to look for evidence of both the initiation of the ice sheet, temperature record and moisture (Fig. 1). While there are several records about the ice sheet's presence, the moisture source evidence during the same intervals remains enigmatic.

5 The Earliest Evidence of Antarctic Ice Sheet—The Ice Rafted Debris

Ehrmann and Mackensen (1992) made sedimentological studies of the Middle Eocene to late Oligocene sediments recovered at ODP Sites 689 and 690 on Maud Rise in the southernmost Atlantic Ocean and Sites 738 and 744 on Kerguelen Plateau in the southernmost Indian Ocean. They found the earliest IRDs in the form of isolated gravel and terrigenous sand grains, which indicate ice-rafting from middle Eocene time at c. 45.5 Ma. This is almost coeval to IRD findings by Tripati et al. (2008)

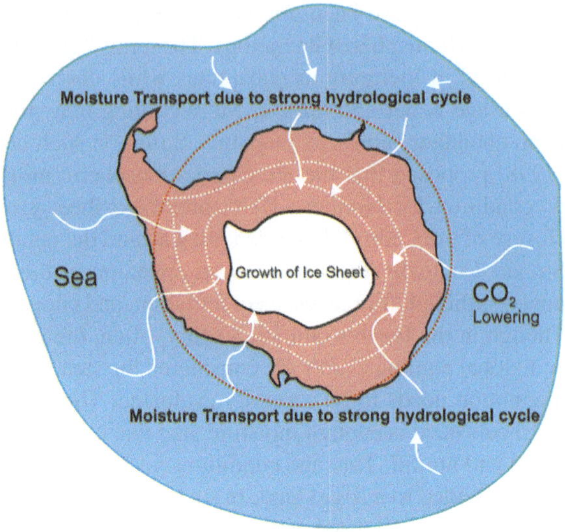

Fig. 1 Essential pre-requisites for ice-sheet growth. Moisture supply through strong hydrological cycle (white arrows) and position of the depositional base (Antarctica) at high latitude (Red dotted line showing 70°S latitude). White dotted lines show successive growth limits of the ice sheet with time. Greenhouse (CO_2) forcing leads to further cooling

and Eldrett et al. (2007) from high northern latitude oceans at 44 Ma. The glacial origin of these sediments was supported by high concentrations of detrital chlorite and reworked kaolinite, probably caused by enhanced physical weathering due to the growth of ice in the interior. Some glaciers reached the sea, while most of the continent remained under the influence of a humid and warm to a temperate climate (Ehrmann and Mackensen (1992). Similar observations were made earlier by Hambrey et al. (1991) and Ehrmann (1991). However, at the same time, Ehrmann (1991) was not convinced with the amount of the ice-rafted material and doubted any continental-scale glaciations. Zachos et al. (2001) imagined small ephemeral ice sheets to have appeared temporarily and were responsible for such low abundance sporadic IRDs and concluded that the first Cenozoic East Antarctic Ice sheet appeared in the Earliest Oligocene.

The arguments for the earliest evidence of an ice sheet in Antarctica were made in yet another study by Hollis et al. (2014). Based on their reviews of the sedimentary basins of New Zealand, they considered the deposition of the organo-facies of the Waipawa Formation due to sea-level fall and climatic cooling. They believed that ice sheets grew on Antarctica in the earliest Late Paleocene (~59 Ma). However, no other record of such cooling and ice–sheet presence in Paleocene is available. The interpretation of Hollis et al. (2014) can be debated considering local tectonism in New Zealand. The lower sea level could not have been Ice-sheet related fall. However, Hollis et al. (2014) argued that based on some geophysical studies (Wilson et al. 2013), that inland regions of Antarctica might have reached higher elevations to allow temporary ice sheets that waxed and waned in response to climatic changes. The logical source of the earliest ice-rafted debris should have been in west Antarctica rather than East Antarctica because the East Antarctic ice sheet was not large enough to reach continental shelves to produce ice-rafted detritus (Fig. 2).

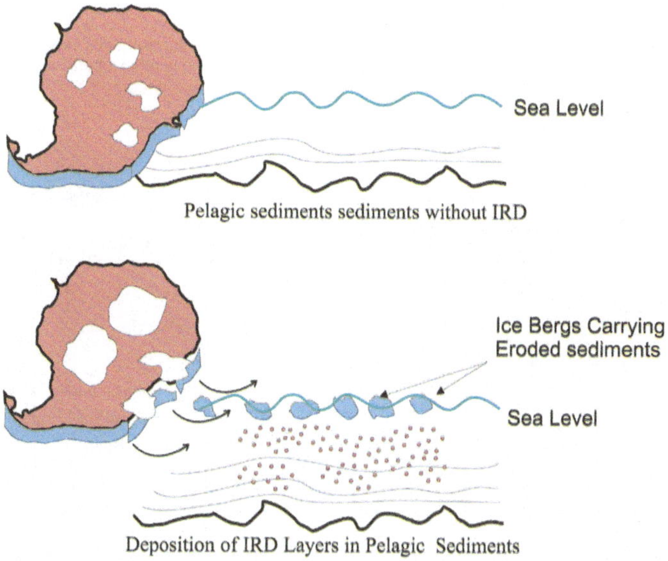

Fig. 2 The first evidence of the presence of the ice sheet in Antarctica comes from Ice Rafted Detritus sediments. The upper figure shows a condition when the Ice sheet might be present inland but have not reached the shelf. This may cause a lowering of sea level as proposed by Hollis et al. (2014) during Paleocene. The lower figure shows Ice sheets reaching shelf areas and depositing IRDs through icebergs in the pelagic sediments presented by Ehrmann and Mackensen (1992) from the ODP Sites

Thus it becomes essential to understand the establishment of the West Antarctic ice sheet. Wilson et al. (2013), based on the new climate-ice sheet model, showed that the West Antarctic Ice Sheet first formed at the Eocene–Oligocene transition with the continental-scale expansion the East Antarctica Ice Sheet. Thus the findings of the earliest Oligocene Ice rafted detritus from marine sediments is a clear indication of shelf ice. The West Antarctic ice sheet history has been more dynamic because of its grounding below sea level (Lurkock and Florindo 2017).

After the findings of the earliest Oligocene ice-rafted material, scientists agreed that Antarctic ice sheets reached sea level in the earliest Oligocene (e.g., Zachos et al. 2001) because the ice-rafted material provided strong evidence of being eroded from the ice sheet near the shelf. Breza and Wise (1992) discovered abundant ice-rafted debris from ODP site 748 on the Central Kerguelen Plateau in the southern Indian Ocean from 35.8–36.0 Ma sediments. This age refers to the latest Eocene as per the Geological Time Scale of Gradstein et al. (2012). This finding of IRDs was not only one of the earliest occurrences, but also spatially the lowest latitudinal event known far from Antarctica. These authors strongly argued for the presence of the earliest Oligocene ice sheet on the Antarctic continent based on the coincidence of this IRD event with the globally recognized shift in $\delta^{18}O$. This $\delta^{18}O$ shift is quite significant (1.2–1.5‰) and is attributed to the ice volume effect (Fig. 3) (Tigchelaar et al. 2011).

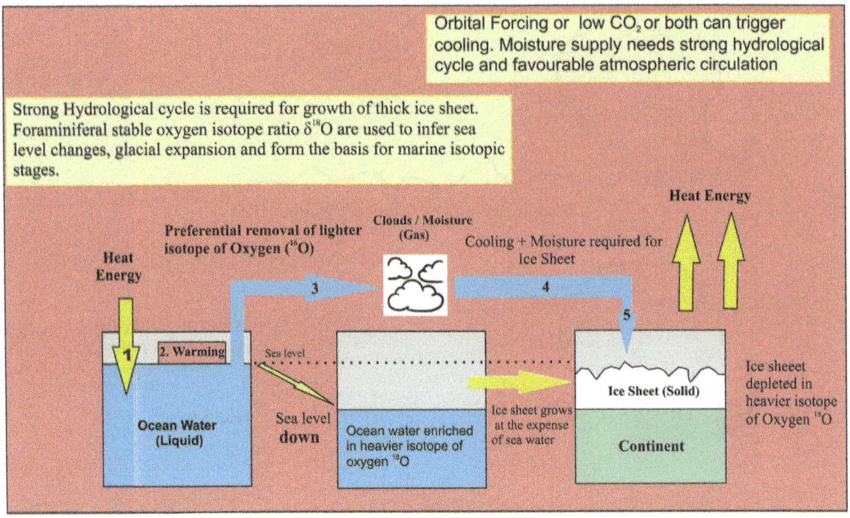

Fig. 3 Conceptual diagram showing necessary conditions for the growth of large ice sheet, the resultant sea-level changes and foraminiferal $\delta^{18}O$. The development of a permanent ice sheet should logically be reflected in an permanent positive shift in foraminiferal $\delta^{18}O$

The above finding was significant as the evidence of the presence of Ice sheet and global cooling, both were coincident.

In yet another study, Carter et al. (2017) found one more IRD event, probably the first evidence for the presence of massive glaciers eroding along the coastline of the southern Weddell Sea. However, the IRD event's age was older than the significant oxygen isotope shift at 34–33.5 Ma. However, if the oxygen isotope shift was due to the ice volume effect, then the IRD event should have been younger than the isotope shift. Nevertheless, the discovery of the IRDs based on petrographic analysis of over 275,000 grains, detrital zircon geochronology, and apatite thermochronometry by Carter et al. (2017) point toward the build-up of the widespread ice sheet on Antarctica even earlier than the significant isotopic shift in $\delta^{18}O$. This IRD event was regarded as evidence of the beginning of the Antarctic glaciation. Considering the above critical studies, one can accept at least the presence of glaciers in Antarctica during the earliest Oligocene.

The first major, well-recorded expansion of Cenozoic glaciation at the start of the Oligocene (~34 Ma), documented by oxygen isotopic record, is matched by ice-rafted debris and other sedimentological evidence from Maud Rise and the Kerguelen Plateau (Ehrmann and Mackensen 1992) and by diamictites deposited in Prydz Bay (Hambrey et al. 1991). The majority of the studies related to sediments derived from glaciers come from East Antarctica. However, there are studies such that Ivany et al. (2006) suggesting that the northern Antarctic Peninsula was glaciated to sea level in the early Oligocene, though the timing of this evidence has been debatable (Marenssi et al. 2010). Further, the deep-sea hiatuses are not physically verifiable unless some biozones are found missing. The difference in ages of the IRDs reported by various authors can be due to biostratigraphic uncertainties. The dating of the CIROS-1 core, the first to capture the Eocene–Oligocene transition, has been revised repeatedly, but the current age model (Wilson et al. 1998; Roberts et al. 2003) confirms a cooling trend during the transition itself, following a relatively warm Eocene interval. The transition is also recorded in the CRP-3 core, which, although difficult to date precisely in this interval (Florindo et al. 2005), also seems to indicate sharp cooling at around the same time, as interpreted from a sharp decrease in magnetite concentration (Sagnotti et al. 2001).

6 Causative factors for the Antarctic glaciations:

Arguing for the causative factors leading to the Oligocene ice sheet, one has to assume some significant warming event that could have supplied moisture to Antarctica, which already reached its present position as the southernmost continent in the cold climatic regime. The views regarding causative factors include:

(a) Thermal isolation of the Antarctic Continent from warm ocean currents due to opening of drake passage and Tasmanian Seaway and establishment of Circum Antarctic Circulation;
(b) Some triggering mechanisms including of atmospheric carbon dioxide levels and Orbital forcing;
(c) Some warm events and favourable ocean and wind circulation moisture supplier through a solid hydrological cycle.

6.1 Opening of the Ocean Gateways and Development of Circum Antarctic Current

The plate tectonic movements that led to Antarctica's positioning as the southernmost continent continued until the final isolation of Antarctica was completed from its neighbours. Important events include.

(a) Opening of Tasmanian Seaway and
(b) Opening of the Drake Passage.

6.1.1 Opening of the Tasmanian Seaway

The knowledge about the spreading history and direction of plate movements are obtained from magnetic stripes preserved on the seafloor and their dating with radiometric methods. The Southwest Pacific ocean provides a magnetic anomaly pattern that reveals the history of spreading between Antarctica and Australia. Christoffel and Falconer (1972) discovered the oldest magnetic anomalies of 80 Ma in the

Southwest Pacific ocean. The separation of New Zealand from Australia and Antarctica formed the Tasman Sea. In the Early Eocene, Australia separated from Antarctica and continued to drift towards lower latitudes, and its journey till recent is well documented by the deep-sea sediments of Leg 90 (Southwest Pacific). Still, since the South Tasman Rise was connected with Victoria Land, Antarctica, Tasmanian Seaway was not fully opened. The South Tasman Rise formed only a shallow water barrier to circum-Antarctic flow (Kennett 1977). The deep-oceanic circulation developed just after the Eocene–Oligocene boundary when South Tasman Rise separated from Victoria Land sufficiently and is supposed to have occurred at some time during the Oligocene (38 to 22 Ma Kennett et al. 1975; Deighton et al. 1976 Kennett 1977). Kennett and Exon (2004) opined that the opening of the Tasmanian Gateway during the Eocene–Oligocene transition (~33.5 Ma), and later Seaway expansion, led to critical changes in Southern Hemisphere Ocean circulation resulting from the development of the Antarctic Circumpolar Current (ACC). This event created a thermal barrier of psychrospheric circulation around Antarctica (Fig. 4), leading to the crossing of a significant global climatic threshold and major ice expansion on Antarctica. The gateway opening was not the reason for this ice sheet but was accompanied by feedback mechanisms associated with ice-sheet expansion. The expansion of the Antarctic ice sheet must have resulted in increased albedo, enhanced production of bottom waters resulting in increased thermohaline circulation and Southern Ocean productivity (Kennett and Exon 2004).

Besides this crucial causative factor related to gateways' opening, Oligocene glaciation recurred at periods corresponding to variation in the eccentricity and obliquity of the Earth's orbit, indicating that the climate system was responding to orbital cycles (Pälike et al. 2006).

6.1.2 Opening of the Drake Passage

The opening of the Drake Passage between South America and Antarctica is another significant plate tectonic event leading to the establishment of the Circum Antarctic Circulation (Fig. 4). Jenkins (1974), based on the migratory route of planktic foraminiferal species *Guembelitria stavensis* from DSDP sites 282 (West of Tasmania), Site 277 (western edge of Campbell plateau) and sections from New Zealand East Coast South Island, concluded that the initiation of proto circum Antarctic current occurred between 27 and 28 Ma. The timing of this initiation was based on numerical age estimates of associated planktic foraminiferal events. Debates are there regarding the timing of the opening of the drake passage. Age estimates for the opening of Drake Passage range from 49 to 17 million years ago (Scher and Martin 2006).

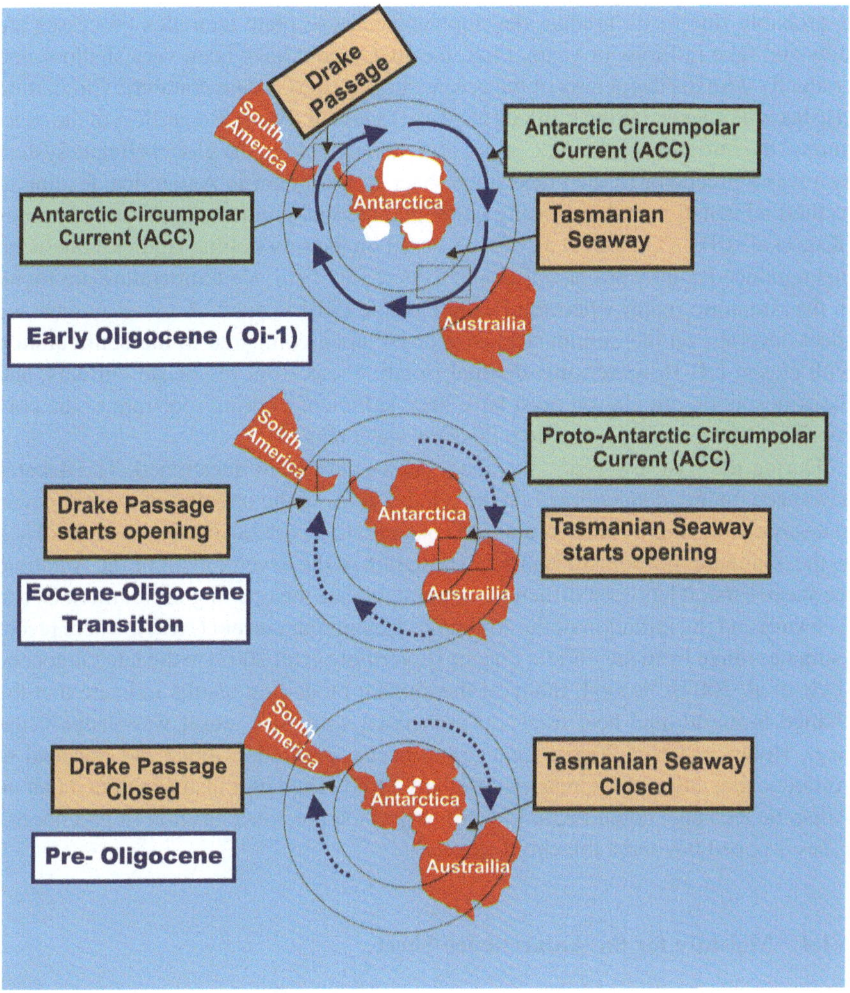

Fig. 4 Cartoon (not to the scale) showing the Drake Passage and Tasmanian Seaway (rectangles with dotted lines), which were earlier closed, not allowing a free Circum Antarctic Current. After these two seaways opened, the Circum Antarctic Circulation was established, leading to Antarctica's thermal isolation from warm ocean currents (Concept from Kennett 1977)

6.1.3 Initiation of the Circum Antarctic Current

With opening of the Drake Passage and Tasmanian Seaway (Fig. 4), the circum Antarctic Circulation developed. However, the exact timing of the opening of the two gateways is not fully resolved. Overall there is an agreement regarding the opening of both seaways during the Oligocene time. Thus it is likely that the Oligocene also witnessed the initial establishment of an Antarctic Circumpolar Current (ACC) (Barker et al. 2007). The uncertainty in fixing the timing of initiation of the ACC is probably due to its gradual development because plate tectonics processes are slow and take millions of years. First, the ACC could have been very shallow, and gradually with the deepening of the ocean gateway, the Circum Antarctic Circulation also became deeper, taking millions of years (Lyle et al. 2007). Regardless of the exact timing, the initiation of the ACC was a critical threshold in the global climate system as it was a precursor to the thickest ice sheet development in Antarctica, leading to its thermal isolation and increasing latitudinal thermal gradients in the world oceans (Katz et al. 2011). Cristini et al. (2012), based on their modelling studies, concluded that a reduced southward heat flux and a decrease of surface temperature are found in the Antarctic realm when the Drake Passage (DP) is open. A more massive ice sheet develops on the continent in case of DP open compared to the configuration with closed DP. However, this thermal isolation was from the ocean currents, and the atmospheric circulation must have been sufficient to bring moisture to the cold continent to eventually turn into a thick ice sheet (Fig. 3).

During the past few years, the above views have been questioned. The debates have been mainly concentrated on the relative roles of the opening of the two ocean gateways, lowering of carbon dioxide concentration, orbital forcing, and feedback as the dominant mechanism leading to initiation of the development of the Antarctic ice sheet. First, it has been difficult to accurately date the opening of Southern Ocean gateways and the initiation of the Antarctic circumpolar current (ACC), where recent estimates range from the middle Eocene (Livermore et al. 2007) to the late Oligocene (Lyle et al. 2007). Second, many of the climate modelling results indicate that the change in meridional heat transport associated with ACC onset was insignificant (e.g., Huber and Nof 2006). Changes in ocean circulation alone did not lead to Antarctic glaciation (De Conto and Pollard 2003). The latter also concluded that an Antarctic ice sheet could begin to form only when the concentration of greenhouse gases dropped beyond a threshold value.

6.1.4 Moisture for the Antarctic Ice-Sheet

Surprisingly all these studies concentrate on the cooling of the climate, and none argues for the moisture factor. As Antarctica contains 90 per cent of the freshwater in the form of its ice sheet, scientists need to bother about transporting this moisture source within a short period (Oligocene to Middle Miocene) rather than only concentrating on the cooling events associated with the Antarctic ice sheet. As stated earlier, various modelling studies on the response of Antarctica to greenhouse warming suggest that the increased accumulation of ice may dominate the ablation under moderate global warming (Gildor 2003; van der Veen 2002). Gildor (2003) also argued about the temperature –precipitation feedback (Le-Treut and Ghil 1983), explaining that the evaporation rate and the strength of the hydrological cycle increase exponentially with temperature due to the Clausius-Clapeyron relation. The argument that extensive ice cover effectively insulates the ocean from the atmosphere and reduces the evaporation rate dramatically (Ruddiman and McIntyre 1981; Gildor and Tziperman 2000) leading to less precipitation more ablation works against the generally believed albedo feedback mechanism of glaciations. The later mechanism only addresses the temperature part and not the source of moisture. The globally warm, intense hydrological cycles are needed for the ice sheet's growth (Gildor 2003), followed by cooling required to sustain it. In this context, more research is needed to understand the moisture source for 90 per cent of the freshwater in the form of the Antarctic ice sheet and its relationship with cooling events. The answers to such questions logically should not lie in Antarctica but in the oceans near and far from the ice continent, which must have been the source of moisture during the Eocene–Oligocene times. Was the Antarctic ice sheet initiated due to a warming event with a robust hydrological cycle followed by cooling? For this reason, it is worthwhile to consider the marine oxygen isotopic record compilations.

6.2 *Carbon Dioxide Concentrations*

Pearson and Palmer (2000) estimated an atmospheric carbon dioxide concentration of 2000 -> 3000 ppm for the earliest Eocene (to 52 Ma) and an erratic decline to less than 800 ppm by 40 Ma (Lawver and Gahagan 2003). They regarded changes in the CO_2 concentration of the atmosphere as a likely forcing mechanism on the global climate because of its predicted effect on temperature overages. A level $pCO_2 > 1000$ ppm is considered a "super greenhouse" condition, so the change from high earliest Eocene values to the late Middle Eocene value of < 800 ppm should have had a significant impact the Earth's climate. There have been some critical studies regarding threshold values of the concentration of atmospheric CO_2. In one such study, Galeotti et al. (2016) observed that about 34 million years ago (Earliest Oligocene), Earth's climate cooled, and an ice sheet formed on Antarctica as atmospheric carbon dioxide (CO_2) fell below ~750 parts per million (ppm). Sedimentary cycles from a drill core in the western Ross Sea provide direct evidence of orbitally controlled glacial cycles between 34 and 31 Ma (Latest Eocene to early Oligocene). The study concluded that initially, under atmospheric CO_2 levels of ≥ 600 ppm, a smaller Antarctic Ice Sheet (AIS), restricted to the terrestrial continent, was highly responsive to local insolation forcing. A more stable, continental-scale ice sheet calving at the coastline did not form until ~32.8 million years ago, coincident with the earliest time that atmospheric CO_2 levels fell below ~600 ppm (Galeotti et al. 2016). Their results provide insight into the AIS potential for threshold behaviour and have implications for its sensitivity to atmospheric CO_2 concentrations above present-day levels. This study is significant as one would wonder whether the present status of CO_2 around 400 ppm would affect the Antarctic ice sheet at all as the threshold is 600 ppm. Another approach to understanding the presence or absence of sufficiently thick Antarctic ice sheets during the Oligocene is to see global sea levels and coastal sedimentary sequences.

7 Oxygen Isotope Studies from Marine Sediments

The stable isotope composition of foraminifera has been extensively used for generating paleoclimatic records from marine cores. The Oxygen isotope composition of foraminiferal tests is controlled by several factors, including temperature, salinity, and composition of the ambient seawater during calcification, besides vital effect specific to the species. The ice volume effect controls the seawater composition. In general, during the glacial ages and expansion and formation of the ice caps, the ocean waters become enriched in ^{18}O. In contrast, the polar ice caps become depleted in this heavier isotope because of preferential removal of the lighter isotope in water vapour, which ultimately precipitates as polar ice (Fig. 3). Thus during the significant ice-sheet expansion, the foraminifera calcifying their tests becomes enriched in the heavier isotope. The Cenozoic glacial and interglacial ages have been identified by the oxygen isotope record mostly from benthic foraminifera as the later's isotopic composition is supposed to have been mainly controlled by the ice volume effect because the bottom water temperatures and salinity do not show much variation owing to the physical properties of water, i.e., densest and cold water occupies the bottom of the sea. Such isotope records have resulted in the establishment of marine isotopic stages (MIS), where the odd number stages represent interglacial, and even number stages represent glacial interval.

7.1 Mid Paleocene-Early Eocene: Was Warming a Precursor to Antarctic Ice Sheet Formation?

As described earlier, the $\delta^{18}O$ record of benthic foraminifera is a more reliable proxy for ice volume change than planktic foraminifera. However, the steps and peaks in the record are also dependent on the sampling interval. Two famous compilations may be mentioned in this regard. One is that of Miller et al. (1987) and the second one is by Zachos et al. (2001). As stated in Fig. 3, the ice sheets have grown at the expense of ocean water. Thus permanent ice sheet development is matched by a permanent shift in the oxygen isotope values (heavier isotope enrichment) and lowered sea level. The compilation of oxygen isotope records by Miller (1987) beautifully displays these aspects.

There are several fluctuations in the $\delta^{18}O$ record; however, few major trends tell us the story of Cenozoic climate and Antarctic ice sheet evolutionary history. The earliest such trend in Cenozoic is observed in a 1.5‰ decrease in δ^{18} O during Mid Paleocene to the Early Eocene (59–52 Ma). The peak of this decrease defines the Early Eocene Climatic Optimum (EECO-52–50 Ma). Hyland et al. (2017) considered EECO to have been caused by perturbation in the global carbon cycle. Payros et al. (2015) stated that EECO was precisely confined to 52.6 Ma and lasted for 2.3 My. The rise in global temperature has been estimated as 4–8 °C. Though none of the studies relate this significant warming event to the formation of the Antarctic ice sheet, yet it seems that this warming event must have introduced a robust hydrological cycle and resulted in moisture generation and transported to Antarctica, which had the perfect latitudinal setting for moisture precipitation and ice (Fig. 3).

The model results show that large ice sheets developed during past ice ages grew when the climate was relatively warm and, therefore, moist (Gildor 2003). This new idea about the origin of the Antarctic ice sheet is supported by climatic modelling, which indicates ice caps on interiors of mountains (DeConto and Pollard 2003) during this time. The study of DeConto and Pollard (2003) was based on atmospheric CO_2 levels. IODP Leg 318 off Wilkes land gives some idea about climatic conditions in Antarctica. The estimated summer temperatures were approximately 25 °C and frost-free winters at 10 °C (Pross et al. 2012). These warm temperatures must have been the causative factors behind the moisture source for the initiation of the Antarctic ice sheet formation.

7.2 Post-Eocene Climate

The EECO is followed by a long trend spanning almost 17 million years when the $\delta^{18}O$ shows a 3‰ rise. This very gradual but prominent cooling trend corresponds to a decline of 7 °C in deep-sea temperatures from 12 °C to 4.5 °C (Zachos et al. 2001). The debates regarding this cooling have been concentrated on the presence/absence of ice sheets during the Earliest Oligocene. If one believes that Ice sheets were present, then the ice volume effect has to be filtered, and the contribution of temperature will be less. If we think that Ice sheet was absent in Oligocene, then the entire $\delta^{18}O$ decline would be attributed to deep-sea/bottom water temperature.

Further, the bottom water temperatures also reflect the surface temperatures at the pole where deep bottom waters are formed. Thus this is a crucial debate whether to attribute Oligocene $\delta^{18}O$ decline to a temperature drop or a combined effect of ice volume and temperature drop. The debate narrows down if one believes EECO to have caused enough evaporation and enrichment of oceans in heavier isotope leading to a gradual 3‰ increase in $\delta^{18}O$. The first detailed $\delta^{18}O$ Cenozoic studies from deep-sea sediments assumed that Earth was substantially ice-free before Middle Miocene (15 Ma) (Shackleton and Kennett 1975; Savin et al. 1975). However, evidence of Oligocene glaciomarine sediments in the Ross sea confirmed glacial ice's presence on the margin of the Antarctic Continent (Barrett et al. 2006). Thus, the presence of Oligocene ice sheet was later accepted by many (Savin and Barrera 1985).

Another aspect related to ice sheet formation is the fluctuation in global sea-level (Fig. 3). But the debates regarding sea-level fluctuations during the early Cenozoic also arises from the fact that this could have also been caused by mantle processes, including an increase in ridge length, the opening of Norwegian –Greenland sea, a significant global reorganization of spreading ridges and extrusion of Brito-Arctic province (Roberts et al. 1984). Thus debate continues whether Oligocene sea-level fluctuations caused by waxing and Waning of the Antarctic ice cap or there was no ice sheet were only due to the reorganization of the spreading ridges (Bond 1979).

High-resolution deep-sea $\delta^{18}O$ record and modelling studies indicate that the growth of the first ice sheet was triggered by orbital forcing combined with a decrease in atmospheric CO_2 levels below a threshold value between 1000 and 750 ppm (Deconto and Pollard 2003; Coxall et al. 2005). Once initiated, the high albedo feedback ensured that the first ice remained a nucleus from which the ice sheet expanded and contracted in response to orbital forcing. But all these studies, unfortunately, seldom speak of the causative factors responsible for moisture and temperature-precipitation feedback. Orbital forcing is, of course, responsible for cold/warm phases, but without an abundant supply of moisture, the ice sheets cannot grow. When the first continent-wide ice sheet formed ~34 Ma, paleotopographic reconstructions reveal that approximately 20% more of the Antarctic continent was above sea level. Virtually all of this additional area was in West Antarctica, which may have allowed the ice sheet to be larger than today, even during the high CO_2 worlds of Paleogene (Wilson et al. 2013).

To summarise the initiation of the Antarctic ice sheet, we can list the following events.

(a) Opening of Tasmanian Seaway
(b) Opening of the Drake passage
(c) Development of free Circum Antarctic circulation resulting in thermal isolation of Antarctica
(d) An Early Eocene Climatic Optimum as the source of moisture
(e) Gradual cooling of the climate from Eocene to Oligocene (in tune with plate movements)
(f) Drop-in CO_2 levels
(g) Orbital forcing
(h) Feedback mechanism related to ice albedo
(i) Development of Permanent Ice Sheet in Antarctica.

Surprisingly seeing the oxygen isotopic record, we once again find a warm event preceding the final drop in global temperature shown by remarkable enrichment in the $\delta^{18}O$ record. Following the cooling and rapid expansion of Antarctic continental ice sheets in the earliest Oligocene, deep-sea $\delta^{18}O$ values remained relatively high (>2.5%), indicating the presence of permanent ice (Hambrey et al. 1991) with a mass as great as 50% of that of the present-day ice sheet and bottom temperatures of ~4 °C (Zachos et al. 1993). These ice sheets persisted until the later part of the Oligocene (26–27 Ma) when a warming trend reduced the extent of the Antarctic ice sheet. This conjecture needs debate once again as the decrease in $\delta^{18}O$ has been attributed to warming and melting of the ice sheet, causing depletion of ocean waters in the heavier isotope. This could have also been due to temperature-precipitation feedback, where ablation might have surpassed precipitation due to continuous warming. From this warming interval until the Middle Miocene, global ice volume remained low, and bottom water temperatures trended slightly higher (Wright et al. 1992; Miller et al. 1991) except for several brief periods of glaciations (e.g., The Mi events) (Wright and Miller 1993). The peak of this warm phase is recorded at the late Middle Miocene Climatic optimum (17–15 Ma) and was followed by a gradual cooling and reestablishment of a major ice sheet on Antarctica. It can be envisaged again that the Middle Miocene Climatic optimum would have resulted in moisture supply for further development of the Antarctic ice sheet by 10 Ma (Vincent et al. 1985; Flower and Kennett 1995). Mean $\delta^{18}O$ values then continue to rise gently through the late Miocene until 6 Ma, indicating additional cooling and small-scale ice sheet expansion in west Antarctica (Kennett and Barker 1990). The Early Pliocene is marked by a subtle warming trend (Poore and Sloan 1996) until 3.2 Ma when $\delta^{18}O$ again increased, reflecting the onset of Northern Hemisphere Glaciation (Maslin et al. 1998).

8 Conclusions

Antarctica's journey to the southernmost position during the Late Cretaceous was the first precursor for developing the Antarctic Ice sheet as it reached the regime of the coldest climate.

Mere reaching the southernmost position in a colder climate was not sufficient to initiate the ice sheet. Various studies point towards multiple causes for the initial ice formation in Antarctica, including a further lowering of the temperature due to decreased atmospheric carbon dioxide concentration coupled with orbital forcing.

The opening of the Drake Passage and Tasmanian Gateway during the late Eocene to Oligocene were significant events that resulted in the development of the Antarctic Circum Polar Current, which made Antarctica thermally isolated from the warm ocean currents. The ACC was, to a great extent, responsible for the further development of the Antarctic Ice sheet. The debates regarding the timing of ACC's establishment are not much valid as the plate tectonics process is prolonged, and the ocean gateways open, become deep very gradually with time taking millions of years. The establishment of the permanent ice sheet in Antarctica is quite later in Miocene, millions of years after the opening of the southern ocean gateways. Thus these events can only be considered as a precursor and not the ultimate and unique cause for the Antarctic Ice sheet development.

The majority of the studies have concentrated on the temperature change and the opening of the gateways. Still, very few discuss the causative factors for moisture transport, which created such large and thick ice sheets on Antarctica, storing 90% of Earth's freshwater. Mere cold temperatures can never result in the growth of such a large ice sheet.

Looking at the Huttonian way, "Present is key to the past," the Antarctic ice sheet must have needed a powerful hydrological cycle and warm event to allow so much moisture to precipitate in the form of snow on its land. The research should be concentrated on the source of moisture and to find out whether the ice sheet has grown gradually with uniform rates or in episodes. Intervals, when there has been excess moisture transport to Antarctica, is another area to explore. Such studies will enlighten us about the punctuated vs gradual growth of the Antarctic Ice sheet. Studies should be integrated with the investigations of coastal marine sequence stratigraphy, which records the eustatic rise and fall of sea levels.

It is proposed that Early Eocene Climatic Optimum, late Middle Miocene Climatic Optimum, Early Pliocene warming might have been events triggering more moisture transport to Antarctica by intricate wind patterns.

9 Future Scope

Future research needs to be concentrated on the nature of thermohaline circulation during Cenozoic, moisture transport, and meridional heat transport to Antarctica in specific time slices.

Modern studies concentrated on the precipitation, ablation, and temperature variations on the ice continent will pave the way for understanding the future of the Antarctic Ice sheet. Ocean circulation, including moisture transport due to poleward currents like Leeuwin Current, East Australian Current, Agulhas Current, and their intensification and resultant shifting of Antarctic Polar Front in geological time, will also be instrumental in understanding the future of the Antarctic ice sheet. Another important aspect being studied is the role of ice sheets in the global carbon cycle. In the last few years, Ice Sheets/ glaciers have been considered active players in the worldwide carbon cycle. This is due to many reasons. The ice sheet/glaciers cover the vegetated areas of the world, isolating the carbon reservoirs from taking part in the carbon cycle. Once they retreat, the vegetated areas are exposed and available for the carbon cycle. So this process works rhythmically with glacial and interglacial ages. The aerosols with carbon fall on the glacier surface and are carried to the ocean/ river by meltwater. Recently discovered rich microbial life beneath ice sheets that respire and release carbon dioxide is another important aspect to be studied (Wadham et al. 2019).

Acknowledgements AKS and DKS thank the Delhi School of Climate Change and Sustainability (DSCCS) of the Institution of Eminence (IoE) and the Department of Geology, University of Delhi, for providing infrastructural support. Financial assistance from the Ministry of Earth Sciences, Govt of India (Sanction No. MoES/CCR/Paleo-4/2019) is thankfully acknowledged. Other authors (KM, VPS, AS, TK) thank their respective institutions for logistic help.

References

Alley RB et al (1993) Abrupt increase in Greenland snow accumulation at the end of the Younger Dryas event. Nature 362:527–529

Barker PF, Filippelli GM, Florindo F, Martin EE, Scher HD (2007) onset and role of the antarctic circumpolar current. Deep-Sea Res Part II: Top Stud Ocean 54(21–22):2388–2398

Barrett PJ, Florindo F, Cooper AK (2006) Introduction to "Antarctic climate evolution: view from the margin." Palaeogeogr Palaeoclim Palaeoecol 231(1–2):1–8

Bleeker W (2003) The late Archean record: a puzzle in ca. 35 pieces. Lithos 71:99–134
Bond GC (1979) Evidence for late Tertiary uplift of Africa relative to North America. S Am Aust Eur: J Geol 86:47–65
Breza JR, Wise S (1992) Lower oligocene ice-rafted debris of the Kerguelen Plateau: evidence for East Antarctic Continental Glaciation. In: Wise SW, Schlich R, et al (eds), Proceedings of the ocean drilling program, scientific results, college station, TX (Ocean Drilling Program), vol 120, 161–178
Carter A, Riley TR, Hillenbrand C-D, Rittner M (2017) Widespread Antarctic glaciation during the Late Eocene. Earth Planet Sci Lett 58(2017):49–57
Cheney ES (1996) Sequence stratigraphy and plate tectonic significance of the Transvaal succession of southern Africa and it's equivalent in Western Australia. Precambr Res 79:3–24
Christoffel D, Falconer R (1972) Marine magnetic measurement in the southwest Pacific Ocean and the identification of new tectonic features. In: Hayes D (ed) Antarctic oceanology.' The Australian-New Zealand Sector, Antarctic Res Ser, vol 19, p 197, AGU, Washington, DC, 1972
Coxall HK, Wilson PA, Pälike H, Lear CH, Backman J (2005) Rapid stepwise onset of Antarctic glaciation and deeper calcite compensation in the Pacific Ocean. Nature 433(7021):53–57
Cristini L, Grosfeld K, Butzin M, Lohmann G (2012) The influence of the opening of the Drake Passage on the Cenozoic Antarctic Ice Sheet: a modelling approach. Palaeogeogr Palaeoclim Palaeoecol 339–341:66–73
Cuffey K, Clow G (1997) Temperature, accumulation, and ice sheet elevation in central Greenland through the last deglacial transition. J Geophys Res 702(26):383–26,396
DeConto RM, Pollard D (2003) Rapid Cenozoic glaciation of Antarctica induced by declining atmospheric CO2. Nature 42:245–249
Deighton l, Falvey DA, Taylor DJ (1976) Depositional environments and geotectonic framework: Southern Australian continental margin. APEA J (16.1. 25–36. 16, I 0976) 25–36. A J. 0976)
Ebbing J, Haas P, Ferraccioli F, Pappa F, Szwillus W, Bouman J (2018) Earth tectonics as seen by GOCE - enhanced satellite gravity gradient imaging. Sci Rep 8(1):34733–34739
Ehrmann WU (1991) Implications of sediment composition on the southern Kerguelen Plateau for paleoclimate and depositional environment. In Proc. ODP sci result, vol 119, pp 185–210
Ehrmann WU, Mackensen A (1992) Sedimentological evidence for forming an East Antarctic ice sheet in Eocene/Oligocene time. Palaeogeogr Palaeoclim Palaeoecol 93:85–112
Eldrett JS, Harding IC, Wilson PA, Butler E, Roberts AP (2007) Continental ice in Greenland during the Eocene and Oligocene. Nature 446:176–179
Elliot DH (1975) The tectonics of Antarctica. Am J Sci A275:45–106
Evans DAD, Mitchell RN (2011) assembly and the breakup of the core of Paleoproterozoic-Mesoproterozoic supercontinent Nuna. Geology 39:443–446
Florindo F, Wilson GS, Roberts AP, Sagnotti L, Verosub KL (2005) Magnetostratigraphic chronology of a late Eocene to early Miocene glacimarine succession from the Victoria Land Basin, Ross Sea. Antarctica. Glob Planet Chang 45(1):207–236
Flower B, Kennett JP (1995) Middle Miocene deepwater paleoceanography in the southwest Pacific: relations with East Antarctic Ice Sheet development. Paleoceano Paleoclimate 10(6):1095–1112
Galeotti S, DeConto R, Naish T, Stocchi P, Florindo F, Pagani M, Zachos JC (2016) Antarctic Ice Sheet variability across the Eocene-Oligocene boundary climate transition. Science 352(6281):76–80
Ghavri S, Catherine JK, Ambikapathy A, Kumar A, Gahalaut V (2017) Antarctica plate motion. Proc Indian Natn Sci Acad 83(2):437–440
Gildor H, Tziperman E (2000) Sea ice as the glacial cycles climate switch: role of seasonal and Milankovitch forcing. Paleoceanography 15(6):605–615
Gildor H (2003) When earth's freezer door is left Ajar Eos, vol 84, No. 2, 3
Gradstein FM, Ogg JG, Schmitz M, Ogg G (2012) The geologic time scale 2012, 1176 p
Grew ES, Manton WI (1979) Archean rocks in Antarctica: 2.5 billion year Uranium led ages of pegmatites in Enderby Land. Science 206(4417):443–445

Grunow AM, Kent DV, Dalziel IWD (1991) New paleomagnetic data from Thurston Island: Implications for the tectonics of West Antarctica and the Weddell Sea opening. J Geophys Res 96:17935–17954
Hambrey MJ, Ehrmann W, Larsen B (1991) Cenozoic glacial record of the Prydz Bay continental shelf, East Antarctica. In: Barron J, Larsen B, et al (Eds), Proceedings of the ocean drilling program: scientific results, vol 119 (pp 77–132). College Station, TX: Ocean Drilling Program
Harley SL, Fitzsimons IC, Zhao Y (2013) Antarctica and supercontinent evolution: historical perspectives, recent advances and unresolved issues. Geol Soc Lond Spec Publ 383:1–34
Haug GH, Tiedemann R (1998) The effect of the formation of the isthmus of Panama on Atlantic thermohaline circulation. Nature 393:673–676
Hollis CJ, Tayler MJS, Andrew B, Taylor KW, Lurcock P, Bijl PK, Kulhanek DK, Crouch EM, Nelson CS, Pancost RD, Huber M, Wilson GS, Todd GV, Crampton JS, Schiøler P, Phillips A (2014) Organic-rich sedimentation in the South Pacific Ocean associated with Late Paleocene climatic cooling. Earth Sci Rev 134:81–97
Huber M, Nof D (2006) The ocean circulation in the Southern Hemisphere and its climatic impacts in the Eocene. Palaeogeogr Palaeoclim Palaeoecol 231:9–28
Hyland EG, Sheldon ND, Cotton JM (2017) Constraining the early Eocene climatic optimum: a terrestrial interhemispheric comparison. Geol Soc Am Bull 129(1–2):244–252
Ivany LC, Van Simaeys S, Domack EW, Samson SD (2006) Evidence for an earliest Oligocene ice sheet Antarctic Peninsula. Geology 34(5):377–380
Jenkins DG (1974) Initiation of the Proto Cicum Antarctic current. Nature 252(5482):371–373
Jones DL, Bates MP, Li Z-X, Corner B, Hodgkinson G (2003) Palaeomagnetic results from the ca. 1130 Ma Borgmassivet intrusions in the Ahlmannryggen region of Dronning Maud Land, Antarctica, and tectonic implications. Tectonophysics 375:247–260
Katz ME, Cramer BS, Toggweiler JR, Esmay G, Liu C, Miller KG, Rosenthal Y, Wade BS, Wright JD (2011) Impact of Antarctic Circumpolar Current development on late Paleogene ocean structure. Science 332:1076–1079
Kennett JP, Exon NF (2004) Paleoceanographic evolution of the Tasmanian Seaway and its climatic implications. Geophys Monogr Ser 345–367
Kennett JP, Houtz RE, Andrews PB, Edwards AR, Gostin VA, Hajos M, Hampton MA, Jenkins DG, Margolis SV, Ovenshine AT, Perch-Nielsen KC (1975) Cenozoic paleoceanography in the southwest Pacific Ocean, Antarctic glaciations and the development of the circum-Antarctic current, in Initial Reports of the Deep Sea Drilling Project, vol 29, p 1155, US Govt Printing Office Washington DC
Kennett JP (1977) Cenozoic evolution of Antarctic glaciation, the circum-Antarctic Ocean and their Impact on Global Paleoceanography. J Geophys Res 82(27):3843–3859
Kennett JP, Barker PF (1990) Latest Cretaceous to Cenozoic climate and oceanographic developments in the Weddell Sea, Antarctica: an ocean-drilling perspective. In: Kennett JP, et al. Proceedings of the ocean drilling program, scientific results, vol. 113. College Station, Texas, Ocean Drilling Program, pp 937–960
Kent DV, Gradstein FM (1986) A Jurassic to recent geochronology, in the geology of North America, vol M. The Western North Atlantic Region, edited by Vogt PR, Tucholke BE, pp 45–50, Geological Society of America, Boulder, Colo., 1986
Lawrence RD, Khan SH, Nakata T (1992) Chaman fault, Pakistan-Afghanistan: Ann. Tectonicae, v. Spec. Issue. - Suppl. to vol. VI, pp 196–223
Lawver LA, Gahagan LM (2003) Evolution of Cenozoic seaways in the circum Antarctic region. Paleogeogr Paleoclimate Paleoecology 198:11–37
Le-Treut H, Ghil M (1983) Orbital forcing, climatic interactions, and glaciation cycles. J Geophys Res: Ocean 88(C9):5167–5190
Livermore R, Hillenbrand C-D, Meredith M, Eagles G (2007) Drake Passage and Cenozoic climate: an open and shut case? Geochem Geophys Geosystems 8(1)
Lourious C, Merlivat L, Jouzel J, Pourchet M (1979) A 30,000-yr isotope climatic record from Antarctic ice. Nature 280:644–648

Lowrie W, Hayes DE (1975) Magnetic properties of oceanic basalt samples in Initial Reports of the Deep Sea Drilling Project, vol 28, p 869, US Government Printing Office, Washington DC
Lurcock P, Florindo F (2017) Antarctic climate history and global climate changes. Oxford Handbooks Online (www.oxfordhandbooks.com). Oxford University Press, 2018
Lyle M, Gibbs S, Moore TC, Rea DK (2007) Late oligocene initiation of the antarctic circumpolar current: evidence from the South Pacific. Geology 35(8):691
Marenssi SA, Casadío S, Santillana SN (2010) Record of Late Miocene glacial deposits on Isla Marambio (Seymour Island) Antarctic Peninsula. Antarct Sci 22(2):193–198
Marschall HR, Hawkesworth CJ, Storey CD, Dhuime B, Leat PT, Meyer HP, Tamm-Buckle S (2010) The Annandagstoppane Granite, East Antarctica: evidence for Archaean intracrustal recycling in the Kaapvaal Grunehogna Craton from zircon O and Hf isotopes. J Petrol 51(11):2277–2301
Maslin MA, Li XS, Loutre M-F, Berger A (1998) The contribution of orbital forcing to the progressive intensification of northern hemisphere glaciations. Quatern Sci Rev 17(4–5):411–426
McElhinny MW (1973) Palaeomagnetism and plate tectonics. Cambridge University Press, New York
Miller KG, Wright JD, Fairbanks RG (1991) Unlocking the Ice House: Oligocene-Miocene oxygen isotopes, eustasy, and margin erosion. J Geophys Res: Solid Earth 96(B4):6829–6848
Miller KG, Fairbanks RG, Mountain GS (1987) Tertiary oxygen isotope synthesis, sea-level history and continental margin erosion. Paleoceanography 2:1–19
Pälike H, Norris RN, Herrle J, Wilson PA, Coxall HK, Lear CH, Shackleton NJ, Tripati AK, Wade BS (2006) The heartbeat of the Oligocene climate system. Science 314:1894–1898
Payne JL, Hand M, Barovich KM, Reid A, Evans DAD (2009) Correlations and reconstruction models for the 2500–1500 Ma evolution of the Mawson Continent. In: Reddy SM, Mazumder R, Evans DAD, Collins AS (eds) Palaeoproterozoic supercontinents and global evolution. Geological Society, London, Special Publications, vol 323, pp 319–355
Payros A, Ortiz S, Millán I, Arostegi J, Orue-Etxebarria X, Apellaniz E (2015) Early Eocene climatic optimum: environmental impact on the North Iberian continental margin. Geol Soc Am Bull 127(11–12):1632–1644
Pearson PN, Palmer (2000) Atmospheric carbon dioxide concentrations over the last 60 million years. Nature 406:695–699
Phillips G, Wilson CJL, Campbell IH, Allen CM (2006) U-Th–Pb detrital zircon geochronology from the southern Prince Charles Mountains, East Antarctica – defining the Archaean to Neoproterozoic Ruker Province. Precambr Res 148:292–306
Poore RJ, Sloan LC (1996) Climates and climate variability of the Pliocene. Mar Micropal 27:326
Pross J, Contreras L, Bijl PK, Greenwood DR, Bohaty SM, Brinkhuis H (2012) Persistent near-tropical warmth on the Antarctic continent during the early Eocene epoch. Nature 488(7409):73–77
Reddy SM, Evans DAD (2009) Palaeoproterozoic supercontinents and global evolution: correlations from the core to atmosphere. In: Reddy SM, Mazumder R, Evans DAD, Collins AS (eds) Palaeoproterozoic supercontinents and global evolution. Geological Society, London, Special Publications, vol 323, pp 1–26
Roberts AP, Wilson GS, Harwood DM, Verosub KL (2003) Glaciation across the Oligocene-Miocene boundary in southern McMurdo Sound, Antarctica: New chronology from the CIROS-1 drill hole. Palaeogeogr Palaeoclim Palaeoecol 198(1):113–130
Roberts DG, Morton AC, Backman J (1984) Late Paleocene-Eocene Volcanic events in the northern North Atlantic ocean Init. Reports. DSDP, 81: Washington (US Govt. Printing Office)
Rogers JJ (1993) India and Ur. Geol Soc India 42(3):217–222
Ruddiman WF, Mclntyre A (1981) Oceanic mechanisms for amplification of the 23,000-year
Sagnotti L, Kenneth V, Andrew R, Wilson GS et al (2001) Environmental magnetic record of the Eocene-Oligocene transition in CRP-3 drillcore. Victoria Land Basin, Antarctica, Terra Antarctica 8(4):507–516
Savin SM, Barrera E (1985) Cenozoic ocean temperatures inferred from 18O/16O ratios of foraminifera abst Geol Soc. Am Abst Programs 17(7):707

Savin SM, Douglas RG, Stehli FG (1975) Tertiary marine paleotemperatures. Geol Soc Am Bull 86(11):1499

Scher HD, Martin EE (2006 Apr 21) (2006), timing and climatic consequences of the opening of drake passage. Science 312(5772):428–430

Shackleton NJ, Kennett JP (1975) Initial Reports of the Deep Sea Drilling Project, vol 29, pp 801–807

Sobotovich EV, Kamenev VN, Kornanstyy AA, Rudnik VA (1976) The oldest rocks of Antarctica (Enderby Land). Int Geol Rev 18(4):371–388. https://doi.org/10.1080/00206817609471218

Srinivasan MS, Sinha DK (1998) Early Pliocene closing of the Indonesian seaway: evidence from the North-East Indian Ocean and Tropical Pacific deep-sea cores. J Asian Earth Sci 16(1):29–44

Stuiver M, Braziunas TF (1985) Compilation of isotopic dates from Antarctica. Radiocarbon 27(2A):117–304. https://doi.org/10.1017/s0033822200007037

Tigchelaar M, von der Heydt AS, Dijkstra HA (2011) (2011) A new mechanism for the two-step δ 18O signal at the Eocene-Oligocene boundary. Clim Past 7:235–247

Tingey RJ (1991) The regional geology of Archaean and Proterozoic rocks in Antarctica. In: Tingey RJ (ed) The Geology of Antarctica. Oxford University Press, Oxford, pp 1–73

Tripati AK, Eagle RA, Morton A, Dowdeswell JA, Atkinson KL, Bahé Y, Thanabalasundaram L (2008) Evidence for glaciation in the Northern Hemisphere back to 44 Ma from ice-rafted debris in the Greenland Sea. Earth Planet Sci Lett 265(1–2):112–122. https://doi.org/10.1016/j.epsl.2007.09.045

van der Veen CJ (2002) Polar ice sheets and global sea level: how well can we predict the future? Global Planet Change 32:165–194

Vaughan APM, Livermore RA (2005) Episodicity of Mesozoic terrane accretion along the pacific margin of Gondwana: implications for superplume plate interactions. In: Vaughan APM, Leat PT, Pankhurst RJ (eds) Terrane processes at the margin of Gondwana. Geological Society, London, Special Publications, vol 246, pp 143–178

Veevers JJ (2012) Reconstructions before rifting and drifting reveal the geological connections between Antarctica and its conjugates in Gondwanaland. Earth Sci Rev 111:249–318

Vincent E, Killinglcy JS, Berger WH (1985) Miocene oxygen and carbon isotope stratigraphy of the tropical Indian Ocean, in The Miocene Ocean: paleoceanography and biogeography, edited by J.P. Kennett. Mem Geol Soc Am 163:103–130

Wadham JL, Hawkings JR, Tarasov L, Gregoire LJ, Spencer RGM, Gutjahr M, Ridgwell A, Kohfeld KE (2019) Ice sheets matter for the global carbon cycle. Nat Commun 10(1)

Webb PN (1962) Isotope dating of Antarctic rocks NZ. J Geol Geophys 5(5):790–796

Wilson GS, Roberts AP, Verosub KL, Florindo F, Sagnotti L (1998) Magnetobiostratigraphic chronology of the Eocene-Oligocene transition in the CIROS-1 core, Victoria Land margin, Antarctica: implications for Antarctic glacial history. Geol Soc Am Bull 110(1):35–47

Wilson DS, Pollard D, DeConto RM, Jamieson SSR, Luyendyk BP (2013) Initiation of the West Antarctic ice sheet and estimation of total Antarctic volume in the earliest Oligocene. Geophys Res Lett 40:4305–4309

Wilson DS, Pollard D, DeConto RM, Stewart SRJ, Luyendyk BP (2013) Initiation of the West Antarctic Ice Sheet and estimates of total Antarctic ice volume in the earliest Oligocene. Geophys Res Lett 40:4305–4309. https://doi.org/10.1002/grl.50797

Wright JD, Miller KG, Fairbanks RG (1992) the evolution of modern deepwater circulation: evidence from the late Miocene Southern Ocean. Paleoceanography 6:275–290

Wright JD, Miller KG (1993) Southern ocean influences late Eocene to Miocene deepwater circulation. Antarct Res Ser 60:1–25

Yakubchuk AS (2019) From Kenorland to modern continents: tectonics and metallogeny. Geotektonika 2:3–32

Zachos J, Pagani M, Sloan L, Thomas E, Billups K (2001) Trends, rhythms, and aberrations in global climate 65 Ma to present. Science 292(5517):686–693

Zachos JC, Lohmann KC, Walker JCG, Wise SW (1993) Abrupt climate change and transient climates during the Paleogene: a marine perspective. J Geol 101:191–213
Zhao G, Sun M, Wilde SA, Li S (2004) A Paleo- Mesoproterozoic supercontinent: assembly, growth and breakup. Earth Sci Rev 67:91–123

Dronning Maud Land (Antarctica) and Reconstruction of Its Glacial History with Cosmogenic Radionuclides

Waseem Ahmad Baba, Pankaj Kumar, Jitendra Kumar Pattanaik, and Neloy Khare

Abstract This chapter deliberates Dronning Maud Land's glacial history (DML) based on available dates obtained using cosmogenic radionuclides. As Dronning Maud Land is a part of the East Antarctica Ice Sheet (EAIS), background information about EAIS and its glacial history is also discussed based on the researchers' various evidence. A comprehensive outline of DML, the basics of cosmogenic radionuclide and its application and major glacial events from DML are presented in this chapter. Further, meltwater's pulse due to deglaciation of EAIS and evidence related to the marine isotope stages were discussed to understand the impact of deglaciation on the global ocean. A very few direct dates were available from Dronning Maud Land to establish the detailed glacial chronology, or some of the results are contradicting. This region shows sparse or no evidence of ice thickening during the last glacial maximum (LGM). Field observations and ice core models show that the ice sheet's interior parts, the ice dome, were possibly 100 m lower during LGM than the present. The results obtained by the various researchers shows that around 600 m high ice sheet existed 4 million years ago, which is decreasing continuously to the present day.

Keywords Dronning maud · East antarctica Ice sheet (EAIS) · Cosmogenic radionuclides · Last glacial maximum (LGM) · Meltwater runoff

W. Ahmad Baba · J. Kumar Pattanaik (✉)
Department of Geology, Central University of Punjab, Punjab 151001, India

P. Kumar
AMS Division & Geochronology Group, Inter-University Accelerator Centre, New Delhi 110067, India

N. Khare
Ministry of Earth Sciences, Lodhi Road, Prithvi Bhawan, New Delhi 110 003, India

N. Khare (ed.), *Assessing the Antarctic Environment from a Climate Change Perspective*, Earth and Environmental Sciences Library,
https://doi.org/10.1007/978-3-030-87078-2_5

1 Introduction

The Continental ice of Antarctica is separated by the Transantarctic Mountains and created two unequal ice sheets. The West Antarctica Ice Sheet is much smaller than the East Antarctica Ice Sheet (EAIS). The EAIS covering a vast tract of a continental area is an enormous ice mass on the earth's surface (Stonehouse 2002). The EAIS is comprised of several domes (e.g. Dome Fuji, Dome Argus, Dome Circe), reaching > 4.8 km of thickness near Dome Circe and estimated total grounded ice volume is 21.76 $\times$ 10^6 km^3 (Lythe and Vaughan 2001; Fretwell et al. 2013; Mackintosh et al. 2014). The EAIS is divided into different parts (e.g., Dronning Maud Land (DML), Enderby Land, Mac. Robertson Land, Wilkes Land, George V Land) based on the continental plateau and regional slope, where mountain chains obstruct large tract of the ice sheet. The EAIS is considered stable than the West Antarctica Ice Sheet and Greenland Ice Sheet; however, recent studies differ (Pingree et al. 2011; Mackintosh et al. 2014). The volume of EAIS is comparable to ~ 53 m of mean seal level (Lythe and Vaughan 2001; Fretwell et al. 2013; Mackintosh et al. 2014); therefore, a minor change in the volume will have a more significant impact on the global sea level. Many unanswered questions about the processes and timescale of the formation and existence of ice sheets in Antarctica. The Continental ice sheet plays a vital role in controlling the earth's climate. Climate modelling suggests that the concentration of CO_2 in the atmosphere is the foremost process for stabilising Antarctica's ice sheet (DeConto and Pollard 2003; Huber et al. 2004; Pollard and DeConto 2009). However, fewer attempts have been made to understand the response of post industrialisation rapid increase of atmospheric CO_2 on ice sheets. Reconstruction of the last 50 years showed significant warming over West Antarctica (0.1 °C per 10 years) and East Antarctica parts (Steig et al. 2009). Paleoglaciation studies on million years' timescale suggest a decrease in the EAIS thickness (Fogwill et al. 2004; Fink et al. 2006; Huang et al. 2008; Strasky et al. 2009; Kong et al. 2010; Di Nicola et al. 2009, 2012; Altmaier et al. 2010; Liu et al. 2010; Lilly et al. 2010). The extent of this decrease and its impact on the global climate are ambiguous. Also, the nature of glacio-eustatic rise, for example, a rapid rise in sea level due to meltwater pulse during the last glacial maximum (LGM), is poorly understood (Clark et al. 2002; Peltier, 2005; Mackintosh et al. 2014). Proxy records from ice cores provide precise and direct methods to analyse Antarctic climate change (Legrand and Mayewski 1997; EPICA Community Members 2006; Mayewski et al. 2009). A continuous record of paleo-temperature and atmospheric compositions is established based on stable isotope study on ice core samples. The most extended history is established up to 800,000 years from Dome Circe (Parrenin et al. 2007). Looking at the vastness of the EAIS and diverse surface and subsurface features, it is difficult to generalise the change observed at one place to the entire ice sheet as the ice sheet's response to the climate change varies with regions. Therefore, each region's glacial history needs to be evaluated using multiple proxies and synthesised for EAIS to have a comprehensive picture at a continental scale. Several studies have been carried out to understand the glacial history based on the available landforms; sediment archives from the lake, coastal and offshore area; ice core data; terrestrial cosmogenic radionuclide (TCN) studies on 'oasis' (ice-free region), nunatak, mountains and glacial debris. In this chapter, the emphasis is given to the Dronning Maud Land region's glacial history based on the study conducted using cosmogenic radionuclides as a proxy.

Cosmogenic radionuclides (CRN) dating techniques have brought a revolution in studying the geomorphic and landscape evolution and the rate at which these processes act on the earth's surface. When the secondary cosmic ray interacts with the rock surface, radionuclides are formed in situ due to spallation reaction. These in situ produced CRNs are used for surface exposure dating. Glacial chronology from thousands to million years is established based on CRN surface exposure ages of glacial landforms, boulders and moraines (Nishiizumi 1993). In a similar timescale, the burial age of sediments or till depositions using CRN provides a chronology for glacial advancement or retreat. Like other methods, this technique has some limitations; however, these limitations can be quickly hindered with detailed field observations, proper sampling strategy and multiple nuclide selections. This age helps to develop the glacial models for ice sheet and valley glaciers. CRN surface exposure dating is only helpful in the ice-free area, mountains, and nunataks present within the ice sheet. In the EAIS, only 1–2% of the site is ice-free and generally found in the coastal zones called 'Oasis' or within Mountain chains. Ages obtained from CRN like ^{14}C from sediment archive also help understand the paleoclimate and glacial history. Several studies were carried out in Dronning Maud Land to understand the glacial history based on CRNs. The literature was reviewed to reconstruct a glacial chronology for the entire region and address a few unanswered questions for future research scope.

1.1 History of Antarctica Glaciation

The ice sheet in Antarctica prevailed since the middle of Tertiary about 35 million years ago, and it is generally agreed that it reached east Antarctica by the Eocene and Oligocene (Barrett et al. 1991, Hambrey et al. 1989; Birkenmayer 1987, 1991; Denton et al. 1991; Prentice and Mathews 1988; Barrera and Huber 1993). However, the commencement and the timing of glaciation required further evaluation. The reconstruction of ice-sheet extent and volume is based on the ocean drilling programmes (ODP), where clay minerals, stable oxygen isotopic concentrations and sediment analysis were carried out on samples collected from the offshore core. Antarctic glaciation's commencement in the middle of Tertiary was possible with the Gondwana breakup, drifting of Antarctica towards poles and formation of ocean gateways or opening of "Darke Passage" (Kennett 1977). Isolation of Antarctica and the development of circumpolar current subsequently led to the cooling and glaciation. Another model suggests that the Antarctica glaciation was initiated due to lower CO_2 concentration in the atmosphere followed by a permanent ice cap due to further lowering CO_2 to a threshold value (DeConto and Pollard 2003; Altmaier et al. 2010). The results from ODP show subtropical to temperate climates on Dronning Maud

Land during Late Cretaceous (Kennett and Barker 1990). Data from the Weddell Sea suggested that Palaeocene's the warmest period (Kennett and Stott 1990; Robert and Kennett 1994). There are no records of ice sheet existence in the Late Cretaceous or Early Tertiary; however, fluctuations of ice sheets in east Antarctica have been reported (Anderson 1999). There was an increase in the ^{18}O concentrations in the deep-sea records during the Eocene period, indicating the growth of ice sheets in Antarctica (Prentice and Mathews 1988, Denton et al. 1991; Abreu and Anderson 1998). Evidence supports ice sheets in east Antarctica during the Oligocene and the spreading of ice in the Ross Sea during Late Oligocene (Denton et al. 1991; Hambrey et al. 1991). However, ice sheet occurrences in west Antarctica are unknown (Birkenmajer 1998; Anderson 1999). Based on the fossil record, earlier studies proposed that temporary large-scale retreat of EAIS during Pliocene (Webb and Harwood 1987); however, the recent studies based on field evidence and numeric modelling do not support the retreat and suggest a stable EAIS during Pliocene (Denton et al. 1984; Sugden et al. 1995; Pollard and DeConto 2009; Altmaier et al. 2010).

1.2 Present-Day Scenario of Antarctica Ice Sheet

The grounding line is an essential indicator of ice sheets' instability, as their changes depict the flow of ice and imbalances with the surrounding ocean. Ocean driven forces have melted various Antarctic glaciers, which have retreated the grounding line rapidly. However, there are limited records to measure imbalance. Between 2010 to 2016, retreat in grounding lines in east Antarctica (3%), Antarctic Peninsula (10%) and West Antarctica (22%) were recorded (Konrad et al. 2018). It has been shown that the retreat has been very swift (25 m yr^{-1}). The loss of grounded ice area has been around 1463 ± 791 km^2 (Konrad et al. 2018). Satellite altimeter to measure the ice elevation and geometry of the ice were combined with tracking the grounding line movement. The fastest rates have been seen in the Amundsen Sea, while in Pine Island, the grounding line has stabilised possibly due to reduced ocean forcing. According to the ice geometry and satellite measurements, the retreat of grounding lines in west Antarctica, East Antarctica and Antarctica peninsula has been faster after the post-glacial event.

Variations in grounded ice sheets appear due to differences in meltwater runoff, discharge of ice in the ocean and snow accumulation at the surface (Rignot et al. 2011). There has been a reduction in ice thickness in recent times, which has disturbed ice's inland flow. Various satellite techniques complemented with the field measurements and mass balance model have been developed to estimate ice sheet masses' variations (Zwally et al. 2012). It has shown that 2720 ± 1390 billion tonnes of ice have been lost from Antarctica during from 1992–2017, increasing the sea level by 7.6 ± 3.9 mm. By this period in west Antarctica, melting has increased from 53 ± 29 to 159 ± 26 Billion tonnes per year. However, there are uncertainties in the models showing again in surface mass balance with an average being 5 ± 46 billion tonnes per year (Shepherd et al. 2018).

1.3 Glacial History: Evidence from Ice Cores

Antarctica's Ice cores indicate changes in the ice volume from the past 4 million years (Petit et al. 1997). Records from the Vostok show a well-built correlation with global ice records. This shows a link between the ice sheets of the northern and southern hemispheres (Petit et al. 1997). However, there are variations in climate and glacial history of Late Quaternary obtained from ice core, terrestrial, offshore records (Jouzel et al. 1987; Petit et al. 1997; Ingólfsson 1998; Anderson 1999; Ingólfsson and Hjort 1999). During the last glacial maxima (LGM), the east Antarctica domes were thinner than the present because the accumulation rates were lower (Jouzel et al. 1989; Siddall et al. 2012). These observations are based on the ice sheet models and the ice core data. However, these changes in ice thickness are poorly known. Ice cores data used to reconstruct elevation using gas content of the bubbles trapped within ice and ice flow models to constraint the accumulation rates. Both the methods have uncertainties in their results; therefore, they should be used carefully for reconstructing glacial event. However, the obtained records of the past ice thickness are consistent with these methods.

2 Dronning Maud Land (DML)

Dronning Maud land is a region in East Antarctica covering an area of 2.7×10^6 km^2 (Fig. 1). In this region, different countries have established permanent research stations operational year-round or sessional to study geology, geodetic, glaciology, geophysics and atmospheric phenomena etc. Some of the essential stations are Maitri (India), Aboa (Finland); Weasands (Sweden); Troll and Tor

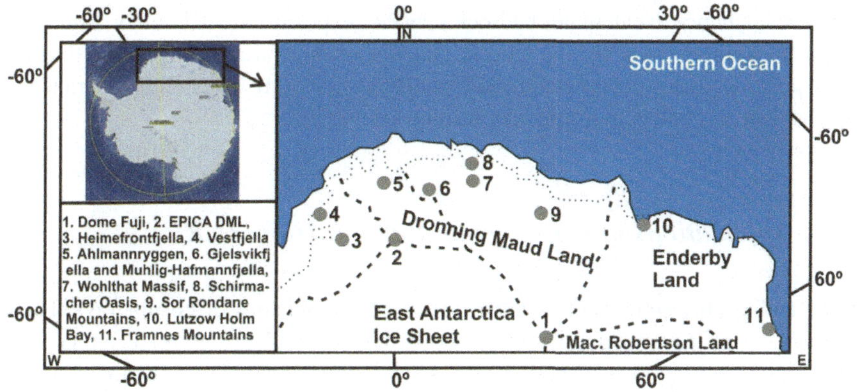

Fig. 1 Location of Dronning Maud Land, East Antarctica and name of the mountain ranges and ice-free regions are given in the diagram (Modified after Mackintosh et al. 2014). Dotted lines indicate the movement of ice mass and slope reduce toward the coastal region

(Norway), Princess Elisabeth Base (Belgium), Neumayer-Station III and Kohnen (Germany), Novolazarevskaya Station (Russia), SANAE IV (South Africa), Asuka, Showa and Dome Fuji Station (Japan).

The Dronning Maud land is dominated by Precambrian gneiss formed between 1 to 1.2 Ga. The mountains in the area are characterised by granitic and crystalline rocks that probably formed 500 to 600 Ma ago during the assemblage of Gondwana land. Younger sedimentary and volcanic rocks are found in the western parts of the region. Borg Massif guards the ice sheet over DML in the west and Yamato Mountain in the east (Pattyn et al. 2010; Mackintosh et al. 2014). The region has thick ice sheets that show downslope towards the coastal part from the continental plateau. Various researchers studied several mountain ranges, nunatak and oasis (ice-free area) located within the DML ice sheet to reconstruct the glacial history of EAIS. The important mountain ranges from west to east are Vestfjella, Heimefrontfjella, Ahlmannryggen, Gjelsvikfjella, Wohlthat Massif (includes Peterman Range, Insel Range, Gruber Mountains and Humboldt Mountains), DallmannBergeandPetermannKetten mountains (south of Wohlthat Massif), and SørRondane Mountains (Fig. 1). Among ice-free areas, Schirmacher oasis, Untersee oasis and coastal oasis in the Lützow-Holm Bay (towards eastern margin of DML ice sheet) were extensively studied for geomorphology, paleoclimate and glacial history. These ice-free regions are home to numerous lakes, Roche Moutonnée, a fossil glacier track filled with boulders, till and moraine deposits.

Striated surfaces and bedrock from nunatak and Roche Moutonnée, erratic's and boulders from fossil glaciers track provide ample opportunities to use cosmogenic radionuclides to measure the surface exposure age, and it can be used for understanding the variation of ice thickness and glacial history. During glacier retreat or thinning and shrinking of the ice sheet, bedrock or boulders are exposed to the cosmic ray, and in situ, cosmogenic radionuclides are produced. Although target nuclides are present in all the rock-forming minerals, quartz is used to measure radionuclides' concentration. The production and decay rate of cosmogenic radionuclides is well established; hence, it calculates bedrock's surface exposure age or boulder. Similarly, sediment archives from the lake deposits and offshore region are used to know the time of deposition using cosmogenic radionuclides, indicating paleoclimate, transport mechanism and paleoenvironmental setting.

2.1 Applications of Cosmogenic Radionuclides (CRN)

Earth and its atmosphere continuously receive solar and galactic cosmic rays. These primary cosmic rays are mainly high-energy (0.01-10^2 GeV) protons and alpha particles. Upon entering into the earth's atmosphere, direct cosmic rays interact with the nuclei of atmospheric elements and produce a cascade of lower energy secondary cosmic rays (mainly neutrons). These secondary cosmic rays further interact with the elements present in the atmosphere and on the earth surface, and in spallation reaction, radioactive isotopes (also called radionuclides) are produced. Radionuclides produced on the earth surface and top layers are called in-situ had CRN (such as ^{10}Be, ^{26}Al, ^{21}Ne etc.) and those made in the atmosphere are called *garden variety* CRN.

Solar cosmic rays are of lower energies and get easily deflected by the geomagnetic field. The atmosphere further causes attenuation. Even at high latitudes, where the geomagnetic deflection is less, solar cosmic rays can penetrate only the topmost layers of the atmosphere and hardly reach the earth's surface their low energy.

Galactic cosmic rays (originating outside the solar system from supernova explosions) are of higher energies and significantly contribute to the in-situ production of ^{10}Be and ^{26}Al. Due to higher energies, GCRs are only partially shielded by the geomagnetic field and reaches earth surfaces after penetrating the whole atmosphere. Major in-situ production channels of ^{10}Be and ^{26}Al on the earth surface are by spallation of neutrons with oxygen (Fig. 2) and silicon, respectively, [$^{16}O\ (n,\ 4p3n)^{10}Be$, $^{28}Si\ (n,\ p\ 2n)\ ^{26}Al$ present in quartz mineral].

In-situ ^{10}Be production rate at sea level and latitude $\geq 60°$, in the rocks having exposure ages ranging from 11 ka to 4 Ma is estimated between 5.8 to 6.4 atoms per year per gm of quartz (Kubik et al 1998). While, the production rate of ^{26}Al in SiO_2 in terrestrial rocks at sea level and latitude > 60° is about 37 atoms per year per gm of quartz (Kubik et al. 1998), and increases rapidly with altitude to 374 atoms per year per gm of quartz at 3.34 km altitude at 44° N (Nishiizumi et al. 1989).

In areas where the inherited and independent ages coexist, magnitude, rate, and spatial patterns can be revealed from single cosmogenic nuclides. However, by applying two radionuclides (^{10}Be and ^{26}Al) with different half-lives on the same samples, uplift rates can be determined with greater accuracy and confidence (Tuniz 2001). These provide a model that links the erosion and ice dynamics processes. The error ranges lie in $\pm\ 5 - 10\%$ for the surface exposure dating, including systematic and analytical error. CRNs like ^{10}Be, ^{26}Al and ^{14}C are used suitably for the burial age of the sediments/pebble/cobbles depending upon the landforms, local setting and type of sediments.

2.2 Measurement of CRNs

Cosmogenic ^{10}Be ($T_{1/2}$ = 1.38 Ma) and ^{26}Al ($T_{1/2}$ = 0.72 Ma) in the geological samples are found at a deficient level, and their isotopic ratios ($^{10}Be/^{9}Be$ and $^{26}Al/^{27}Al$) ranges between 10^{-11} to 10^{-15}. The measurement of such trace CRNs is challenging and beyond the measurement capabilities of conventional mass spectrometric methods insensitivities and isobaric interferences. Accelerator Mass Spectrometry, in which an individual atom of CRN is counted after removing isobaric and isotopic interferences, is the technique, which can perform such ultrasensitive measurements with very high precision (Kumar et al. 2011, 2014, 2015). The sample processing and AMS measurement methods are described in various references (Kumar et al. 2011, 2014, 2015).

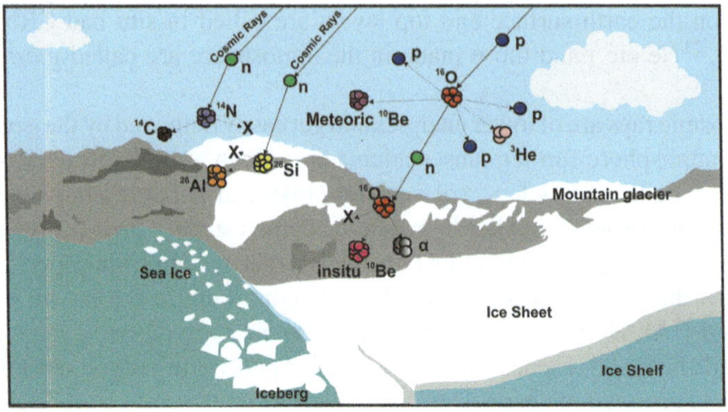

Fig. 2 Showing the in-situ production of Beryllium-10 and Aluminium-26 and production of carbon-14 in the atmosphere (Tuniz 2001; Willenbring and von Blanckenburg 2010)

3 Glacial History of Dronning Maud Land (DML)

Various studies like field observations, ice thickness measurement based on echo sounder, GPR, gravimeter; proxy-based analysis, ice core data, modelling and simulations were used to understand the glacial history of the Droning Maud Land. Most of the field-based studies and data collections were conducted in the continental and ice-free regions as oasis located in the low-lying coastal areas and marginal mountainous areas (Pattyn et al. 1992). The region's present glacial geomorphology is developed due to polycyclic glaciation and deglaciation phases, where deglaciation occurred frequently and for longer durations (Pattyn et al. 1992). Base on-field evidence around the Sor Rondane Mountain region and flow line model, Pattyn et al. (1992) reported that ice thickness was increased by 300–400 m during the last glacial maximum. The previous study based on the ice age model using glaciological and geomorphological evidence also suggests a similar increase in ice thickness of 400 m (Hirakawa and Moriwaki 1990). Based on the ice age simulation, it has been reported that the Grounding line of the Antarctica ice sheet was advanced by 20 km and 100 m thickening of polar ice plateau during that period (Pattyn et al. 1992). The ice cover of DML and EAIS responded slowly to climate change, as reported by Pollard and Deconto (2009) and Huybrechts (1993). In the longer time scale, the recent study using surface exposure dating of nunatak (south-east of Wohlthat Massif) suggest that between 0.75 to 3.57 Ma, the ice surface lowered at a rate less than 1 mm/year (Strub et al. 2015). The authors also did not rule out the possibility of exhumation as it can bring the nunatak above the ice surface. Previous surface exposure study from the same region suggests that Wohlthat Massif was exposed between 1 to 4 Ma ago (Altmaier et al. 2010). However, Strub et al. (2015) argued that this could be due to the thickness of the ice sheet as it was thicker (200–400 m) than today until ~ 0.5 Ma ago or due to the upliftment of that region. Many instance results from recent studies on other parts of the EAIS and for a different period do not converge. Therefore, the glacial history of EAIS is debatable. The effort was made to reconstruct the deglaciation history of Antarctica ice sheet since the Last Glacial Maximum based on the available data from different part of Antarctica by 'The RAISED Consortium' et al. (2014), where a change of the ice thickness and the grounding line position for the different period were discussed. Whereas Mackintosh et al. 2014, described the glacial history for a different location and summarised the changes of EAIS since the Last Glacial Maximum. Previous studies based on the surface exposure ages and burial age of sediments using cosmogenic radionuclides like ^{14}C, ^{10}Be and ^{26}Al were compiled (Table 1), and a sequence of glacial events was established.

3.1 Holocene Period

From the DML region, sparse records from a few oases and nunataks projecting from the ice sheet provides limited information about glacial fluctuation during the Holocene period. Like most Antarctica parts, the grounding line of ice sheets in the DML region was either on the inner shelf or close to the present-day position by 5 ka (The RAISED Consortium et al. 2014). However, in the west part of DML, the grounding line was seaward at the Weddell Sea compared to the present-day position. In the Heimefrontfjella region snow, petrel nests position is lying between 30 and 230 m above the present-day surface, and ^{14}C date of mumiyo sample from basal layer indicate that the since 8700 ± 40 yr B.P. (Lintinen and Nenonen 1997), top of the ice sheet was ~ 30 m above the present surface (Mackintosh et al. 2014). Similarly, mumiyo samples from Ahlmannryggen and Gjelsvikfjella ice-covered ridges in the western part of DML show ^{14}C dates of Holocene age with oldest dates recorded 8330 ± 70 yr B.P. and 3730 ± 80 yr B.P., respectively (Steele and Hiller 1997). These dates considered to be the minimum age of deglaciation of that region. However, other ^{14}C dates from Muhlig-Hafmannfjella, a nearby region, provides older dates and suggesting the thickness of EAIS is close to the present position before LGM (Steele and Hiller 1997; Mackintosh et al. 2014). In the Lützow-Holm Bay region (eastern part of DML), marine fossil samples from raised beach provided two clusters of ^{14}C age, where a sample from Ongul island situated towards north shows ~33–50 ka age, and sample from Skarvsnes and Skallen peninsula towards south shows <7 ka age (Miura et al. 1998; Takada et al. 2003; Mackintosh et al. 2014).

Similarly, samples from other islands in the south show Holocene ages. It was inferred that during the late Pleistocene, EAIS was retreated from the northern part and did not advance during LGM. However, the ice sheet has withdrawn from the southern part after the LGM and region were ice-free during the Holocene. The difference in surface weathering in this area's northern and southern region supports the relative variation in the glaciation history. Another cosmogenic radionuclide dating confirms that the Skarvsnes peninsula had ~ 360 m ice sheet, and it has retreated between 10 and 6 ka ago (Yamane et al. 2011; Mackintosh et al. 2014). Erratic boulders from east Antarctica show that ice sheets reached the present configuration by

Table 1 Cosmogenic radionuclide dating from Dronning Maud Land used for the chronological constraint of glacial history

S.N	Location	Latitude	Longitude	Method	Age (ka)	Reference
1	Dronning Maud Land	69°54.10'S	11°29.61'E	^{14}C	1.55 ± 0.07	Gingele et al. (1997)
2	Gjelsvikfjella	72°09′ S	2° 36′ E	^{14}C	3.73 ± 0.08	Steele and Hiller (1997)
3	Ahlmannryggen	71°50′ S	2° 25′ W	^{14}C	8.33 ± 0.07	
4	Heimefrontjella	74°35'S	11° 0'W	^{14}C	8.7 ± 0.04	Lintinen and Nenonen (1997)
5	Lazarv Sea	70°0.78'S	11°45.30'E	^{14}C	11.14 ± 0.12	Gingele et al. (1997)
6	Schirmacher Oasis	70°45.87' S	11°51.40' E	^{10}Be, ^{26}Al	21 ± 3	Altmaier et al. (2010)
7		70°45.73' S	11°50.79' E	^{10}Be,	22 ± 3	
8	Sør Rondane Mountains	72.2° S	27.8° E	^{10}Be	30 ± 2	Yamane et al. (2015)
9	Gjelsvikfjella	72°9′ S	2°36′ E	^{14}C	31.962 ± 1.7	Steele and Hiller (1997)
10	Lake Unterse (Wohlthat Massif)	71°20′ S	13°27′ E	^{14}C	33.9 ± 3.02	Hiller et al (1988)
11	Schirmacher Oasis	70°45.78' S	11°50.90' E	^{10}Be, ^{26}Al	35 ± 4	Altmaier et al. (2010)
12	Sør Rondane Mountains	72.2° S	27.8° E	^{10}Be, ^{26}Al	35 ± 3	Yamane et al. (2015)
13	Heimefrontjella	74°34′36"S	11°13′24"W	^{14}C	37.4 ± 1.5	Thor and Low (2011)
14	Vestfjella			^{14}C	38.7 ± 1.5	
15	Dallmann Berge	71°45.37'S	10°11.07'W	^{10}Be, ^{26}Al	70 ± 8	Altmaier et al. (2010)
16	Sør Rondane Mountains	72° S	24.9° W	^{10}Be	73 ± 5	Yamane et al. (2015)
17	Dallmann Berge	71°44.4' S	10°9.20' E	^{10}Be, ^{26}Al	81 ± 5	Altmaier et al. (2010)
18	Wohlthat Massif	70°0.78'S	11°45.30'E	TCN	100	

(continued)

Table 1 (continued)

S.N	Location	Latitude	Longitude	Method	Age (ka)	Reference
19	Dallmann Berge	71°45.55' S	10°11.08' E	^{10}Be, ^{26}Al	112 ± 7	
20		71°45.3' S	10°8.6' E	^{10}Be, ^{26}Al	135 ± 8	
21		71°45.3' S	10°9.10' E	^{10}Be, ^{26}Al	166 ± 17	
22		71°45.47' S	10°11.07' E	^{10}Be, ^{26}Al	192 ± 11	
23	Schussel/Eckhorner	71° 31.8' S	11°24.55' E	^{10}Be, ^{26}Al	196 ± 9	
24	Dallmann Berge	71°45.3' S	10°10.9' E	^{26}Al	202 ± 22	
25	Schussel/Seitental	71°36.95' S	11°28.8' E	^{10}Be, ^{26}Al	227 ± 11	
26	Untersee Westgrat	71°20' S	13°24.6' E	^{10}Be, ^{26}Al	235 ± 11	
27	Dallmann Berge	71°45.86' S	10°09.78' E	^{10}Be, ^{26}Al	239 ± 13	
28		71°45.36' S	10°10.9' E	^{10}Be	283 ± 15	
29	Untersee Ostgrat	71°21' S	13°31' E	^{10}Be, ^{26}Al	339 ± 11	
30		71° 21' S	13°31' E	^{10}Be, ^{26}Al	388 ± 14	
31	Sør Rondane Mountains	71.6° S	25.4° E	^{10}Be, ^{26}Al	428 ± 30	Yamane et al. (2015)
32	Dallmann Berge	71°45.42' S	10°10.9' E	^{10}Be, ^{26}Al	462 ± 18	Altmaier et al. (2010)
33	Sør Rondane Mountains	72° S	26° E	^{10}Be, ^{26}Al	476 ± 34	Yamane et al. (2015)
34		72° S	24.9° E	^{10}Be	594 ± 43	
35	Untersee Ostgrat	71° 21' S	13°31' E	^{10}Be, ^{26}Al	631 ± 19	Altmaier et al. (2010)
36		71° 20' S	13°24.7' E	^{10}Be, ^{26}Al	725 ± 35	
37	Sør Rondane Mountains	72° S	26.4° E	^{10}Be	745 ± 57	Yamane et al. (2015)
38	Schussel/Seitental	71°36.95' S	11°28.8' E	^{10}Be, ^{26}Al	912 ± 37	Altmaier et al. (2010)

(continued)

Table 1 (continued)

S.N	Location	Latitude	Longitude	Method	Age (ka)	Reference
39	PetermannKetten	71°52.782'S	11°57.364'E	^{10}Be, ^{26}Al	1050 ± 30	
40		71°30.765'S	12°34.481'E	^{10}Be	1150 ± 40	
41		71°53.741'S	11°57.661'E	^{10}Be, ^{26}Al	1180 ± 30	
42		71°24.67' S	12°48.193'E	^{10}Be, ^{26}Al	1530 ± 40	
43	Schussel / Eckhorner	71°31.75' S	11°25.25' E	^{10}Be, ^{26}Al	1760 ± 40	
44		71°31.75' S	11°25.25' E	^{10}Be, ^{26}Al	1760 ± 40	
45	Sør Rondane Mountains	72° S	24.3° E	^{10}Be	1851 ± 116	Yamane et al. (2015)
46	Petermann Ketten	71°53.3' S	11°57.5' E	^{10}Be, ^{26}Al	1920 ± 40	Altmaier et al. (2010)
47	Sør Rondane Mountains	72° S	24.3° E	^{10}Be, ^{26}Al	2980 ± 180	Yamane et al. (2015)
48	Petermann Ketten	71°25.934'S	12°43.88' E	^{10}Be	3000 ± 60	Altmaier et al. (2010)
49		71°50.149'S	12°17.734' E	^{10}Be, ^{26}Al	3770 ± 70	
50	Sør Rondane Mountains	72° S	25° E	TCN	4000	Moriwaki et al. (1991)
51	Petermann Ketten	72°8.362' S	11°30.961' E	^{10}Be	4110 ± 150	Altmaier et al. (2010)

this time. There was considerable thinning of ice sheets between 10 and 5 ka. The Lazarev Sea of east Antarctica has unearthed the processes that controlled the sedimentation over the past 10,000 yr during deglaciation. Lazarev Sea has distinct facies which reveal the environment of deposition with glaciomarine processes. These depositional sequences preserve the retreat of the ice history in this part of the continent. The minimum age of retreat of glaciers obtained from ^{14}C dating is 1550 ± 70 yr B.P. (Gingele et al. 1997).

The dates were obtained from carbonate particles terrestrial areas were exposed by 5 ka. Some areas also indicate fluctuations during Holocene (Gingele et al. 1997). The Nivl Ice Shelf of the Lazarev Sea is situated north of Schirmacher Oasis (Fig. 1), central DML. Laminated sediments from the Nivl Ice Shelf, dated to be 11,140 ± 120 ^{14}C yr B.P. (Gingele et al. 1997) and suggesting deglaciation of continental shelf during early Holocene. Subsequently, proglacial lakes were formed in the Schirmacher oasis (Mackintosh et al. 2014). This ice retreat was continued, and further, Schirmacher Oasis was becoming ice-free, and the proglacial lake became landlocked lakes around 3 ka (Schwab 1998; Phartiyal et al. 2011). However, another study based on surface exposure dating using cosmogenic radionuclide dating and lake sediment dating suggest that Schirmacher oasis was ice-free before LGM (Altmaier et al. 2010).

The substantial recession of ice sheets in east Antarctica arose around 13 cal yrs B.P., and the retreat was swift in Holocene. In West Antarctica, the retreat began at 10 ka. The ice sheets retreated significantly in the eastern and western Antarctica peninsula by the 15 and 10 ka and reached the present state during the Holocene middle. While on the east side, it would have gone by 10 ka (Ingólfsson et al. 2004).

3.2 LGM and Post LGM

The time duration of LGM (Last Glacial Maximum) varies from place to place, and the global chronostratigraphy refers to the time of the event from c. 26.5 to 19 ka B.P. (Clark et al. 2009). The literature term LLGM (Local Last Glacial Maximum) is being used for a specific location (Clark et al. 2009) to explain the last glacial maxima that differ widely. However, global LGM is considered roughly around 20 ka. To avoid such ambiguity, 'The RAISED Consortium et al. (2014) explain Antarctica's glacial history in the different time slices such as 20 ka, 15 ka, 10 ka and 5 ka. Available data shows Antarctica Ice sheet was not at its maximum extent during LGM (Anderson et al. 2002) and shows local variations. The DML region of EAIS shows the variation in glacial history during the LGM period. The glacial-geological data and ice sheet model contradict EAIS elevation changes around the Weddell Sea region during LGM. This region's overall glacial history is diverse compared to other DML sites (Hillenbrand et al. 2012). As per the Weddell Sea marine sediment record, the grounding line's extent is nearly 100 km seaward during 21 ka than the present (Elverhøi 1981; Mackintosh et al. 2014). Several glacio-geomorphological studies were conducted in the Vestfjella, and Heimefrontfjella mountain ranges and some of the result on the past ice thickness and its timing are contradicting (Jonsson 1988; Lintinen 1996; Lintinen and Nenonen 1997; Hattestrand and Johansen 2005). As no chronological constraints are available from the region, based on the field observation like the position of striations and till depositions in the Vestfjella region, the thickness of the ice sheet during LGM was estimated to be 700 m thicker than the present (Hattestrand and Johansen 2005; Mackintosh et al. 2014). However, these results are not supported by the ^{14}C dating of mumyio sample from this region and indicate that since 38,700 ± 1500 yr B.P., the region was ice-free (Thor and Low 2011). Based on the glacio-geomorphological evidence and surface weathering analysis in the Heimefrontfjella area, it has been contended that 100–200 m thicker ice sheet was present during LGM than today. The sediment core samples collected from few lakes situated on the Schirmacher Oasis shows that the region was covered with an ice sheet during LGM (Schwab 1998; Phartiyal et al. 2011); however, it is contradicted by other results. Based on the surface exposure dating using cosmogenic radionuclides in the SorRondane Mountain, it is inferred that the region had ~ 100 m thicker ice sheet during LGM than present-day. Studies on the nunataks from DML shows significantly less or no ice sheet thickening. Additionally, evidence from the ice-core and the ice-sheet model offers a thickness of the ice sheet was 100 m lower than the present during LGM (The RAISED Consortium et al. 2014) in the interior part of the ice sheet.

3.3 *Pre LGM*

Surface exposure dating using cosmogenic radionuclide from the Wohlthat Massif suggested that the thickness of EAIS has not changed significantly since 100 ka (Altmaier et al. 2010). Similarly, studies indicate that Sor-Rondane Mountain was ice-free since 4 Ma, and however, nunataks from the peripheral parts of this mountain range have become ice-free since 200 ka. Five phases of this region's deglaciation were established based on the surface weathering, where the last stage was dated using cosmogenic radionuclide to be 25 ka (Moriwaki et al. 1991; Nishiizumi et al. 1991; Moriwaki 1992; Ishizuka et al. 1993). There are no advances seen in Schirmacher and Untersee; however, dating and grain size distribution from Lazarev suggests that it may have advanced between 82 ka B.P. (Gingele et al. 1997). Ice sheets in Queen Maud Land were not stable and linked to the ice sheet's interior. There is evidence of changes in the central Antarctic ice sheet during this time scale. The maximum advancement in ice has been sampled from the high altitude Petermann Ketten Mountains. Areas like Untersee, Schussel, on the other hand, suggest that ice sheets were 400 m higher before 0.5 million years. From the data, the impression we get is the steady thinning of the ice sheets; this may be related to the global cooling, which began at the end of the Pliocene. This cooling would have lowered the precipitation rates and, subsequently, Antarctica's ice thickness (Raymo 1992). Other workers Welten et al. 2008 and Höfle 1989, have also supported this.

The evidence from east Antarctica indicates no advance in the ice thickening during the LGM, as is evident from the Sorrondane and Wohlthat Massif. There was a decrease in ice elevation in these areas, as supported by the ice core records. The ice coverage in the last 8 million years is exposed at the high altitude Petermann Ketten Mountains. On the other hand, the ice sheet in Wohlthat Massif had been 200–400 m higher, as shown by the exposure ages of Schussel, Dallmann Berge and Untersee samples. The Eckhorner indicates ice coverage did not exceed 300 m. In general, there is a decrease in ice thickness and exposure of rocks that were buried in ice. This is interpreted as a result of global cooling, which ended in the Pliocene (Raymo 1992). The retreat in the present ice level culminated approximately 0.1 Ma ago. During LGM, the ice level increase was enough to cover the Schirmacher Oasis, as evident from Dallmannberge. A higher concentration of cosmogenic nuclides represents lower rates of erosion from the Peterman Ketten Mountains. These low erosion rates are found in hyper-arid and cold climates. The dates obtained from this area were the first attempt.

4 Melt Water Pulse (MWP)

Several meltwater pulses (MWP) were recorded since the LGM period. There was about an 18 m increase in sea level due to MWP1a during 14.7 to 13.3 ka (Deschamps et al. 2012). The pattern in sea level rise indicates a considerable contribution from the Southern hemisphere. However, the recent data from ice sheet models and ice core records show that only 10 m of ice was locked equally to the eustatic level during LGM (Golledge et al. 2012; Whitehouse et al. 2012; Mackintosh et al. 2011). Mackintosh et al. 2011 suggest that the volume of EAIS increased by 1 m, excluding the embayments of Weddell and Ross seas which is equal to the eustatic level of LGM. This indicates a total 10% contribution from Antarctic ice sheets as shown by the ice sheet models (Golledge et al. 2012; Mackintosh et al. 2011; Whitehouse et al. 2012; Pollard and DeConto 2009). It is not straight forward to understand the contribution of EAIS to the MWP1a, as the volume of ice is small and deglaciation was slow and late. There is evidence of a small donation to MWP1a from Amery and Lambert (Verleyen et al. 2005). There is no clear evidence on how significantly EAIS contributed to MWP1a due to the lack of data or insufficient data constraining or modelling techniques, which needs to be assessed.

5 Marine Isotope Records

Marine isotope records are essential for understanding the Quaternary climate globally (Lisiecki and Raymo 2005; Golledge et al. 2012). Although isotopic records are also affected by the deep-water temperatures (Shackleton 1967), these records are used as a proxy since 1960 to decipher global ice volume. The LGM in Antarctica is not well established; however, it is assumed that it may have occurred during the marine isotope stage 2 (MIS-2). In east Antarctica, maximum extension occurred by 17 ka and 10.7 ka B.P. at Prydz and Mac's coasts. Robertson Land. In the Antarctic Peninsula, the LGM occurred after 30 ka B.P. (Sugden and Clapperton 1980). Recent studies suggest that the Wisconsin ice sheet would have formed by 20 to18 ka (Bentley and Anderson 1998), indicate ice sheets were higher before 35 ka.

In the same way, ice sheets in the Weddell Sea were higher before 26 ka. However, Hjort et al. (2003) indicate the maximum extension in ice occurred before MIS-3 in the western part of the Weddell Sea. Late Quaternary ice distribution suggests Antarctic sea ice in winter advanced towards the present polar zone by MIS-2. During MIS-3, there were several climatic warmings known as Dansgaar and Oeschgerevents (D.O.), between 60 and 27 ka. D.O. events are a period of transition from cold to mild conditions followed by the return of stadial conditions (Dansgaard et al. 1993).

In MIS-3, D.O. events were regular is unclear why they were so frequent. These events were absent during the LGM. Ice cores from Greenland show the rise of 8–16 °C followed by a cooling period before the temperatures returned to stadial values. These transitions have also been indicted by the North Atlantic Ocean (Huber et al. 2006; North GRIP-Members 2004). Marine records suggest that interstadials had higher sea temperature and ocean deep ventilation than stadia. There is a scarcity of information on the east Antarctic ice sheet during MIS-3. Ice models suggest that EAIS expanded during MIS-3 in comparison to the Holocene. However, the field evidence contradicts the modelling evidence and indicates that ice sheets did not advance than the present. Cosmogenic results show that there were areas, which were ice-free for most of the marine isotope stage-3. The last glacial cold cycle has the most prolonged period at around 118 ka. Interglacial period MIS 5-e is at 115 ka (Shackleton et al. 2002 and 2003). The cold cycle of the last glacial had two various complex stages during MIS 4 and 2. Temperature and dust records of Antarctica also indicate this. The average temperature in Antarctica reached −10.2 °C and −10.6 °C for marine isotope stage 4 and 2. These two are divided by the warm interglacial of MIS-3. The millennial-scale variability indicated by Antarctic and Greenland ice core records (Blunier et al. 1998; Markle et al. 2017). Hughes et al. 2013 reported the asynchronies in the glacial cycles, mainly in Asia, where they advanced during the yearly glaciation cycle (Astakhov 2018; Larsen et al. 2006; Svendsen et al. 2004; Vorren et al. 2011). There is evidence of thicker ice sheets in Antarctica before LGM, while at the centre of east Antarctica, there was no thicker ice at LGM than at present (Lilly et al. 2010). Some marine oxygen records indicate the global volume of ice was higher in marine isotope stage 6 than MIS-2 (Roucoux et al. 2011; Shackleton 1987). This is supported by Shakun et al. (2015) for global sea level. However, data obtained by Lisiecki and Raymo (2005) show that MIS-2 has higher ^{18}O records than MIS-6. However, temperature effects may hide the ice volume because, during MIS-6, sea surface temperatures were warmer than MIS-5d-2. This has allowed the supply of moisture to drive the extent of ice masses more in other glacial periods. The distribution of ice before LGM was different from LGM. Eurasia had more ice masses before LGM. Similarly, North America had smaller ice masses before LGM compared to the LGM (Rohling et al. 2017). The pre-glacial maximum peak occurred around 140 ka (Colleoni et al. 2016; Stirling et al. 1998; Raisbeck et al. 2014).

6 Summary

Grounding line in most Antarctica parts was near the present shelf before 20 ka except in the Ross and the Weddell Sea. Besides, the extent of ice reached before 20 ka in some places while recession had started at this time (Hillenbrand et al. 2014). The geological and marine data shows considerable ice by 20 ka in the Weddell Sea's continental shelf. From the east and west Antarctica, the data is limited. Half of the ice that has been grounded in the Ross Sea comes from east Antarctica (Anderson et al. 2014). Radiocarbon dates show that the recession of ice sheets started before the LGM. Some parts of East Antarctica reached the present shelf margin by this time (Mackintosh et al. 2014). The sediments of the MIS-3 suggest that several areas were ice-free at this time. Likewise, in Dronning Maud land, there are sparse or no evidence of ice thickening during LGM. Ice sheet and ice core models show the ice domes were possibly 100 m lower than at present. Striated bedrock, till, and organic deposits like mumiyo provide evidence for changes in the past and the dynamics of EAIS. Various workers' analysis was combined, which showed that a 600 m high ice sheet existed 4 million years ago, decreasing continuously to the present day. The question is how much mean sea level is expected to rise under different climatic scenario's (Bindoff et al. 2007). By reconstructing the glacial history of EAIS could help address these questions and understand the more direct dates required from this region. Surface exposure dating using cosmogenic radionuclide proved to be a potential technique to date the deglaciation phase. From Dronning Maud Land, more dates need to be generated to establish a detailed glacial chronology. The work is in progress to establish chronology in DML with the sample collected during 36th Indian Scientific Expedition to Antarctica by one of the authors.

Acknowledgements JKP is thankful to the present and former Vice-chancellor, Central University of Punjab, Bathinda and the Director, NCPOR, Goa, Ministry of Earth Sciences, for support and help during the 36th Indian Scientific expedition to Antarctica. JKP also thankful to Ms. Amrutha K. for her help to prepare diagram for this chapter.

References

Abreu VS, Anderson BJ (1998) Glacial eustasy during the Cenozoic: sequence stratigraphic implications. Am Asso Petrol Geol Bull 82(7):1385–1400

Altmaier M, Herpers U, Delisle G, Merchel S, Ott U (2010) Glaciation history of Queen Maud Land (Antarctica) reconstructed from in-situ produced cosmogenic ^{10}Be, ^{26}Al and ^{21}Ne. Polar Sci 4(1):42–61

Anderson JB (1999) Antarctic marine geology. Cambridge University Press, Cambridge

Anderson JB, Shipp SS, Lowe AL, Wellner JS, Mosola AB (2002) The Antarctic Ice Sheet during the Last Glacial Maximum and its subsequent retreat history: a review. Quat Sci Rev 21:49–70

Anderson JB, Conway H, Bart PJ, Witus AE, Greenwood SL, McKay RM, Hall BL, Ackert RP, Licht K, Jakobsson M, Stone JO (2014) Ross Sea paleo ice sheet drainage and deglacial history during and since the LGM. Quat Sci Rev 100:31–54

Astakhov VI (2018) Late Quaternary glaciation of the northern Urals: a review and recent observations. Boreas 47(2):379–389

Barrera E, Huber BT (1993) Eocene to Oligocene oceanography and temperatures in the Antarctic Indian Ocean. In: The Antarctic paleoenvironment: a perspective on global change: American Geophysical Union Antarctic Research Series, vol 60, pp 49–65

Barrett PJ, Hambrey MJ, Robinson PR (1991) Cenozoic glacial and tectonic history from CIROS-1, McMurdo Sound. In: International symposium on antarctic earth sciences, vol 5, pp 651–656

Bentley MJ, Anderson, JB (1998) Glacial and marine geological evidence for the ice sheet configuration in the weddell sea–antarctic peninsula region during the last glacial maximum. Antarct Sci 10(3):309–325

Bindoff NL, Willebrand J, Artale V, Cazenave A, Gregory JM, Gulev S, Shum CK (2007) Observations: oceanic climate change and sea level. ClimateChange 385–432
Birkenmajer K (1991) Tertiary glaciations in the South Shetlands Islands, West Antarctica: evaluation of data. In: International symposium on Antarctic earth sciences, vol 5, pp 629–632
Birkenmajer K (1998) Geology of Volcanic Rocks (? Upper Cretaceous-Lower Tertiary) at Potter Peninsula, King George Island (South Shetland Islands, West Antarctica. Bull Pol Acad Sciences Earth Sci. Earth Sciences 46(2):147–155
Birkenmajer KRZYSZTOF (1987) Oligocene-Miocene glaciomarine sequences of King George Island (South Shetland Islands), Antarctica. PalaeontologiaPolonica 49(1):9-36
Blunier T, Chappellaz J, Schwander J, Dällenbach A, Stauffer B, Stocker TF, Raynaud D, Jouzel J, Clausen HB, Hammer CU, Johnsen S (1998) Asynchrony of antarctic and greenland climate change during the last glacial period. Nature 394(6695):739–743
Clark PU, Mitrovica JX, Milne GA, Tamisiea ME (2002) Sea-level fingerprinting as a direct test for the source of global meltwater pulse I.A. Science 295(5564):2438–2441
Clark PU, Dyke AS, Shakun JD, Carlson AE, Clark J, Wohlfarth B, McCabe AM (2009) The last glacial maximum. Science 325(5941):710-714
Colleoni F, Wekerle C, Näslund JO, Brandefelt J, Masina S (2016) Constraint on the penultimate glacial maximum Northern Hemisphere ice topography (~140 kyrsBP). Quatern Sci Rev 137:97–112
Dansgaard W, Johnsen SJ, Clausen HB, Dahl-Jensen D, Gundestrup NS, Hammer CU, Bond G (1993) Evidence for general instability of past climate from a 250-kyr ice-core record. Nature 364(6434):218-220
DeConto RM, Pollard D (2003) Rapid Cenozoic glaciation of Antarctica induced by declining atmospheric CO_2. Nature 421(6920):245–249
Denton GH, Prentice ML, Kellogg DE, Kellogg TB (1984) Late Tertiary history of the Antarctic ice sheet: evidence from the Dry Valleys. Geology 12(5):263–267
Denton GH, Prentice ML, Burckle LH (1991) Cainozoic history of the Atlantic ice sheet. The geology of Antarctica, pp 365–433
Deschamps P, Durand N, Bard E, Hamelin B, Camoin G, Thomas AL, Yokoyama Y (2012). Ice-sheet collapse and sea-level rise at the Bølling warming14,600years ago. Nature 483(7391):559-564
Di Nicola L, Strasky S, Schlüchter C, Salvatore MC, Akçar N, Kubik PW, Baroni C (2009). Multiple cosmogenic nuclides document complex Pleistocene exposure history of glacial drifts in Terra Nova Bay (northern Victoria Land, Antarctica). *Quaternary Research*, *71*(1), 83-92
Di Nicola L, Baroni C, Strasky S, Salvatore MC, Schlüchter C, Akçar N, Wieler R (2012). Multiple cosmogenic nuclides document the East Antarctic Ice Sheet's stability in northern Victoria Land since the Late Miocene (5–7 Ma). *Quat Sci Revs*, *57*, 85-94
Elverhøi A (1981) Evidence for a late Wisconsin glaciation of the Weddell Sea. Nature 293(5834):641–642
EPICA Community Members (2006) One-to-one coupling of glacial climate variability in Greenland and Antarctica. Nature 444(7116):195
Fink D, McKelvey B, Hambrey MJ, Fabel D, Brown R (2006) Pleistocene deglaciation chronology of the Amery Oasis and Radok Lake, northern Prince Charles Mountains, Antarctica. Earth Planet Sci Lett 243(1-2):229-243
Fogwill CJ, Bentley MJ, Sugden DE, Kerr AR, Kubik PW (2004) Cosmogenic nuclides 10Be and 26Al imply limited Antarctic Ice Sheet thickening and low erosion in the Shackleton Range for> 1. Geology 32(3):265–268
Fretwell P, Pritchard HD, Vaughan DG, Bamber JL, Barrand NE, Bell R, Catania G (2013) Bedmap2: improved ice bed, surface and thickness datasets for Antarctica. The Cryosphere 7(1):375–393
Gingele FX, Kuhn G, Maus B, Melles M, Schöne T (1997) Holocene ice retreat from the Lazarev Sea shelf, East Antarctica. Cont Shelf Res 17(2):137–163
Golledge NR, Fogwill CJ, Mackintosh AN, Buckley KM (2012) Dynamics of the last glacial maximum Antarctic ice-sheet and its response to ocean forcing. Proc Natl Acad Sci 109(40):16052–16056

Hambrey MJ, Larsen B, Ehrmann WU (1989) Forty million years of Antarctic glacial history yielded by Leg 119 of the Ocean Drilling Program. PolarRecord 25(153):99–106

Hambrey MJ, Ehrmann W, Larsen B (1991) Cenozoic glacial record of the Prydz Bay continental shelf, East Antarctica. In: Barron J, Larsen B et al (eds.) Proc. ODP, Sci. Results, College Station, TX. (Ocean Drilling Program), vol 119, pp 77–132; vol 119, pp 77–132

Hättestrand C, Johansen N (2005) Supraglacial moraines in Scharffenbergbotnen, Heimefrontfjella, Dronning Maud Land, Antarctica-significance for reconstructing former blue ice areas. Antarct Sci 17(2):225

Hillenbrand CD, Melles M, Kuhn G, Larter RD (2012) Marine geological constraints for the Antarctic Ice Sheet's grounding-line position on the southern Weddell Sea shelf at the Last Glacial Maximum. Quatern Sci Rev 32:25–47

Hillenbrand CD, Bentley MJ, Stolldorf TD, Hein AS, Kuhn G, Graham AG, Larter RD (2014) Reconstruction of changes in the Weddell Sea sector of the Antarctic Ice Sheet since the Last Glacial Maximum. Quat Sci Revs 100:111-136

Hiller A, Wand U, Kampf H, Stackebrandt W (1988) Occupation of the Antarctic continent by petrels during the past 35000 years - inferences from a ^{14}C study of stomach oil deposits. Polar Biol 9:69–77

Hirakawa K, Moriwaki K (1990) Former ice sheet based on the newly observed glacial landforms and erratics in the central SørRondane Mountains, East Antarctica. Proc NIPR Symp Antarct Geosci 4:41e54

Hjort C, Ingólfsson Ó, Bentley MJ, Björck S (2003) The late pleistocene and holocene glacial and climate history of the antarctic peninsula region as documented by the land and lake sediment records-a review. Antarct Penins Clim Var: Hist Paleoenviron Perspect 79:95–102

Höfle HC (1989) The glacial history of the Outback Nunataks area in western North Victoria Land. Geologisches Jahrbuch. Reihe e, Geophysik 38:335–355

Huang F, Liu X, Kong P, Fink D, Ju Y, Fang A, Na C (2008) Fluctuation history of the interior East Antarctic Ice Sheet since mid-Pliocene. Antarct Sci 20(2):197

Huber C, Leuenberger M, Spahni R, Flückiger J, Schwander J, Stocker TF, Jouzel J (2006) Isotope calibrated Greenland temperature record over Marine Isotope Stage 3 and its relation to CH4. Earth Planet Sci Lett 243(3-4):504-519

Huber M, Brinkhuis H, Stickley CE, Döös K, Sluijs A, Warnaar J, Williams GL (2004) Eocene circulation of the Southern Ocean: was Antarctica kept warm by subtropical waters? Paleoceanography 19(4):1-12

Hughes PD, Gibbard PL, Ehlers J (2013) Timing of glaciation during the last glacial cycle: evaluating the concept of a global 'Last Glacial Maximum (LGM). Earth Sci Rev 125:171–198

Huybrechts P (1993) Formation and disintegration of the AntarcticIce Sheet. Ann Glaciol 20:336–340

Ingólfsson Ó, Hjort C (1999) The Antarctic contribution to Holocene global sea-level rise. Polar Res 18(2):323–330

Ingólfsson Ó (2004) Quaternary glacial and climate history of Antarctica. In: Developments in Quaternary sciences, vol 2, pp 3–43. Elsevier

Ingólfsson Ó, Hjort C, Berkman PA, Björck S, Colhoun E, Goodwin ID, Prentice ML (1998) Antarctic glacial history since the Last Glacial Maximum: an overview of the record on land. Antarct Sci 10:326-344

Ishizuka H, Shiraishi K, Moriwaki K (1993) Geological Map of Bergersenfjella. In: AntarcticGeological Map Series. National Institute for Polar Research, Tokyo

Jonsson S (1988) Observations on the physical geography and glacial history of the Vestfjellanunataks in western Dronning Maud Land. Stockholms Universitet, Naturgeografiska Institutionen, Antarctica

Jouzel J, Lorius C, Petit JR, Genthon C, Barkov NI, Kotlyakov VM, Petrov VM (1987) Vostok ice core: a continuous isotope temperature record over the last climatic cycle (160,000 years). Nature 329(6138):403–408

Jouzel J, Raisbeck G, Benoist JP, Yiou F, Lorius C, Raynaud D, Kotlyakov VM (1989) A comparison of deep Antarctic ice cores and their implications for climate between 65,000 and 15,000 years ago. Quat Res 31(2):135-150

Kennett JP (1977) Cenozoic evolution of Antarctic glaciation, the circum-Antarctic ocean, and their impact on global paleoceanography. J Geophys Res 82(27):3843–3860

Kennett JP, Barker PF (1990) Latest Cretaceous to Cenozoic climate and oceanographic developments in the Weddell Sea, Antarctica: an ocean-drilling perspective. In: Proceedings of the ocean drilling program, scientific results, vol 113, pp 937–960

Kennett JP, Stott LD (1990) Proteus and Proto-Oceanus: ancestral Paleogene oceans as revealed from Antarctic stable isotopic results; ODP Leg 113. In: Proceedings of the ocean drilling program, scientific results, vol 113, pp 865–880). Ocean Drilling Program College Station, TX

Kong P, Huang F, Liu X, Fink D, Ding L, Lai Q (2010) Late Miocene ice sheet elevation in the Grove Mountains, East Antarctica, inferred from cosmogenic ^{21}Ne–^{10}Be–^{26}Al. Global Planet Change 72(1–2):50–54

Kubik PW, Ivy-Ochs S, Masarik J, Frank M, Schlucher C (1998) ^{10}Be and ^{26}Al production rates deduced from an instantaneous event within the dendro-calibration curve, the landslide of Kofels, Otz Valley, Austria. Earth Planet Sci Lett 161(1–4):231–241

Kumar P, Pattanaik J, Ojha S, Gargari S, Joshi R, Roonwal G, Kanjilal D (2011) ^{10}Be measurements at IUAC-AMS facility. J Radioanal Nucl Chem 290(1):179-182

Kumar P, Pattanaik JK, Khare N, Chopra S, Yadav S, Balakrishnan S, Kanjilal D (2014) Study of ^{10}Be in the sediments from the Krossfjorden and Kongsfjorden Fjord System, Svalbard. J Radioanal Nucl Chem 302(2):903–909

Kumar P, Chopra S, Pattanaik JK, Ojha S, Gargari S, Joshi R, Kanjilal D (2015) A new AMS facility at Inter-University Accelerator Centre, New Delhi. Nucl Instrum Methods Phys Res Sect B 361:115–119

Konrad H, Shepherd A, Gilbert L, Hogg AE, McMillan M, Muir A, Slater T (2018) The net retreat of Antarctic glacier grounding lines. Nat Geosci 11(4):258–262

Larsen E, KJaeR KH, Demidov IN, Funder S, Grøsfjeld K, Houmark-Nielsen MICHAEL, Lysa A (2006) Late Pleistocene glacial and lake history of northwestern Russia. Boreas 35(3):394-424

Legrand M, Mayewski P (1997) Glaciochemistry of polar ice cores: a review. Rev Geophys 35(3):219–243

Lilly K, Fink D, Fabel D, Lambeck K (2010) Pleistocene dynamics of the interior East Antarctic ice sheet. Geology 38(8):703–706

Lintinen PETRI (1996) Evidence for the former existence of a thicker ice sheet on the Vestfjella-nunataks in western Dronning Maud Land, Antarctica. Bull-Geol Soc Finl 68:85–98

Lintinen P, Nenonen J (1997) Glacial history of the Vestfjella and Heimefrontfjellanunatak ranges in western Dronning Maud Land, Antarctica. Geological Evolution and Processes. Terra Antarctica Publications, Sienna, The Antarctic Region, pp 845–852

Lisiecki LE, Raymo ME (2005) A Pliocene-Pleistocene stack of globally distributed benthic δ18O records, Paleoceanography, 20, PA1003

Liu X, Huang F, Kong P, Fang A, Li X, Ju Y (2010) History of ice sheet elevation in East Antarctica: Paleoclimatic implications. Earth Planet Sci Lett 290(3–4):281–288

Lythe MB, Vaughan DG (2001) BEDMAP: a new ice thickness and subglacial topographic model of Antarctica. J Geophys Res: Solid Earth 106(B6):11335–11351

Mackintosh A, Golledge N, Domack E, Dunbar R, Leventer A, White D, …& Gore D(2011) The retreat of the East Antarctic ice sheet during the last glacial termination. Nat Geosci 4(3):195–202

Mackintosh AN, Verleyen E, O'Brien PE, White DA, Jones RS, McKay R, Miura H (2014) Retreat history of the East Antarctic Ice Sheet since the last glacial maximum. Quat Sci Revs 100:10-30

Markle BR, Steig EJ, Buizert C, Schoenemann SW, Bitz CM, Fudge TJ, Sowers T (2017) Global atmospheric teleconnections during Dansgaard–Oeschger events. Nat Geosci 10(1):36-40

Mayewski PA, Meredith MP, Summerhayes CP, Turner J, Worby A, Barrett PJ, Bromwich D (2009) State of the Antarctic and Southern Ocean climate system. Rev Geophys 47(1):RG1003

Miura H, Maemoku H, Seto K, Moriwaki K (1998) Late Quaternary East Antarctic melting event in the Soya Coast region based on stratigraphy and oxygen isotopic ratio of fossil molluscs. Polar Geosci 11:260–274

Moriwaki K (1992) Late Cenozoic glacial history in the Sφr-Rondane Mountains, East Antarctica. In: Recent progress in Antarctic earth science, 661–667

Moriwaki K, Hirakawa K, Matsuoka N (1991) Weathering stage of till and glacial history of the central SørRondane Mountains, East Antarctica. Proc NIPR Symp Antarct Geosci 5:99–111

Nishiizumi K, Winterer EL, Kohl CP, Klein J, Middleton R, Lal D, Arnold JR (1989) Cosmic ray production rates of 10Be and 26Al in quartz from glacially polished rocks. J Geophys Res: Solid Earth 94(B12):17907–17915

Nishiizumi K, Kohl CP, Arnold JR, Klein J, Fink D, Middleton R (1991) Cosmic ray produced 10Be and 26Al in Antarctic rocks: exposure and erosion history. Earth Planet Sci Lett 104(2–4):440–454

Nishiizumi K, Kohl CP, Arnold JR, Dorn R, Klein I, Fink D, Lal D (1993) Role of in situ cosmogenic nuclides ^{10}Be and ^{26}Al in the study of diverse geomorphic processes. Earth Surf Process Landfs 18(5):407-425

North Greenland Ice-Core Project (NorthGRIP) Members (2004) High-resolution climate record of the Northern Hemisphere reaching into the last Glacial Interglacial Period. Nature 431:147–151

Parrenin F, Barnola JM, Beer J, Blunier T, Castellano E, Chappellaz J, Kawamura K (2007) The EDC3 chronology for the EPICA Dome C ice core. Clim Past 3(3):485–497

Pattyn F, Decleir H, Huybrechts P (1992) Glaciation of the central part of the SørRondane, Antarctica: glaciological evidence. In: Recent progress in Antarctic earth science. Tokyo: Terra Scientific Publishing Company, pp 669–678

Pattyn F, Matsuoka K, Berte J (2010) Glacio-meteorological conditions in the vicinity of the Belgian Princess Elisabeth Station, Antarctica. Antarct Sci 22(1):79-10

Peltier WR (2005) On the hemispheric origins of meltwater pulse 1a. Quatern Sci Rev 24(14–15):1655–1671

Petit JR, Basile I, Leruyuet A, Raynaud D, Lorius C, Jouzel J, …& Davis M(1997) Four climate cycles in Vostok ice core. Nature 387(6631):359–360

Pingree K, Lurie M, Hughes T (2011) Is the East Antarctic ice sheet stable? Quatern Res 75(3):417–429

Phartiyal B, Sharma A, Bera SK (2011) Glacial lakes and geomorphological evolution of Schirmacher Oasis, East Antarctica, during late quaternary. Quatern Int 235(1–2):128–136

Pollard D, DeConto RM (2009) Modelling West Antarctic ice sheet growth and collapse through the past five million years. Nature 458(7236):329–332

Prentice ML, Matthews RK (1988) Cenozoic ice-volume history: development of a composite oxygen isotope record. Geology 16(11):963–966

Raisbeck LB, Xiao H, Liang F, Akers PD, Brook GA, Dennis WM, Edwards RL (2014) A stalagmite record of abrupt climate change and possible Westerlies-derived atmospheric precipitation during the Penultimate Glacial Maximum in northern China. Palaeogeogr Palaeoclim Palaeoecol 393:30–44

Raymo ME (1992) Global climate change: a three million year perspective. In: Start of a glacial, pp 207–223. Springer, Berlin, Heidelberg

Rignot E, Mouginot J, Scheuchl B (2011) Ice flow of the Antarctic ice sheet. Science 333(6048), 1427–1430

Robert C, Kennett JP (1994) The antarctic subtropical humid episode at the Paleocene-Eocene boundary: clay-mineral evidence. Geology 22(3):211–214

Rohling EJ, Hibbert FD, Williams FH, Grant KM, Marino G, Foster GL, Webster JM (2017) Differences between the last two glacial maxima and implications for ice-sheet, δ^{18}O, and sea-level reconstructions. Quatern Sci Rev 176:1–28

Roucoux KH, Tzedakis PC, Lawson IT, Margari V (2011) Vegetation history of the penultimate glacial period (Marine isotope stage 6) at Ioannina, north-west Greece. J Quat Sci 26(6):616–626

Schwab MJ (1998) Reconstruction of the Late Quaternary climatic and environmental history of the Schirmacher Oasis and the Wohlthat Massif (East Antarctica). BerichtezurPolarforschung 293:128

Shackleton NJ (1987) Oxygen isotopes, ice volume and sea level. Quatern Sci Rev 6(3–4):183–190

Shackleton NJ, Chapman M, Sánchez-Goñi MF, Pailler D, Lancelot Y (2002) The classic marine isotope substage 5e. Quatern Res 58(1):14–16

Shackleton NJ, Sánchez-Goñi MF, Pailler D, Lancelot Y (2003) Marine isotope substage 5e and the Eemian interglacial. Global Planet Change 36(3):151–155

Shackleton N (1967) Oxygen isotope analyses and Pleistocene temperatures re-assessed. Nature 215(5096):15–17

Shakun JD, Lea DW, Lisiecki LE, Raymo ME (2015) An 800-kyr record of global surface ocean δ18O and implications for ice volume-temperature coupling. Earth Planet Sci Lett 426:58–68

Shepherd A, Ivins E, Rignot E, Smith B, Van Den Broeke M, Velicogna I, Nowicki S (2018) Mass balance of the Antarctic Ice Sheet from 1992 to 2017. Nature 558:219-222

Siddall M, Milne GA, Masson-Delmotte V (2012) Uncertainties in elevation changes and their impact on Antarctic temperature records since the last glacial period. Earth Planet Sci Lett 315:12–23

Steig EJ, Schneider DP, Rutherford SD, Mann ME, Comiso JC, Shindell DT (2009) Warming of the Antarctic ice-sheet surface since the 1957 International Geophysical Year. Nature 457(7228):459–462

Steele WK, Hiller A (1997) Radiocarbon dates of snow petrel (Pagodromanivea) nest sites in central Dronning Maud Land, Antarctica. Polar Record 33(184):29-38

Stirling CH, Esat TM, Lambeck K, McCulloch MT (1998) Timing and duration of the Last Interglacial: evidence for a restricted interval of widespread coral reef growth. Earth Planet Sci Lett 160(3–4):745–762

Stonehouse B (2002) Encyclopaedia of Antarctica and the southern oceans. Wiley

Strasky S, di Niocola L, Baroni C, Salvatore MC, Baur H, Kubik PW, Wieler R (2009) Surface exposure ages imply multiple low-amplitude Pleistocene variations in the East Antarctic ice sheet, Ricker Hills, Victoria Land. Antarct Sci 21(1):59-69

Strub E, Wiesel H, Delisle G, Binnie SA, Liermann A, Dunai TJ, Coenen HH (2015) Glaciation history of Queen Maud Land (Antarctica)–New exposure data from nunataks. Nucl Instrum Methods Phys Res Sect B: Beam Interact Mater Ats 361:599-603

Sugden DE, Clapperton CM (1980) West Antarctic ice sheet fluctuations in the Antarctic Peninsula area. Nature 286(5771):378–381

Sugden DE, Denton GH, Marchant DR (1995) Landscape evolution of the Dry Valleys, Transantarctic Mountains: tectonic implications. J Geophys Res: Solid Earth 100(B6):9949–9967

Svendsen JI, Alexanderson H, Astakhov VI, Demidov I, Dowdeswell JA, Funder S, Hubberten HW (2004) Late Quaternary ice sheet history of northern Eurasia. Quatern Sci Rev 23(11–13):1229–1271

Takada M, Tani A, Miura H, Moriwaki K, Nagatomo T (2003) ESR dating of fossil shells in the Lützow-Holm Bay region, East Antarctica. Quatern Sci Rev 22:1323–1328

The RAISED Consortium, Bentley MJ, Cofaigh CO, Anderson JB, Conway H, Davies B, Graham AG, Zwartz (2014). A community-based geological reconstruction of Antarctic Ice Sheet deglaciation since the Last Glacial Maximum. Quat Sci Revs 100:1-9

Thor G, Low M (2011) The snow petrel's persistence (Pagodromanivea) in Dronning Maud Land (Antarctica) for over 37,000 years. Polar Biol 34(4):609–613

Tuniz C (2001) Accelerator mass spectrometry: ultra-sensitive analysis for global science. Radiat Phys Chem 61(3–6):317–322

Verleyen E, Hodgson DA, Milne GA, Sabbe K, Vyverman W (2005) Relative sea-level history from the Lambert Glacier region, East Antarctica, and its relation to deglaciation and Holocene glacier readvance. Quatern Res 63(1):45–52

Vorren TO, Landvik JY, Andreassen K, Laberg JS (2011) Glacial history of the Barents Sea region. In: Developments in quaternary sciences, vol 15, pp 361–372. Elsevier

Webb PN, Harwood DM (1987) Terrestrial flora of the Sirius Formation: its significance for Late Cenozoic glacial history. Antarct J Unit States 22(4):7–11

Welten KC, Folco LUIGI, Nishiizumi K, Caffee MW, Grimberg A, Meier MMM, Kober F (2008) Meteoritic and bedrock constraints on the glacial history of Frontier Mountain in northern Victoria Land, Antarctica. Earth Planet Sci Lett 270(3-4):308-315

Whitehouse PL, Bentley MJ, Le Brocq AM (2012) Antarctica's deglacial model: geological constraints and glaciological modelling as a basis for a new Antarctic glacial isostatic adjustment model. Quatern Sci Rev 32:1–24

Willenbring JK, von Blanckenburg F (2010) Meteoric cosmogenic Beryllium-10 adsorbed to river sediment and soil: Applications for earth-surface dynamics. Earth-Sci Rev 98(1–2):105–122

Yamane M, Yokoyama Y, Miura H, Maemoku H, Iwasaki S, Matsuzaki H (2011) The last deglacial history of Lützow-Holm Bay, East Antarctica. J Quat Sci 26(1):3–6

Yamane M, Yokoyama Y, Abe-Ouchi A, Obrochta S, Saito F, Moriwaki K, Matsuzaki H (2015) Exposure age and ice-sheet model constraints on Pliocene East Antarctic ice sheet dynamics. Nat Commun 6. https://doi.org/10.1038/ncomms8016

Zwally HJ, Giovinetto MB, Beckley MA, Saba JL (2012) Antarctic and Greenland drainage systems. In: GSFC cryospheric sciences laboratory. http://icesat4.gsfc.nasa.gov/cryo_data/ant_grn_drainage_systems.php

Late Quaternary Climate Change in Schirmacher Region, East Antarctica: As Revealed from Terrestrial Diamicts and Lacustrine Sediments

Prakash Kumar Shrivastava, Rajesh Asthana, and Sandip Roy

Abstract The multi-proxy data, generated from the moraines and sediment cores of variety of lakes from Schirmacher region, cDML, East Antarctica has provided better insight into the Late Pleistocene to Holocene paleoclimatic evolution of the region during Late Quaternary. This chapter highlights the glacial signatures which are very well preserved in all kinds of sediments of this region. The clay minerals indicate a gradual shift in the weathering regime and therewith in climate from strongly glacial to fluvio-glacial during Late Quaternary. The results of surface textures of quartz grains have been discussed depth wise and in the same samples. In general it shows dominant glacial and glaciofluvial actions. The OSL chronology on moraines has provided information on different events of deglaciation in Schirmacher region, East Antarctica. The retreat of EAIS initiated much earlier than it was thought (sometime prior to 171 ka). There is no evidence of morainic deposits belonging to the Last Glacial Maximum. Formation of lake sediments started before 42 ka marks the beginning of climate change from glacial to glaciofluvial.

Keywords Surface textures · Lake sediments · Late Quaternary · Clay minerals · Climate change

1 Introduction

Climate change is a very complex phenomenon and involves terrestrial as well as extra-terrestrial variables. Earth was subjected to drastic climate changes since its birth, which has been well recorded in geological formations. The time between Proterozoic to Recent has marked by different phases of evolution and extinction. Many of these phases are directly related to climate changes. The latest among them, the Late Quaternary time, has witnessed culmination in all modern-day species. Although many places worldwide have preserved paleoclimatic records in their

P. Kumar Shrivastava · R. Asthana (✉) · S. Roy
Geological Survey of India, Kolkata, India

N. Khare (ed.), *Assessing the Antarctic Environment from a Climate Change Perspective*, Earth and Environmental Sciences Library,
https://doi.org/10.1007/978-3-030-87078-2_6

natural formations, the higher latitudes areas serve best proxies to study paleoclimatic changes due to their undisturbed nature.

Antarctica is an island continent with about 13.66 million sq. km., almost entirely covered by ice. It's total area changes on an annual basis and also known as the 'pulsating continent'. The freezing of the surrounding ocean adds up another 18.8 million sq. km. The ice sheet over the continent is average ~3500 m thick, with the maximum thickness being 4776 m at Terra Adelie in East Antarctica. This enormous ice cap locks up about 90% of ice of the earth, which is equivalent to ~70% of the freshwater of our planet. Exposed rocks in Antarctica are less than 2% of the area, confined chiefly to its peripheral outcrops, Dronning Maud Land and the Trans-antarctic mountain range.

It is the continent of extremes, being coldest, windiest, driest, highest and virtually uninhabited. The coldest temperature ever recorded on earth was in Antarctica −94.7 °C (−135.8°F), which happened in August 2010. The old record had been −89.2 °C (−128.6°F) in Vostok (Russian Station) in July 1983. The temperature on the polar plateau ranges from −5 °C to −80 °C, while in the coastal areas, the climate remains in between 0 °C and −35 °C. Such an extreme cold freezes most of the moisture in the atmosphere, which gets precipitated as snow, having annual precipitation as low as less than 5 cm in the interior areas making Antarctica a 'cold desert', and coupled with the polar day and night cycles, giant low-pressure systems moving around the continent from west to east cause extreme weather conditions giving rise to unique set-up on the earth carving unique geomorphological milieu. Its unique geographical position with all its extremities has made it very special for scientists and researchers worldwide and provides valuable information on various aspects of the scientific knowledge base. Far away from the significant pollution sources, these regions also act as natural laboratories, rendering sharper easier to interpret scientific observations.

Antarctica has seen glaciations and deglaciations throughout geological time on a different scale (Pollard and DeConto 2009). The most continuous data has been retrieved from the Vostok ice core, indicating fluctuating climate during the past 800 ka with eight glacial and interglacial cycles (Lambert et al. 2008; Cortese et al. 2007; Petit et al. 1999). Such periodic changes in Antarctica's climatic condition determined the evolution of different geomorphic landforms and the overall landscape.

Cosmogenic isotope-based evidence of minor changes in the ice thickness over the past 100 ka has been reported from different central Dronning Maud Land (cDML), including the Gruber, Petermann Humboldt mountains (Altmaier et al. 2010) areas. Optically stimulated luminescence dating of glacio-fluvial and glacial lake shoreline sediments indicates that in the Bunger Hills and cDML regions, the most recent large-scale retreat of EAIS began around 40 ka (Gore et al. 2001). Post 40 ka till the present (including the period of last glacial maximum), these areas were never covered with ice sheet (Hodgson et al. 2001; Gore et al. 2001).

The grounding line zone of EAIS in the north of the cDML area has been nearly stationary over the past 30 ka (Anderson et al. 2002; Livingstone et al. 2012; Mackintosh et al. 2011, 2014). Mackintosh et al. (2011; 2014) suggested that in this interval, the ice sheet was restricted to 800 m in Lambert/Amery Glacier system, and Larsemann Hills and Bunger Hills did not carry signatures glaciations. The presence of nunataks and striations in the cDML region suggested thinning the ice sheet by ~100 m. This estimate arrived based on the nature of indentation on the bedrock. A consensus also exists that the retreat of EAIS at the Lambert/Amery Glacier system began at around 18 ka, at ~14 ka in Mac Robertson Island and during ~12 ka to 6 ka at other places (Mackintosh et al. 2014).

Schirmacher Oasis (70°44′30° S-70°46′30"S & 11°22′04" E-11°22′04"E) is a ~17 km long and ~3 km wide semi-linear exhumed Neo-Proterozoic terrain (Fig. 1) and is amongst few regions in Antarctica that have preserved sedimentary records of different stages of deglaciation. Baalsrudfjellet is located in the South-East direction of Schirmacher Oasis. The sediment cover in this oasis and the adjoining areas are thin because of low debris volume and weak fluvial activities (Fitzsimons 1996). Schirmacher Oasis was exposed and evolved through the continental ice sheet retreat, leading to different glacial and glaciofluvial geomorphic features.

In coastal Antarctica, few landmasses are exposed to freshwater lakes. These landmasses are known as Oases. With the climatic fluctuations in the past, other than glaciological agents, the oases have experienced other active geomorphic agents, viz. fluvial and aeolian processes, in reshaping the existing landforms factors to the geomorphic evolution of the area. The Schirmacher Oasis is an exhumed terrain among few Antarctica regions that have preserved sedimentary records of different deglaciation stages. Schirmacher Range is Antarctica's few areas that provide distinct geomorphic features (Fig. 2) formed due to the deglaciation process. This oasis in the rocky area elongated in the west-northwest to east-southeast direction, dotted with more than 100 pro-glacial, landlocked and epi-shelf lakes (Ravindra et al. 2002; Verlecar et al. 1996). This oasis is located on the Princess Astrid Coast in Queen Maud Land, East Antarctica (Fig. 1) and is bound by the Antarctic ice sheet on the south and epi-shelf lakes and ice shelf in the north.

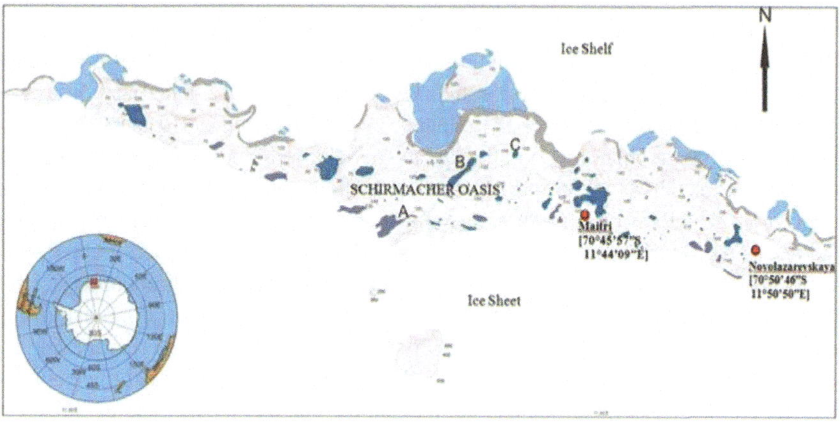

Fig. 1 Location map of Schirmacher Oasis, East Antarctica

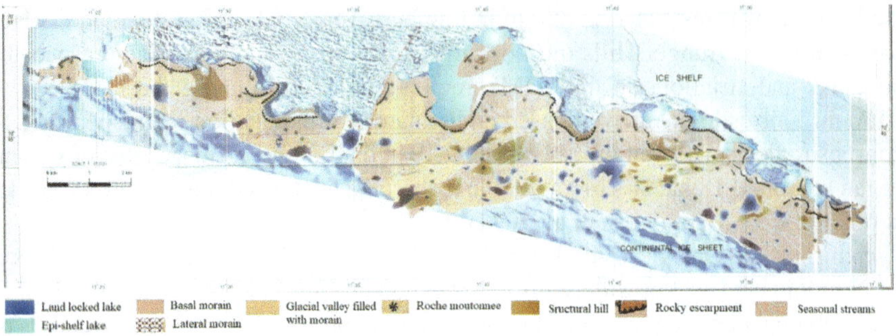

Fig. 2 Geomorphological map of Schirmacher Oasis (*modified after GSI map 2006*)

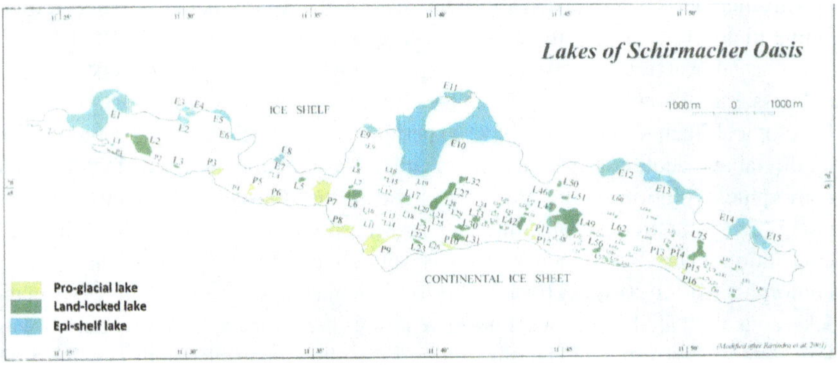

Fig. 3 Epi-shelf, Land-locked and Pro-glacial Lakes of Schirmacher Oasis, East Antarctica

The Schirmacher region in East Antarctica is characterized by a very distinct periglacial environment having typical geomorphological units. The oasis is dotted with more than 100 lakes of Pro-glacial, Land-locked and Epi-shelf nature (Fig. 3). The landscape, including erosional and depositional features (Fig. 2), offers a scope to study paleoclimatic changes significantly during the Late Quaternary time. The recent works of Warrier et al. (2014), Mahesh et al. (2015) and (2017) have explained the climate change phenomena in this area. The lake sediments offer an opportunity to study residues' hydrodynamics, resulting in a change in environmental parameters. The immature and unsorted sediments in this periglacial environment are transported and deposited by meltwater (Srivastava et al. 2018). The visible layers in lake sediments are not present in the entire Schirmacher region. But the indirect approach led to the identification of different layers, and these sedimentary layers contain chronologically ordered information on the paleoclimatic evolution of the region. The ice sheet recession deposited various moraines, helping in understanding climate change in a better way. Understanding the geomorphology and paleoclimatic evolution of

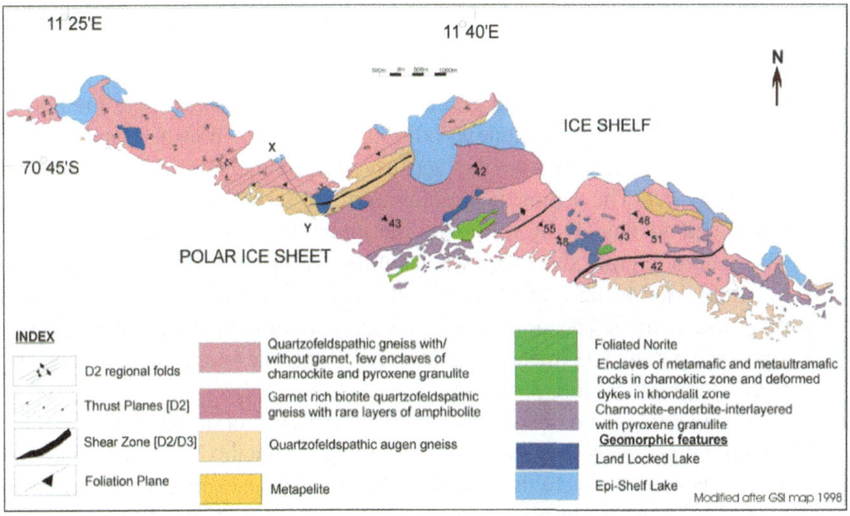

Fig.4 Geological map of the Schirmacher Oasis (modified after GSI map, 1998)

the Schirmacher region will contribute significantly to the present knowledge of palaeoclimatic information and sedimentation characteristics.

2 General Geology of Schirmacher Oasis

The Schirmacher Oasis has undergone complex tectono-metamorphic evolution and emplacement of various intrusives. The rocks of the Schirmacher Oasis were subjected to multiple events of metamorphism and deformation. Several earlier workers (Sengupta 1986, 1988; Peach and Stackebrandt 1995; Ravikant et al. 2004; Bose and Sengupta 2003) studied the detailed geological history of the area. The modified geological map of GSI (1998), Fig 4, shows NE-SW trending litho-units represented mainly by orthogneiss, paragneiss, mafic granulite, ultramafic enclaves and lamprophyre dykes. These rocks later underwent a subsequent amphibolite facies equilibration along a retrogressive P-T path resulting in symplectitic textures in various litho-units (Ravikant and Kundu 1998; Bose and Sengupta 2003).

3 Methodology

The Late Quaternary paleoclimatic history of the Schirrmacher region has been explained with the multi-proxi approach's help. This includes sedimentological studies, sediment geochemistry, OSL geochronology etc. A brief description of these methods are as follows;

(i) Granulometric analysis: The granulometric study is an excellent method to understand the hydrodynamic condition of sediment transport and deposition. The sediment core from the lacustrine environment provides regular data on this aspect. The sediment samples from the lacustrine environment were macerated adequately by following the methods given by Jacksons (1979). The pattern of grain size distributions of the macerated sediment samples has been obtained using a mechanical automatic sieve shaker and the standard pipetting method. The sieve analysis results are plotted to determine various statistical parameters (mean particle size, sorting, skewness, and kurtosis) and facilitate graphical interpretation.

(ii) Clay mineralogy: For clay mineralogical analysis, oriented slides for clay minerals have been prepared after following the methods given by Jackson 1979. These slides were then allowed to dry in air and subsequently scanned from 3 to 30° 2θ on a PANalytical X-ray diffractometer (X'Pert PRO) using Ni-filtered Cu Kα radiation. To know the presence of expandable clay minerals, the Ca-saturated slide of clay & silt fraction is exposed to ethylene glycol vapours and then scanned again with the same setting. The crystallinity of biotite is measured as a half-height width.

(iii) Scanning Electron Microscopy (SEM): SEM can obtain the quartz grain's high-resolution surface texture. For this, about 10 g of each sample was collected after coning and quartering and treated with dilute HCL and $SnCl_2$ (5M). From these chemically cleaned sediment samples, medium to refined sand-sized representative quartz grains was randomly picked for detailed surface texture studies. The grains were first mounted on specially designed aluminium stubs and then coated with 150 Å gold-palladium film before being studied under an SEM (EVO 40). The surface textures of quartz grains were classified in mechanical and chemical features to describe the processes which had acted on them. The work of Krinsley and Doornkam (1973), Krinsley and Funnell (1965), Krinsley and Margolis (1969), Krinsley and Smith (1981), Mahaney (1990), Mahaney and Kalm (1995) and Mahaney et al. (1996) have been used in explaining the surface texture of quartz grains.

(iv) Optically Simulated Luminescence (OSL) dating: OSL dating of glacial sediments from this type of Antarctic environment proved somewhat tricky as the luminescence sensitivity of the samples was low and many samples contained feldspars (possibly as inclusions) that could not be removed entirely even after repeated quartz-feldspar separation and treatment with mild HF. Therefore, an infrared stimulation step was introduced in the SAR protocol to optically remove the feldspar signal (Jain and Singhvi 2001). The samples were processed with 10% HCl and 30% H_2O_2 to remove carbonates, and organic matter under subdued red LED light (~650nm). These were then dried and sieved to obtain a 90–210 μm grain size fraction. Quartz and feldspar mineral fractions were isolated using sodium-polytungstate heavy liquid ($\rho =$

2.58 g/cm^3). Magnetic grains were separated using Franz magnetic separator. Quartz grains were etched using 40% HF for 80 minutes with continuous magnetic stirring (followed by a treatment of 12N HCl for 30 minutes, to remove insoluble fluorides. The resultant fraction was dried and checked for purity using IRSL stimulation. Samples with finite IRSL signal were re-etched for 10 min with HF followed by 12N HCl, re-sieved and retested with IRSL.

The annual radiation dose rate was determined by measuring the radioactivity concentration (Singhvi et al. 2001; Murray and Olley 2002). Dose-rate estimates were based on either or a combination of thick source ZnS (Ag) alpha counting for Uranium and Thorium's elemental concentrations using the Daybreak-582 alpha counters and K using gamma-ray spectrometry.

(v) Total Organic Carbon (TOC): Sub-samples from the untreated bulk sediment samples were analyzed for TOC by heat treatment.

4 Results

The multi-proxy data generated from the moraines and sediment cores of various lakes from the Schirmacher region, cDML, East Antarctica has provided better insight into the Late Pleistocene Holocene paleoclimatic evolution of the area during the Late Quaternary.

The glacial signatures are very well preserved in all kinds of sediments of this region. The sediments from the lacustrine environment are immature, chemically unaltered, have microscopic drainage pattern (Srivastava et al. 2018). The provenance of these sediments is from medium to high-grade metamorphic terrain. The patterns of fluctuations in different granulometric and statistical parameters (Fig. 5) show

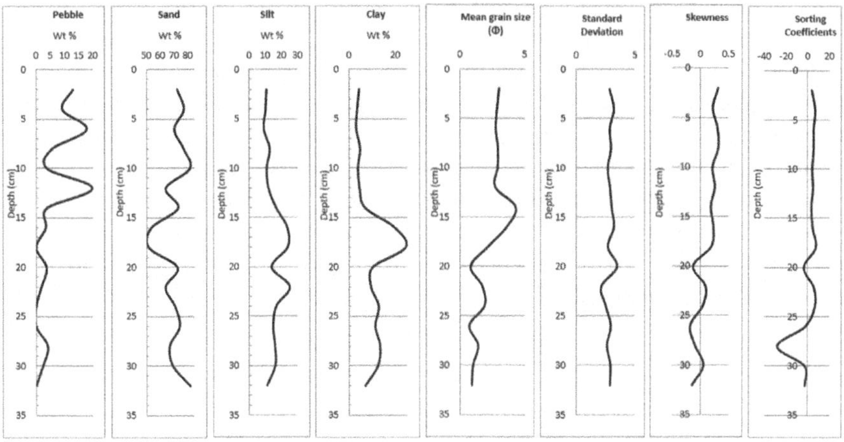

Fig. 5 Granulometric variation in lake sediments from Schirmacher Oasis, East Antarctica

alternate warmer and cooler phases at different intervals. At 15–20 cm depth, a noticeable change in the pattern (a dominant warming phase) has been observed.

The clay minerals indicate a gradual shift in the weathering regime and that in climate from strongly glacial to fluvioglacial during Late Quaternary. But the warm period has not altered the overall clay chemistry. The effect of warming is visible on the sediments of the upper horizon and in pro-glacial sediments. Exclusively physical weathering has controlled the widespread sediments and composition of the clay fraction. The mixed layers of biotite have evidenced the evidence of gradual warming.

The results of surface textures of quartz grains have been discussed depth-wise and in the same samples. In general, it shows dominant glacial and glaciofluvial actions. The more refined quartz grains have shown maximum new growths and silica precipitation, while the coarse grains are primarily fresh, representing sub-glacial to supraglacial transportation. The quartz grains show wide variations in their surface textures when compared with depth and size. New growths, straight grooves and arcuate steps show a positive correlation with depth in refined quartz grains (63 μm) and are dominant in older glacial sediments. A similar result has also been demonstrated by medium quartz grains (250 μm). The role of glacial crushing and grinding is more pronounced on the coarser quartz grains (250–500 μm). The fluvioglacial system's effect is evident with a mix population of these surface textures, especially in the middle horizon lake sediments. This effect becomes more pronounced with an increase in size. The younger quartz grains show a dominance of fluvial impact and lacustrine environment, especially in the coarser grains. Edge abrasion and surface abrasion are high in older sediments as well as in coarser grains. The coarser quartz grains have preserved glacial action better than, the more refined grains (Srivastava et al. 2018).

The OSL chronology on moraines has provided information on different events of deglaciation in the Schirmacher region, East Antarctica. The retreat of EAIS initiated much earlier than it was thought (sometime before 171 ka). Since then, it has receded in different phases. In the first phase of retreat, the exposure of some of the highest points (>200 m msl) of Schirmacher Oasis took place. The ice sheet behaves partially as a valley glacier by moving along the pre-existing structural and topographical valleys. This happened in the second phase of retreat (~95 ka to 65 ka). Deglaciation of the area was almost complete by the end of this phase. Few localized remnants of ice sheet remained in the exposed valleys' lower reaches, which vanished around 22 ka. There is no evidence of morainic deposits belonging to the Last Glacial Maximum. Formation of lake sediments started before 42 ka marks the beginning of climate change from glacial to glaciofluvial.

References

Altmaier M, Herpers U, Delisle G, Merchel S, Ott U (2010) Glaciation history of Queen Maud Land (Antarctica) reconstructed from in-situ produced cosmogenic 10Be, 26Al and 21Ne. Polar Sci 4:42–61

Anderson JB, Shipp SS, Lowe AL, Wellner JC, Mosola AB (2002) The Antarctic Ice sheet during Last Glacial Maximum and its subsequent retreat history: a review. Quat Sci Rev 21, 1–3, 49–70

Bose S, Sengupta S (2003) High-temperature mylonitization of quartzofeldspathic gneisses: Example from the Schirmacher Hills, East Antarctica. Gondwana Res 6(4):805–816

Cortese G, Abelmann A, Gersonde R (2007) The last five glacial-interglacial transitions: a high-resolution 450,000-year record from the subantarctic Atlantic. Paleoceanography 22(PA4203):1–14. https://doi.org/10.1029/2007PA001457

Fitzsimons SJ (1996) Paraglacial redistribution of glacial sediments in the Vestfold Hills, East Antarctica. Geomorphology 15:93–108

Gore DB, Rhodes EJ, Augustinus PC, Leishman MR, Colhoun EA, Rees-Jones J (2001) Bunger Hills, East Antarctica: ice-free at the last glacial maximum. Geology 29:1103–1106

Hodgson DA, Noon PE, Vyverman W, Bryant CL, Gore DB, Appleby P, Gilmour M, Verleyen E, Sabbe K, Jones VJ, Ellis-Evans JC, Wood PB (2001) Were the Larsemann Hills ice-free through the Last Glacial Maximum? Antarct Sci 13(4):440–454

Jacksons Ml (1979) Soil chemical analysis—advanced course, 2nd edn, 11th printing. Madison, Wisconsin, pp 169–251

Jain M, Singhvi AK (2001) Limits to depletion of green light stimulated luminescence in feldspars: implication for quartz dating. Rad Meas 33:883–892

Krinsley DH, Doornkam JCP (1973) Glossary of' Quartz Sund Grain Textures. Cambridge University Press, Cambridge, England

Krinsley DH, Funnell BM (1965) Environmental history of sand grains from the Lower and Middle Pleistocene of Norfolk. England Q Jlgeol Soc Land 121:435–461

Krinsley DH, Margolis S (1969) A study of quartz sand surface texture with the scanning electron microscope. Trans N Y Acad Sci Ser II 31:457–477

Krinsley DH, Smith DB (1981) A selective SEM study of grains from Permian yellow sands of northeast England. Proc Geol Assoc Engl 92:242–247

Lambert F, Delmonte B, Petit JR, Bigler M, Kaufmann PR, Hutterli MA, Stocker TF, Ruth U, Steffensen JP, Maggi V (2008) Dust-climate couplings over the past 800,000 years from the EPICA Dome C ice core. Nature 3:452. https://doi.org/10.1038/nature06763

Livingstone SJ, Cofaigh CO, Stokes CR, Hillenbrand C-D, Vieli A, Jamieson SSR (2012) Antarctic palaeo-ice streams. Earth Sci Rev 111(1–2):90–128

Mackintosh AN, Verleyen E, O'Brien PE, White DA, Jones RS, McKay R, Dunbar R, Gore DB, Fink D, Post AL, Miura H, Leventer A, Goodwin I, Hodgson DA, Lilly K, Crosta X, Golledge NR, Wagner B, Berg S, Ommeno T, Zwartz D, Roberts SJ, Vyverman W, Masse G (2014) Retreat history of the East Antarctic Ice Sheet since the Last Glacial Maximum. Quatern Sci Rev 100:10–30

Mackintosh A, Golledge N, Domack E, Dunbar R, Leventer A, White D, Pollard D, DeConto R, Fink D, Zwartz D, Gore D, Lavoie C (2011) The retreat of the East Antarctic ice sheet during the last glacial termination. Nat Geosci 4:195–202. https://doi.org/10.1038/ngeo1061

Mahaney WC, Kalm V (1995) Scanning electron microscopy of Pleistocene tills in Estonia. Boreas 24:13–29

Mahaney WC (1990) Glacially-crushed quartz grains in late Quaternary deposits in the Virunga Mountains, Rwanda—indicators of wind transport from the north. Boreas 19:81–84

Mahaney WC, Claridge G, Campbell I (1996) Microtextures on quartz grains in tills grains, Antarctica. Palaeogeogr Palaeoclim Palaeoecol 121:89–103

Mahesh BS, Warrier AK, Mohan R, Tiwari M, Babu A, Chandran A, Asthana R, Ravindra R (2015) The response of Long Lake sediment to Antarctic climate: a perspective gained from sedimentary organic geochemistry and particle size analysis. Polar Sci 1–9

Mahesh BS, Warrier AK, Mohan R, Tiwari M, Asthana R, Ravindra R (2017) The response of Sandy Lake in Schirmacher Oasis, East Antarctica, to the glacial-interglacial climate shift. J Palaeomnol. Springer. https://doi.org/10.1007/s10933-07-9977-8

Murray AS, Olley JM (2002) Precision and accuracy in the optically stimulated luminescence dating of sedimentary quartz. Geochronometria 21:1–16

Peach HJ, Stackebrandt W (1995) Geology (of the Schirmacher Oasis). In: Bormann P, Fritzsche D (eds) The Schirmacher Oasis, Queen Maud Land, East Antarctica, and Its Surroundings. PetermannsGeographischeMitteilungen, Erganzungsheft, vol 289, pp 63–165

Petit JR, Jouzel J, Raynaud D, Barkov NI, Barnola J-M, Basile I, Bender M, Chappellaz J, Davis M, Delaygue G, Delmotte M, Kotlyakov VM, Legrand M, Lipenkov VY, Lorius C, Pepin L, Ritz C, Saltzman E, Stievenard M (1999) Climate and atmospheric history of the past 420,000 years from the Vostok ice core, Antarctica. Nature 399:429–436

Pollard D, DeConto RM (2009) Modelling West Antarctic ice sheet growth and collapse through the past five million years. Nature 458:329–332. https://doi.org/10.1038/nature07809

Ravikant V, Kundu A (1998) Reaction texture of retrograde pressure-temperature deformations concerning metamorphism – migmatite events in the Schirmacher Hills, East Antarctica. J Geol Soc India 32:295–315

Ravikant V, Bhaskar Rao YJ, Gopalan K (2004) Schirmacher Oasis as an extension of the East African Orogen into Antarctica: new Sm-Nd isochron age constraints. J Geol 112:607–616

Ravindra R, Chaturvedi A, Beg MJ (2002) Meltwater lakes of Schirmacher Oasis – their genetic aspects and classification. In: Sahoo DB, Pandey PC (eds) Advances in marine and antarctic science. APH Publishing Corporation, New Delhi, pp 301–303

Sengupta S (1988) History of deformation about metamorphism-migmititic events in the Schirmacher Hills, Queen Maud Land, East Antarctica. J Geol Soc Ind 32:295–319

Sengupta SM (1986) Geology of Schirmacher range (DakshinGangotri), East Antarctica. Sci. Rep., 3rd Indian Sci. Exp. Antarctica. DOD, Govt. of India Publ., Technical Publ. no.3, pp 187–217

Singhvi AK, Bluszcz A, Bateman MD, Someshwarrao M (2001) Luminescence dating of loess-paleosol sequences, methodological aspects and paleoclimatic implications. Earth Sci Rev 54(1–3):193–221

Srivastava HB, Shrivastava PK, Roy SK, Beg MJ, Asthana R, Govil P, Verma K (2018) Transition in late quaternary paleoclimate in Schirmacher region, East Antarctica as revealed from Lake sediments. J Geol Soc India 91:651–663

Verlecar XN, Dhargalkar VK, Matondkar SGP (1996) Ecobiological studies of the freshwater lakes at Schirmacher Oasis, Antarctica. Twelfth Indian Expedn Antarct Sci Rep 10:233–257

Warrier AK, Mahesh BS, Mohan R, Shankar R, Asthana R, Ravindra R (2014) Glacial–interglacial climatic variations at the Schirmacher Oasis, East Antarctica: the first report from environmental magnetism. Palaeogeogr Palaeoclim Palaeoecol 412:249–260

A Review of the paleoclimatic Studies from Lake Sediments of Schirmacher Oasis, East Antarctica

Pawan Govil and Abhijit Mazumder

Abstract Paleoclimate of Antarctica has remained a prime subject of study throughout the past half of the last century to understand the global perspective of temperature variability during the late Quaternary. Of the few restricted rocky regions within the Antarctica ice sheet, Schirmarcher Oasis attracts scientific workers for its immense potential for paleoclimatic studies because of several lakes within this area and the different meltwater sources feeding these lakes. Major paleoclimatic/paleoenvironmental studies from this region reveal several episodes of alternating warm-cold events during Holocene and even beyond, based on a large number of various proxies, mainly biological, geochemical and sedimentological parameters. The morphological evolution of the lakes and paleoclimatic reconstruction of the Schirmacher Oasis has been deciphered based on these data. These works collectively offer enormous scope to explore further the high resolution paleoclimatic/paleoenvironmental work of Schirmacher Oasis.

Keywords Schirmacher oasis · East Antarctica · Dronning maud land · Paleoclimate

1 Introduction

Antarctica and its surrounding Ocean are crucial in regulating the earths' climate system. The climate system of the earth is a complex system in view of the exchange of heat between the atmosphere and Ocean in effect of the oceanic current systems and trade winds (Massom et al. 2010). The oceanic currents, trade winds, snow cover, vegetation etc. affect the climate system and vice versa. Antarctica contributes around 90% of the world's freshwater due to its ice-sheets. It also plays a crucial role in the high southern latitudes' radiative forcing and is one of the major driving components for the circulation systems (Turner et al. 2009).

P. Govil (✉) · A. Mazumder
Birbal Sahni Institute of Palaeosciences, Lucknow, UP 226007, India

N. Khare (ed.), *Assessing the Antarctic Environment from a Climate Change Perspective*, Earth and Environmental Sciences Library,
https://doi.org/10.1007/978-3-030-87078-2_7

The climate system of Antarctica as well as surrounding Southern Ocean waters form due to the interaction of the ice sheets, oceanic circulation, sea ice extent, and atmospheric circulation, which altogether respond in the past and present climate forcing (Mayewski et al. 2009). Antarctic region acts as the Southern Hemisphere atmospheric heat sink, which is critical for both the air and sea-surface temperatures. The floating ice, which calves from ice shelves near the Antarctic coastal regions and also sea-ice are the main contributor to climate change. Taken together, both the oceanic and atmospheric mechanisms are connected for the climate system (Bargagli 2005).

The anthropogenic activities in the Antarctic region are least thereby Antarctic pristine is the best for deciphering the natural (un-biased) climatic changes in the past. Such changes are helpful for understanding how biotic components were affected by climate and how the permanent glaciers retreated or advanced in Antarctica along with how it affects the climate change of the earth. The past climatic changes of Antarctica are reconstructed using long ice-cores to better understand the air temperature changes in the polar region (Jouzel et al. 1989, 2007). Although ice-core records presented long-term glacial-interglacial climatic variability, ice core studies' within the Holocene from East Antarctica's coastal areas has not delivers much promising results. Also, the minor variations in isotopic results suggest less climatic variations and therefore needs to be resolved. The inland records from Antarctic regions were collected mostly from the inshore sites (Bromwich et al. 1998; Verleyen et al. 2011), and a also from the high central plateau (Masson et al. 2000). Theice cores from the coastal areas of Antarctica helped reconstructing the dynamic nature of the ice sheet at a higher resolution (Hodgson et al. 2005). It transforms lake and coastal marine sediment into valuable archives for the paleoclimatic study. Therefore, lake sediments are best suite of collections for targeting the Holocene climatic variability by using different proxies. Moreover, the Pleistocene-Holocene boundary in East Antarctic sediments recorded the data to exhibit the deglaciation initiation after the Last Glacial Maximum (LGM) in the coastline areas (CLIMAP 1981; Huybrechts 1990; Hughes 1998), making lakes present in those regions favourable sites for paleoclimatic studies.

The present chapter reviews all the literature that addresses paleoclimatic studies over the Pleistocene and Holocene in the Schirmacher Oasis region. However, paleoclimatic interpretations, specifically in Antarctica, are greatly influenced by the location, geology, geomorphology, and glaciology of the sampling location. Thus, a brief on these have also been included in the chapter.

2 The Geographical Extent of Schirmacher Oasis

The Schirmacher Oasis (SO) (70°44′21″–70°46′04″ S to 11°26′03″–11°49′54″ E) is an ice-free vast plateau along the Princess Astrid Coast in Dronning Maud Land, East Antarctica (Fig. 1), which has around 118 freshwater lakes. These lakes are characterized as epishelf, land-locked, and pro-glacial types. The oasis is mainly situated between the Antarctic Ice Sheet and the NovolazarevskayaNivl Ice Shelf. As SO is a part of Central Dronning Maud Land in eastern Antarctica, it is a rocky terrain of about 17 km long and around 0.7–3.3 km wide. The vertical elevation of the area varies between 0 and 228 m above the mean sea level. Topographically, SO comprises four distinct units, namely, the southern continental ice sheet, rocky hill slopes, a vast central pro-glacial lake, and northern undulatory shelf ice.

3 Geology

The Schirmacher Oasis, Queen Maud Land, a part of the east Antarctic Precambrian shield, is composed of Precambrian polymetamorphic gneisses. The main rock types of this area are charnockites, underbites, garnet-sillimanite gneisses, garnet-biotite gneisses, quartzofeldspathicaugen gneisses including some foliated lamprophyres, amphibolites, dolerite, metagabbro, and metabasalt. The gneisses show mylonitization, polyphase deformation and have undergone upper amphibolite to granulite facies metamorphism (Sengupta 1988). The ages of these gneisses range from 460 to 845 Ma (Grew and Manton 1983; Ravich and Krylov 1964; Kampf and Stackebrandt 1985). The gneissic basement is intruded by pegmatites basic sills, lamprophyres, gabbro, dolerite and basaltic dykes (Sheraton and England 1980; Shiraishi et al. 1988). Lamprophyres dykes in Schirmacher Oasis are 1 km in length and 10–15 m in width, trending NW-SE. These are coarse-grained, having a panidiomorphic texture with phlogopite and pyroxene phenocrysts in a groundmass of K-feldspar, diopside, amphibole, quartz, plagioclase, appetite, and calcite. Based on the mineralogy, lamprophyres are classified as Minette. Lamprophyres dykes intrude on the late Proterozoic metamorphic terrain of Schirmacher Oasis. The study also reveals that the main fault is situated in the north-east corner, intersecting the geomorphological units, namely, shelf ice, rocks, and continental ice sheet. Several lamprophyre dykes of calc-alkaline (shoshonitic) nature are reported from Schirmacher Oasis, Queen Maud Land, in Eastern Antarctica. These dykes are characterized by high K2O/Na2O (avg. 5.81) ratios, high Rb, Sr, Zr, and Th abundances. REE distribution patterns show significant enrichment of LREE against HREE (avg. Lan/Ybn = 27.51) without any Eu anomaly. It appears that these ultrapotassic dykes have been derived from partial melting of a thickened lithosphere (Jafri 1997). The studies from Sangewar (1987) recorded the surface features and temporal changes in different Antarctic ice types and carried out glacial-geomorphological studies in Schirmacher Oasis. Geological studies were also carried out, which revealed the existence of olivine basalt, hornfels, pyroxenite, and skarn rocks in the Skaly IGA nunatak area.

The rock units predominantly represent Grenvillean (1000 Ma), and Pan-African (550 Ma) events. Rb-Sr whole-rock/mineral isochrone ages of two lamprophyres dykes are 450 ± 12 Ma (Sr_i = 0.70886 ± 5) and 458 ± 6 Ma (Sr_i = 0.71388 ± 98). This Lamprophyres dyke's Schirmacher Oasis activity may be interpreted as an appearance of post-orogenic alkaline magmatism related to the Ross orogeny of the Trans-Antarctic Mountains (Dayal and Hussain 1997). The whole-rock Rb-Srisochron age of 514 ± 59 Ma has been obtained for granites from the Nordevestoya area, Humboldt Mountains (Ravindra and Pandit 2000). The minettes were found to be abundant with the compatible elements as well as high mg values (58–75), which indicates that these rocks' petrogenesis is mainly derived from mantle (Hoch and Tobschall 1998). The higher values of incompatible elements, especially LREE, and Ba, Rb, K, Sr were also found. The Chondrite-normalized REEs' distribution patterns exhibit differentiated enrichments and high concentration in the LREE (240–530 × chondrite) in comparison to low and nearly stable enrichments of the La/Yb-ratios from 28 to 52 and HREE (11–15 × chondrite). The samples also show higher 87 Sr/86 Sr (0.70775–0.71337) and low 143 Nd/144 Nd (0.51229–0.51135) with εNd = -6.5 to −21.5. It shows 207 Pb/204 Pb ranging between 15.46–15.57 together with high 208 Pb/204 Pb (38.06–39.79) and unradiogenic 206 Pb/204 Pb (16.77–18.08). Rb/Sr-dating yields two different age groups of the time (430 and 703 Ma) of intrusion of the lamprophyre dikes.

Joshi (1997) delineated tectonic features of Schirmacher Oasis using total field geomagnetic suggests that the exposed rocks have little magnetic contrast. The exposed rock has undergone multiple episodes of metamorphism, migmatization, and deformation. However, nine profiles were obtained crossing different geological units, which was subjected to identify the faults and shear zones by Bormann et al. (1986) from the photogeological map. It is observed that the rock type gradation varies from east to west. The same banded and leucocratic rocks show different magnetic characteristics in Central Schirmacher Oasis for that in the eastern region. The absence of foliation in the banded gneiss in Central Schirmacher Oasis has been observed.

Scheinert (2001) investigated geodynamic phenomena in Schirmacher Oasis using a variety of geodetic observation techniques. Satellite-based techniques (Interferometric SAR) were combined with terrestrial methods (for instance, surveying, static, and kinematic GPS) for inferring the ice dynamics, especially the horizontal velocity field of the inland ice and the dynamics of the shelf ice. GPS epoch observations were used to describe the ocean, ice, and solid earth dynamics and interactions. New gravity field data of the planet, and investigation of ice sheet will be provided by future satellite missions (e.g. CHAMP, GRACE, GOCE, ICESat/GLAS, CRYOSAT, and ENVISAT). Using satellite data and airborne data, namely, gravimetry, laser altimetry, and radio-echo sounding helped to explore geodetic, geodynamic, and glaciological features for extended regions in Antarctica.

The data (during January–March 2005) of eight MT stations studied by Murthy et al. (2012) showed high resistive upper crust followed by a comparatively conductive lower crust comparable to the southern Granulite Terrain with a lateral heterogeneity in the crust's electrical conductivity across the Schirmacher Oasis.

Siddiqui et al. (1988) carried out the study of seismic reflection, higher energy sparker source (405 line km), bathymetric (458 line km), and magnetic (452 line km) studies of the Astrid ridge, a comprehensive bathymetric feature was recorded in the water depths of about 1520–2900 m towards the continental margin, Dronning Maud Land.

4 Glaciology

The Schirmacher Oasis observation and data suggests the southern part of the ice sheet (blue ice) is uncovered. The contour line starting from 180 to 10 m shapes at the snout of glacier, which is intersected by NE-SW to NNE-SSW trending several fractures. Two transient supraglacial streamlets derive towards an NNE direction. Overall meteorological data shows that the area falls under a dry polar climate. Snowfall is very common throughout the winter periods. However, one way snow deposition is regular over the leeward hillock's side and the other way strong winds clean the rock surfaces.

According to observations made in to the glaciology between 1983 and 1996, surveys from two fixed points using electronic distance measurement (EDM) or theodolite, the glacier is continuously retreating annually with an average retreat at the rate of 70 cm/annum. The moderate annual recession during 1997–2002 was in the range of (~48.7–74.9 cm), with an average of 65.3 per/annum during 1997–2002, and it is in agreement with the interpretation for the preceding years from1983 to 1996 with a decline rate of 7 m/decade (Chaturvedi et al. 2004).

The study of the snow accumulation/ablation over the ice shelf near the first Indian station Dakshin Gangotri (70°05′37" S Lat.; 12°00′00" E long.) was launch by fixing a 100 m by 100 m network of wooden stakes in 1982–1983. Consequently, much-reduced accumulation has been noticed during austral summer (November-January) in the last twenty years compared to the rest of the year. The polar winter shows the maximum snow accumulation. Vaikmäe (1991) reported that the isotopic composition of ice, fire, and snow data decipher, recent structures and Holocene evolution of this marginal area of East Antarctica from northern Queen Maud Land near 12° E. Isotopic analysis from the ice cover samples situated south of the Schirmacher Oasis suggests that under the Holocene ice, the thick relics of the Pleistocene ice cover with $\delta^{18}0$ values of −38 to −48‰.

Analysis of thermal structure and estimation of heat budget for lake Priyadarshini, during December'96–March'97, reveal that weak thermal stratification prevailed in the lake for the significant part of the period of observation during mid-summer. However, the lake becomes unstratified as the winter approaches. The heating cycle of the lake is a function of the climate and morphometry of the lake itself. The National Geophysical Research Institute (NGRI), India in 1997, initiated the studies on seismotectonic and geodynamical processes by using the seismology and GPS-Geodesy. The scientific proposals to study the crustal deformation and geodynamical processes between Antarctica and India were continued up to 2005 and achieved the prime objectives. NGRI launched several programes to improve the outcomes on the interplate and intraplate strain accumulation and the seismicity within the Antarctica by setting up the seismic stations and using densifying GPS near Maitri and third station (named Bharati) at Larsemann Hills. The critical evaluation of the deformation processes examined by the IGS GPS stations, Casey and Davis which are far away from Maitri. To understand the Antarctica and the Southern Indian Peninsula's geodynamical system, Antarctica's internal deformation processes has been estimated very precisely using the planned array of 5 GPS stations Near to Maitri. Similarly, a Seismic array with 5 Broadband seismometers has been installed around Maitri. To achieve the long-term objectives of the program, this Seismic array helped to identify the major and minor magnitudes of the earthquakes occurred in and around Antarctica and to assess, the velocity and crustal structure below Maitri and the nearby region.

Detailed analysis of the measurement of the ozone hole depth, during the 16th and consecutive three (21st, 22nd, and 23rd) Indian Antarctic Scientific Expedition was done from a high latitude isolated area near Indian Station Maitri, Schirmacher Oasis of East Antarctica. The lowest ozone value recorded over Maitri was 135 (±9.3) DU in 1997, 185 (±12) DU in 2002, 126 (±9.7) DU in 2003 and 159.8 (±8.8) DU in 2004 spring. During Antarctic spring (day 225–365), concentration of Ozone below 220 DU shows 45.1%, 20.7%, 62.7% and 60% in 1997, 2002, 2003 and 2004, respectively. The Ozone loss (below 220 DU ozone values) during days from 225 to 365 in 2003 was observed to be increased by a factor of 0.4 compared to 1997 and 2 compared to 2002. The data collected from Maitri also exhibited the significant event of stratospheric warming during 2002. In 2002, the ozone hole was not severe and improved early, in contrast to 1997, 2003 and 2004. Experimental columnar ozone values has been used to decipher the increase in UV irradiance theoretically during the deep Ozone Hole period (30-09-2004 to 10-10-2004). During the intense ozone hole period, the maximum increase in UV irradiance theoretically was estimated to be an average loss of 44% total ozone with 403% at 305 nm, 84% at 312 nm, and 24% at 320 nm. To understand and calculate the variation in direct UV irradiance per DU change (range 320–100 DU) in columnar ozone at different solar zenith angle, the TUV model investigated the relationship between the columnar ozone and direct UV irradiances.

Survey of India (SOI) has been participating in the Indian Scientific Expedition to Antarctica since 10th Expedition (1991–92). The SOI set the objective to carry out the survey and respective mapping of the area and finally to provide technical assistance to the different participating organisations. Originally, conventional triangulation and Global Positioning System (GPS) techniques were employed to deliver major outcome. Later, with contour interval 1 m, the mapping on scale 1:1000 and with contour interval 5 m, the mapping on scale 1:5000 has been commenced. The SOI has already established Geodetic and Geophysical Control in and around Schirmacher Oasis for better scientific investigations in the field of glaciology and climatology. In this process, Control points was provided for India's Geological Survey, the Indian Meteorological Department and assistance was given to the survey work to the National Environmental Engineering Research Institute and National Physical Laboratory, Defense Electronics Application Laboratory. From 2003–2004, neotectonic and glacier movement assessments has been completed with the aim of undertaking the varied geodetic works, such as establishment of 40 nos. of GPS control points (with permanent documentation). Later on, these data connect High Precision Leveling, and provide precise gravity values, repetition of levelling, gravity, and GPS observations for at least three epochs. As much as 28 stations have been established for this studies along with seven days GPS observation at Maitri station

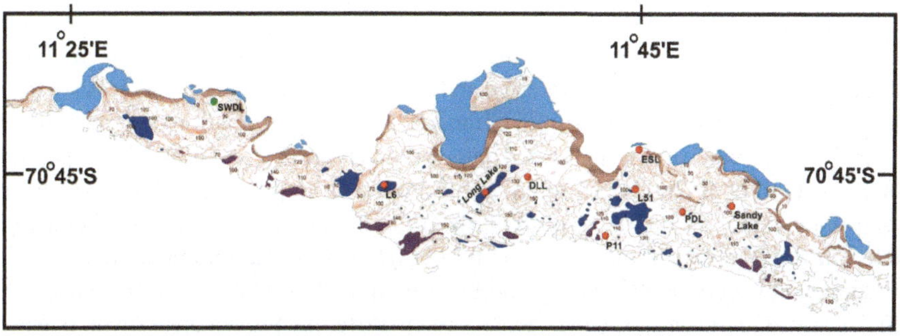

Fig. 1 Sediment core locations in Schirmacher Oasis, East Antarctica (after Govil et al. 2018)

from 23rd Expedition till date. In and around Schirmacher Oasis, a area of 5.7 sq km of the has been surveyed. This Survey has published in form of large scale maps (both in analogue and digital format) covering most of the area of interest.

5 A Review of the Paleoclimatic Reconstructions at Schirmacher Oasis

5.1 Pleistocene Paleoclimatic Record

An early attempt to study freshwater lakes in East Antarctica was made in 1984–85 to describe the freshwater microfaunal components in Schirmacher Oasis (Ingole and Parulekar 1993). This study was followed by very few investigations on lake sediments from this region (Sharma et al. 2000; Sinha et al. 2000a; Sinha and Chatterjee 2000; Sharma et al. 2007a) to reconstruct the palaoclimate, transportation of palynodebris and pollen (Sharma et al. 2002; Bera 2004) and report of the presence of freshwater protozoans—Arcellaceans (Mathur et al. 2006). Hydrogen and oxygen isotopic studies has been carried out from water samples of different lakes in Schirmacher Oasis (Sinha et al. 2000b); and preliminary study of quartz grains morphology and microtextures from surface sediments from lakes in Schirmacher Oasis was done to recognize their provenance and depositional environments (Asthana and Chaturvedi 1998). Furthermore, Schirmacher Oasis, being one of the larger coastal oases in eastern Antarctica, remains a significant region for paleoenvironmental and paleoclimatic studies., Lacustrine sediments should be addressed to understand the paleoclimatic variations as they preserve the sediments reflecting long term and abrupt changes through the geological period. However, Schirmacher Oasis has remained a less worked region in Antarctica regarding paleoclimatic studies of lake sediments (Bera 2004; Sharma et al. 2007b; Phartiyal et al. 2011; Phartiyal 2014; Warrier et al. 2014, 2015; Mahesh et al. 2015; Govil et al. 2016; Govil et al. 2018).

Nevertheless, the rapid climatic changes affirm the internal climate changes involving the interaction between the atmosphere and the ocean during a stable climatic condition and possible internal and external forcing of earth, such as solar or volcanic effects (Stuiver et al. 1995). Climate variability information can be obtained from these forcings and transformed into the paleoclimatic data based on spectral analysis (Ghil et al. 2002). Sediment core samples from pro-glacial lake P-11 have been analyzed for the quartz grains. Quartz grains has been used over any other available minerals present in the lakes due to its higher preservation potential and their resistance to weathering (Krinsley and Doornkamp 1973; Mahaney 1995; Mahaney 2002). The Quartz grains have the record of different provenances, which were derived mechanically or chemically along with through wind-driven aeolian activities (Stanley and DeDeckker 2002). Quartz grain microtextures and morphology, such as shape of grain, different patterns, such as step patterns, fracture patterns etc., principally depend on the transportational conditions and depositional environment and hence provide data of the paleoenvironmental as well as paleoclimatic conditions (Whalley and Krinsley 1974; Mahaney et al. 1996; Helland and Holmes 1997; Hart 2006; Mathur et al. 2009).

To decipher the provenance of clay minerals in the sediments and reconstruct Schirmacher Oasis's paleoclimate, a detailed clay mineralogical study has been carried out from glacial deposits found near the lakes (Srivastava et al. 2011). This study exhibits prominent peaks of five clay minerals: kaolinite, illite, chlorite, smectite, and vermiculite. These clay minerals were formed under cold climatic settings from the weathered and altered rocks, such as igneous or metamorphic rocks. Further study has been carried out on the sediment samples collected from surface of two epi-shelf lakes in Schirmacher Oasis to investigate the sedimentological and mineralogical properties of surficial sediments (Shrivastava et al. 2012). The study based on the quartz grains concluded that the grains were deposited in these epi-shelf lakes by different physical weathering and chemical precipitation processes. Chemical analysis revealed that the sediments were gneissic in composition, and saltation and suspension were two main modes of transportation of the poorly sorted glaciofluvial sediments. Sediment samples from different regions in the Schirmacher Oasis, such as ice-free lakes, polar ice, and the coastal shelf area, reveal that the Schirmacher Oasis was formed predominantly by glacial processes, with some glaciofluvial and aeolian influences (Srivastava et al. 2012). Glacial sediments from different Schirmacher Oasis locations were investigated to assess the palaeodepositional environment (Srivastava et al. 2013). Mineralogically, the rare earth and trace elemental data indicates that heavy minerals enriched in the coarse fraction over finer one. The transportation of sediments happened mostly by fluvial as well as aeolian agents, including the minor influence of gravity and glacier actions. The overall conclusion was that almost all the deposits collected show similar geochemical and mineralogical characteristics.

5.2 *Holocene Paleoclimatic Evidence*

The evolutional history of Schirmacher Oasis (SO) from 13 kyr BP to Recent was reconstructed based on the data of loss-on-ignition (LOI) and magnetic susceptibility (MS) from seven sediment profiles of five palaeo-lake deposits (Phartiyal et al. 2011). It has been deciphered that glaciers dominated the entire SO from 13 to 12.5 kyr BP and the today's land-locked lakes were designated as the glacial lakes. In the beginning of Holocene warming (11.5 kyr BP), five large pro-glacial lakes situated the low-lying valleys of SO were formed due to the recession of glaciers. On the other hand, the landlocked lakes got dried up due to the absence of water influx during summer months because of the retreat of glaciers, diminished meltwater inflows, lesser snow accumulation, stronger winds, and sublimation of the lake due to ice-cover for longer period. However, both the pro-glacial and the epishelf lakes were fed by the continuous melt-water flux from the ice shelf and continental ice sheet. An analysis of remnant magnetism in the lake sediments of SO reconstructed the paleoclimatic condition of this area (Phartiyal 2014). In this study, a total of six phases of climatic variation were recorded from 13 to 3 kyr BP, and Phase 1 is a short one of climatic event occurred at ca. 12.9 kyr BP, which was followed by Phase 2 at 12.5 kyr BP. Both phases correspond to comparatively warmer climatic condition. At ca. 11.8–11 kyr BP, Phase 3 records similar climatic conditions to those of Phase 1. After that, a better and longer Phase 4 was recorded from 11 to ca. 8.7 kyr BP, which can corresponds to worldwide Holocene Optimum event, and a significant period of shelf areas deglaciated in the Antarctic in this period. Lastly, Phase 5 shows an arid and cold climatic condition with the period of 8.7 to 4.4 kyr BP. Two shorter and comparatively warmer climatic periods during 7.9–7.1 and 6.4–5.6 kyr BP were sandwiched within Phase 5. From 5.3 kyr BP onwards, the climatic condition became increasingly warmer recorded in most Antarctica areas till 3 kyr BP, which is comparable to mid-Holocene Hypsithermal (Bentley et al. 2005). In SO, the time period of 4.6–4.4 kyr BP was reported as a time of intermediate climatic conditions. The period of the mid-Holocene Hypsithermal shows a considerable variation in different regions of Antarctica as compared to ca. 4.4–3 kyr BP in SO; such as 4–2.7 kyr BP (Bentley et al. 2005, 2009) to 4.5–2.8 kyr BP on the Antarctic Peninsula, and 3.8–1.7 kyr BP in maritime Antarctica (Hodgson and Convey 2005; Jones and Bowser 1978).

Sedimentological parameters along with quartz grain morphology and microtextures of surface sediments from Schirmacher Oasis have been analysed to comprehend the transportation and depositional processes in this region (Asthana et al. 2013). It has been concluded that fluvioglacial activity with some variations of energy conditions played a vital role in this area. It has also been surmised that the deposition of sediments in periglacial lakes of SO is dominated by sedimentary and glacial processes.

Paleoclimatic reconstruction for the past 42.5 kyr BP has been performed using the environmental magnetic properties of sediments deposited in Sandy Lake, SO

(Warrier et al. 2014). Coldest periods of the surrounding region were recorded consecutively during 40.7, 36.0, 34.5, 29.0 and, 28.0–21.4 kyr BP. Intermittent relatively warm periods were intercalated during 38.4–39.2 kyr BP, 33.7–29.8 kyr BP and 28.5 kyr BP. The LGM in the SO has described as a widespread glacial condition. The following Holocene period was also intersected by changing phases of relatively warmer (12.5–9.8 kyr BP and 4.2–~2 kyr BP) to colder (9.2–4.2 kyr BP and from ~2 kyr BP onwards) periods.

Most of these alternative warm and cold periods can be easily correlated with the climatic events recorded from other lake sediments and ice-core records from the SO as well as other East Antarctic areas. Overall, it has been surmised that the SO was not affected by complete glaciation during the past 40 kyr BP. Study of sedimentological parameters along with quartz grains morphology and microtextures from the sediments of Sandy Lake were examined to infer the provenance of the sediment and the transport mechanism during the past 42.5 kyr BP (Warrier et al. 2016). It was concluded that the sediments were mainly transported by glaciers and meltwater inflows. During the period of last glaciation, the climate was colder with the prominent aeolian transportation activity. On the other hand, post-LGM enhanced the energy for the transportion, hence meltwater stream played an active role under glaciogenic conditions. The study also surmised that the various kinds of physical and chemical weathering, mostly of glacial conditions during the transportation affected Quartz grains A study of organic geochemistry and sedimentology from sediments from Long Lake, SO, depicted a history of the paleoclimatic condition during the last 48 kyr BP (Mahesh et al. 2015a). The extended ice-cover period for a freezing environment lowered the productivity of the last glacial period as well as a significant part of Holocene. The Holocene warm period around the lake would have resulted a extended ice-free environment starting around 6 kyr BP. Collectively, the productivity and sedimentation of the Long Lake related to the response of the glaciated conditions controlled by Antarctic climate.

The presence of *Pohlianutans* moss species was the first ever record of this subfossil from the central Dronning Maud Land region dated 10.6 kyr BP in a sediment core from Lake L6 in SO (Singh et al. 2012). This is one of the common species found even in the other part of Antarctica (Singh et al. 2012). The moss population data can assess the variation of its distribution and diversity in the SO region and decipher the paleoclimatic and palaeoenvironmental conditions. Another study from Lake L-6 within the SO was carried out extensively with a multiproxy approach to reconstruct the paleoproductivity and paleoclimatic changes during Holocene (Govil et al. 2016). This study exhibits a warm period during the Pleistocene-Holocene boundary (~11.6 to ~10 cal kyr BP) with the record of elevated paleoproductivity and reducing environment as organic matter decayed in the bottom water of the lake. Later, a rapid shift happened from deglaciation to glaciation during the commencement of the Holocene (~10 kyr BP). A relatively colder phase has been recorded from the entire stretch of Early to Late Holocene (~10 to ~3.1 kyr BP) due to the extension of continental ice sheet from the southward to the northward in this area. This phase was characterized by lowering of paleoproductivity within the lake and lessening of the rate of sedimentation (~2.9 cm/kyr). The Late Holocene period (~3.1 kyr BP to Recent) experienced a stepwise warming that corresponds to the winter and summer solar insolation. This stepwise warming event in East Antarctica has ultimately affected the paleoproductivity in the lake. During this period, lakes had received an adequate amount of meltwater from the continental ice sheet and snow around the lake that increased the lake water level. Another study reveals the paleoclimatic shift from the last glacial period to the Holocene (Govil et al. 2018), which was substantially juxtaposed with the changes that occurred during Holocene (Johnsen et al. 1992a).

Paleoenvironmental studies from a pro-glacial lake in SO were performed based on quartz grain morphology and microtextures (Mazumder et al. 2017). This study exhibits the predominantly colder climatic condition at the advent of the Holocene. During the beginning of the Holocene, melting of the glaciers started in this region. Middle Holocene is characterized by relatively warmer conditions, while in Late Holocene the reappearance of a somewhat colder climatic condition occurred in this region. Thus, this study suggest an interchanging glacial-interglacial environmental phase during Holocene.

Sediment cores from few landlocked lakes and a grab sample from the pro-glacial lake in the central part of SO have been analysed using multi-proxy sedimentological approach to reconstruct the paleoclimatic history of the region (Srivastava et al. 2018). The lake sediments exhibit a glacial signature derived from medium to high-grade metamorphic terrain under restricted drainage pattern. The different sedimentological parameters show alternate warmer and colder phases at different time intervals. This study concluded that the gradual warming in SO began much before 42 kyr BP. During the early stage of the Late Quaternary, this area was covered by ice, which gradually waned to expose an ice-free area in the Late Holocene. The dominance of sand from sediment cores of Antarctic lakes GL-1, V-1, and L-6 deciphered the high degree of mechanical weathering, which weathered coarse-grained particles from the provenance to the depositional area transported by the fluvioglacial process (Choudhary et al. 2018). The grain size variation of lake system was controlled by many factors, such as amount and energy of meltwater, aeolian action, the surface area of the lake, geology of the catchment area, and local hydrodynamic settings.

5.3 A Case Study from Schirmacher Oasis (SO)

The recent study from the L-51 landlocked lake focused on two observations. The first observation to distinguish the effect of the reservoir age in the lake sediment from entire Schirmacher Oasis (SO). The second observationis based on one of the well-documented events of Antarctic Cold Reversal (ACR) (Jouzel et al. 1989; Blunier et al. 1997; Pedro et al. 2011). The 2.2 kyr age was subjected to subtract from all radiocarbon ages individually and the corrected reservoir ages for the core L-51 was obtained in SO. The considered corrected reservoir age was finalized after calculating the mean of all ages obtained from four lake sediment top samples within SO. The core tops from the lake L-6, lake P-11, epishelf lake (ESL) and sandy lake, situated within 3 km of present southern boundary glacier in the SO, interpret the ages of 640 years (Govil et al. 2016), 3320 years (Mazumder et al. 2017), 3685 years (unpublished data) and 1245 years (Warrior et al. 2014), respectively. Earlier, the reservoir age correction of 640 years has been standerdizedfrom the lake L-6 (Govil et al. 2016) because of the depositionof the old carbon to the core site through the glacial meltwater. If the glacial meltwater inflows with depleted ^{14}C move lesser distance in the catchment area of the lake and suggests the equilibrium condition within the atmospheric gas exchanges (Doran et al. 1999; Wagner et al. 2004). The reservoir age effect can be neglected if temporary ice covers the lake, but it will gradually enhance during the colder periods with the enhancement of meltwater influx from surrounding glaciers (Wagner et al. 2004). However, measurement of this transported carbon, both incoming and outgoing, along with the mobility and the meltwater flux is difficult to obtain accurately. Hence, to attain the corrected reservoir ages, 2.2 kyr was subtracted from all radiocarbon ages individually.

Moreover, detailed work from SO samples obtained from trenches and dry lakes also confirms comparatively older ages at the top 0–10 cm as the raw radiocarbon dates show a rangebetween 3.11 and 11.1kyr BP (Phartiyal et al. 2011). But these radiocarbon dates have not been used in calculating the average value in reservoir age due to the collection of these samples and some sediment core samples collected from pits, channels and dry lakes. However, the upper part of sediment collected from the entire SO contains older carbon in it (Phartiyal et al. 2011). The large and small lakes situated in SO receive sedimentation through the meltwater flux of glaciers from the surrounding regions during SO warm summer period. The top sediments from lakes ESL and P-11 represent the old raw radiocarbon ages in the present scenario. However, L-51 sediment core should precisely be assessed by the incurred chronology before insuring further inferences. Hence, the reservoir age and its corrections should be established to apprehendpaleoclimatic and palaeoenvironmental settings accurately from the SO lake sediment cores.

A multiproxy study from the different sediment proxies (percentage of sand-silt-clay, mean diameter of the sediment and roundness), percentage of biogenic silica (BSI %) as well as spectral examination of the available data to determine the significant periodicities (at >95% significance) deciphers the presence of Antarctic Cold Reversal (ACR) around 14–11.8 kyr BP, which was characterized by the increase of freshwater runoff in the lake followed by early-Holocene climate optimum and Holocene climate optimum around 11.8–7.2 kyr BP (Govil et al. 2018). However, the Antarctic ice sheet mass loss, Melt-water Pulse 1A (WMP1A), and enhanced iceberg flux during the Antarctic cold reversal (ACR) in the Southern Ocean were documented (Weber et al. 2014; Fogwill et al. 2017), though the effect of ice of Antarctica on the MWP1A during the ACR is questionable (Liu et al. 2016). The ACR timing was reported to be during 14.7–13 kyr BP, which corresponds with the major warming event, namely Bølling–Allerødevent from the north Atlantic region (Pedro et al. 2016). A detailed comparative study between Antarctica and Greenland ice core record depicts an unsynchronized relationship between the two poles, designated as 'bipolar seesaw' (Blunier et al. 2007). Hence, there is ambiguity regarding the meltwater flux during ACR that questions the intensity of the ACR signals in the SO in East Antarctica. Furthermore, in present study the sedimentological proxies has been used on the sediments from the L-51 lake of SO. The analysis of grain size infers that during the 18.5–14.4 kyr BP, the clay variesfrom 3.6 to 2.5% (Fig. 2c), silt ranges between 52.9 and 40.0% (Fig. 2b), and sand ahows a range of 57.4–43.6% (Fig. 2a) along with the mean diameter changes from 143.65 to 102.61 μm (Fig. 2d) and sorting varies from 28.0 to 25.7% (Fig. 2e). However, the BSI percentage ranges from 1.2 to 0.5% from 18.5 to 14.4 kyr BP (Fig. 2f). Within the 14.4–12.6 kyr BP period, the clay varies from 2.9 to 1.0% (Fig. 2c), silt ranges between 42.6 and 25.3% (Fig. 2b), and sand varies between 73.5 and 54.3% (Fig. 2a) along withthe mean diameter varies from 256.8 to 136.3 μm (Fig. 2d) and the values of sorting range from 36.9 to 26.9% (Fig. 2e). On the other hand, the BSI percentage shows a range from 1.8 to 0.6% within 14.4–12.6 kyr BP (Fig. 2f). During Holocene (~11.7 kyr BP to Recent), the data from the analysis of grain size shows that the clay varies between 5.3 and 1.4% (Fig. 2c), silt ranges from 76.5 to 35.2% (Fig. 2b), sand shows a range between 61.2 and 20.5% (Fig. 2a) as well asthe values of mean diameter range between 228.5 and 44.5 μm (Fig. 2d) and sorting shows a variation between 38.3 and 24.2 (Fig. 2e). During the Holocene the BSI percentage varies from 3.6 to 0.4% (Fig. 2f). The paleoclimatic data obtained from the ice cores from Antarctic decipher that the Antarctic cold reversal (ACR) interrupted significantly post-LGM warming in the Southern Hemisphere around 14.5 to 13 kyr BP (Blunier et al. 1997; Pedro et al. 2011, 2016; Jouzel et al. 1989). In the present study, the increased sand percentage by ~20%, mean diameter by ~164 μm, sorting by ~10% and BSI percentage by ~0.9% along with the decrease in clay percentage ~2% and silt percentage by ~17% have been noticed during a short time span after 14.7–12.2 kyr BP (Fig. 2). Simultaneously, increase in sand percentage, mean diameter and sorting indicates that coarser grain was transported through enhanced meltwater flux in the result of the retreat of ice cover and hence deposited at the centre of the lake (Fogwill et al. 2017). This hypothesis also supported the reason of the possible initiation of the ACR signal (14.6–12.7 ka BP) due to sufficient meltwater flux into the Southern Ocean. During MWP1A a significant quantity of meltwater has been released into the Ocean that contributed to maintain the stratification in ocean and hence enhance the ice cover retreat. Moreover, the probable methods of the retreat of grounded ice may be attributed to the subsurface ocean warming (Golledge et al. 2014) and substantially reduction of Antarctic bottom water (AABW) formation in the Southern Ocean resulting to the enhancement of the cooling in atmosphere over Antarctica during this period (Parrenin et al. 2007). Thereafter it reduced the sea surface temperature of the Southern Ocean (Siani et al. 2013; Nielsen 2004) and finally happened to increase the sea ice extension. Furthermore, the sea ice extension affected on the reduction in the ventilation from surface water to deeper water and enhanced subsurface temperature in the ocean as well as the Antarctic coast (Menviel et al. 2010). The grounded ice melting resulted due to this warming lowered the ice sheet surface that enhanced the horizontal release (Fogwill et al. 2017). Finally, the feedback mechanism over the climate system of Antarctica seems to be a conflict between the differences of ocean and atmospheric heat fluxes, which trigger the regional climate change of the SO.

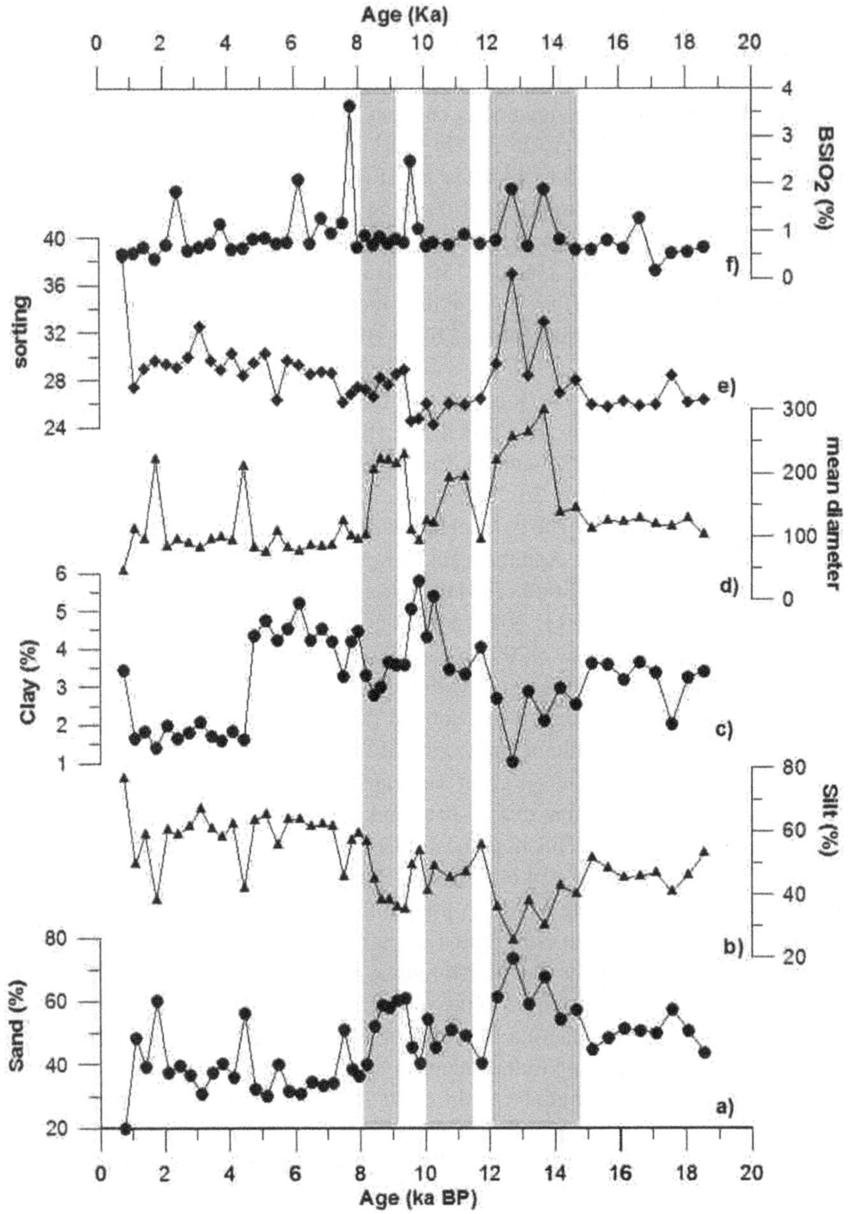

Fig. 2 Grain size distribution, Biogenic silica and sorting variation with the chronological sequence of lake L-51 **a** Sand (%). **b** Silt (%). **c** Clay (%). **d** Mean diameter (μm). **e** Sorting (%). **f** Biogenic Silica (BSI %). Three dark band shows the meltwater flux episodes in SO (also refer Govil et al. 2018)

Although limitations arise in the present study due to the radiocarbon dating uncertainties, it was obvious that increase of meltwater flux correlated with the event of ACR. However, the data of grain size distribution along with biogenic silica demarcate three prominent episodes of meltwater flux (Fig. 2). They may trigger to the recession of the southern glacier boundary towards more south in SO. During pre-Holocene deglaciation period, the episodes of glacier meltwater flux corresponds to the event of ACR. This time, the lake L-51 have received maximum meltwater derived from ice cover and hence increased the deposition of coarse sediments. The Holocene climate optimum (11.2–8.4 kyr) exhibits a mixed signals of aeolian deposition and surroundings snow meltwater flux as well as ice sheet meltwater flux. Later, from 7.2 kyr BP to Recent core site received the majority of the sediments mainly because of surrounding snow meltwater flux along with aeolian deposition rather than ice sheet meltwater flux.

6 Conclusions and Recommendations

This chapter deals with review work by many Institute/Universities research workers in the Schirmacher Oasis over the period or the initiation of the Indian Expedition to East Antarctica (Schirmacher Oasis). The geographical location, the geology of the area, and paleoclimatic work show the research work's progression in the Schirmacher Oasis. The main emphasis is on the high-resolution paleoclimate reconstruction based on the lake sediments and the multiproxy dataset. But, the work on biological proxies such as Diatom or any other biota is less explored. The relation between the biological component with geochemistry requires more attention in the SO.

Furthermore, the radiocarbon dating in the lake sediment core is required more attention as the reservoir ages have to be calculated based on other dating techniques. One of the case studies shows 640–2200 years; radiocarbon ages may be used as a reservoir age correction within the present lakes system and used to interpret the paleoclimatic/paleoenvironmental conditions.

Acknowledgements PG and AM are indebted to the Director BSIP for her support and permission for the present chapter. We are grateful to the National Centre for Polar and Ocean Research for providing logistics and financial support during the three Indian Scientific Antarctic Expedition. The assistance by summer Indian Scientific Antarctic Expedition members to retrieve the sediment core is gratefully acknowledged. This is the BSIP publication number 40/2020-21.

References

Asthana R, Beg MJ, Swain AK, Dharwadkar A, Roy SK, Srivastava HB (2013) Sedimentary processes in two different polar periglacial environments: examples from Schirmacher Oasis and Larsemann Hills. East Antarct Geol Soc Lond Spec Publ 381:411–427

Asthana R, Chaturvedi A (1998) The grain-size behaviour and morphoscopy of supraglacial sediments, south of Schirmacher Oasis, East Antarct J Geol Soc India 52(5):557–568

Bargagli R (2005) Antarctic ecosystems. Ecological studies series 175, pp 1–395. Springer, Berlin

Bentley MJ, Hodgson DA, Smith JA, Cofaigh CO, Domack EW, Larter RD, Roberts SJ, Brachfeld S, Leventer A, Hjort C, Hillenbrand CD (2009) Mechanisms of Holocene palaeoenvironmental change in the Antarctic Peninsula region. The Holocene 19(1):51–69

Bentley MJ, Hodgson DA, Sugden DE. Roberts SJ, Smith JA, Leng MJ, Bryant C (2005) Early Holocene retreat of the George VI ice shelf, Antarctic Peninsula. Geology 33:173e176

Bera SK (2004) Late Holocene Palaeo winds and climatic changes in Eastern Antarctica as indicated by long-distance transported pollen spores and local microbiota in polar lake core sediments. CurrSci 86:1485–1488

Blunier T, Schwander J, Stauffer B, Stocker T, Dällenbach A, Indermühle A, Tschumi J, Chappellaz J, Raynaud D, Barnola JM (1997) Timing of the Antarctic cold reversal and the atmospheric CO2 increase with respect to the younger Dryas event. Geophys Res Lett 24(21):2683–2686

Blunier T, Spahni R, Barnola JM, Chappellaz J, Loulergue L, Schwander J (2007) Synchronization of ice core records via atmospheric gases. Clim past 3:325–330

Bormann P et al (1986) Structure and development of the passive continental margin across the Princess Astrid Coast, East Antarctica. J Geodyn 6(1–4):347–373

Bromwich D, Zhong L, Rogers AN (1998) Winter atmospheric forcing of the Ross Sea polynya. In: Jacobs SS, Weiss RF (eds) Ocean, ice and atmosphere, interactions at the Antarctic continental margin, vol 75, pp 101–133. American Geophysical Union, Antarctic Research Series

Chaturvedi A, Beg MJ, Keshavprasad AV (2004) Monthly patterns of advance and retreat of Dakshin Gangotri glacier snout in Schirmacher range. In: Nineteenth Indian expedition to Antarctica, Scientific report, DOD, Tech Publ No 17, pp 21–31

Choudhary S, Tiwari AK, Nayak GN, Bejugam P (2018) Sedimentological and geochemical investigations to understand source of sediments and processes of recent past in Schirmacher Oasis, East Antarctica. Polar Sci 15:87–98

CLIMAP Project Members (1981) Seasonal reconstructions of the Earth's surface at the last glacial maximum. Geol Soc Am Chart Ser MC-36

Dayal AM, Hussain SM (1997) Rb-Sr ages of lamprophyre dykes from Schirmacher Oasis, Queen Maud Land, East Antarctica. J Geol Soc India 50(4):457–460

Doran PT, Berger GW, Lyons WB, Wharton RA, Davisson ML, Southon J, Dibb JE (1999) Dating quaternary lacustrine sediments in the McMurdo dry valleys Antarctica. Palaeogeogr Palaeoclimatol Palaeoecol 147(3):223–239

Fogwill CJ, Turney CSM, Golledge NR, Etheridge DM, Rubino M, Thornton DP, Baker A, Woodward J, Winter K, Van Ommen TD, Moy AD (2017) Antarctic ice sheet discharge driven by atmosphere-ocean feedbacks at the Last Glacial Termination. Sci Rep 7:39979

Ghil M, Allen MR, Dettinger MD, Ide K, Kondrashov D, Mann ME, Robertson AW, Saunders A, Tian Y, Varadi F, Yiou P (2002) Advanced spectral methods for climatic time series. Rev Geophys 40(1)

Golledge NR, Menviel L, Carter L, Fogwill CJ, England MH, Cortese G, Levy RH (2014) Antarctic contribution to meltwater pulse 1A from reduced Southern Ocean overturning. Nat Commun 5:5107

Govil P, Mazumder A, Asthana R, Tiwari A, Mishra R (2016) Holocene climate variability from the lake sediment core in Schirmacher Oasis region, East Antarctica: multiproxy approach. Quat Int 425:453–463

Govil P, Mazumder A, Ram R, Singh DS, Azharuddin S (2018) Meltwater flux and climate change record of last 18.5 ka from Schirmacher Oasis East Antarctica. Polar Sci 18:135–141

Grew ES, Manton WI (1983) Geochronologic studies in East Antarctica: reconnaissance uranium/thorium/lead data from rocks in the Schirmacher Hills and Mount Stinear. Geology 77:1405–1418
Hart JK (2006) An investigation of subglacial processes at the microscale from Briksdalsbreen Norway. Sedimentology 53(1):125–146
Helland PE, Holmes MA (1997) Surface textural analysis of quartz sand grains from ODP Site 918 off the southeast coast of Greenland suggests glaciation of southern Greenland at 11 Ma. Palaeogeogr Palaeoclimatol Palaeoecol 135:1–4; 109–121
Hoch M, Tobschall HJ (1998) Minettes from the Schirmacher Oasis, East Antarctica—indicators of an enriched mantle source. Antarct Sci 10:476–486
Hodgson DA, Convey P (2005) A 7000-year record of oribatid mite communities on a maritime-Antarctic island: responses to climate change. Arct Antarct Alp Res 37(2):239–245
Hughes TJ (1998) Ice sheets: New York, Oxford University Press, 343 p
Huybrechts P (1990) The Antarctic ice sheet during the last glacial-interglacial cycle: a three-dimensional experiment. Ann Glaciol 14:115–119
Ingole BS, Parulekar AH (1993) Limnology of freshwater lakes of Schirmacher Oasis, East Antarctica
Jafri S (1997) Calc-alkaline Lamprophyres from Schirmacher Oasis, Queen Maud Land, East Antarctica. In: Thirteenth Indian Expedition to Antarctica, Scientific report, Department of Ocean Development, Technical Publication No 11, pp 227–236
Johnsen SJ, Clausen HB, Dansgaard W, Fuhrer K, Gundestrup N, Hammer CU, Iversen P, Jouzel J, Stauffer B (1992) Irregular glacial interstadials recorded in a new Greenland ice core. Nature 359(6393):311–313
Johnsen SJ, Clausen HB, Dansgaard W, Fuhrer K, Gundestrup N, Hammer CU, Iversen P, Jouzel J, Stauffer B, Steffensen J-P (1992) Irregular glacial interstadials recorded in a new Greenland ice core. Nature 359:311–313
Jones BF, Bowser CJ (1978) The mineralogy and related chemistry of lake sediments. In: Lakes, pp 179–235.Springer, New York, NY
Joshi MS (1997) Delineation of tectonic features of Schumacher Oasis, East Antarctica using total field geomagnetic profiling. In: Thirteenth Indian Expedition to Antarctica, Scientific eeport, Department of Ocean Development, Technical Publication Scientific Report 11, pp 237–244
Jouzel J, Raisbeck G, Benoist JP, Yiou F, Lorius C, Raynaud D, Petit JR, Barkov NI, Korotkevitch YS, Kotlyakov VM (1989) A comparison of deep Antarctic ice cores and their implications for climate between 65,000 and 15,000 years ago. Quatern Res 31(2):135–150
Jouzel J, Masson-Delmotte V, Cattani O, Dreyfus G, Falourd S, Hoffmann G, Minster B, Nouet J, Barnola JM, Chappellaz J, Fischer H (2007) Orbital and millennial Antarctic climate variability over the past 800,000 years. Science 317(5839):793–796
Kampf H, Stackebrandt W (1985) Crustal evolution of the Eastern Antarctic Craton (Schirmacher Oasis, Dronning-Maud Land). GerlandsBeitragezurGeophysik 94(4–6):251–258
Krinsley DH, Doornkamp JC (1973) Atlas of Quartz Sand Surface Textures. Cambridge University Press, 91 pp
Liu J, Milne GA, Kopp RE, Clark PU, Shennan I (2016) Sea-level constraints on the amplitude and source distribution of Meltwater Pulse 1A. Nat Geosci 9(2):130
Mahaney WC, Kalm V (1995) Scanning electron microscopy of Pleistocene tills in Estonia. Boreas 24:13–29
Mahaney WC et al (1996) Geochemistry and clay mineralogy of termite mound soil and the role of geophagy in chimpanzees of the Mahale Mountains, Tanzania. Primates 37(2):121–134
Mahaney WC (2002) Atlas of Sand Grain Surface Textures and Applications. Oxford University Press, 237 pp
Mahesh BS, Warrier AK, Mohan R, Tiwari M, Babu A, Chandran A, Asthana R, Ravindra R (2015) Long Lake sediments to Antarctic climate: a perspective gained from sedimentary organic geochemistry and particle size analysis. Polar Sci 9(4):359–367

Mahesh BS, Warrier AK, Rahul M, Tiwari M, Babu A, Chandran A, Asthana R, Ravindra R (2015) Response of Long Lake sediments to Antarctic climate: a perspective gained from sedimentary organic geochemistry and particle size analysis. Polar Sci 9:359–367
Massom RA, Stammerjohn SE (2010) Antarctic sea ice change and variability–physical and ecological implications. Polar Sci 4(2):149–186
Masson V, Vimeux F, Jouzel J, Morgan V, Delmotte M, Ciais P, Hammer C, Johnsen S, Lipenkov VY, Mosley-Thompson E, Petit JR (2000) Holocene climate variability in Antarctica based on 11 ice-core isotopic records. Quatern Res 54(3):348–358
Mathur AK, Asthana R, Ravindra R (2006) Arcellaceans (thecamoebians) from core sediments of Priyadarshini Lake, Schirmacher Oasis. East Antarct Curr Sci 90:1603–1605
Mathur AK, Mishra VP, Singh J (2009) Study of quartz grain surface texture by electron microscopy–a tool in evaluating palaeoglacial sediments in Uttarakhand. Curr Sci 1377–1382
Mayewski PA et al (2009) State of the Antarctic and Southern Ocean climate system. Rev Geophys 47:1
Mazumder A, Govil P, Kar R, Gayathri NM (2017) Paleoenvironments of a proglacial lake in Schirmacher Oasis, East Antarctica: Insights from quartz grain microtextures. Polish Polar Res 38(1)
Menviel L, Timmermann A, Timm OE, Mouchet A (2010) Climate and biogeochemical response to rapid melting of the West Antarctic Ice Sheet during interglacials and implications for future climate. Paleoceanography 25(4)
Murthy DN, Veeraswamy K, Harinarayana T, Singh UK (2012) Electrical structure beneath the Schirmacher Oasis, East Antarctica, from Magnetotelluric measurements. In: Asthana R (ed) Twenty fourth Indian expedition to Antarctica. Technical Publication No 22, pp 207–226. National Institute of Science Communication and Information Resources, Council of Scientific and Industrial Research, New Delhi
Nielsen S (2004) Deglacial and Holocene Southern Ocean climate variability from diatom stratigraphy and paleoecology. PhD thesis, University of Tromsø
Parrenin F, Barnola JM, Beer J, Blunier T, Castellano E, Chappellaz J, Dreyfus G, Fischer H, Fujita S, Jouzel J, Kawamura K (2007) The EDC3 chronology for the EPICA Dome C ice core. Clim past 3:485–497
Pedro JB, Bostock HC, Bitz CM, He F, Vandergoes MJ, Steig EJ, Chase BM, Krause CE, Rasmussen SO, Markle BR, Cortese G (2016) The spatial extent and dynamics of the Antarctic Cold Reversal. Nat Geosci 9(1):51–55
Pedro JB, Rasmussen SO, van Ommen TD, Morgan VI, Chappellaz J, Moy AD, Masson-Delmotte V, Delmotte M (2011) The last deglaciation: timing the bipolar seesaw. Clim Past 7
Phartiyal B (2014) Holocene paleoclimatic variation in the Schirmacher Oasis, East Antarctica: a mineral magnetic approach. Polar Sci 8:357–369
Phartiyal B, Sharma A, Bera SK (2011) Glacial lakes and geomorphological evolution of Schirmacher Oasis, East Antarctica, during late quaternary. Quatern Int 235(1–2):128–136
Ravich MG, Krylov AJ (1964) Absolute ages of rocks from East Antarctica. Antarct Geol 579–590
Ravindra R, Pandit MK (2000) Geochemistry and geochronology of A-type granite from northern Humboldt Mountain, East Antarctica: a vestige of Pan-African event. J Geol Soc India 56(3):253–262
Sangewar CV (1987) Some geological and glaciological observations during reconnaissance of the terrain South of Dakshin Gangotri Station, Antarctica. In: Fourth Indian Expedition to Antarctica, Department of Ocean Development, Technical Publication 4, pp 99–104
Scheinert M (2001) Geodynamic Investigations in Dronning Maud Land/Antarctica. JoumeesLuxembourgeoises De Geodynamique 89:12–14
Sengupta S (1988) Precambrian rocks of the Schirmacher range, east Antarctica. ZietschriftfürGeologieWissenschaften 16:647e660
Sharma C, Bera SK, Upreti DK (2002) Modern pollen-spore rain in Schirmacher Oasis. East Antarct CurrSci 82:88–91

Sharma C, Chauhan MS, Sinha R (2007) Studies on Holocene climatic changes from Priyadarshini Lake sediments East Antarctica: the Palynological Evidence. J Geol Soc India 69:92–96
Sharma C, Chauhan MS, Sinha R (2007) Studies on Holocene climatic changes from Priyadarshini Lake sediments, East Antarctica: the palynological evidence . J Geol Soc India 69:92–96
Sharma C, Chauhan MS, Bera SK, Sinha R, Upreti DK (2000) Early Holocene sedimentological and palynological studies from Lake Priyadarshini, Eastern Antarctica. In: Volume of Abstract, 10th International Palynological Congress, Nanjing, China, pp 151–152
Sheraton JW, England RN (1980) Highly potassic mafic dykes from Antarctica. J Geol Soc Aust 27(1–2):129–135
Shiraishi K, Kanisawa S, Ishikawa K (1988) Geochemistry of post-orogenic mafic dike rocks from the eastern Queen Maud Land, East Antarctica
Shrivastava PK, Asthana R, Roy SK, Swain AK, Dharwadkar A (2012) Provenance and depositional environment of epishelf lake sediment from Schirmacher Oasis, East Antarctica, vis-a-vis scanning electron microscopy of quartz grain, size distribution and chemical parameters. Polar Sci 6:165–182
Siani G, Michel E, De Pol-Holz R, DeVries T, Lamy F, Carel M, Isguder G, Dewilde F, Lourantou A (2013) Carbon isotope records reveal precise timing of enhanced Southern Ocean upwelling during the last deglaciation. Nat Commun 4(1):1–9
Siddiquie HN, GopalaRao D, Ramana MV, (, (1988) Geology and structure of the Astrid Ridge, Dronning Maud Land Antarctica. Paleoceanography 3(5):583–599
Singh SM, Ochyra R, Pednekar SM, Asthana R, Ravindra R (2012) A Holocene moss species preserved in lake sediment core and the present moss diversity at Schirmacher Oasis Antarctica. Antarct Sci Inst Subscr 24(4):353
Sinha R, Chatterjee A (2000) Lacustrine sedimentology in the Schirmacher Range Area, East Antarctica . J Geol Soc India 56:39–45
Sinha R, Sharma C, Chauhan MS (2000a) Sedimentological and pollen studies of Lake Priyadarshini. East Antarct Palaeobot 49:1–8
Sinha R, Navada SV, Chatterjee A, Kumar S, Mitra A, Nair AR (2000b). Hydrogen and oxygen isotopic analysis of Antarctic lake waters. Curr Sci 992–995
Srivastava AK, Randive KR, Khare N (2013) Mineralogical and geochemical studies of glacial sediments from Schirmacher Oasis, East Antarctica. Quatern Int 292:205–216
Srivastava AK, Ingle PS, Khare N (2018) Controlling factor for nature, pattern and accumulation of the glacial sediments of Schirmacher Oasis, East Antarctica: comments on paleoclimatic condition. Polar Sci 18:113–122
Srivastava AK, Khare N, Ingle PS (2011) Characterization of clay minerals in Schirmacher Oasis's sediments, East Antarctica: their origin and climatological implications. Curr Sci 363–372
Stanley S, Dedeckker P (2002) A Holocene record of allochthonous, aeolian mineral grains in an Australian alpine lake; implications for the history of climate change in southeastern Australia. J Paleolimnol 27:207–219
Stuiver M, Grootes PM, Braziunas TF (1995) The GISP2 δ18O climate record of the past 16,500 years and the sun, ocean, and volcanoes' role. Quat Res 44(3):341–354
Turner J (2009) Non-annular atmospheric circulation change induced by stratospheric ozone depletion and its role in the recent increase of Antarctic sea ice extent. Geophys Res Lett 36(8)
Vaikmae R, Hermichen W, Kowski P, Strauch G, Savatyugin L (1991) Glaam-Ocean-Aoncaphmbumctions. In: Proceedings of the International Symposium held at St Petersburg, September 1990. IAHS Publ no 208, 1991. Deciphering recent structures and Holocene evolution of the marginal East Antarctic ice cover in Queen Maud Land. In Glaciers--ocean-atmosphere interactions: proceedings of the International Symposium held at St Petersburg, 24–29 September 1990 (No. 208, p. 3). IAHS Press, Institute of Hydrology
Verleyen E, Hodgson DA, Sabbe K, Cremer H, Emslie SD, Gibson J, Hall B, Imura S, Kudoh S, Marshall GJ, McMinn A (2011) Post-glacial regional climate variability along the East Antarctic coastal margin—evidence from shallow marine and coastal terrestrial records. Earth Sci Rev 104(4):199–212

Wagner B, Cremer H, Hultzsch N, Gore DB, Melles M (2004) Late Pleistocene and Holocene history of lake terrasovoje, Amery Oasis, East Antarctica, and its climatic and environmental implications. J Paleolimnol 32(4):321–339
Warrier AK, Mahesh BS, Mohan R, Shankar R, Asthana R, Ravindra R (2014) Glacial-interglacial climatic variations at the Schirmacher Oasis, East Antarctica: the first report from environmental magnetism. Palaeogeogr Palaeoclimatol Palaeoecol 412:249–260
Warrier AK, Pednekar H, Mahesh BS, Mohan R, Gazi S (2015) Sediment grain size and surface textural observations of quartz grains in late quaternary lacustrine sediments from Schirmacher Oasis, East Antarctica: paleoenvironmental significance. Polar Sci 10:89–100
Warrier AK, Pednekar H, Mahesh BS, Rahul M, Gazi S (2016) Sediment grain size and surface textural observations of quartz grains in late quaternary lacustrine sediments from Schirmacher Oasis, East Antarctica: paleoenvironmental significance. Polar Sci 10:89–100
Weber ME, Clark PU, Kuhn G, Timmermann A, Sprenk D, Gladstone R, Zhang X, Lohmann G, Menviel L, Chikamoto MO, Friedrich T (2014) Millennial-scale variability in Antarctic ice-sheet discharge during the last deglaciation. Nature 510(7503):134
Whalley WB, Krinsley DH (1974) A scanning electron microscope study of surface textures of quartz grains from glacial environments. Sedimentology 21(1):87–105

A Synthesis of Glacial-Interglacial Paleoenvironmental Records from Lake Sediments of Schirmacher Oasis, East Antarctica

Anish Kumar Warrier, B. S. Mahesh, Joju George Sebastian, A. S. Yamuna Sali, and Rahul Mohan

Abstract Antarctica plays a significant role in regulating the global climate, mainly due to its geographic position, characterized by the freezing climate and high albedo. Antarctica attributes this attribute amid several paleoclimatic questions ranging from global warming to Antarctic ice sheet melting associated with sea-level rise. Thus, reconstructing the Antarctic past-climate is of prime importance in understanding and modelling future climatic changes. Many lakes in Antarctica remained free from the continental ice sheet's influence during the last glacial maxima. Hence, their sedimentary archives are a repository of paleoclimatic evidence for the Late Quaternary. Paleoclimatic studies using lake sediments drew scientific attention due to their efficiency to record long, high-resolution climate records. Recent studies have employed multiple proxies like environmental magnetism, isotope geochemistry, petrography, sedimentology, and geochronology on lake sediments of Schirmacher Oasis to decipher the past climate and the prevailing ecological conditions. The existing studies poorly record climatic events such as the Mid-Holocene Hypsithermal and neoglacial cooling. Despite better chronometric control in these studies, coarse temporal resolution and sparsely documented finer-scale climatic variations place the need for future high resolution works in the East Antarctic region.

Keywords Lake sediments · Environment · Organic matter · Magnetic minerals · Late quaternary · East Antarctica

---

A. K. Warrier (✉) · J. G. Sebastian · A. S. Y. Sali
Centre for Climate Studies / Department of Civil Engineering (Manipal Institute of Technology), Manipal Academy of Higher Education, Manipal, Karnataka 576104, India
e-mail: anish.warrier@manipal.edu

B. S. Mahesh · R. Mohan
National Centre for Polar and Ocean Research, Ministry of Earth Sciences, Headland Sada, Vasco-da-Gama, Goa 403804, India

N. Khare (ed.), *Assessing the Antarctic Environment from a Climate Change Perspective*, Earth and Environmental Sciences Library,
https://doi.org/10.1007/978-3-030-87078-2_8

1 Introduction

Antarctica's lakes, which dote the ice-free regions along the continental margin, are significant archives of past-climate records and the best source to monitor changes in modern climate that respond to the increasing global temperature. The Polar Regions are the first respondents of climate change. Due to their limited spatial entity, the lakes respond to subtle changes in the global weather system and act as good thermometers to measure human-induced climate change.

During recent decades, the global mean temperatures around the world have registered an increase. During the period January–October 2019, the global mean temperature was around 1.1 ± 0.1 °C above pre-industrial levels (1850–1900) (WMO Provisional Statement of the State of the Climate 2019). To forecast the future climatic scenario, we must have a fair understanding of the modern-day weather and study the past climate. This way, one can explore the interconnections that exist between these changes. The Southern Hemisphere, especially the Antarctic continent, plays a vital role in modulating the global climate during the present and the past. Antarctica is particularly crucial to the worldwide climate system due to its high albedo, low thermal conductivity, and contribution to deep ocean water circulation. Most of the global climatic signatures are also found pronounced in this continent due to its pristine conditions. Many important questions regarding the paleoclimate and environment of Antarctica are yet to be answered even though research in this area has flourished well in the last couple of decades (Verleyen et al. 2011).

A novel finding that many existing ice-free regions of Antarctica escaped the last glaciation (Gore et al. 2001) awaken paleoclimatologists' enthusiasm. These ice-free regions have numerous pristine lakes, the repository to sedimentary archives that offer high-resolution climate records and environmental changes. The Antarctic summer in these ice-free regions is generally warmer (above freezing point). The lakes become ice-free, responding well to local and regional climatic changes. These changes are preserved as additional biochemical and physical evidence (Govil et al. 2016). A varied number of techniques are used in past-climate reconstruction using lake sediments, viz. environmental magnetism, organic and inorganic geochemistry, quartz petrography, sedimentology, geochronology.

Schirmacher Oasis, one of the many ice-free regions, is a 25 km long and up to 3 km wide ice-free region on the Princess Astrid Coast in Queen Maud Land in East Antarctica is on average 100 m above sea level (Fig. 1). It has close to 118 lakes (proglacial, epi-shelf and periglacial lakes; Ravindra 2001). The majority of the lakes' surface remain ice-free during austral summer, and the ice cover rarely exceeds 2 m thickness (Hermichen et al. 1985). The lakes receive melt-water contributed by the snow and continental ice sheet. The sediments deposit into these lakes via fluvioglacial, in-situ productivity and aeolian activities ((Warrier et al. 2021a, b; Govil et al. 2016). Several of these lakes are found in rocks eroded by glaciers, and some are closed by moraines or ice (Bormann and Fritzsche 1995). The plant life of Schirmacher Oasis is limited to lichens and mosses, which thrive in rocky soils.

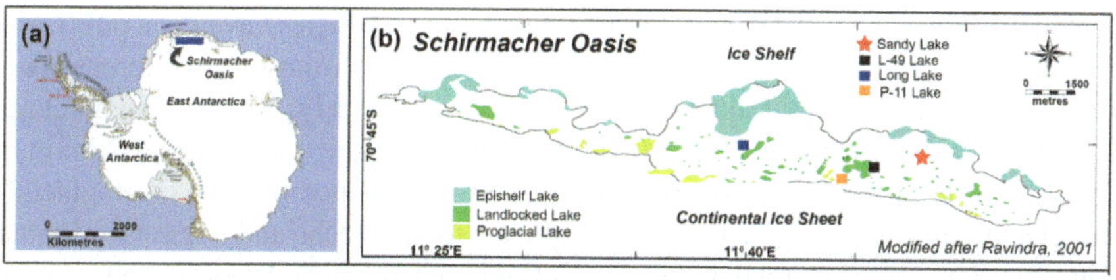

Fig. 1 **a** General map of Antarctica showing the location of Schirmacher Oasis. **b** Map of Schirmacher Oasis showing the distribution of epi-shelf lakes, periglacial (land-locked) lakes and proglacial lakes (modified after Ravindra 2001)

Over the last couple of decades, several paleoclimate records have been reconstructed using lacustrine sedimentary archives from various Antarctic lakes to understand local and regional climate. These records were often correlated with global climatic events and probed for their linkages and their processes. This chapter summarizes a review of past-climate records reconstructed from sediment cores from Schirmacher Oasis using different proxies.

2 Organic Geochemistry and Variations in Productivity

The sedimentary organic matter (OM) deposited in lakes acts as a source for elemental analysis (e.g. carbon, nitrogen, phosphorous) and stable isotopes ($\delta^{13}C$ and $\delta^{15}N$) along with C/N ratios to understand the provenance of organic matter (Meyers 2003; Talbot 2001). These multi-proxy records help understand long-term climatic changes and evolution during the last glacial stage through deglaciation and Holocene. This approach will also help minimize bias in interpreting the measured data and ascertain OM's source and possible variations concerning climate-induced environmental changes.

The type and amount of sedimentary OM can be used to reflect past fluctuations in lake productivity and organic matter source perhaps linked to climate-forced environmental changes (Leng and Marshall 2004; Meyers 1997; Talbot and Johannessen 1992). The utility of the C/N ratio is well-studied to identify the source of organic matter in sediments (Meyers 2003; Talbot 2001). The relative proportion of aquatic versus terrestrial OM where phytoplankton and aquatic macrophytes exhibit C/N ratio between 6 and 12 while terrestrial derived OM, between 14 and 20 (Meyers 1994, 2003; Meyers and Teranes 2001). The $\delta^{13}C$ and $\delta^{15}N$ of bulk sedimentary OM is a good indicator of past environmental changes in lacustrine systems (e.g. Talbot and Johannessen 1992; Engel and Macko 1993), which forges a basis for further studies. When used in conjunction, they (C_{org}, C/N, $\delta^{13}C$ and $\delta^{15}N$) provide information on past-changes in (a) productivity (e.g. Hodell and Schelske 1998), (b) nutrients (Meyers 1997, 2003), (c) environmental variation (e.g., Meyers 2003; Talbot and Johannessen 1992) and supply of organic material (Hedges and Keil 1995).

Records based on sedimentary organic geochemical proxies have been published (Mahesh et al. 2015, 2017, 2019) for Schirmacher Oasis from three different lakes viz., Sandy Lake, Long Lake and Zub Lake, respectively. The sediment cores (<1 m in length) span the last 36–43 kyr spanning the Last Glacial Maxima. These records provide overall productivity and provenance patterns over glacial-interglacial timescale and the lakes' response to the local and regional climate (Fig. 2). The organic matter in high-latitude (Antarctic) lakes widely varies from that of low-latitude lakes, wherein the former is dominated by aquatic macrophytes (autochthonous: cyanobacterial algal mats) and terrestrial bryophytes (allochthonous: lichens and moss). In the modern period, i.e., Holocene, algae flourish (diminish) during periods of ice-free (ice-cover) conditions, i.e., during Austral summer (winter) owing to sustained warm (cold) conditions. During the Holocene summer, the temperature is warm enough to melt the lake cover rendering them ice-free for 3–5 months. The melting facilitates increased productivity in the lake through algae production. However, productivity is hindered when the lake surface is covered with ice as there is no exchange of gases with the atmosphere and the limitation of sunlight. The sedimentary organic matter in Schirmacher Oasis shows, in general, less than 1% during the glacial stage, indicating prolonged and intense winter and mild summer, which would have led to the existence of ice-cover for a more extended period. This would have hindered any exchange of gases between the lake and the atmosphere limiting sunlight and extremely low productivity within the lake. The ice-cover would also have restricted the influx of sediments either from melt-water or wind-blown.

Interestingly, SO does not record higher productivity at the beginning of Antarctic deglaciation (~ 17 cal ka BP: Petit et al. 1999). The ice-cover conditions continued

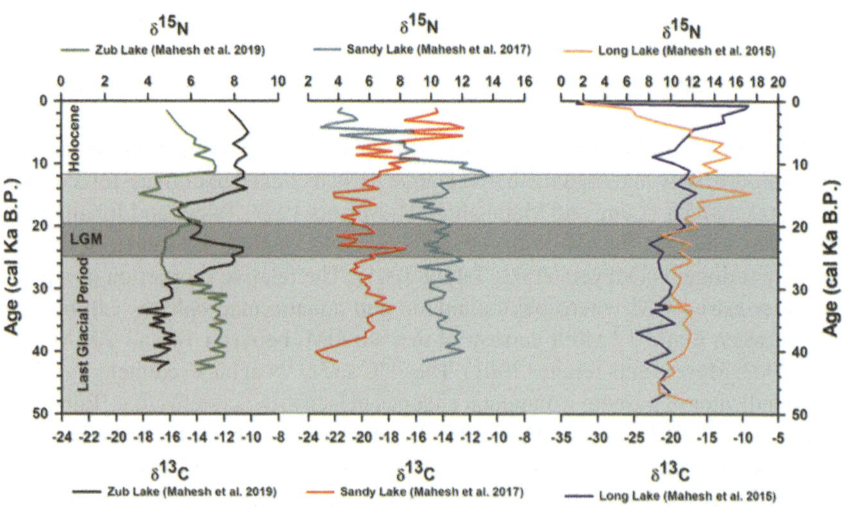

Fig. 2 A comparison of down-core variations of $\delta^{13}C$ and $\delta^{15}N$ for Zub (L-49) Lake, Sandy Lake and Long Lake sediment cores from Schirmacher Oasis

to persist through deglaciation in all three lakes indicating cold summer conditions, i.e., sub-zero conditions. The $\delta^{13}C$ values for non-marine aquatic plants and algae are between −26‰ to −12‰ (Fry and Sherr 1989; Farquhar et al. 1989), while for terrestrial plants, it varies between −24 and −32‰ (Trumbore and Druffel 1995). The $\delta^{15}N$ values of terrestrial plants which use atmospheric N_2 are lower (0‰) as compared to the nitrate incorporating algae ($\delta^{15}N$ values of 7–10‰: Peters et al. 1978).

However, the Holocene's productivity in these three lakes widely varies even though they are spatially within the same oasis. The Long Lake records an increase in C_{org}% (1.2%) only in the core-top (0–3 cm). In comparison, Sandy Lake shows an increase in C_{org}% at the beginning of the Holocene. The Zub Lake offers increasing productivity (C_{org}%: 1–7%) beginning at 18 cal ka BP and attains Holocene optimum conditions at ~11 cal ka BP. Such contrasting variation in C_{org}% suggests the response of these lakes to climate. Under current conditions, it can be noted that the Sandy Lake and Long Lake (periglacial lakes) are located significantly farther from the continental ice sheet and are very unlikely to receive melt-water from the ice-sheet. At the same time, Zub Lake is connected with numerous lakes along the continental ice-sheet edge and gets a significantly large amount of melt-water from the other lakes fed by the melting ice-sheets. Hence, factors such as the lake's proximity to the ice-sheet or the ice-shelf play a significant role in the lake ecosystem. The nitrogen content (N%) also shows similar variations to that of the C_{org}%, suggesting no selective loss of both carbon and nitrogen with time (Talbot and Johannessen 1992).

The C/N ratios (<10) for all the three sedimentary records indicate that the source of organic matter during the glacial stage is autochthonous. In contrast, the Holocene records higher C/N ratios (>10), suggesting an admixture of aquatic macrophytes and terrestrial bryophytes. The input of terrestrial organic matter would be possible when the lakes are ice-free. The information on terrestrial organic matter from the catchment area is aided by fluvial input from the snow and ice-sheet melt-water. Higher C/N ratios throughout the Holocene suggests steady warmer conditions in the Schirmacher Oasis. The provenance of the organic matter can be further understood by studying the bi-plots of C/N rates vs $\delta^{13}C$ values (Mahesh et al. 2015, 2017, 2019). The bi-plots from these three records distinctly indicates that the organic matter is sourced from both aquatic macrophytes (bacteria and freshwater algae: autochthonous) and terrestrial bryophytes (lichens and moss: allochthonous).

The down-core $\delta^{13}C$ values in these lakes vary between −10 and −24‰ (Fig. 2). The glacial stage records depleted values (~−16 ± 4‰). This is most likely due to the utilization of CO_2 by aquatic organisms resulting in a deposition of ^{13}C-poor organic matter in the sediments (Meyers and Terranes 2001). The terrestrial input matter can also result in depleted values, but this is very unlikely as the $\delta^{15}N$ values have recorded enriched values indicating the predominance of aquatic organisms during the same period. Low productivity under consistently ice-covered lake surface during the cold glacial period is documented in these lakes. Enriched values during the Holocene indicates a shift in the balance of supply/demand on DIC with possible enhanced contribution from the terrestrial OM and hence suggest a dominance of in-situ productivity.

The $\delta^{15}N$ values are enriched during the glacial stage for the lakes in Schirmacher Oasis. Prolonged enrichment of $\delta^{15}N$ values during the glacial phase suggest that the lake was ice-covered, limiting lake-to-atmosphere exchange of gases leading to low productivity. Such common productivity conditions have been recorded in sub-glacial lakes (Smith et al. 2006; Hodgson et al. 2009). Sustained enrichment of $\delta^{15}N$ values generally reflects ammonification and denitrification/nitrification, which occurs under ice-cover and mineralization of organic matter by sedimentary and suspended bacteria (Talbot 2001; Meyers and Teranes 2001). The $\delta^{15}N$ values exhibit depleted values during the Holocene. This shift in $\delta^{15}N$ values from enriched (glacial stage) to exhausted (Holocene) values in the sedimentary record suggest a change of lake environment from an oxygen-depleted (ice-cover) to an oxygen-rich (ice-free) condition. In summary, the sedimentary organic proxies (C_{org}%, N%, $\delta^{13}C$, $\delta^{15}N$ and C/N ratios) have been well utilized to understand the climate trends by reconstructing the productivity trends provenance in the lake ecosystem.

3 Grain Size Variation and Climate Change

The input of terrestrial derived erosional products to the lake system helps understand the transferring agent's energy, which depends on the climatic conditions that prevail in Antarctica. The input of erosional products is predominant in the austral summer (Simmons et al. 1986), indicating the significant role of melt-water in transporting to the lake (Mora et al. 1994; Retelle and Child 1996; Spaulding et al. 1997). Westerlies-borne aeolian dust from mid-latitude deserts also contributes to erosional products to the lakes (Sugden et al. 2009; Petit et al. 1990) along with dust deposited by the wind from the surrounding area. The glacier action also contributes to the lake sedimentary deposits (Squyres et al. 1991; Hendy et al. 2000).

The sand-clay-silt content measured for the three lakes provides an overview of the depositional pattern of terrigenous sediments in Schirmacher Oasis over the glacial-interglacial stage. The down-core variations for the three lakes show sand dominance (Fig. 3), followed by silt and clay. The content of sand is lower during the glacial stage, while silt content is higher. The increased contribution of mud to the lake during the glacial phase suggest enhanced wind delivered detritus to the lake. During the warm Holocene period, more increased sedimentation is primarily due to enhanced sediment load delivery to the lake through increased frozen/snow melt-water to an open lake. Warmer summer resulting in higher melt-water (high energy) would bring in coarser sediments into the lake. In comparison, a milder summer would lead to low melt-water (low power) with deposition of clayey particles.

Even though the sediment records' resolution is low (200–600 years), major climatic events such as Antarctic Cold Reversal, Antarctic Warming Event and Antarctic Isotope Maxima were recorded in the down-core variations of the proxy records. This suggests that the climate in Schirmacher Oasis responded well to the

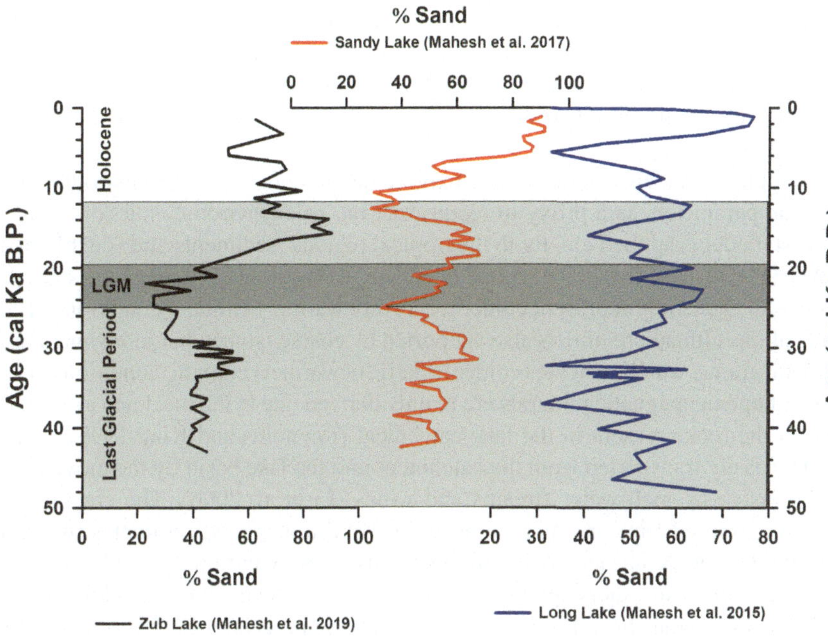

Fig. 3 Down-core variations of sand content in the sediment cores of Zub (L-49) Lake, Sandy Lake and Long Lake, Schirmacher Oasis

general Antarctic climate. The beginning of deglaciation and attainment of Holocene optimum is inconsistent between the lake records suggesting that regional factors such as the geomorphology plays an essential role in influencing the lake ecosystem.

4 Magnetic Mineral Records of Glacial-Interglacial Climatic Changes

Iron oxide minerals are omnipresent and can be effectively used to reconstruct past variations in the environment (Evans and Heller 2003). The environmental magnetic method deals with the systematic study of concentration, grain size and mineralogy of magnetic minerals present in soils and sediments (Walden et al. 1999; Thompson and Oldfield 1986). These minerals, which may be formed, carried away and deposited in different deposition basins, can be accurately studied using environmental magnetism. These iron oxides exhibit specific properties under a particular climatic phase, such as abundance, grain size, and mineralogy. However, as the climate changes, these properties also vary. Sediments deposited in the lakes and oceans record this signal through temporal variations in the concentration, grain size, and iron oxide carriers' mineralogy. Due to its several advantages (like rapid measurements, sensitive, inexpensive, non-destructive), an environmental magnetic technique has found applications in different facets of earth sciences like paleoclimate/paleoceanography (Sandeep et al. 2015), archaeology (Warrier et al. 2011), environmental pollution (Warrier et al. 2014a), soil genesis and erosion (Sandeep et al. 2012).

A couple of models can be used to explore the potential of using environmental magnetic parameters as a proxy to reconstruct the paleoenvironmental conditions. The first model is applied chiefly to the tropical regions' sediments and is explained by Warrier et al. (2014b). In the second model, high (low) values of magnetic susceptibility (X_{lf}) generally represent colder (relatively warmer) climatic conditions. The more callous climatic regime is also supported by coarse (somewhat more nuanced) magnetic grains, which indicates colder (relatively warmer) climatic conditions. The crude lithogenic magnetic minerals are mainly derived due to the mechanical weathering of the rocks present in the lake catchment (Reynolds and King 1995). They could be easily transported from the catchment into the lake basin by the movement of glaciers or by melt-water streams and winds (Li et al. 2006). The cracks that develop on the lake-ice are suitable hosts for wind-transported sediments that fall on to the lake ice's surface. When the ice melts, these wind-transported materials fall through the lake waters and get mixed with the lake sediments (Spaulding et al. 1997). A few studies have been made on the lake sediments of Schirmacher Oasis by using environmental magnetic techniques.

Warrier et al. (2014b) reconstructed the paleoenvironmental conditions in the Schirmacher Oasis on glacial-interglacial timescales (past 43,000 cal years BP) They studied the abundance, grain size, and mineral assemblage of the iron oxide minerals present in Sandy Lake's sediments and reconstructed periods of relatively warm and icy climatic conditions. During the glacial period, frost weathering gave rise to catchment-derived, primary, coarse grains of ferrimagnetic minerals later deposited within the lake sediments. These coarse-grained ferrimagnets' presence was evident by the increased values of X_{lf} (Fig. 4) and reduced X_{ARM}/X_{lf}. During the interglacial period, a relatively warm and wet climate accelerated the catchment rocks' chemical weathering, leading to enhanced soil formation. The environmental magnetic parameters showed an opposite trend when compared to the glacial period. From 42.5 to 11.86 cal ka BP., the glacial period was recorded within the Schirmacher Oasis, which alternated between exceedingly colder (40.78, 36.08, 34.51, 29.03, 28.02–21.45 cal ka BP) and relatively warmer climates (38.44–39.22, 33.73–29.81 and 28.52 cal ka BP; Fig. 4). Deglaciation commenced at about 20 cal Ka BP followed by the early Holocene optimum at around 12.55 cal Ka BP. In the Holocene, alternating warm (12.55–9.88, 4.21–2 cal ka BP) and cold events (9.21–4.21 cal ka BP and from 2 cal ka BP onwards) were also observed. Most of these out and warm phases showed a broad correlation with significant climatic events seen in lake-sediment records and ice-core records from different East Antarctica regions (Fig. 4). However, circumstances such as the Medieval Warm Period and the Little Ice Age were absent due to the core-top loss during coring operation. Phartiyal et al. (2011) reconstructed the climate history for the past ~13 ka of the Schirmacher Oasis based on the analysis of magnetic concentration, mineralogy and grain-size dependent parameters. The

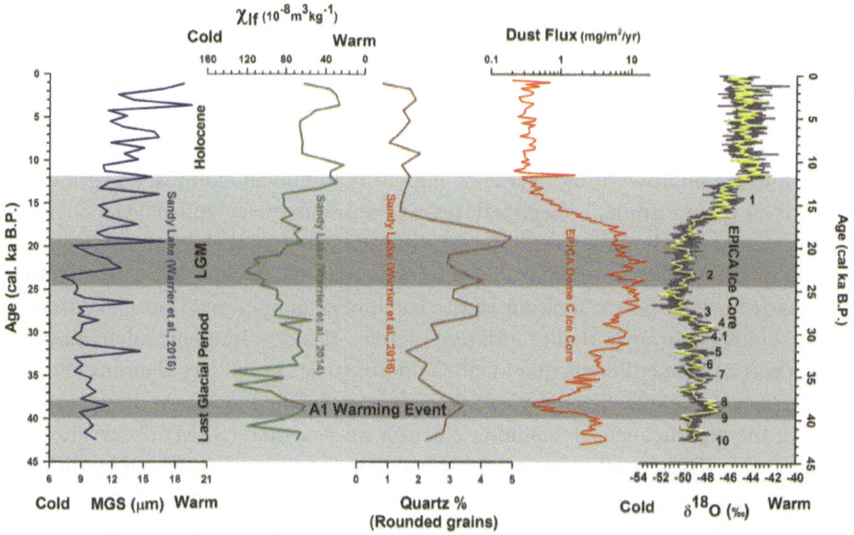

Fig. 4 A comparison of multi-proxy data like mean grain size, magnetic susceptibility, rounder quartz grains (%) of Sandy Lake with dust (Lambert et al.2012) and oxygen isotopic data (EPICA Community Members 2006) of EPICA ice core. The A1 event is one of the seven warming events during the past 90 cal ka BP documented from Antarctica (Blunier and Brook 2001)

study reported that colder periods recorded higher magnetic concentration-dependent values and relatively warmer periods show reduced values. Based on multivariate cluster analysis, the paleoclimate record (13–3 ka BP) of the Schirmacher Oasis was divided into six phases. The periods 12.5 ka BP (late Pleistocene), 11–8.7 ka BP (Holocene Optimum) and 4.4–3 ka BP (Mid-Holocene Hypsithermal) represented phases of warmer climatic conditions. Higher values of magnetic mineral concentrations (X_{lf}, SIRM) were observed for colder glacial periods and lower values for warmer interglacial periods (Phartiyal 2014). The studies show that environmental magnetic techniques can be successfully used to reconstruct the past variations in the climate and the associated changes in the lake catchment that bring about a difference in the magnetic concentration, grain size and mineralogy.

5 Petrographic Studies

Detrital minerals such as quartz, feldspar, mica, garnet have been successfully used to reconstruct the past climate in the polar regions as they are mainly derived from physical weathering (Pistolato et al. 2006; Chamley 1989). According to Bowen's reaction series, quartz (SiO_2) is the last mineral to crystallize at lower temperatures. Due to this, it is very stable and found in greater abundance in all the rock types. Due to its hardness (7 according to Moh's scale), it is a mechanically and chemically resistant mineral. Therefore, it has higher chances of getting preserved in the sediments (Krinsley and Doornkamp 1973; Mahaney 1995, 2002). Surface texture observations can also be made on the quartz grains that will reveal the type of transportation and deposition environment the mineral was subjected to in the previous sediment cycle. Surface microtextures, degree of angularity, chemical features, and grain-size analysis of quartz grains collectively reveal the sedimentary and physicochemical processes that acted on the grains during different geological phases.

The quartz grains present in the sediments may be transported by melt-water streams from the snow's melting in the catchment or may also be picked up by strong winds and transported to different regions. Besides, biogenic silica may also form a part of the sediments due to silica microfossils' burial like diatoms (Stanley and DeDeckker 2002). Several works have been published in recent decades, highlighting the significance of scanning electron microscopic (SEM) observations of quartz grains. These studies have established that quartz grains affected by different geological processes exhibit distinct grain surface features and micro-textural characteristics (Helland and Holmes 1997; Mahaney 1995; Strand et al. 2003). Although several studies have been made on quartz grains in Schirmacher Oasis, they are primarily on surface sediments and soils. Only a couple of studies have been reported from Schirmacher Oasis wherein the quartz grains deposited in the lake sediment cores have been used to perceive the modes of transportation, weathering processes, sediment maturity and its relation to regional climate (Warrier et al. 2016). Statistical parameters of particle size data indicated a fluvioglacial deposition of sediments.

Despite the coarse resolution of rounded quartz data, a significant correlation between the round quartz data and the high-resolution dust-flux data of EPICA ice-core suggested a peak in aeolian transport sediments during the Last Glacial Period (LGP) (Fig. 4). The low value of mean grain size of deposits from the LGP (colder climate) showed an increasing trend after the LGM corresponding to increased energy of transporting agent (melt-water streams) and temperature. The mean grain size of sediments showed peaks at 38 and 32 cal ka BP when warmer climatic conditions prevailed, which might have corresponded to the Antarctic Warming Event A1 (Fig. 4). The Holocene epoch was characterized by alternating periods of more or less warm and cold climate as evidenced by a cyclic variability of mean grain size of sediments. Mazumder et al. (2017) studied the quartz petrography and sedimentological analysis on a sediment core from a proglacial lake (P 11) in Schirmacher Oasis. The microtextural and morphological observations of quartz grains revealed a combination of the glacial, aeolian and fluvioglacial mode of transport in the study area. Three prominent climatic zones were identified from the study in the period spanning from 13.9 to 3.3 ka BP alternating between relatively warmer and colder phases.

6 Scope for Future Work

Several studies have been made on lake sediments from the Schirmacher Oasis to reconstruct the paleoclimate and the paleoenvironmental changes recorded in the region. However, only a few of them report the environmental variations on glacial-interglacial timescales. Many of these studies are based on a coarse temporal resolution and, as a result, do no record the finer scale climatic variations. Events like the Mid Holocene Hypsithermal and neoglacial cooling are poorly represented in the current studies. Future coring operations should be carried out to obtain the longest possible sediment cores to get a detailed paleoclimate record of the SO. Higher resolution studies will allow us to better constrain the small scale climatic fluctuations and correlate with other archives such as ice-core records that produce high-resolution climate data.

Acknowledgements We thank the Secretary-MoES and the Director, ESSO-NCPOR, for their encouragement and support under the project "Past Climate and Oceanic Variability". The authors thank the Logistics Division and members of the 28th Indian Scientific Expedition to Antarctica for their help. AKW acknowledges the financial support provided by the National Centre for Polar and Ocean Research, Ministry of Earth Sciences, in the form of a research project (Sanction: NCPOR/2019/PACER-POP/ES-02 dated 05/07/ 2019) under the PACER Outreach Programme (POP) initiative. This is NCPOR contribution no. B-3/2021-22.

References

Blunier T, Brook EJ (2001) Timing of millennial-scale climate change in Antarctica and Greenland during the last glacial period. Science 291:109–112

Bormann P, Fritzsche D (1995) The Schirmacher Oasis, Queen Maud Land, East Antarctica, and its surroundings. J Perthes, Gotha

Chamley H (1989) Clay sedimentology. Springer-Verlag, Berlin/Heidelberg

Community Members EPICA (2006) One-to-one coupling of glacial climate variability in Greenland and Antarctica. Nature 444:195–198

De MSJ, Whitehead RF, Gregory M (1994) The chemical composition of glacial melt-water ponds and streams on the McMurdo Ice Shelf, Antarctica. Antarct Sci 6:17–27. https://doi.org/10.1017/S0954102094000039

Engel MH, Macko SA (1993) Organic geochemistry. Springer, US, Boston, MA

Evans M, Heller F (2003) Environmental magnetism: principles and applications of enviromagnetics. Academic Press, San Diego, USA, Int. Geophys. Ser

Farquhar GD, Ehleringer JR, Hubick KT (1989) Carbon isotope discrimination and photosynthesis. Annu Rev Plant Physiol Plant Mol Biol 40:503–537. https://doi.org/10.1146/annurev.pp.40.060189.002443

Fry B, Sherr EB (1989) δ13C measurements as indicators of carbon flow in marine and freshwater ecosystems BT—stable isotopes in ecological research. In: Rundel PW, Ehleringer JR, Nagy KA (eds) Springer. New York, NY, New York, pp 196–229

Gore DB, Rhodes EJ, Augustinus PC et al (2001) Bunger Hills, East Antarctica: ice-free at the last glacial maximum. Geology 29:1103. https://doi.org/10.1130/0091-7613(2001)029%3c1103:BHEAIF%3e2.0.CO;2

Govil P, Mazumder A, Asthana R et al (2016) Holocene climate variability from the lake sediment core in Schirmacher Oasis region, East Antarctica: multiproxy approach. Quat Int 425:453–463. https://doi.org/10.1016/j.quaint.2016.09.032

Hedges JI, Keil RG (1995) Sedimentary organic matter preservation: an assessment and speculative synthesis. Mar Chem 49:81–115. https://doi.org/10.1016/0304-4203(95)00008-F

Helland PE, Holmes MA (1997) Surface textural analysis of quartz sand grains from ODP Site 918 off the southeast coast of Greenland suggests glaciation of southern Greenland at 11 Ma. Palaeogeogr Palaeoclimatol Palaeoecol 135:109–121. https://doi.org/10.1016/S0031-0182(97)00025-4

Hendy CH, Sadler AJ, Denton GH, Hall BL (2000) Proglacial lake-ice conveyors: a new mechanism for deposition of drift in polar environments. Geogr Ann Ser A, Phys Geogr 82:249–270. https://doi.org/10.1111/j.0435-3676.2000.00124.x

Hermichen W-D, Kowski P, Wand U (1985) Lake Untersee, a first isotope study of the largest freshwater lake in the interior of East Antarctica. Nature 315:131–133. https://doi.org/10.1038/315131a0

Hodell DA, Schelske CL (1998) Production, sedimentation, and isotopic composition of organic matter in Lake Ontario. Limnol Oceanogr 43:200–214. https://doi.org/10.4319/lo.1998.43.2.0200

Hodgson DA, Roberts SJ, Bentley MJ et al (2009) Exploring former subglacial Hodgson Lake, Antarctica. Paper-II: paleolimnology. Quat Sci Rev 28:2310–2325. https://doi.org/10.1016/j.quascirev.2009.04.014

Krinsley DH, Doornkamp JC (1973) Atlas of quartz sand surface textures. Cambridge University Press, Cambridge

Lambert F, Bigler M, Steffensen JP, Hutterli M, Fischer H (2012) Centennial mineral dust variability in high-resolution ice core data from Dome C. Antarctica Clim past 8:609–623

Leng MJ, Marshall JD (2004) Palaeoclimate interpretation of stable isotope data from lake sediment archives. Quat Sci Rev 23:811–831. https://doi.org/10.1016/j.quascirev.2003.06.012

Li Y, Yu Z, Kodama K, Moeller R (2006) A 14,000-year environmental change history revealed by mineral magnetic data from White Lake, New Jersey, USA. Earth Planet Sci Lett 246:27–40. https://doi.org/10.1016/j.epsl.2006.03.052

Mahaney WC (2002) Atlas of sand grain surface textures and applications. Oxford University Press, New York

Mahaney W (1995) Glacial crushing, weathering and diagenetic histories of quartz grains inferred from scanning electron microscopy. Glacial Environ—Process Sediments Landforms 487–506

Mahesh BS, Warrier AK, Mohan R et al (2015) Response of Long Lake sediments to Antarctic climate: a perspective gained from sedimentary organic geochemistry and particle size analysis. Polar Sci 9:359–367. https://doi.org/10.1016/j.polar.2015.09.004

Mahesh BS, Warrier AK, Mohan R et al (2017) Response of Sandy Lake in Schirmacher Oasis, East Antarctica to the glacial-interglacial climate shift. J Paleolimnol 58:275–289. https://doi.org/10.1007/s10933-017-9977-8

Mahesh BS, Warrier AK, Mohan R, Tiwari M (2019) Impact of Antarctic climate during the Late Quaternary: records from Zub Lake sedimentary archives Schirmacher Hills, East Antarctica. Palaeogeogr Palaeoclimatol Palaeoecol 514:398–406. https://doi.org/10.1016/j.palaeo.2018.10.029

Mazumder A, Govil P, Kar R, Gayathri NM (2017) Paleoenvironments of a proglacial lake in Schirmacher Oasis, East Antarctica: insights from quartz grain microtextures. Polish Polar Res 38:1–19. https://doi.org/10.1515/popore-2017-0002

Meyers PA (1994) Preservation of elemental and isotopic source identification of sedimentary organic matter. Chem Geol 114:289–302. https://doi.org/10.1016/0009-2541(94)90059-0

Meyers PA (1997) Organic geochemical proxies of paleoceanographic, paleolimnological, and paleoclimatic processes. Org Geochem 27:213–250. https://doi.org/10.1016/S0146-6380(97)00049-1

Meyers PA (2003) Applications of organic geochemistry to paleolimnological reconstructions: a summary of examples from the Laurentian Great Lakes. Org Geochem 34:261–289. https://doi.org/10.1016/S0146-6380(02)00168-7
Meyers PA, Teranes JL (2001) Tracking environmental change using lake sediments. In: Last WM, Smol JP (eds) Tracking environmental change using lake sediments, vol 2. Physical and geochemical methods. Kluwer Academic Publishers, Dordrecht, pp 239–269
Peters KE, Sweeney RE, Kaplan IR (1978) Correlation of carbon and nitrogen stable isotope ratios in sedimentary organic matter 1. Limnol Oceanogr 23:598–604. https://doi.org/10.4319/lo.1978.23.4.0598
Petit JR, Mourner L, Jouzel J et al (1990) Palaeoclimatological and chronological implications of the Vostok core dust record. Nature 343:56–58. https://doi.org/10.1038/343056a0
Petit JR, Jouzel J, Raynaud D et al (1999) Climate and atmospheric history of the past 420,000 years from the Vostok ice core, Antarctica. Nature 399:429–436. https://doi.org/10.1038/20859
Phartiyal B (2014) Holocene paleoclimatic variation in the Schirmacher Oasis, East Antarctica: a mineral magnetic approach. Polar Sci 8:357–369. https://doi.org/10.1016/j.polar.2014.06.001
Phartiyal B, Sharma A, Bera SK (2011) Glacial lakes and geomorphological evolution of Schirmacher Oasis, East Antarctica, during Late Quaternary. Quat Int 235:128–136. https://doi.org/10.1016/j.quaint.2010.11.025
Pistolato M, Quaia T, Marinoni L et al (2006) Grain size, mineralogy and geochemistry in late quaternary sediments from the Western Ross sea outer slope as proxies for climate changes. Antarctica. Springer-Verlag, Berlin/Heidelberg, pp 423–432
Ravindra R (2001) Geomorphology of Schirrmacher oasis, East Antarctica. In: Proceedings, a symposium on snow, ice and glacier. Geol Sur Ind. Spl Pub pp 379–390
Retelle M, Child J (1996) Suspended sediment transport and deposition in a high arctic meromictic lake. J Paleolimnol 16:151–167. https://doi.org/10.1007/BF00176933
Reynolds RL, King JW (1995) Magnetic records of climate change. Rev Geophys 33:101. https://doi.org/10.1029/95RG00354
Sandeep K, Shankar R, Warrier AK et al (2015) The environmental magnetic record of palaeoenvironmental variations during the past 3100years: a possible solar influence? J Appl Geophys 118:24–36. https://doi.org/10.1016/j.jappgeo.2015.03.023
Sandeep K, Warrier AK, Harshavardhana BG, Shankar R (2012) Rock magnetic investigations of surface and sub-surface soil samples from five lake catchments in tropical southern India. Int J Environ Res 6:1–18. https://doi.org/10.22059/ijer.2011.467
Simmons GM Jr, Wharton RA Jr, McKay CP et al (1986) Sand/ice interactions and sediment deposition in perennially ice-covered Antarctic lakes. Antarct J US 21:217–220
Smith JA, Hodgson DA, Bentley MJ et al (2006) Limnology of two Antarctic Epishelf lakes and their potential to record periods of ice shelf loss. J Paleolimnol 35:373–394. https://doi.org/10.1007/s10933-005-1333-8
Spaulding SA, McKnight DM, Stoermer EF, Doran PT (1997) Diatoms in sediments of perennially ice-covered Lake Hoare, and implications for interpreting lake history in the McMurdo Dry Valleys of Antarctica. J Paleolimnol 17:403–420. https://doi.org/10.1023/A:1007931329881
Squyres SW, Andersen DW, Nedell SS, Wharton RA (1991) Lake Hoare, Antarctica: sedimentation through a thick perennial ice cover. Sedimentology 38:363–379. https://doi.org/10.1111/j.1365-3091.1991.tb01265.x
Stanley S, De Deckker P (2002) A Holocene record of allochthonous, aeolian mineral grains in an Australian alpine lake; implications for the history of climate change in southeastern Australia. J Paleolimnol 27:207–219. https://doi.org/10.1023/A:1014249404845
Strand K, Passchier S, Näsi J (2003) Implications of quartz grain microtextures for onset Eocene/Oligocene glaciation in Prydz Bay, ODP Site 1166, Antarctica. Palaeogeogr Palaeoclimatol Palaeoecol 198:101–111. https://doi.org/10.1016/S0031-0182(03)00396-1
Sugden DE, McCulloch RD, Bory AJ-M, Hein AS (2009) Influence of Patagonian glaciers on Antarctic dust deposition during the last glacial period. Nat Geosci 2:281–285. https://doi.org/10.1038/ngeo474

Talbot MR (2001) Nitrogen Isotopes in Paleolimnology. In: Last WM, Smol JP (eds) Tracking environmental change using lake sediments: physical and chemical techniques. Kluwer Academic Publishers, Dordrecht, pp 401–439

Talbot MR, Johannessen T (1992) A high-resolution palaeoclimatic record for the last 27,500 years in tropical West Africa from the carbon and nitrogen isotopic composition of lacustrine organic matter. Earth Planet Sci Lett 110:23–37. https://doi.org/10.1016/0012-821X(92)90036-U

Thompson R, Oldfield F (1986) Environmental magnetism. Allen & Unwin, London

Trumbore SE, Druffel ERM (1995) Carbon isotopes for characterizing sources and turnover of nonliving organic matter. In: Zepp RG, Sonntag C (eds) The role of nonliving organic matter in the earth's carbon cycle. John Wiley & Sons, New York, pp 7–21

Verleyen E, Hodgson DA, Sabbe K et al (2011) Post-glacial regional climate variability along the East Antarctic coastal margin—evidence from shallow marine and coastal terrestrial records. Earth Sci Rev 104:199–212. https://doi.org/10.1016/j.earscirev.2010.10.006

Walden J, Oldfield F, Smith J (1999) Environmental magnetism: a practical guide. Quaternary Research Association, London

Warrier AK, Sandeep K, Harshavardhana BG et al (2011) A magnetic rock record of Pleistocene rainfall variations at the Palaeolithic site Attirampakkam, Southeastern India. J Archaeol Sci 38:3681–3693. https://doi.org/10.1016/j.jas.2011.08.039

Warrier AK, Mahesh BS, Mohan R et al (2014) Glacial–interglacial climatic variations at the Schirmacher Oasis, East Antarctica: The first report from environmental magnetism. Palaeogeogr Palaeoclimatol Palaeoecol 412:249–260. https://doi.org/10.1016/j.palaeo.2014.08.007

Warrier AK, Shankar R, Manjunatha BR, Harshavardhana BG (2014) Mineral magnetism of atmospheric dust over the southwest coast of India: impact of anthropogenic activities implications to public health. J Appl Geophys 102:1–9. https://doi.org/10.1016/j.jappgeo.2013.11.013

Warrier AK, Joju GS, Amrutha K, Yamuna Sali AS, Mahesh BS, Mohan R (2021a) Magnetic properties of surface sediments in Schirmacher Oasis, East Antarctica: Spatial distribution and controlling factors. Jour. Soils Sed. 21:1206-1221. https://doi.org/10.1007/s11368-020-02824-8

Warrier AK, Mahesh BS, Mohan R Shankar R (2021b) A 43-ka mineral magnetic record of environmental variations from lacustrine sediments of Schirmacher Oasis, East Antarctica. Catena, 202:105300. https://doi.org/10.1016/j.catena.2021.105300

WMO Provisional Statement of the State of the Climate (2019) Weather Climate Water - A handbook by the World Meteorological Organization, 34 p. https://library.wmo.int/doc_num.php?explnum_id=10108. Accessed 29 Feb 2020

Nutrient Cycling and Productivity in Antarctic Lakes

Shabnam Choudhary, G. N. Nayak, and Neloy Khare

Abstract Sedimentary organic matter from the Antarctic lakes is the source of various proxies used to study productivity changes. A total of three sediment cores (GL-1, V-1, and L-6) collected from the lakes of Schirmacher Oasis, East Antarctica, were analysed for total organic carbon (TOC), total nitrogen (TN), total phosphorus (TP), biogenic silica (BSi) and their ratios were computed to understand the nutrient cycling and productivity in Antarctic lakes. In core GL-1 and V-1, high TOC and high clay in the upper section of the core indicated high primary productivity due to the lakes' exposure to the ice meltwater influx. The C/N ratio of substances GL-1, V-1, and L-6 varied from 2.72 to 8.52, indicating the source of organic matter as autochthonous exclusively derived from algae (C/N < 10). N/P ratio is <7.81 in all three lakes, meaning a potential limitation of N in all the lakes. In cores GL-1, V-1, N/Si ratio is lower than 1, indicating N limitation, while in core L-6, N/Si ratio is higher than 1, i.e. 1.53, indicating Si limitation. Si/P ratio is found to be greater than 3 in all the cores, indicating P limitation. Deviation from the Redfield ratio suggested that the lakes are oligotrophic.

Keywords Nutrients · Algae · Carbon · Nitrogen · Phosphorus · Sediment · Redfield ratio

S. Choudhary (✉)
National Centre for Polar and Ocean Research, Vasco da Gama, Goa-403802, India

G. N. Nayak
Marine Sciences, School of Earth, Ocean and Atmospheric Sciences, Goa University, Taleigao, Goa 403206, India

Present Address:
N. Khare
Ministry of Earth Sciences, Lodhi Road, Prithvi Bhawan, New Delhi 110 003, India

N. Khare (ed.), *Assessing the Antarctic Environment from a Climate Change Perspective*, Earth and Environmental Sciences Library,
https://doi.org/10.1007/978-3-030-87078-2_9

1 Introduction

Antarctica, the southernmost continent on planet earth, played an essential role in its climate system. It consists of a large number of glacial landforms such as mountain tops and nunataks. Along with these landforms, ice-free areas known as "dry valleys or Antarctic oasis" occurs commonly in the Antarctic region and are surrounded by the Antarctic ice sheet. These valleys or oasis are situated between the Antarctic ice shelves and ice sheet. These ice-free regions of Antarctica viz. McMurdo dry valleys, Bunger hills, Vestfold Hills, Larsemann Hills, and Schirmacher Oasis occupies about 2% of the Antarctic landmass and consist of numerous lakes. In recent years, these lakes have gained importance as they present pristine conditions, act as an essential source for paleo-archives, and are easily accessible in ice-free areas.

In general, lake sediments are ideal repositories of eolian and fluvial materials. The sediment input through glacial meltwater and biological productivity is the primary source of sedimentation in Antarctic lakes. The biological productivity is confined primarily to algae and cyanobacteria in the Antarctic lakes (Yoon et al. 2006; Smith et al. 2006; Hodgson et al. 2009; Choudhary et al. 2018b). During the austral summer, lake sedimentation is predominant (Simmons et al. 1986); hence, transferring the detrital matter from the Antarctic landmass to its lakes, meltwater seems to play an important role. Significantly lower sediment accumulation rates are observed in high-latitude lakes than temperate lakes (Wolfe et al. 2004).

The source and accumulation of organic matter can be identified by studying the abundance of total organic carbon and total nitrogen in lacustrine sediments. The type and amount of sedimentary organic matter can reflect the past fluctuations in lake's productivity and terrestrial inputs influenced by climate-induced environmental changes (Talbot and Johannessen 1992; Meyers 1997; Leng and Marshall 2004). Carbon, Nitrogen, and Phosphorus are the primary archives extracted from lake sedimentary organic matter to understand the limiting factor affecting algal growth. These robust proxies are indicators of organic matter's provenance, the type and amount of organic matter that has been deposited in the lake over a while (Talbot and Johannessen 1992; Meyers 1997; Leng and Marshall 2004). The C/N ratio of organic matter is also used as an indicator of the source of organic matter (Talbot 2001; Meyers 2003). Past environmental conditions (redox) in the lacustrine systems due to climate change can be deciphered from the TOC and TN of bulk sedimentary organic matter (Talbot and Johannessen 1992). The molar concentrations of C, N, and P have been used to estimate which of these nutrients is limiting the growth of algae in aquatic systems compared to the Redfield ratio. Redfield observed that phytoplankton contains a molecular C:Si:N:P ratio of 106:15:16:1 (Harrison et al. 1977). A departure from this ratio has been assumed to imply nutrient deficiency. The use of elemental ratios has become widespread in marine and freshwater phytoplankton studies. In the present study, three lake sediment cores from the Schirmacher Oasis, East Antarctica, have been studied for understanding the cycling of the nutrients controlling the primary productivity in the lakes. As in the lakes, nutrients rather than physical conditions tend to limit primary productivity.

2 Nutrient Cycling in Lakes

In an ecosystem, nutrient cycling is an essential process that describes the usage of the nutrients, their movement, and the methods and their recycling in the environment (Fig. 1). For an organism's existence, nutrients like carbon, oxygen, hydrogen, phosphorus, and nitrogen are essential and are termed macronutrients. Nutrient cycles involve living organisms and non-living components and biological, geological, and chemical processes. Thus, these nutrient cycles are also known as biogeochemical cycles. Biogeochemical cycles can be categorised into two main types: global cycles and local cycles. Elements such as carbon, nitrogen, oxygen, and hydrogen are recycled through abiotic environments, including the atmosphere, water, and sediment.

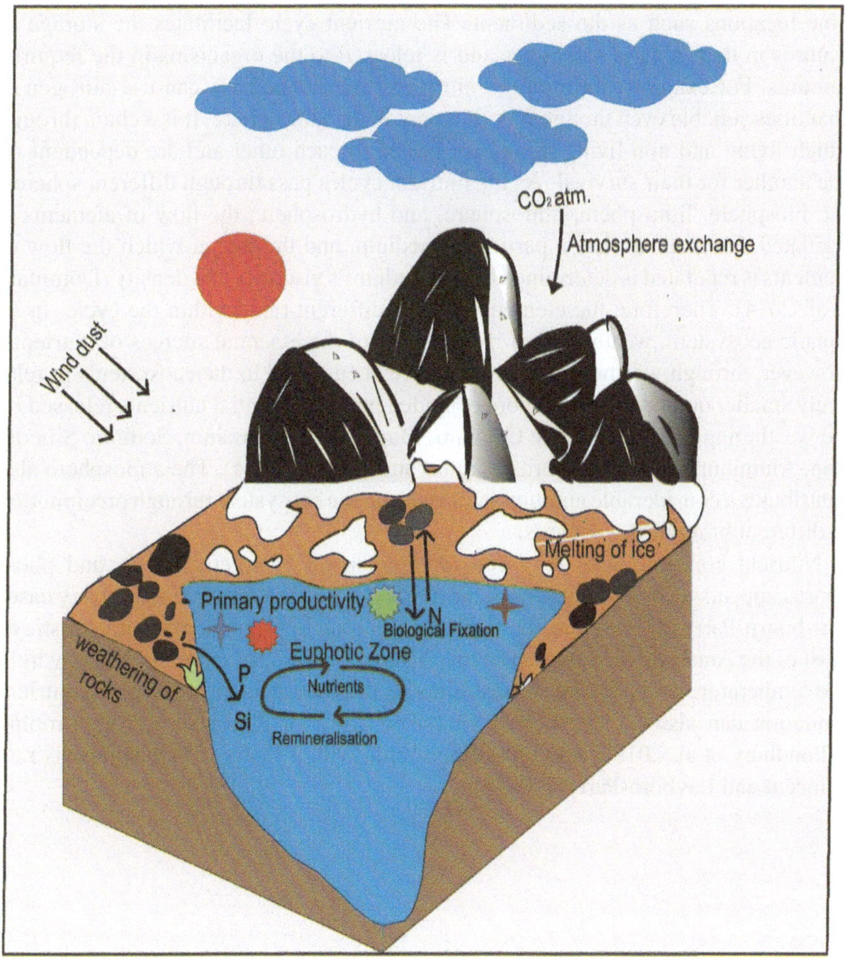

Fig. 1 Nutrient cycling in lakes

The atmosphere is the primary abiotic environment. Most of the elements enter the cycle from the atmosphere and travel long distances before the organisms' uptake in a cycle (Inagaki and Ishizuka 2011). The sediment is the primary abiotic environment for recycling elements such as phosphorus, calcium, and potassium. As such, their movement is typically over a local region.

The nutrient cycle allows the transformation of the element into different specific forms that enable utilising other organisms' particular component. Example, although nitrogen is abundant in the atmosphere, plants can only uptake nitrogen in two solid forms, viz. ammonium and nitrate. Without the transformation of nitrogen into these usable forms, the growth of an organism would be limited. The transfer of elements from one location to another is also aided by nutrient cycles. Some features are highly concentrated in an inaccessible area to most living organisms, such as nitrogen in the atmosphere. Through the nutrient cycle, these elements are transferred to accessible locations such as the sediment. The nutrient cycle facilitates the storage of features in their natural reservoirs and is released to the organisms in the required amounts. For example, through the nitrogen cycle, organisms can use nitrogen in quantities suitable even though it is abundant in the atmosphere. It is a chain through which living and non-living things are linked to each other and are dependent on one another for their survival. As the nutrient cycles pass through different spheres, viz. biosphere, lithosphere, atmosphere, and hydrosphere, the flow of elements is regulated. Each sphere has a particular medium, and the rate at which the flow of elements is regulated is determined by the medium's viscosity and density (Dommain et al. 2014). Therefore, the elements flow at different rates within the cycle. In an aquatic ecosystem, weathering of rocks is one of the essential sources of nutrients. However, through weathering, nutrients have been added to the ecosystems in relatively smaller quantities over a more extended period. Essential nutrients released by the weathering of rocks include Calcium, Magnesium, Potassium, Sodium, Silicon, Iron, Aluminum, and Phosphorus (Kumar and Sekaran 2014). The atmosphere also contributes a considerable amount of nutrients to the ecosystem through precipitation or different biological processes.

Nutrient concentrations vary considerably in the Antarctic Lakes, and phosphorus appears to be a limiting nutrient for phytoplankton production in many cases (Laybourn-Parry 2003). Thus, the nutrient limitation is a significant characteristic of most of the Antarctic lacustrine systems. Although physical factors such as low light and temperature are substantial constraints on production in most lakes, the nutrient limitation can also be significant. Therefore, most of the lakes are oligotrophic (Choudhary et al. 2018b) and ultraoligotrophic with a shallow photosynthesis rate (Vincent and Layborn-Parry 2008).

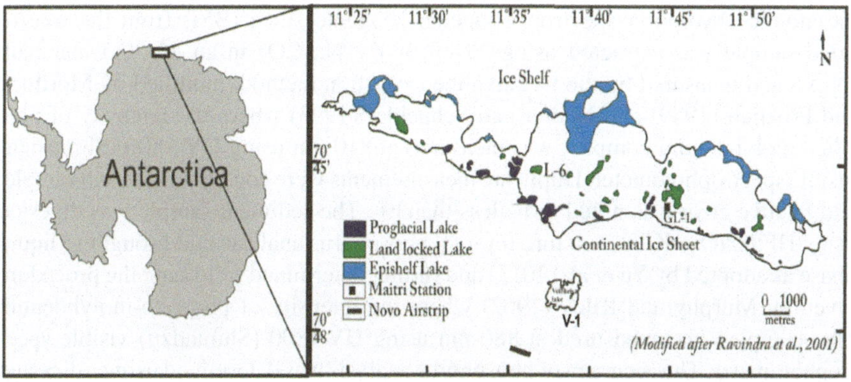

Fig. 2 Map of Antarctica showing the location of Schirmacher oasis and map of Schirmacher Oasis (modified after Ravindra et al. 2001) showing the sampling locations in the study area

3 Materials and Methodology

3.1 Study Area and Sample Collection

The samples were collected from Schirmacher Oasis (Fig. 2) during the 31st Indian Scientific Expedition to Antarctica, January 2012. Sediment core samples of varying length 40 cm (GL-1), 32 cm (V1), and 58 cm (L-6) were retrieved manually from near the periphery of the lake when the lakes were ice-free. A PVC handheld corer was inserted by hammering manually into the lake sediment bed and then retrieved. Further, the cores were labelled, packed, and stored in a deep freeze at <4 °C. The cores were transported to the laboratory, core GL-1 and V-1 subsampled at 4 cm interval while L-6 was subsampled at 2 cm and later on dried at 60 °C.

3.2 Sample Analysis

For determination of total organic carbon (TOC) small portion of each subsample was powdered and homogenised in an Agate mortar. TOC was determined using the Walkey Black method (Walkey 1947), adopted and modified by Jackson (1958), which utilised exothermic heating and oxidation with potassium dichromate ($K_2Cr_2O_7$) and sulfuric acid (H_2SO_4). The freeze-dried sediment sample was also analysed for total nitrogen (TN) concentration in the Marine Stable Isotope Lab (MASTIL) at National Centre for Antarctic & Ocean Research, Goa, India, using an EA (Isoprime, Vario Isotope Cube). The precision for N% was ±0.63% (1σ standard deviation) obtained by repeatedly running sulfanilamide as the standard. Calcium carbonate was computed using the values of Ca analysed through the atomic absorption spectrophotometer. Biogenic silica (BSi) from the freeze-dried sample was extracted using 25 ml of 1% Na_2CO_3 in an 85 °C water bath for 5 h and measured by the wet alkaline extraction method, modified by Mortlock and Froelich (1989) and Muller and Schneider (1993) where the intensity of blue silico molybdenum complex was measured at 810 nm using UV-1800 (Shimadzu) visible spectrophotometer. Duplicate measurements were conducted on each sample, and relative error was noted to be less than3%. The sediment sample was digested using $HF:HNO_3:HClO_4$ mixture for total phosphorus analysis and brought to liquid phase as adopted by Yu et al. (2013) and further determined following the procedure given by Murphy and Riley (1962) where the intensity of phospho-molybdenum blue complex was measured at 880 nm using UV-1800 (Shimadzu) visible spectrophotometer. The accuracy of phosphorus analysis was determined using a digested sample of JLK-1, and relative error was noted to be less than 4%.

4 Source of Sedimentary Organic Matter in Lacustrine Sediments

Sedimentary organic matter offers different proxies that can be used to reconstruct past environmental changes as preservation and production of organic matter is affected by the environmental changes to a more considerable extent (Meyers 1997). Organic matter preserved in lake sediments reflects the ecological changes. In core GL-1 and V-1 (Fig. 3), high TOC along with high clay in the upper section of the core indicated high primary productivity due to the exposure of lakes to the ice melt-water influx (Choudhary et al. 2018a). C/N ratios have been used often to identify the source of organic matter in lake sediments (Talbot and Johannessen 1992; Meyers 1997). Algae and cyanobacteria typically have an atomic C/N ratio between 4 and 10, while the terrestrial organic matter is above 20 (Meyers 1994, 2003; Meyers and Teranes 2001). The C/N ratio of cores GL-1, V-1, and L-6 varied from 2.72 to 8.52, indicating the source of organic matter as in situ exclusively derived from algae (C/N < 10) as per the classification of Meyers (1994). The C/N ratio for all three cores was less than 10 for the entire core length, indicating that the significant organic matter source was autochthonous. However, a high C/N ratio in core GL-1 at a depth of 16 cm and the surface showed prolonged ice-free conditions. It increased meltwater influx which must have delivered terrestrial organic matter to the lake, possibly from lichens and mosses (Choudhary et al. 2018a). Also, the loss of N from the sediments during diagenesis or nitrogen limitation in the surface water due to high primary production must have resulted in the relatively large range of the C/N ratio. Partial degradation of algal organic matter can selectively diminish proteinaceous components and thereby raise C/N ratios. During early diagenesis, selective degradation of organic matter components can modify C/N ratios of organic matter in sediments (Meyers 1997).

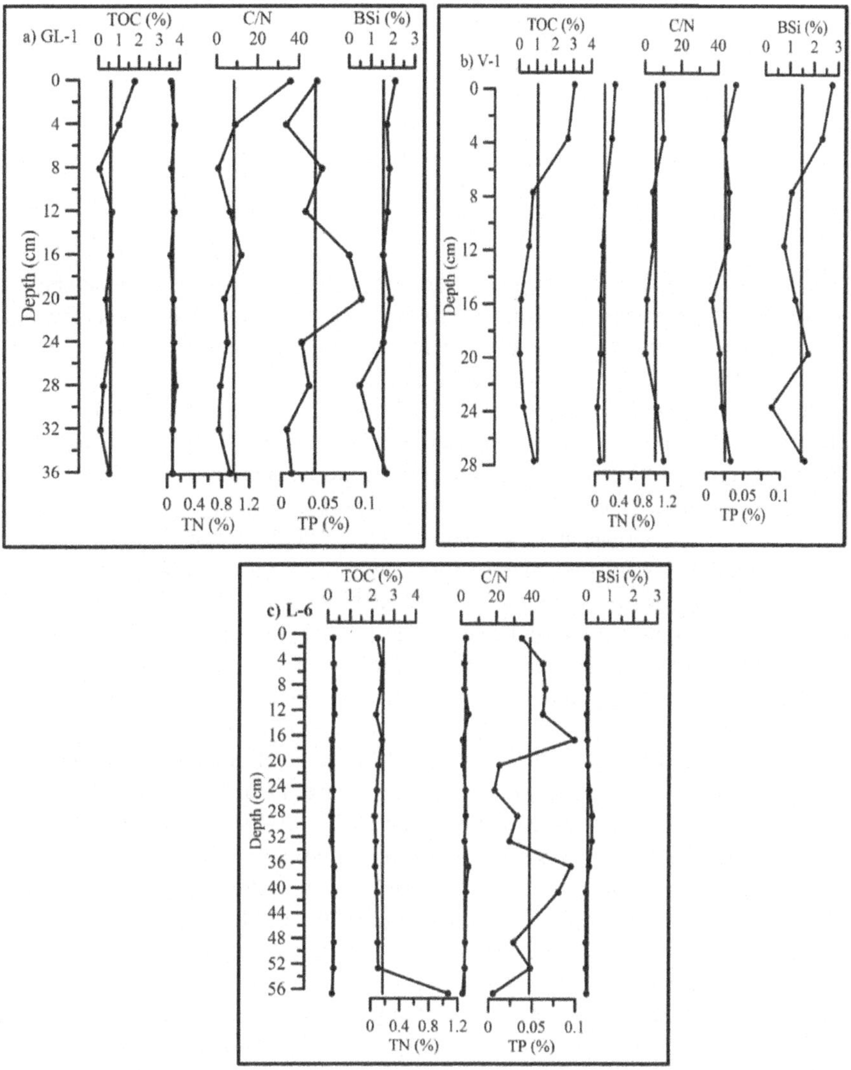

Fig. 3 Carbon, nitrogen, phosphorus and biogenic silica concentration in the core **a** GL-1. **b** V-1. **c** L-6

4.1 Nutrient Limitations in Lakes

In lakes where nutrients rather than physical conditions tend to limit algal growth leading to changes in productivity, the molar concentrations of carbon, nitrogen, and phosphorus (Table 1) have been used to estimate nutrient limitation (Choudhary et al. 2018b). The least available nutrient in any system is considered to be the limiting nutrient for the total amount of photosynthetic C-fixation that a system can sustain.

Table 1 Molar ratios of carbon/nitrogen, nitrogen/phosphorus, nitrogen/silica, and silica/phosphorus contents of cores (a) GL-1 (b) V-1 (c) L-6

Cores	TOC/TN	TN/TP	TOC/TP	TN/Si	Si/P
GL-1	7.30	2.47	14.03	0.03	80.46
V-1	5.14	4.05	31.49	0.05	73.21
L-6	2.21	7.81	4.04	1.53	4.71

This is a stoichiometric concept that presumes that one nutrient is consumed before other nutrients. It is considered that P is limiting in lakes, while N is usually limiting in the marine environment. However, there are exceptions; among P, N and Si, any of these nutrients can be limiting (Hecky and Kilham 1988).

Redfield surmised that the C:N:P ratio is 106:16:1 under ordinary conditions, when neither of the nutrients is limiting (Redfield 1934, 1958; Sterner and Elser 2002; Choudhary et al. 2018b). A deflection from the particular ratio attributes to the deficiency of C, N, or P in an aquatic system. Harrison et al. (1997) suggested that siliceous microorganisms like diatoms need silicate to form the hard part of their shell known as frustule, and an optimal C:Si:N:P ratio is 106:15:16:1. When the C/N molar ratio was compared with the Redfield ratio, it was lower in all three lakes. C/N ratios typically <10 indicate algal growth suggesting autochthonous lacustrine organic matter, whereas ratios above 20 may be >200 suggested allochthonous organic matter (Talbot and Johannessen 1992). In all three cores, as stated earlier, the C/N ratio is lower than 10, indicating algal growth and the autochthonous nature of organic matter. Further, the N/P ratio above 17 indicates P limitation; a ratio below 10 suggests N limitation, and values between 10 and 17 indicate that either of the nutrients may be limiting suggested by Ulen (1978) and Hellstrom (1996). In the present study, the N/P ratio is <7.81 in all the three lakes, namely core GL-1, V-1, and L-6, which is much lower than 16, indicating a potential limitation of N in all the lakes. This limitation of N may be either due to phosphorus incorporation into the sediments, reducing the N/P ratio or removing nitrogen by denitrification in water column or sediments (Tyrrell and Law 1997; Downing and McCauley 1992; Choudhary et al. 2018b). Further, the C/P molar ratio varied from 4.04 to 31.49 in these lakes, higher than the Redfield ratio observed by Dore and Priscu (2001) in McMurdo dry valley lakes Antarctica. According to Harrison et al. (1977), N/Si > 1 and Si/P < 3 are indicative of Si limitation. In cores GL-1, V-1, N/Si ratio is lower than 1, indicating N limitation, while in core L-6, N/Si ratio is higher than 1, i.e. 1.53, indicating Si limitation. Si/P ratio is found to be much higher than 3 in all the cores that indicated P limitation. Priscu's (1995) experimental work demonstrated that Lake Vanda and Lake Bonney in McMurdo dry valleys are phosphorus-deficient while Lake Fryxell and Lake Hoare are nitrogen deficient. The ratio of nutrients C, N, P, and Si deviated from the Redfield ratio and showed low concentrations suggesting that the lakes are oligotrophic, leading to low rates of plankton biomass and low primary production despite relatively high temperature, ice-free conditions of the lake, and high influx of sediment. The nutrient concentrations are low in all three lakes due to less organic matter concentration as it might have been diluted by coarse-grained sediment. They might also be related to glacial meltwater being low in nutrient concentrations.

Although data generated from the limnetic ecosystems does not fit the Redfield paradigm very well as marine data, because in the open ocean, the particulate organic carbon of the surface oceans is dominated by phytoplankton that follows Redfield stoichiometry, the terrigenous organic matter with higher C/P and C/N can contribute significantly to the pool of organic carbon available for remineralisation to offset the C:N:P ratio in some lakes. Also, bacterial denitrification and N-fixation can affect N/P ratios.

5 Factors Affecting Nutrient Concentrations and Primary Productivity

Climate change has directly affected the Antarctic aquatic systems primarily. During the austral summer, with warming conditions in the region due to the retreat of glaciers in the study area, higher melting occurs, leading to a sizeable freshwater influx to the lakes. As the lakes are ice-free, they are getting exposed to the atmosphere. Atmosphere exchange enhancing CO_2 input decreased ice cover (improving photosynthetically active radiation). High temperature has increased nutrient influx and increased primary productivity in lakes (Lyons and Finlay 2008; Choudhary et al. 2018b).

Nutrient chemistry is modified to a more considerable extent by the streams coming out from the melting glaciers and draining into the lakes. The chemistry of the nutrients can also be affected by supraglacial processes (Vincent and Laybourn-Parry 2008). Recent studies in the Taylor Valley (MacMurdo dry valley) aquatic systems suggested that the C:N:P stoichiometry is affected primarily as water flows through the hydrological sequence from snow/glacier/ice to the closed basins, i.e. lakes. Streams flowing on younger surfaces exposed in the catchment area provide more phosphorus relative to nitrogen, and streams with high abundances of algal mats have a much lower N:P ratio (Lyons and Finlay 2008). Nitrogen fixation, i.e. conversion of atmospheric N_2 to NO^{3-} occurs in the microbes' catchment area while weathering of the continental rocks contributes to phosphate. Therefore, nitrogen and phosphate concentrations are regulated by runoff from the land (Meyers 1997).

The age and history of a landform and the gradient and geomorphology of the streams also play an essential role in regulating the nutrient input into the lakes and ultimately primary productivity (Lyons and Finlay 2008).

6 Conclusions

In the studied Antarctic lakes, organic matter is found to be autochthonous exclusively derived from algae, with significant terrestrial organic matter contribution at some intervals derived from the lichens and mosses growing around the lake. The ratio of nutrients C, N, P, and Si deviated from the Redfield ratio showed low concentrations suggesting that the lakes are oligotrophic, leading to low plankton biomass rates and low primary production despite relatively high temperature, ice-free conditions of the lake, and high influx of sediment. Low nutrient concentrations in all three lakes may be due to inadequate organic matter concentration. The terrestrial material might have diluted it and might also be related to glacial meltwater, devoid of nutrient concentrations.

Acknowledgements The authors place on record to thank the Director, National Centre for Polar and Ocean Research (NCPOR), Goa, for providing the opportunity to one of the authors (SC) to participate in the Indian Antarctic programme and Ministry of Earth Sciences (MoES) for providing the logistic support required for the collection of samples. The authors, thank Dr. Manish Tiwari, Scientist and Mr. Siddesh Nagoji, ESSO-NCAOR, Goa, are thanked for kindly extending the instrumental facility of the elemental analyser. Dr. Anoop K. Tiwari, Scientist -E, ESSO-NCAOR, Goa, is thanked for providing the samples. One of the authors (SC) thanks to the University Grant Commission (UGC) for awarding a national fellowship (F1-17.1/2015-16/MANF-2015-17-UTT-51739). Another author (GNN) thanks to the Council of Scientific and Industrial Research (CSIR) for awarding the Emeritus Scientist position.

References

Choudhary S, Tiwari AK, Nayak GN and Bejugam P (2018a) Sedimentological and geochemical investigations to understand the source of sediments and recent past processes in Schirmacher Oasis, East Antarctica. Polar Sci 15:87–98

ChoudharyS NGN, Tiwari AK, Khare N (2018b) Sediment composition and its effect on productivity in Larsemann Hills East Antarctica. Arab J Geosci 11(15):416

Dommain R, Couwenberg J, Glaser PH, Joosten H, Suryadiputra INN (2014) Carbon storage and release in Indonesian peatlands since the last deglaciation. Quatern Sci Rev 97:1–32

Dore JE, Priscu JC (2001) Phytoplankton phosphorus deficiency and alkaline phosphatase activity in the McMurdo Dry Valley lakes Antarctica. Limnol Oceanogr 46(6):1331–1346

Downing JA, McCauley E (1992) The nitrogen: phosphorus relationship in lakes. Limnol Oceanogr 37(5):936–945

Harrison PH, Conway H, Holmes W, Davis D (1977) Marine diatoms were grown in chemostats under silicate or ammonium limitation. III. Cellular chemical composition and morphology of *Chaetoceros debilis*, *Skeletonemacostatum*, and *Thalassiosira gravida*. Mar Biol 43:19–31

Harrison PJ, Khan N, Yin K, Saleem M, Bani N, Nisa M, Ahmed SI, Rizvi N, Azam F (1997) Nutrient and phytoplankton dynamics in two mangrove tidal creeks of the Indus River Delta, Pakistan. Mar Ecol Prog Ser 157:13–19

Hecky RE, Kilham P (1988) Nutrient limitation of phytoplankton in freshwater and marine environments: a review of recent evidence on the effects of enrichment. Limnol Oceanogr 33(4part2):796–822

Hellstrom T (1996) An empirical study of nitrogen dynamics in lakes. Water Environ Res 68(1):55–65
Hodgson DA, Verleyen E, Vyverman W, Sabbe K, Leng MJ, Pickering MD, Keely BJ (2009) A geological constraint on relative sea level in marine Isotope stage 3 in the larsemann hills, lambert glacier region, East Antarctica (31 36633 228 cal yr BP). Quat Sci Rev 28:2689–2696
Inagaki M, Ishizuka S (2011) Ecological impact on nitrogen and phosphorus cycling of a widespread fast-growing leguminous tropical forest plantation tree species Acacia Mangium. Diversity 3(4):712–720
Jackson ML (1958) Soil chemical analysis. Prentice-Hall, New York
Johanna L-P (2003) Polar limnology-the past, the present and the future
Kumar S, Sekaran V (2014) Nutrient cycles in lakes. Int J Lakes Rivers 7(1):11–24
Leng MJ, Marshall JD (2004) Palaeoclimate interpretation of stable isotope data from lake sediment archives.Quat Sci Rev 23:811–831
Lyons W, Finlay J (2008) Biogeochemical processes in high-latitude lakes and rivers. Oxford University Press, Oxford
Meyers PA (1994) Preservation of elemental and isotopic source identification of sedimentary organic matter. Chem Geol 114(3–4):289–302
Meyers PA (1997) Organic geochemical proxies of paleoceanographic, paleolimnology, and paleoclimatic processes. Org Geochem 27(5):213–250
Meyers PA (2003) Applications of organic geochemistry to paleolimnological reconstructions: a summary of examples from the Laurentian Great Lakes. Org Geochem 34(2):261–289
Meyers PA, Teranes JL (2001) Sediment organic matter. In: Last WM, Smol JP (eds) Tracking environmental changes using lake sediments—volume II: physical and chemical techniques. Springer, Dordrecht, pp 239–269
Mortlock RA, Froelich PN (1989) A simple method for the rapid determination of biogenic opal in pelagic marine sediments. Deep-Sea Res Part A Oceanogr Res Papers 36(9):1415–1426
Muller PJ, Schneider R (1993) An automated leaching method for the determination of opal in sediments and particulate matter. Deep-Sea Res I Oceanogr Res Papers 40(3):425–444
Murphy JAMES, Riley JP (1962) A modified single solution method for determining phosphate in natural waters. Anal Chim Acta 27:31–36
Priscu JC (1995) Phytoplankton nutrient deficiency in lakes of the McMurdo Dry Valleys Antarctica. Freshwater Biol 34(2):215–227
Ravindra R, Chaturvedi A, Beg MJ (2001) Meltwater lakes of Schirmacher Oasis -their genetic aspects and classification. In: Sahu DB, Pandey PC (eds) Advances in marine and Antarctic sciences, pp 301–313. Dariyaganj, New Delhi
Redfield AC (1934) On the proportions of organic derivatives in seawater and their relation to the composition of plankton. University Press of Liverpool, James Johnstone memorial volume, pp 176–192
Redfield AC (1958) The biological control of chemical factors in the environment. Am Sci 46(3):205–221
Simmons GM Jr, Wharton RA Jr, McKay CP, Nedell S, Clow G (1986) Sand/ice interactions and sediment deposition in perennially ice-covered Antarctic lakes. Antarct J US 21(5):217–220
Smith JA, Hodgson DA, Bentley MJ, Verleyen E, Leng MJ, Roberts SJ (2006) Limnology of two Antarctic epishelf lakes and their potential to record periods of ice shelf loss. J Paleolimnol 35(2):373–394
Sterner RW, Elser JJ (2002) Ecological stoichiometry: the biology of elements from molecules to the biosphere, pp 80–134. Princeton University Press
Talbot MR, Johannessen T (1992) A high-resolution palaeoclimatic record for the last 27,500 years in tropical West Africa from the carbon and nitrogen isotopic composition of lacustrine organic matter. Earth Planet Sci Lett 110(1):23–37
Talbot MR (2001) Nitrogen isotopes in paleolimnology. In: Last WM, Smol JP (eds) Tracking environmental change using lake sediments. Physical and geochemical methods

Tyrrell T, Law CS (1997) Low nitrate: phosphate ratios in the global ocean. Nature 387(6635):793–796

Ulen B (1978) Seston and sediment in Lake Norrviken. Seston composition and sedimentation. Schweiz Z Hydrol 40:262–286

Vincent WF, Laybourn-Parry J (eds) (2008) Polar lakes and rivers: limnology of Arctic and Antarctic aquatic ecosystems. Oxford university press

Walkey A (1947) A critical examination of a rapid method for determining organic carbon in soils: effects of variations in digestion conditions and organic soils constituents. Soil Sci 63:251–263

Wolfe AP, Miller GH, Olsen CA, Forman SL, Doran PT, Holmgren SU (2004) Geochronology of high latitude lake sediments. In: Long-term environmental change in Arctic and Antarctic lakes. Springer, Dordrecht. Mm W, Morgan JJ (1996) Aquatic chemistry, John Wiley & Sons Inc, New York

Yoon H, Khim B, Lee K, Park Y, Yoo K (2006) Reconstruction of postglacial palaeoproductivity in Long Lake, King George Island West Antarctica. Polish Polar Res 27(3):189–206

Yu Y, Song J, Li X, Yuan H, Li N, Duan L (2013) Environmental significance of biogenic elements in surface sediments of the Changjiang estuary and its adjacent areas. J Environ Sci 25(11):2185–2195

Chemical and Isotopic Characterization of Lakes in the Larsemann Hills, East Antarctica

T. R. Resmi, Girish Gopinath, P. S. Sunil, M. Praveenbabu, P. Arjun, and Rahul Rawat

Abstract The ionic and isotopic ratios of some lakes in the Larsemann Hills in East Antarctica are determined to assess the lakes' chemical and isotopic evolution. The lakes occupied the natural depressions carved by glacial abrasions forming small basins, thus providing suitable water accumulation locales under lacustrine conditions. Lakes in the Grovness Peninsula, Stornes Peninsula, Brokenness Peninsula, Sigdoy and McLeod Islands were sampled. Weathering and reverse ion exchange reactions are found to regulate the ionic makeup of the lakes. The isotopic ratios were relatively enriched in the sampled lakes than the typical glacial meltwater fed lakes in other parts of the East Antarctica. Kinetic controlled ice-water fractionation, and evaporation processes are found to affect the isotopic evolution of the lake water in the Larsemann Hills region.

Keywords Ionic chemistry · Isotopic ratio · East Antarctic lakes · Larsemann Hills

1 Introduction

Antarctica, the remote continent, has been the fulcrum of climate studies in recent years. Oceanographers, geologists, geophysicists, marine biologists and climatologists are fascinated by the past climate records embedded in the ice cores, landforms and special features of the continent. Antarctica can be considered composed of two

T. R. Resmi · G. Gopinath · M. Praveenbabu · P. Arjun
Centre for Water Resources Development and Management, Kozhikode, Kerala 673 571, India

G. Gopinath
Kerala University of Fisheries and Ocean Studies, Kochi, Kerala 682 508, India

P. S. Sunil (✉)
Department of Marine Geology and Geophysics, CUSAT-NCPOR Centre for Polar Sciences, School of Marine Sciences, Cochin University of Science and Technology, Kochi, Kerala 682 016, India

R. Rawat
Indian Institute of Geomagnetism, Mumbai, Maharashtra 410 218, India

N. Khare (ed.), *Assessing the Antarctic Environment from a Climate Change Perspective*, Earth and Environmental Sciences Library,
https://doi.org/10.1007/978-3-030-87078-2_10

units, West Antarctica and East Antarctica, making up two-thirds of the landmass. A tiny fraction of this landmass is ice-free, which mainly consists of Nunataks. Among the ice-free regions, 1–2% contains coastal oases that have been exposed by the post-glacial retreat of the ice cap and isostatic rebound of the Earth's crust following the recent deglaciation (Hodgson 2012). Large oases are found in East Antarctica like the Vestfold Hills, Larsemann Hills, Bunger Hills, Schirmacher Oasis, Syowa Oasis etc. During the austral summer, meltwater accumulates in the land depressions, and lakes and ponds are formed in these oases. The Larsemann Hills, one of the significant oases in East Antarctica, sustains many lakes varying in depth, quality, water spread area, and water availability (seasonal or perennial). As these lakes are fed by glacial meltwater, the chemical composition depends mainly on the precipitation and regional bedrock geology. However, continuous cycles of freezing, thawing and evaporation alter the primary composition giving distinct chemical signatures. Similar to the chemical composition, stable isotope ratios are also varied in these lakes. Earlier studies have focussed on many aspects of lakes in the Larsemann Hills (Gasperon et al. 2002; Kaup and Burgess 2002; Hodgson 2012; Wand et al. 2011; Verleyen et al. 2012; Bharti and Niyogi 2015; Nakai et al. 1975; Horita 2008; Quayle 2002; Gasperon 2002; Beg and Asthana 2002; Verleyen 2004; Sabbe et al. 2004; Shrivastava et al. 2012). This study presents the ionic and isotopic characterization of selected lakes in the Larsemann Hills of East Antarctica.

2 Materials and Methods

The Larsemann Hills is the ice-free coastal oasis in the Ingrid Christensen Coast of Princess Elizabeth Land in East Antarctica (69°30'S, 76°19′58″ E). It is spread in approximately 50 km^2 and includes about 150 lakes with different ionic character. Stornes and Broknes are the two significant peninsulas in the Larsemann Hills. There are other minor peninsulas and many coastal islands in the area. The Indian research station 'Bharti' is located in the central part of the Larsemann Hills, which is about ~12 km^2 in the area.

The introductory geology of the Larsemann Hills consists of supracrustal volcanogenic and sedimentary rocks metamorphosed under granulite facies conditions. The supracrustal rocks are intruded by several generations of pegmatites and granites and are underlain by and possibly derived from the proterozoic orthopyroxene-bearing orthogneiss basement (ATCM 2014). Lakes are formed in the land depressions caused by glacial erosion.

We sampled water to form 16 lakes from the Broknes Peninsula (BP), Stornes Peninsula (SP), Grovnes Peninsula (GP, near Bharti Promontory), Sigdoy Island (SI) and McLeod Island (MI) in the Larsemann Hills during the 29th Indian Antarctic Scientific Expedition. The study area's location map is provided in Fig. 1, and a few photographs of water sampling are given in Plate 1. Samples were collected in pre-cleaned plastic bottles, and physicochemical parameters were determined in situ, and standard procedures were followed to find out the ionic composition (APHA

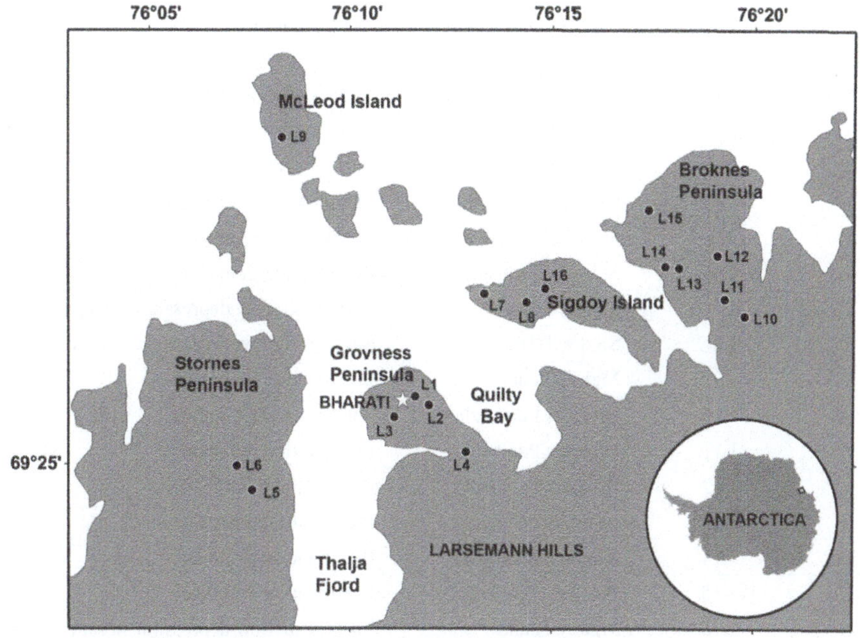

Fig. 1 Sampling locations of the lakes in Larsemann Hills, East Antarctica

Plate 1 Water sampling from the lakes

1995). For isotopic measurements, water samples were collected in High-density polyethylene (HDPE) bottles, and the ratios were determined in an off-axis integrated cavity output spectroscopy system (IWA-45EP, Los Gato's research), with an external precision of ±0.1‰ for $\delta^{18}O$ and ±0.5‰ for δD. The results were expressed in δ as,

$$\delta = \left[\frac{\text{R sample}}{\text{R standard}} - 1\right] \times 1000$$

Table 1 Location details of the sampled lakes in Larsemann Hills, East Antarctica

Sl. no	Code	Location name	Type
1	L1	Grovness Peninsula	Surface shallow depression storage
2	L2	Grovness Peninsula	Surface shallow depression storage
3	L3	Grovness Peninsula	Surface shallow depression storage
4	L4	Grovness Peninsula	Surface shallow depression storage
5	L5	Stornes Peninsula	Precipitation dominant
6	L6	Stornes Peninsula	Precipitation dominant
7	L7	Sigdoy Island	Surface shallow depression storage
8	L8	Sigdoy Island	Surface shallow depression storage
9	L9	McLeod Island	Precipitation dominant
10	L10	Broknes Peninsula	Surface shallow depression storage
11	L11	Broknes Peninsula	Precipitation dominant
12	L12	Broknes Peninsula	Precipitation dominant
13	L13	Broknes Peninsula	Precipitation dominant
14	L14	Broknes Peninsula	Precipitation dominant
15	L15	Broknes Peninsula	Precipitation dominant
16	L16	Sigdoy Island	Surface shallow depression storage

Where R represents either $^{18}O/^{16}O$ or D/H ratio.

Details of the lakes selected for the study are provided in Table 1. For isotopic measurements, water was collected from all the 16 lakes, whereas 13 lakes were chosen for ionic determination. All the lakes were shallow and contained freshwater.

3 Results and Discussion

3.1 *Hydrochemical Composition of the Lakes*

Table 2 illustrates the hydrochemical data of the lakes in the study area. pH varied from slightly acidic (6.2) to alkaline (7.7) in the lake water. pH > 7 was noted in the Sigdoy and McLeod island lakes compared to the peninsular lakes (Groveness, Broknes and Stornes). Electrical conductivity (EC) showed large variation with a minimum of 13.8 μS/cm at L6 (Stornes Peninsula) to the maximum of 1330 μS/cm at L9, McLeod Island. Based on EC, the lakes can be categorized into three, such as EC < 500 μS/cm (L1–L7, L12–L13) and EC = 500–1000 μS/cm (L11) and EC > 1000 μS/cm (L8–L10). Lakes in the GP and SP belonged solely to the first category, whereas the lakes in the MI and one each from SI and BP belonged to the higher EC category. Only one lake in BP was in the second category. The lakes in the BP showed a large variation in ionic content falling in the three categories. Similarly,

Table 2 Hydrochemical composition of the selected lake water in the study area

Code	pH	EC (μS/cm)	TDS (ppm)	Na (mg/L)	K (mg/L)	Ca (mg/L)	Mg (mg/L)	Cl (mg/L)	HCO_3 (mg/L)	SO_4 (mg/L)
L1	6.38	360	192	3.4	5.1	48.5	2.1	80.2	12.9	9.7
L2	6.99	383	203	5.0	3.1	53.5	2.9	84.0	13.1	19.3
L3	6.86	205	112	1.7	4.1	29.4	1.9	47.7	10.6	14.5
L4	7.27	234	127	6.7	5.1	29.9	1.8	51.5	9.8	29.0
L5	6.29	274	145	5.0	3.1	32.4	1.9	57.3	6.7	9.7
L6	6.22	13.8	7	1.7	0.0	1.5	1.2	1.9	4.0	4.8
L7	6.4	141	72	3.4	1.0	17	1.5	22.9	3.3	14.5
L8	7.66	1210	610	25.2	23.5	130	7.3	274.8	3.9	120.8
L9	7.27	1330	690	10.1	21.5	170	8.7	324.5	35.9	33.8
L10	7.06	1210	650	15.1	21.5	140	6.6	284.4	39.9	38.6
L11	6.41	532	284	3.4	5.1	75	3.8	110.7	19.0	24.2
L12	6.75	373	195	6.7	5.1	43.5	2.3	61.1	33.0	24.2
L13	6.65	163	87	3.4	2.0	21.2	1.6	26.7	7.3	4.8

the SI lakes belonged to one of the lowest and highest ionic content categories simultaneously. Overall, the islands' lakes showed higher conductivity, perhaps due to the sea's proximity or increased evaporation as being closed basins. Among the major ions, calcium and chloride ions dominated invariably in all the sampled lake water. In the cationic part, either K or Na was present next to Ca; and among the anions, sulphate was present substantially than bicarbonate ions in most lakes.

3.2 Major Hydrochemical Interactions

The lake water's inter-ionic relations were determined with Pearson's product momentum correlation (Table 3). Most of the ions in the study area's lakes were correlated with each other either strongly or moderately. Bicarbonate ion was the only ion that did not show any relationship with any other except for Ca. The strong correlation among the ions of the lakes can be attributed to their common origin.

To find out the origin of ions in the lake water, Gibb's classification was used (Fig. 2), which can differentiate between the three significant sources such as (1) evaporation/seawater dominant, (2) rock-water interaction/weathering dominant and (3) precipitation prevalent.

Though the concentration of ions in these lakes was low, except for one lake in the Stornes Peninsula (SP), all other lakes showed the predominance of weathering reactions. As these lakes are shallow and primarily temporary, the host rocks' interactions are significant. As obtained in Piper's diagram, the major geochemical facies present in the lake water were Ca–Cl, pointing to lake water's strong interaction with the host rocks. It can be considered that in the pristine waters of Antarctica, mineralogy defines the ionic character of the lakes (Fig. 3).

Table 3 Correlation coefficients of the hydrochemical components of lakes in the Larsemann Hills *(Significant coefficients are given in bold and italics)*

	pH	EC	TDS	Na	K	Ca	Mg	Cl	HCO_3	SO_4
pH	1									
EC	***0.69***	1								
TDS	***0.68***	**1.00**	1							
Na	***0.77***	**0.81**	**0.80**	1						
K	***0.75***	**0.98**	**0.98**	**0.88**	1					
Ca	***0.66***	**0.99**	**0.99**	***0.74***	**0.95**	1				
Mg	***0.69***	**0.99**	**0.98**	***0.78***	**0.96**	**0.99**	1			
Cl	***0.69***	**1.00**	**1.00**	***0.79***	**0.98**	**0.99**	**0.99**	1		
HCO_3	0.27	0.61	0.63	0.22	0.53	***0.65***	0.57	0.60	1	
SO_4	***0.77***	***0.70***	***0.68***	**0.94**	***0.77***	0.63	***0.68***	***0.67***	0.03	1

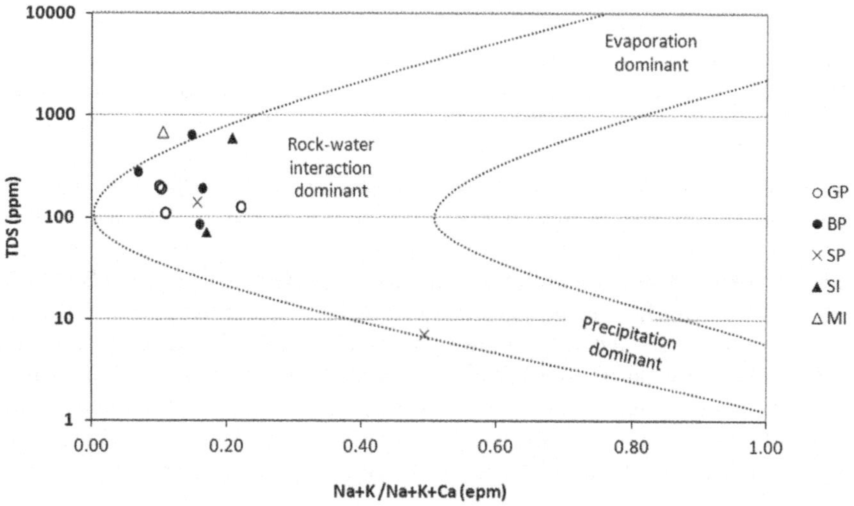

Fig. 2 Gibb's classification of lake water in the Larsemann Hills

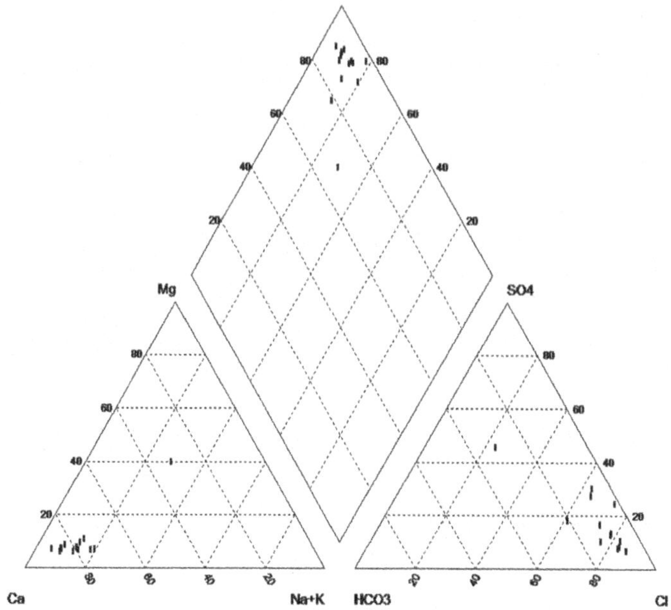

Fig. 3 Piper diagram of lake water of the Larsemann Hills, East Antarctica

Different ionic relations were plotted to understand the significant weathering reactions responsible for the lake water's ionic composition (Fig. 4a–f).

The relation of the lakes' total alkalinity, which is mainly due to the bicarbonate alkalinity, with total cations (TZ^+) (Fig. 4a) showed an excess of cations, which cannot be balanced by the bicarbonate ion alone. It has been reported earlier that the rate of weathering reactions is very high in Larsemann Hills's peninsulas (Kiernen et al. 2009). Consequently, the dissolved carbon dioxide was used up, and accordingly, bicarbonate ion concentration is lesser in these lakes. Ca and Mg ions were plotted with bicarbonate and sulphate ions (Fig. 4b), and as can be seen, there is an excess of Ca and Mg, which points to additional sources of these ions. Since the geology of the area is determined mainly by silicate rocks, silicate weathering can be the lake's ions source. To find out the silicate weathering, Na + K and Ca + Mg were plotted against total cations separately (Figs. 4c, d). Na and K were not varying linearly with total cations or did not fall on the $0.5TZ^+$ line, and there can be a possibility of Ca/Na exchange reactions. However, as the Ca and Mg ions were

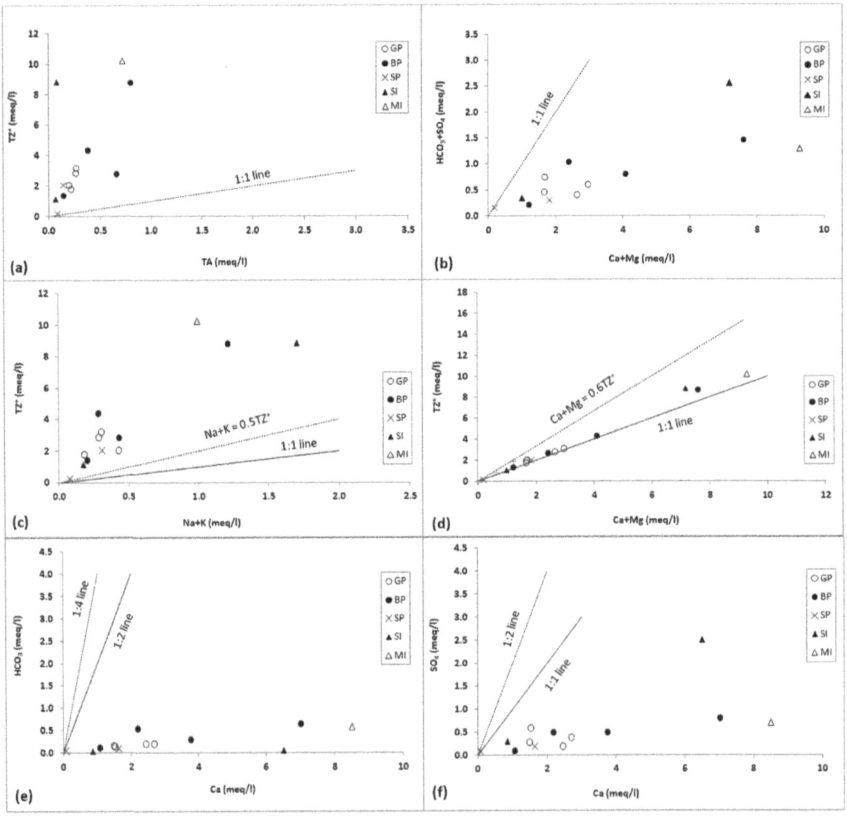

Fig. 4 Variation of **a** Total alkalinity with total cations **b** Calcium and Magnesium with Bicarbonate and Sulphate **c** Sodium and Potassium with total cations **d** Calcium and Magnesium with total cations **e** Calcium with Bicarbonate **f** Calcium with Sulphate

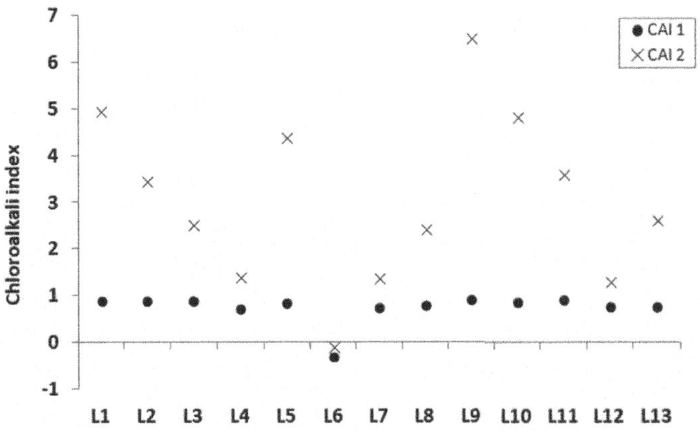

Fig. 5 Chloro alkali indices in the lake water of Larsemann Hills

varying linearly with total cations in most of the lakes, these ions may have originated by silicate weathering reactions. Ca to Mg molar ratios were >2 in most of the lakes, indicating the predominance of silicate weathering reactions (Katz et al. 1998). Moreover, the plot of Ca with bicarbonate and sulphate (Figs. 4e, f) showed excess calcium in the lake water due to different geochemical processes such as exchange reactions contributing these ions.

Chloroalkali indices (CAI 1 & 2) devised by Schoeller (1965, 1967) were used to determine the lakes' ion exchange reactions. If both the indices are negative, there is an exchange of Ca and Mg of the water with Na and K of the earth material. If reverse ion-exchange reactions are occurring, the indices will be positive and accordingly, Ca and Mg of the rock material will get exchanged with Na and K ions in water.

Except for L6 of SP, all the lakes showed reverse ion-exchange reactions, contributing Ca and Mg to the lake water (Fig. 5). Hence apart from silicate weathering, Ca ions are added to the water by the reverse ion exchange process accounting for the excess Ca present in the water. A direct ion-exchange reaction was observed in L6, in Stornes Peninsula, and the primary source of Na and Ca into this lake water is the weathering of silicate minerals.

Table 4 shows the principal component analysis results of the lake water in the study area. Two major factors were extracted with varimax rotation, which could explain 93.7% of the total variance of data. Eigen values >7 were selected in this analysis. In factor 1, pH, SO_4, Na, and K showed strong loadings, and EC, TDS, Mg and Cl showed moderate loading. This factor could alone explain 80.7% of the total variance. EC, TDS, Ca, Mg, Cl and HCO_3 loaded strongly and K moderately in Factor 2.

From the PCA, two dominant origins of ions can be discerned: seawater and, second, weathering reactions. The entrapped seawater in the closed basins or the marine aerosols transported by the solid Katabolic winds may be responsible for the lakes' ionic composition. As seen in the previous sections, weathering and exchange

Table 4 Varimax rotated factor loadings of PCA analysis of lake water of Larsemann Hills

	Factor 1	Factor 2
pH	**0.804**	0.283
EC	*0.637*	**0.765**
TDS	*0.616*	**0.782**
Na	**0.918**	0.303
K	**0.725**	*0.673*
Ca	0.563	**0.814**
Mg	*0.630*	**0.751**
Cl	*0.624*	**0.770**
HCO_3	−0.106	**0.926**
SO_4	**0.970**	0.105
Eigenvalue	8.068	1.301
% total variance	80.7	13.0
Cumul. %	80.7	93.7

reactions form the second component. As the lakes are shallow and fleeting, evaporation can be another process that increases the ionic concentration in lake water. Since Lake L6 showed clear precipitation origin, to check the evaporation effect, this lake was omitted. Since the Na/Cl molar ratio remains constant even when EC is increased due to evaporation, these two variables' covariance was checked (Fig. 6).

In Fig. 6, the dotted line gives the linearity of lakes without L6 and the solid line with L6. If L6 is considered an outlier, the line's slope is very close to 0, i.e. evaporation is a dominant process in these lake systems. To get a clear picture of these

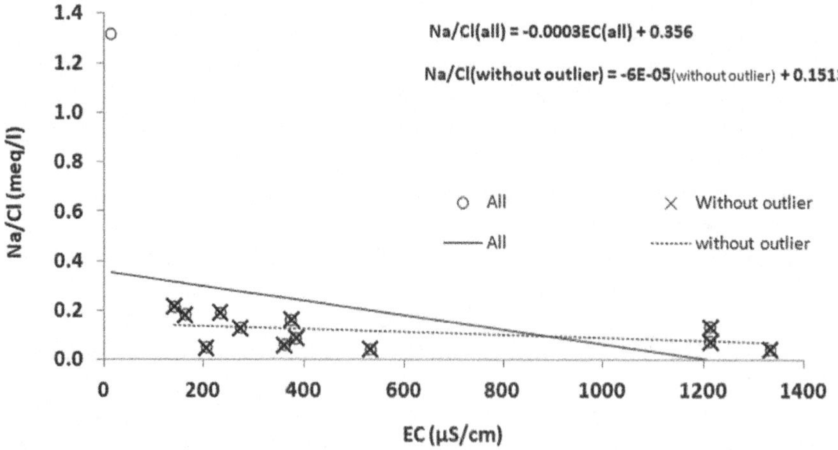

Fig. 6 Variation of EC with Na/Cl molar ratio in the lake water of Larsemann Hills

processes, isotopic ratios were determined, the observations of which are provided in the subsequent sections.

3.3 *Isotopic Characterization of Lakes of Larsemann Hills*

The stable isotope composition of 16 lake samples of the study area was determined, and the results are depicted in Table 5. Similar to the chemical composition, a considerable variation is observed in these lake waters' isotopic ratios. $\delta^{18}O$ ranged between –7.6 ‰ and −17.9 ‰ and δD between −108.7 ‰ and −146.5 ‰.

Compared to other lakes, such as in the Schirmacher Oases in East Antarctica ($\delta^{18}O = -30$ ‰, Gopinath et al. 2020), Larsemann Hills's lakes have an enriched isotopic composition. L9, located in the McLeod island, was the most enriched ($\delta^{18}O = -7.6$‰). The lakes in the Groveness Peninsula (GP) has an average isotopic composition of ($\delta^{18}O = -14.6$‰, $\delta D = -121.3$‰); lakes in Stornes Peninsula (SP) has $\delta^{18}O = -16.8$‰, $\delta D = -141.7$‰, Sigdoy Island (SI) has $\delta^{18}O = -12.5$‰, $\delta D = -110.6$‰ and the Broknes Peninsula (BP) has $\delta^{18}O = -14.3$‰, $\delta D = -121.9$‰. Lakes located in the islands were enriched, and, on average, a difference of 5‰ was noted as that of the peninsular lakes. We separated the lakes into precipitation dominant and surface depression storage and compared the isotopic composition (Fig. 7).

Table 5 Stable isotope data of lake waters of Larsemann Hills, East Antarctica

Sl. no	Code	Location name	$\delta^{18}O$ (‰)	$\delta^{2}H$ (‰)	d-excess
1	L1	GP	−17.4	−138.7	0.08
2	L2	GP	−12.7	−110.6	−9.08
3	L3	GP	−13.4	−113.3	−6.02
4	L4	GP	−14.8	−122.4	−4.09
5	L5	SP	−17.9	−146.5	−3.12
6	L6	SP	−16.8	−141.7	−7.06
7	L7	SI	−15.0	−130.2	−10.4
8	L8	SI	−11.8	−111.0	−16.3
9	L9	MI	−7.60	−108.7	−48.3
10	L10	BP	−12.7	−112.0	−10.5
11	L11	BP	−12.8	−117.0	−14.6
12	L12	BP	−11.0	−104.7	−16.6
13	L13	BP	−17.3	−138.7	−0.66
14	L14	BP	−17.9	−142.5	0.98
15	L15	BP	−14.0	−116.5	−4.19
16	L16	SI	−13.3	−110.3	−4.31

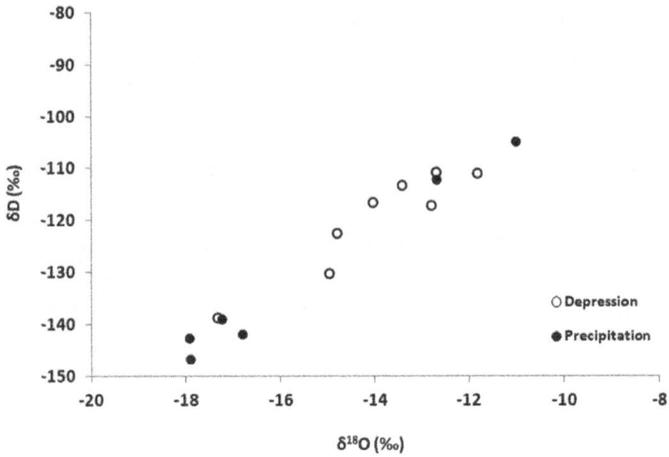

Fig. 7 Isotopic composition of the precipitation dominant lakes and shallow surface depression lakes in Larsemann Hills, East Antarctica

As expected, most of the lakes in the precipitation-dominant category were depleted in heavier isotopes than the shallow depression storage lakes. However, Lake L1 (GP), though being depression storage, was depleted in heavier isotopes may be because of the inflow of isotopically depleted melt eater. Similarly, L9 (MI) and L12 (BP) had isotopic enrichment though they were precipitation dominant. The contrasting behaviour can be attributed to the evaporation processes occurring in the lakes.

3.4 $\delta^{18}O$-δD Covariance in the Lake Water

Figure 8 shows the covariation of δ^{18}O and δD in the lake waters. The lake water line (LWL) has a slope of 4.54, close to that observed for systems dominant in evaporation. Horita (2009) have shown that in a naturally freezing method, the remaining water will be depleted in heavier isotopes and will have a slope of 5–6 in the regression plot of δ^{18}O-δD that are similar to the slopes due to evaporation processes.

The lake water line's intersection point (LWL) with the global meteoric water line represents the isotopic composition of the region's precipitation. Thus $\delta^{18}O = -20‰$ and $\delta D = -144‰$ can be considered as the initial isotopic composition of precipitation. A few lakes in the BP and SP have δ values close to this value but are more enriched, probably due to the kinetically controlled ice-water fractionation processes. However, the slope of LWL is mainly defined by other lakes in which evaporation can be the dominant process. The d-excess values obtained also are in accordance. The evaporation process enriches the water body lowering the d-excess

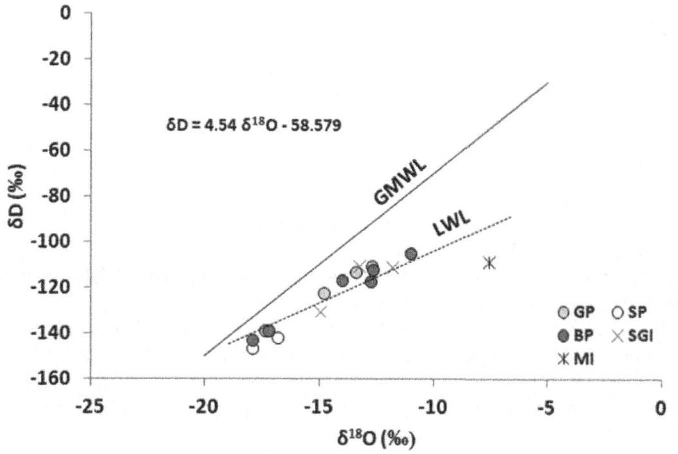

Fig. 8 Regression plot $\delta^{18}O$ versus δD of the lake waters. The evaporation line or lake water line is also given

of the residual water. The island lakes have the lowest d-excess values being shallow basins with entrapped seawater.

4 Conclusions

Hydrochemical and isotopic characteristics of the lakes in the Larsemann Hills in East Antarctica were determined. Lakes in the Grovness Peninsula, Stornes Peninsula, Broknes Peninsula, Sigdoy and McLeod Islands were selected for the study. The island lakes' pH was slightly alkaline, and the major hydrochemical facies identified in the lake water was Ca–Cl. There was strong interaction of lake water with the host rocks, and weathering and reverse ion exchange processes were the source of ions. The lakes have an enriched isotopic composition than the lakes in the Schirmacher Oases. The isotopic ratios could identify the lakes as precipitation dominant and evaporation dominant. The kinetic controlled ice-water fractionation and evaporation processes control the lake water's isotopic composition in the Larsemann Hills.

Acknowledgements The authors gratefully acknowledge the support rendered by the Director, Indian Institute of Geomagnetism (IIG) and Director, National Centre for Polar and Ocean Research (NCPOR) for the permission and support to carry out this study. The authors express their sincere thanks to the Executive Director, CWRDM (Centre for Water Resources Development and Management), Kozhikode, for its facilities.

References

ATCM XXXVII Final Report (2014) Larsemann Hills, East Antarctica Antarctic Specially Managed Area Management Plan

APHA (1995) Standard methods for the analysis of water and wastewater, 19th edn. American Public Health Association, Washington, D.C.

Beg JM, Asthana R (2002) Geological studies in the Larsemann Hills, Ingrid Christensen Coast, East Antarctica. In: 24th Indian Antarctic expedition 2003-2005, Ministry of Earth Sciences, Technical Publication No. 22, pp 363–367

Bharti PK, Niyogi UK (2015) Assessment of pollution in a freshwater lake in Fisher Island, Larsemann Hills over East Antarctica. Sci Int 3(1):25–30

Gasperon M, Lanyon R, Burgess JS, Sigurdsson IA (2002) Freshwater lakes in Larsemann Hills, East Antarctica: chemical characteristics of the water column. ANARE Report 147

Gopinath G, Resmi TR, Praveenbabu M, Pragath M, Sunil PS, Rawat R (2020) Isotope hydrochemistry of the lakes Schirmacher Oasis, East Antarctica. Indian J Geo Mar Sci 49(6):947–953

Hodgson DA (2012) Antarctic lakes. In: Bengtsson L, Herschy RW, Fairbridge R (eds) Encyclopedia of lakes and reservoirs. Springer, Berlin, pp 26–31

Horita J, Rozanski K, Cohen S (2008) Isotope effects in water evaporation: a Craig-Gordon model's status report. Isot Environ Health Stud 44(1):23–49

Katz BG, Coplen TB, Bullen TD, Davis JH (1998) Use chemical and isotopic tracers to characterize the interaction between groundwater and surface water in mantled karst. Ground Water 35:1014–1028

Kaup E, Burgess JS (2002) Surface and subsurface flow of nutrients in natural and human-impacted lake catchments on Broken, Larsemann Hills Antarctica. Antarct Sci 14(4):343–352

Kiernen K, Gore DB, Fink D, White DA, McConnel A, Sigurdsson IA (2009) Deglaciation and weathering of Larsemann Hills, East Antarctica. Antarct Sci 21:373–382

Nakai N, Wada H, Kiyosu Y, Takimoto M (1975) Stable isotope studies on the origin and geological history of water and salts in the Lake Vanda area, Antarctica. Geochem J 9:7–24

Quayle WC, Peck LS, Peat H, Ellis-Evans JC et al (2002) Extreme responses to climate change in Antarctic lakes. Science 295:645

Sabbe K, Hodgson DA, Verleyen E et al (2004) Salinity, depth and the structure and composition of microbial mats in continental Antarctic lakes. Freshwater Biol 49:296–319

Schoeller H (1965) Qualitative evaluation of groundwater resources. In: Methods and techniques of groundwater investigations and development, UNESCO, pp 54–83

Schoeller H (1967) Geochemistry of groundwater—an international guide for research and practice, UNESCO, pp 1–18

ShrivastavaP K, Asthana R, Roy SK, Swain AK, Dharwadkar A (2012) Provenance and depositional environment of epi-shelf lake sediment from Schirmacher Oasis, East Antarctica, via vis scanning electron microscopy of quartz grain, size distribution and chemical parameters. Polar Sci 6(2):165–182

Verleyen E, Hodgson DA, Sabbe K, Vyverman W (2004) Late Quaternary deglaciation and climate history of the Larsemann Hills (East Antarctica). J Quart Sci 19(4):361–375

Verleyen E, Hodgson DA, Gibson J et al (2012) Chemical limnology in coastal east Antarctic lakes: monitoring future climate change in endemism and biodiversity centres. Antarc Sci 24(1):23–33

Wand U, Hermichena W-D, Brüggemannb E, Zierath R, Klokov VD (2011) Stable isotope and hydrogeochemical studies of Beaver Lake and Radok Lake, MacRobertson Land, East Antarctica. Isot Environ Health Stud 47(4):407–414

Effect of Ionosphere Scintillations on the Loss of Lock-In GPS Signals at Antarctica Region

A. K. Gwal, Suryanshu Choudhary, and Ritesh Yadav

Abstract We studied the effect of amplitude and phase scintillations on the loss of lock during five most disturbed days of the December 2006. During all the disturbed days weak, moderate or intense geomagnetic storms were observed. The main findings of the study are following; during all the disturbed days the amplitude and phase scintillations of moderate intensity were observed. The amplitude and phase scintillation become frequently, the visible PRNSs does not remains stable for longer periods. The loss of lock occurs frequently whenever the GPS signals scintillate. In present studies the diurnal variation of TEC goes higher during the daytime but it decreases gradually in the Antarctica region.

Keywords Scintillations · Loss of lock · Geomagnetic storms · GPS · Radio wave

1 Introduction

Antarctica is a land of extremes; it is the highest, driest and coldest continent. In many ways, it is a Paradise of Science, and many scientific phenomena have happened, having weird and unknown facts about earth and sun relation. In between the Magnetosphere and Ionosphere, incoming solar wind particles disturb the earth ionospheric condition. Ionospheric scintillation studies in the polar region are also fascinating and provide scientific results for space weather phenomena. Many instruments and scientific payloads are capable of delivering valuable and real-time information of the polar ionosphere. The Global Positioning System (GPS) has accuracy and reliability, and its GPS receiver's performance gives a perfect image of the earth ionosphere. When a radio wave (either from the satellite or the cosmic noise, especially from the

Present Address:
A. K. Gwal (✉) · S. Choudhary · R. Yadav
Department of Physics, Rabindranath Tagore University, Raisen 464993, India

Present Address:
Department of Physics, Dr A P J Abdul Kalam University, Indore 452016, India

N. Khare (ed.), *Assessing the Antarctic Environment from a Climate Change Perspective*, Earth and Environmental Sciences Library,
https://doi.org/10.1007/978-3-030-87078-2_11

radio star) traverses in an irregular ionosphere (i.e. ionospheric irregularities of electron density), it develops a random phase of fluctuations across the wave-front. As the wave-front travels towards the ground, phase mixing occurs. Due to relative motion between the satellite, ionospheric irregularities and the receiver on the ground, the spatial pattern of amplitude and phase variations sweeps past the receiver and the temporal variations of phase and amplitude, which is known as scintillations. It is the single most crucial deleterious factor in utilising the earth-space propagation path.

The occurrence of scintillation occurs when the earth's ionised upper atmosphere often becomes turbulent and develops electron density. These irregularities scatter radio waves from satellites in the frequency range of 100 MHz–4 GHz (Basu et al. 1988; Aarons 1993). Scintillations are strong at high latitudes, weak at mid-latitudes and intense in the equatorial region. Scintillation at all margins attains its maximum value during the maximum solar period when the F-region ionisation density increases and the irregularities occur in a background of enhanced ionisation density. Variability of ionospheric irregularities is of grave concern to the GPS because these irregularities affect the amplitude and phase of trans-ionospheric radio (satellite) signals. Amplitude scintillation may induce signal fading, and when the depth of fading exceeds the fade margin of a receiving system, message errors are introduced in satellite communication systems. It leads to loss of lock by degrading the carrier-to-noise ratio (C/N_0) to below the receiver lock threshold (De Oliviera Moraes et al. 2011), requiring the receiver to attempt reacquisition of the satellite signals. Losing signal is a significant concern in GPS receiver navigation performance (Doherty et al. 2000; Phoomchusak et al. 2003; Aquino et al. 2005; Kintner et al. 2007). If navigation is Global Positioning System (GPS) dependent, then amplitude scintillations may lead to data loss and cycle slips (Aarons and Basu 1994). Phase scintillation, characterised by rapid fluctuations of carrier-phase, can be a source of cycle slips and sometimes affect the receiver's ability to hold a lock on a signal. Fortunately, many of the essential characteristics of scintillation are already well known (Basu et al. 1988). These studies revealed that scintillation activity varies with Season, local time, operating frequency, geographic location, magnetic activity and 11-year solar cycle. Most severe scintillations occur for a few hours after sunset in the months of equinoxes at equatorial latitudes during the peak years of the solar cycle.

It is essential to clearly understand the location, magnitude, and frequency of scintillation effects on GPS. It is desirable to have statistical tools to recognise scintillation occurrence and classify its characteristics as effectively as possible. During the high solar activity period, equatorial scintillation is adequately severe, capable of disabling many communication and navigation systems (Groves et al. 1997). The effects of small-scale plasma density irregularities on trans-ionospheric radio signals constitute a problem for a wide range of military and civilian users. The WBMOD computer code was developed beginning in the early 1970s to model these effects to provide planners and system operators a tool for assessing these irregularities' impact on their systems. The WBMOD program uses a collection of empirically derived models to describe the global distribution and behaviour of naturally occurring ionospheric irregularities and a power-law phase screen propagation model to calculate the intensity and phase scintillation estimates irregularities would impose on a user-defined system and geometry. The model outputs are estimates of intensity and phase scintillation levels and occurrence statistics for the user-defined scenario. A report on improvements made to the equatorial sections of the WBMOD model was presented at the 1993 Ionospheric Effects Symposium (Secan et al. 1995). Thus to determine the role of space weather events on scintillations is essential. Many types of research (Aarons et al. 1980; Rastogi et al. 1981,1990; Pathan et al. 1991; Kumar et al. 1993; Kumar and Gwal 2000; Banola et al. 2001; Biktash 2004; Li et al. 2006; Bhattacharya et al. 2010, 2011) have studied the geomagnetic activity effects on the occurrence of scintillation over equatorial and low latitude. A moderate amplitude scintillation (S4 ~ 0.6) may cause more than 10 m error in GPS C/A code positioning (Phoomchusak et al. 2003). Severe scintillation (S4 ~ 0.7) can lead to up-to 22 m latitude error and 14 m longitude error in GPS C/A code positioning (Dubey et al. 2006). Single point precise positioning error may reach several meters in vertical and tens of centimetres horizontally during intense ionospheric scintillation events (Moreno et al. 2010). Scintillation is severe in the equatorial region, strong at high latitudes and weak at mid-latitudes (Basu et al. 1988). High latitude aurora irregularities are formed from the precipitation of energetic electrons along terrestrial magnetic field lines into at high latitude ionosphere. These electrons are energised through a complex interaction between the solar wind and the earth magnetic field, resulting in optical and UV emissions commonly known as the auroras. This phenomenon characterises the magnetosphere sub-storm, where associated irregularities in electron density lead to scintillations (Aarons 1982). Ionosphere irregularities at high latitude are of interest for users of transmissions and those studying aeronomy. With the advent of GPS and GPS use for geodesy, a network of stations has continuously reported data.

In the polar region, irregularities are typical. These irregularities on different scale cause fluctuations in a signal whose scale is more extensive than 100–300 km and occur as deep spatial variations of TEC. High latitude and polar cap scintillation are mainly produced by geomagnetic storms associated with Coronal Mass Ejections (CME) and coronal hole. Compared to equatorial scintillation, high latitude scintillations show slight diurnal variation in their rate of occurrence. They can last from a few hours to days, and they can begin at any time (Klobuchar 1991). The development of TEC fluctuations over the Antarctica region of the earth has been studied with different satellite transmissions over several years. It is found that the intermediate phase and amplitude scintillation occurrence at high latitudes while working with three stations. He found maximum phase scintillation at Churchill studied amplitude and phase scintillation at high latitudes for seasonal patterns and distinct storms. These studies have shown correction between amplitude and TEC fluctuation occurrence. High margin and polar cap scintillation also offer a seasonal variation opposite to that observation at low latitude region, being peaked during autumn equinox through winter to the vernal equinox and found maximum during summer (Since the occurrence of geomagnetic storms is solar activity-dependent through sunspot numbers, solar flares and coronal mass ejections, aurorally and polar cap scintillation also enormously varies with the 11-year solar cycle, being on the peak during solar maxima and least occurred during solar minima.

Recently, several types of research used GPS permanent observations to study irregularities in the auroral region and utilised the GPS data at 30 s intervals to study ionospheric irregularities of electron density by computing the time rate of change of the differential carrier phase. This is equivalent to the quality of change of the total electron content, termed ROT, in units of TEC/min have demonstrated the utility of such dataset for studying the evolution of different scale irregularities during magnetic storms at high latitude. A rate of change of TEC index (ROTI) based on a standard deviation of ROT over a 5 min period. This index statistically quantifies the ROT measurements (Pi et al. 1997). The intensive phase fluctuations observed along GPS satellite passes are caused by dramatic changes in total electron content (TEC) and a solid horizontal gradient of TEC. Fluctuation effects and TEC gradient can have a different impact on GPS measurements and data processing for high-precision GPS positioning. They affect phase ambiguity resolution, increases the number of undetected and uncorrected cycle slips and loss of signal lock (Wanninger 1993; Krankowski et al. 2002). In the high latitudes phase, scintillation can be more severe than amplitude scintillation.

At low latitudes, both amplitude and phase scintillations may occur, but in general, amplitude scintillation is more severe than phase scintillation (Gwal et al. 2006; Forte 2012). Both amplitude and phase scintillations can degrade the GPS positioning performance by increasing the tracking error, number of cycle slips and the probability of losing lock. In precise GPS positioning, based on double differenced (DD) carrier-phase observables, the vital issue is to resolve the ambiguities to their integer values and derive an improved estimator of the baseline coordinates (i.e., fixed solution). Therefore, the ambiguity resolution (AR) performance determines the quantity of the resulting position coordinates. The development of TEC fluctuations and their impact on GPS signal loss of lock is presented from the data collected at Indian Antarctic Base Station, Maitri, during the solar activity period 2006.

2 Datasets and Methodology

This study evaluates GPS performance during extreme or disturbed ionospheric conditions; consequently, we have chosen five troubled days of December 2006. At the same time, we have also selected the five quietest days of the same month. During these disturbed days, the GPS performance was evaluated by considering the loss of GPS signals during the scintillation events. Irregularities of different scales characterise the disturbed ionosphere. When the GPS signals encounter these irregularities, significant changes occur in their phase and amplitude, commonly known as scintillations. The ionospheric scintillations were monitored using NovAtel's dual-frequency GISTM- (GPS Ionospheric Scintillation and TEC Monitor) based GPS receiver GSV4004A. The receiver has been widely used to monitor ionospheric scintillations and TEC in the past as well. The receiver performs amplitude and phase measurements at a 50 Hz sampling rate and measures carrier-code divergence at the 1 Hz rate for each satellite tracked on L1. It computes TEC from combined L1 L2 pseudorange and carrier phase measurements. The 12 channel is used to measure noise for C/N_0 as well as scintillations computations. The receiver was installed in the sub auroral region at the Indian Antarctic Base station, Maitri (70.45°S, 11.45°E), Antarctica, during the low solar activity period 2006. Disturb and quiet days are considered for investigation and calculation of Vertical Total Electron Content (VTEC), measurement of phase and amplitude scintillation during the loss of lock for each PRN, carrier to noise ratio (C/N_0) is also measured.

3 Results

The small-scale irregularities in the electron density that usually occur during disturbed solar and geomagnetic conditions can diffract the signal, leading to rapid fluctuations in signal intensity and phase, known as amplitude and phase scintillations, respectively. Amplitude and phase scintillations can be severe enough for the received GPS signal intensity to drop below a receiver's lock threshold, forcing the receiver to lose lock. We have considered five geomagnetically disturbed days of December 2006 and investigated the effect of scintillations on lock loss during high geomagnetic activity.

3.1 15th December 2006

The 15th of December 2006 was the most disturbing day in December, as indicated by the geomagnetic indices. On this day, an intense geomagnetic storm was observed. The DST index's maximum or peak value occurred at 07:00 hrs UT with a peak value of −162 nT. The other geomagnetic indices like Kp and AE also underwent a significant increase in achieving peak values of 8.3 and 1372 nT. Therefore, this day was identified as a disturbing day. We now show the occurrence and effect of scintillation during this event on the loss of lock. Figure 1 represents the 15 min Averaged Total Electron Content's hourly variation using all visible PRNs during the 15th of December 2006. It means a usual diurnal patter with TEC achieving a peak of about 28 TECU around 12:00 hrs UT.

3.2 Average VTEC (15 Min) 15 Dec 2006

Figure 2 shows the temporal variation of amplitude and phase scintillation indices along with the VTEC observed by all visible PRNs at Maitri, Antarctica, on the 15th of December 2006. The S4 represents the amplitude scintillation, and the Phi

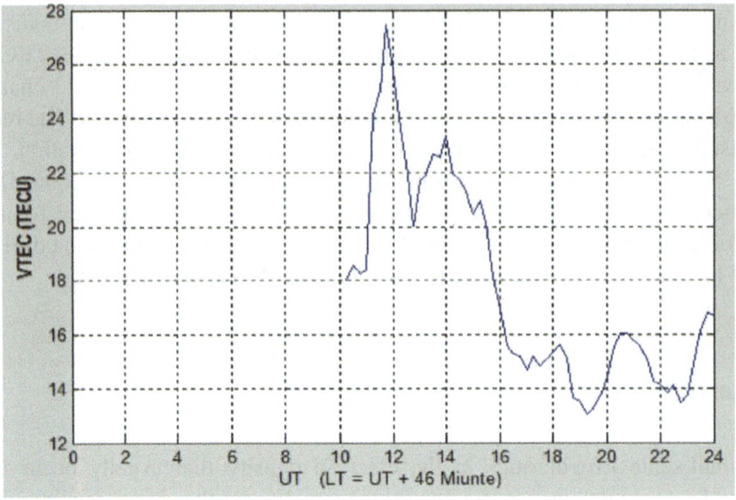

Fig. 1 The diurnal variation of total electron content (VTEC) on the 15th of December 2016

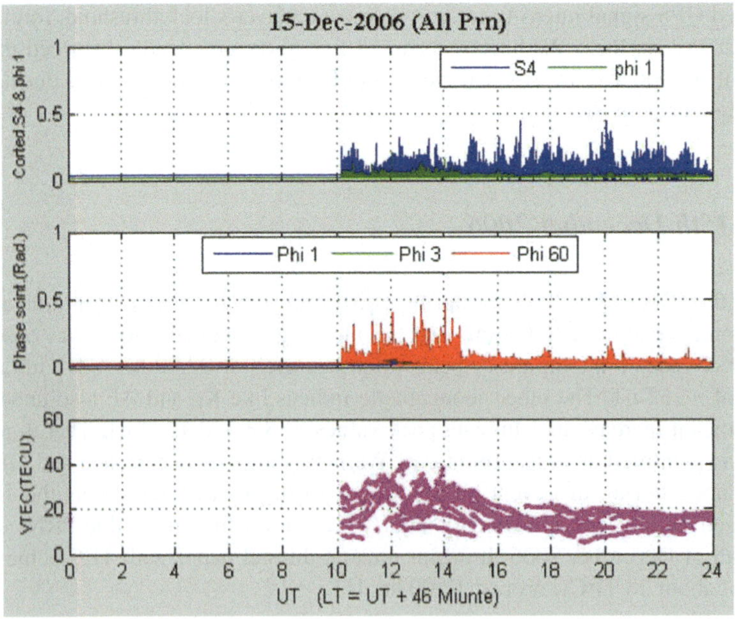

Fig. 2 Temporal variation of amplitude and phase scintillation indices along with VTEC

1, Phi 3 and Phi 60 represent the phase scintillation at 1, 3 and 60 s, respectively. From the figure, we find that all the S4 index undergoes rapid fluctuation, and it reaches a threshold of about 0.5 between 16:00 to 20:00 hrs UT. Similarly, the phase scintillation indices also reach the point of 0.5 during 10:00 to 15:00 hrs UT, indicating moderate scintillation during this period. The bottom panel of the figure shows the variation of VTEC calculated from different PRNs visible at Maitri. The value of VTEC is highest during 10:00 to 14:00 hrs UT with an apparent dip around 12:30 hrs. During this dip of VTEC, the scintillation activity increases.

3.3 15 Dec, 2006 (All PRN)

The relevant effect of these scintillations on the loss of lock can be seen in Fig. 3, where the product is described by the C/N_0 ratio and safety time for different visible PRN's. The C/N_0 ratio goes less than 40 dB and remains disturbed during the time of occurrence of scintillations. The locked PRN's do not seem to be stable for more extended periods, as shown in blue colour and breaking of lock occurs continuously during the amplitude and phase scintillation of the signals.

Different PRNs showing loss of lock (in TEC curve) with the rise in scintillation have been demonstrated in Fig. 4. Their Azimuth and Elevation angle has also been shown in Figs. 5a, b, c, d, e, respectively.

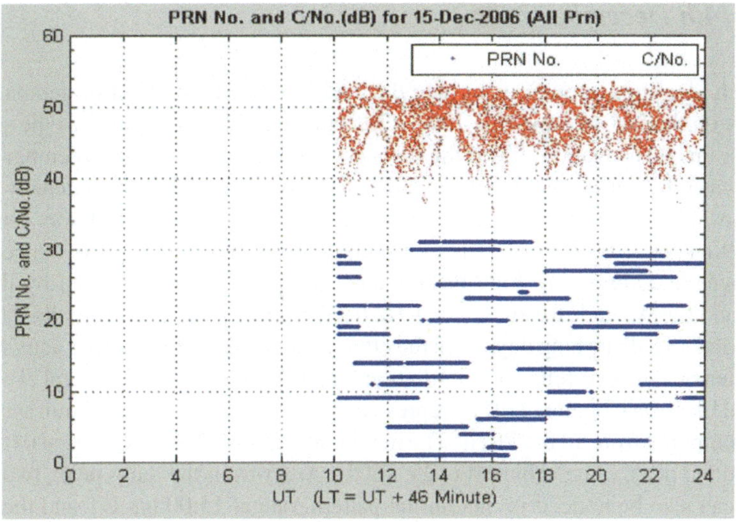

Fig. 3 The C/N0 ratio and lock time for different visible PRNs

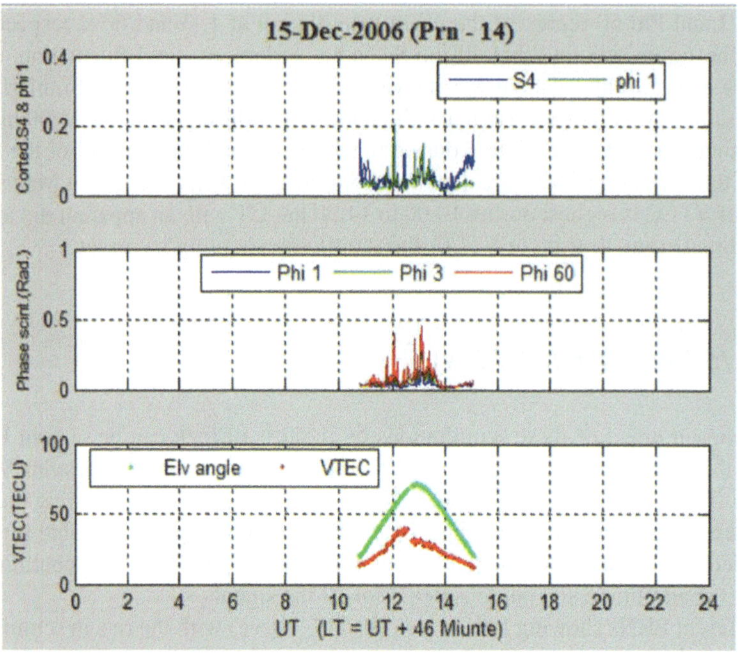

Fig. 4 RNs showing loss of lock-in TEC curve with the rise scintillation

3.4 14th December 2006

We then identified another disturbing day and presented the effect of geomagnetic activity on the loss of lock. The 14th of December 2014 was selected as the second case for this investigation. It was the second most disturbing day of December 2006, as indicated by various geomagnetic indices. The variation of Storm intensity index Dst showed a moderate geomagnetic storm. The peak value of Dst was found to be −69 nT at 23:00 hrs UT. The interpretation of other indices also showed that geomagnetic activity was high on this day. The peak values achieved by the Kp index and the AE index were 5.3 and 1616 nT, respectively. Therefore the day was designated as a disturbing day. Then the effect of increased geomagnetic activity was evaluated on the occurrence of scintillations and loss of lock. The temporal evolution of Total Electron Content (TEC) is represented in Fig. 6. It shows the 15 min averaged TEC computed from all the PRNs. The maximum value of TEC occurred at 08:00 hrs UT and the peak value achieved is 25 TECU. Apart from the daily peak, two other peaks can also be noticed in the diurnal pattern, one at 15:00 hrs UT and the other at 20:00 hrs UT. Moreover, a steep and sharp decrease in TEC can also be noticed around 18:00 hrs UT.

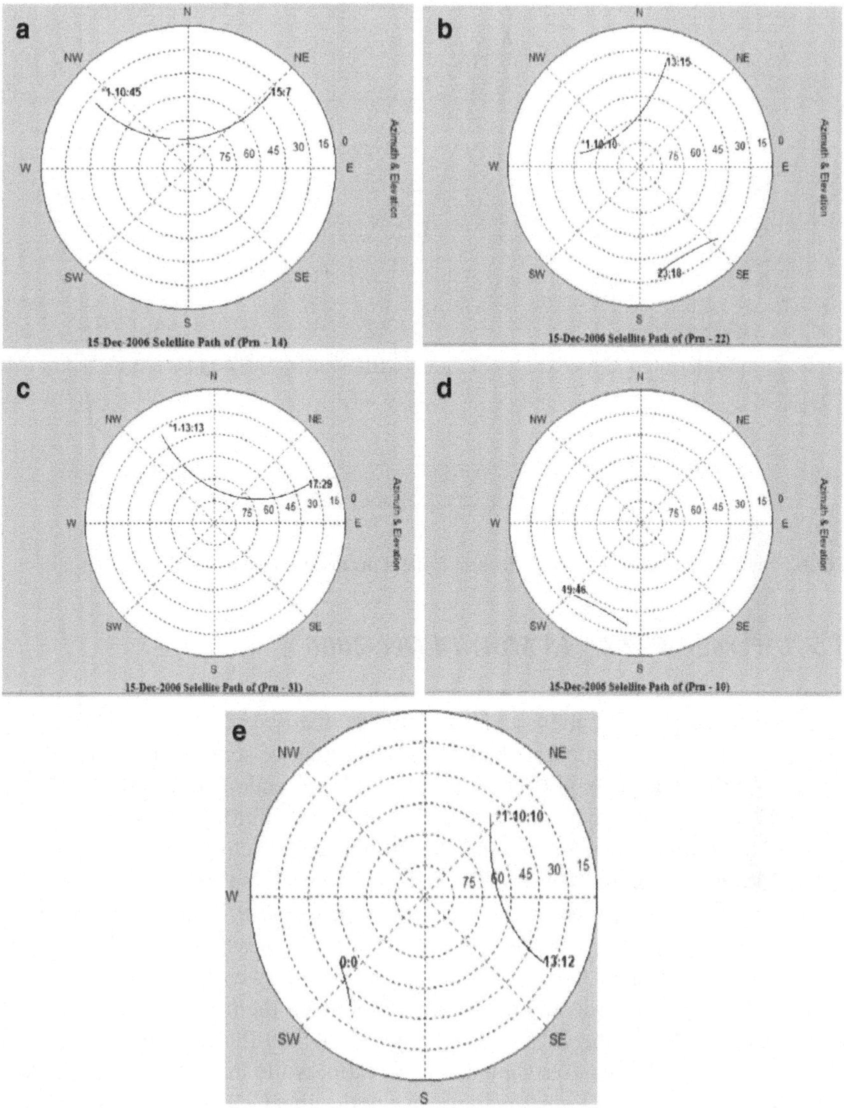

Fig. 5 **a**: Polar plot showing Azimuth and Elevation for PRN-14 during the 15th of December 2006. **b**: Azimuth and Elevation for PRN-22 during the 15th of December 2006. **c**: Azimuth and Elevation for PRN-31 during the 15th of December 2006. **d**: Azimuth and Elevation for PRN-10 during the 15th of December 2006. **e**: Azimuth and Elevation for PRN-09 during the 15th of December 2006

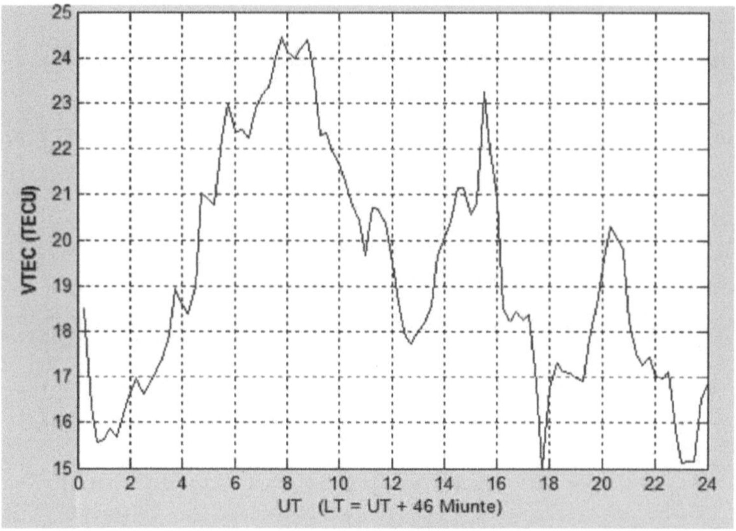

Fig. 6 The diurnal variation of total electron content during the 14th of December 2006

3.5 Average VTEC (15 Min) 14 Dec 2006

The occurrence of scintillations and the VTEC from different visible PRNs is shown in Fig. 7. It shows the temporal changes in the amplitude scintillation index S4 and phases scintillation index phi. The phase scintillation index phi is calculated at 1, 3 and 60 s. From the figure, we notice that both the scintillation indices increase around 18:00 hrs UT to about 0.5 thresholds indicating the occurrence of moderate amplitude and phase scintillations. The scintillation was also observed around 23:00 hrs UT when the geomagnetic storm was in its peak phase. The occurrence of this scintillation corresponds to the depletion in TEC. The values of the S4 index fluctuate continuously throughout the day, indicating the event of weak scintillations.

The effect of amplitude and phase scintillations on the loss of lock is shown in Fig. 7 and changes in the C/N_0 ratio along with locked PRN's during the 14th of December 2006. The red dots show the C/N_0 ratio, while the blue diamond's show the lock of locked PRNs. From the figure, we notice that C/N_0 falls below 40 dB and shows rapid fluctuations during the scintillations.

The lock time of the visible PRN's at Maitri during the time of scintillation, particularly during the main phase of the geomagnetic storm, is not stable for more extended periods. The lock brakes regularly. Therefore, we conclude that the occurrence of amplitude and phase scintillation during the 14th of December 2006 resulted in the loss of locks of the visible PRNs at Maitri, Antarctica, and at different PRN's their Azimuth and Elevator angle has been shown in Figs. 8a, b, c, d, e respectively.

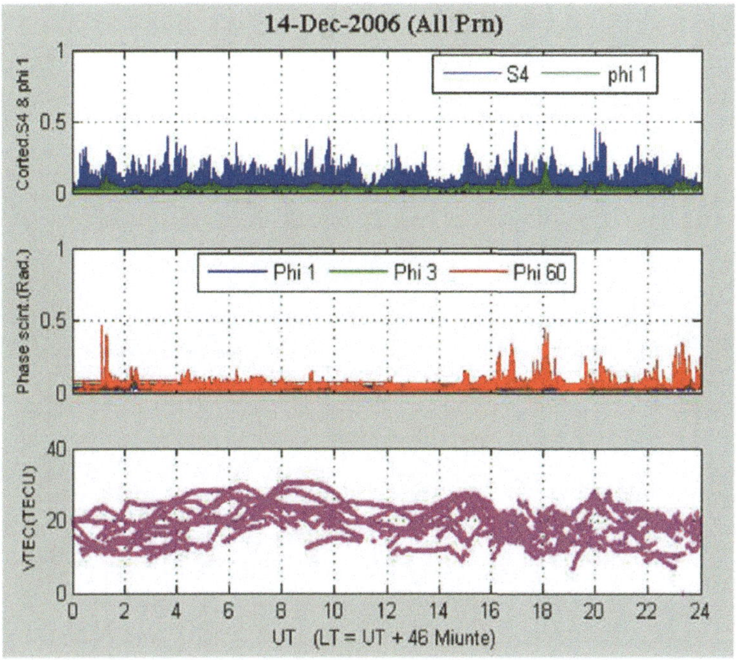

Fig. 7 The C/N_0 ratio and lock time for different visible PRN

3.6 12th December, 2006

The occurrence of amplitude and phase scintillation is shown in Fig. 9, along with the time profile of VTEC of the different PRNs visible at PRNs during the 12th of December 2006. The S4 index describes the occurrence and intensity of amplitude scintillations while the phase scintillations are described by Phi 1, Phi 3 and Phi 60 at 1, 3 and 60 s, respectively. We can quickly notice amplitude scintillations of moderate intensity of 0.5 around 17:00 hrs UT and 21:00 hrs UT from the figure. Similarly, phase scintillations can also be found to occur between 20:00 hrs UT to 23:00 hrs UT. The occurrences of these scintillations correspond to the recovery phase of the geomagnetic storm. Around the same time, the AE index, specific to high latitude regions, peaked at its peak value. Thus during increased geomagnetic activity, particularly at high latitudes, the occurrence of scintillations is quite frequent. At the same time, we found that there is a decrement in VTEC.

The relevant effect of the scintillations of moderate-intensity is evaluated in Fig. 10 and demonstrates the impact of the C/N_0 ratio and the loss of lock. The red dots represent the C/N_0 ratio in decibels, while the blue diamond's show the lock time of different PRN's visible on the 12th of December 2006 at Maitri station. From the figure, we find that the C/N_0 ratio falls below 40 dB and fluctuates rapidly around the time when the amplitude and phase scintillation become frequent. We also notice

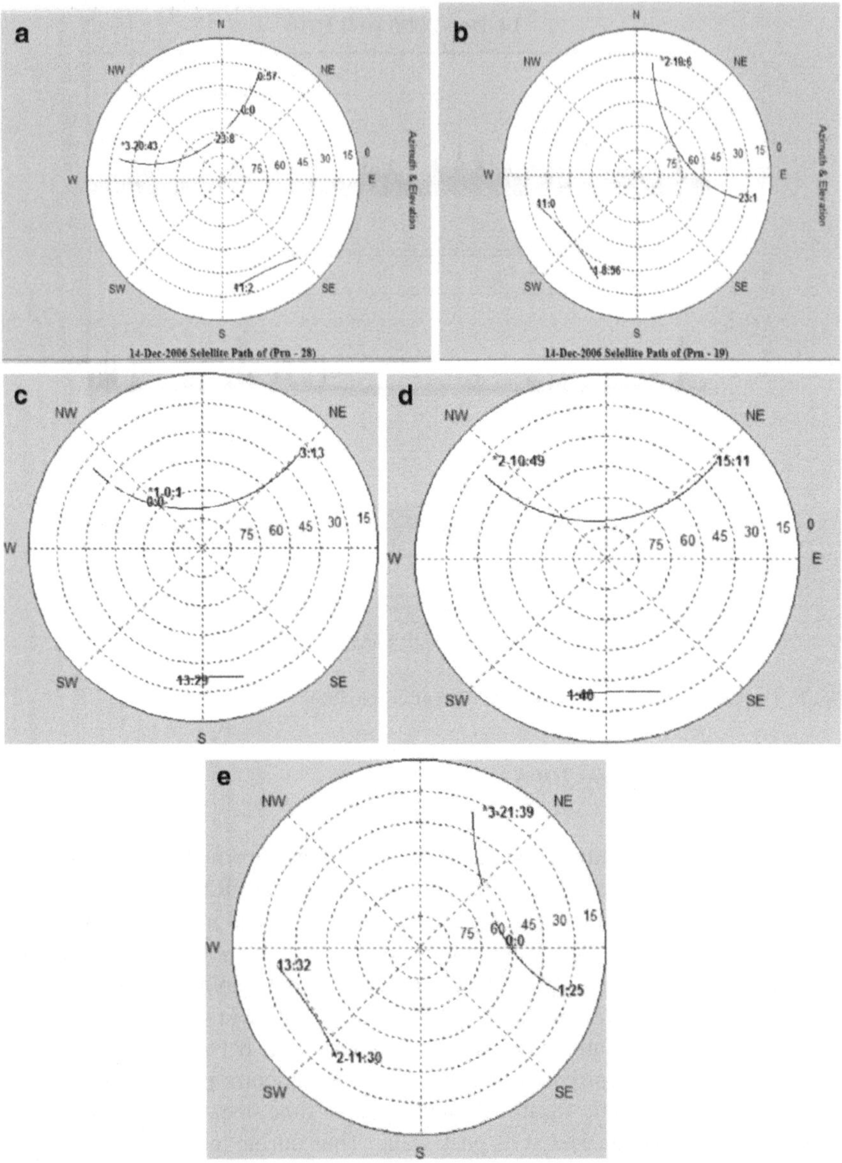

Fig. 8 **a**: Azimuth and Elevation for PRN-28 during the 14th of December 2006. **b**: Azimuth and Elevation for PRN-19 during the 14th of December 2006 **c**: Azimuth and Elevation for PRN-17 during the 14th of December 2006. **d**: Azimuth and Elevation for PRN-14 during the 14th of December 2006. **e**: Azimuth and Elevation for PRN-11 during the 14th of December 2006

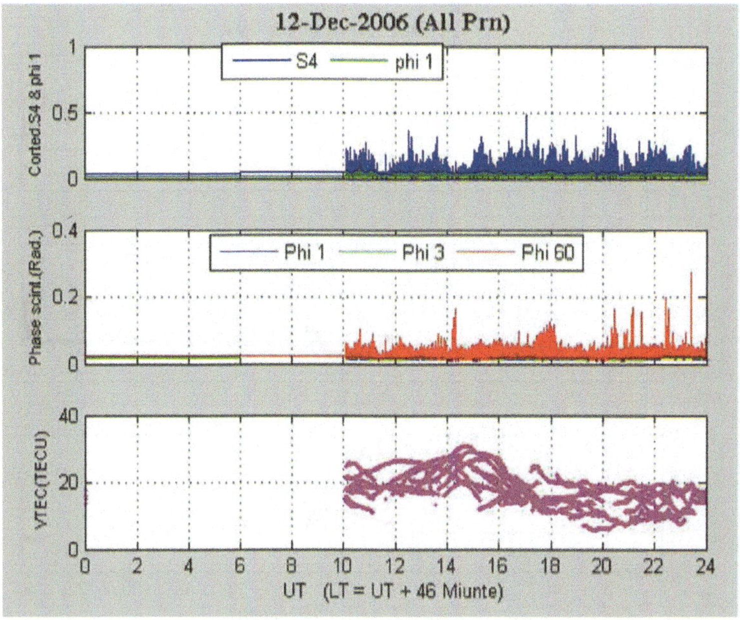

Fig. 9 Temporal variation of amplitude and phase scintillation indices along with VTEC

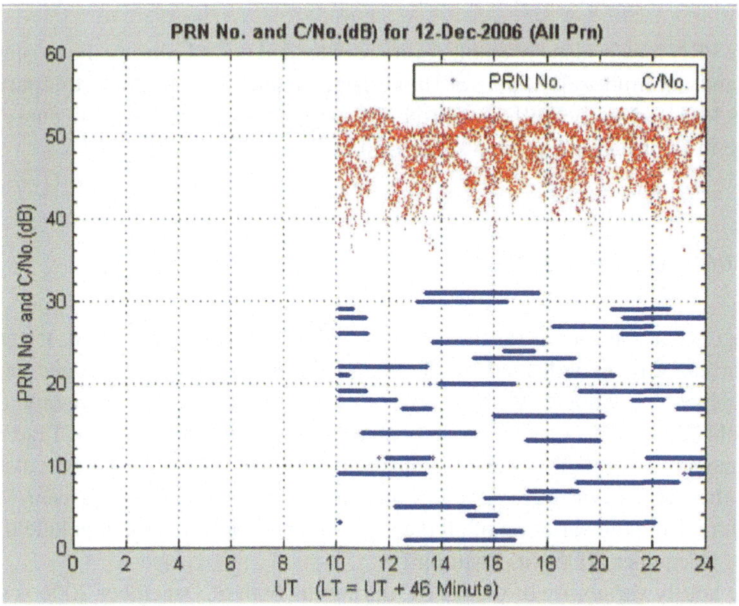

Fig. 10 The C/N_0 ratio and lock time for different visible PRN's

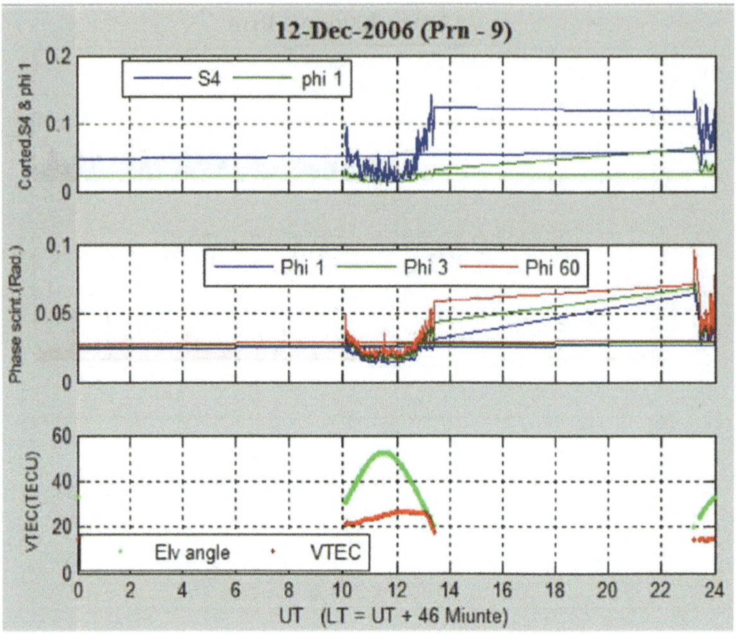

Fig. 11 PRN's showing loss of lock-in TEC curve with the rise scintillation

that the PRN lock is also not stable for more extended periods during the existence of phase and amplitude scintillations. The certainty of the PRN's visible at this particular time is lost frequently. Different PRN's showing loss of lock (in TEC curve) with the rise in scintillation has been shown in Fig. 11.

3.7 06th December 2006

The last event considered for this study is the 06th of December 2006. It was also a disturbing day, as indicated by the temporal variation of different geomagnetic indices like Dst, Kp and AE. It was identified as the fifth most worrisome day of December 2006. The Dst achieved its peak value of −55 nT at 12:00 hrs UT indicating a geomagnetic storm of moderate intensity. Kp and AE indices' value was also well above their typical or average values and recorded 4.7 and 810 nT's peak values. Therefore, from the variation of all these geomagnetic indices, we conclude that the 06th of December 2006 was disturbing.

The hourly variability of the VTEC during the 06th of December 2006 is shown in Fig. 12. The figure describes the behaviour of 15 min averaged TEC calculated from the observations of all the visible PRNs at Maitri, Antarctica. From the model, we find that the VTEC follows a regular diurnal pattern with a daily peek at 14:00 hrs

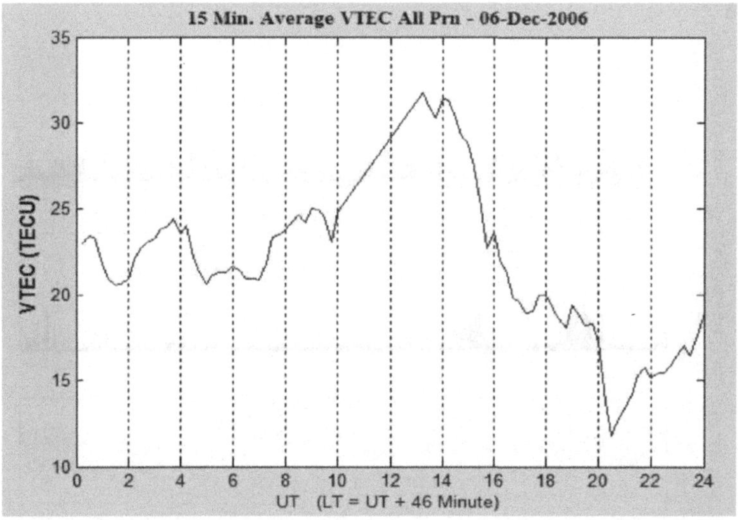

Fig. 12 The diurnal variation of total electron content during the 06th of December 2006

UT and takes a peak value of 32 TECU, which is comparatively higher than observed in previous events. A sharp decrement can also be observed in VTEC around 20:00 hrs UT with TEC's value decreasing to 12 TECU. The Azimuth and Elevator angle has been shown in Figs. 13a, b, respectively.

3.8 PRN No. and C/No. (dB) for 07-Dec-2006 (All PRN)

The occurrences of scintillations are shown in Fig. 14. It describes the amplitude scintillation by the S4 index while phi1, phi 3 and phi 60 relate phase scintillations at 1, 3 and 60 s, respectively. From the figure, we notice that the phase scintillations of moderate-intensity at all three thresholds occur around 20:00 hrs UT as well as 04:00 hrs UT. It is the same time at which the decrement is VTEC can be observed from the bottom panel. The weak amplitude scintillations are frequent throughout the day; however, amplitude scintillation of moderate to vigorous intensity can occur between 18:00–22:00 hrs UT. Different PRNs showing loss of lock (in TEC curve) with the rise in scintillation has been demonstrated in Figs. 15a, b, c, d, e.

4 Conclusions and Discussion

We studied the effect of amplitude and phase scintillations on lock loss during the five most disturbing days of December 2006. During all the hectic days, weak,

a

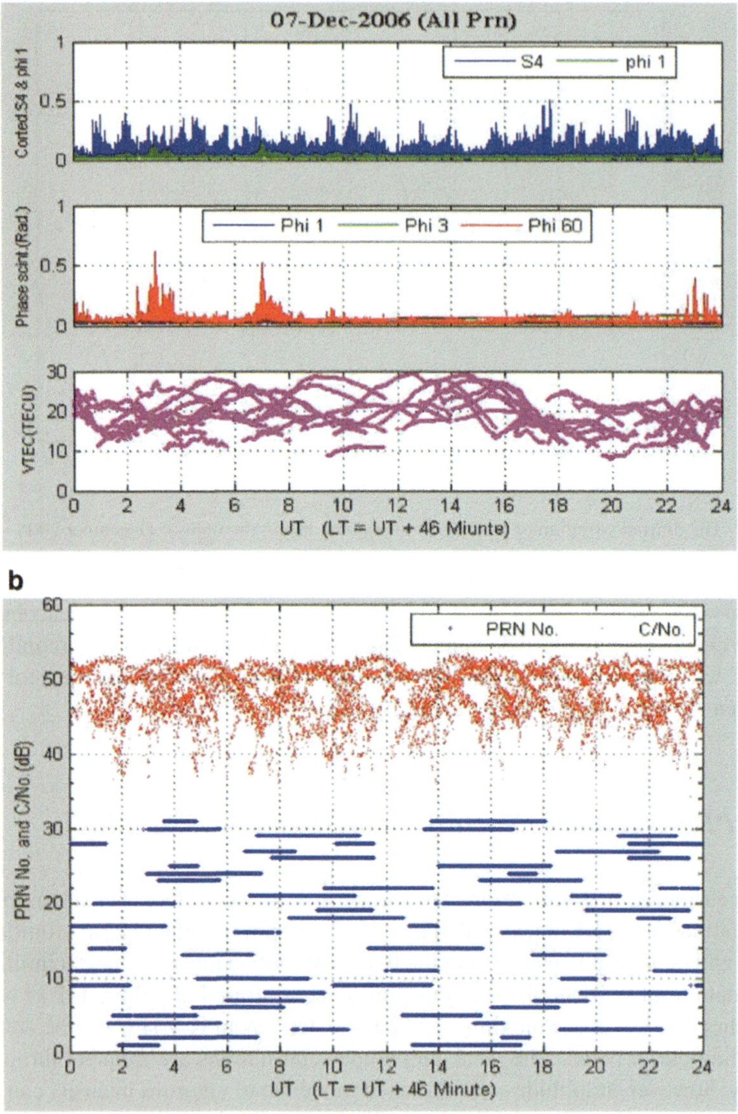

b

Fig. 13 **a**: Temporal variation of amplitude and phase scintillation indices along with VTEC. **b**: The C/N_0 ratio and lock time for different visible PRN's

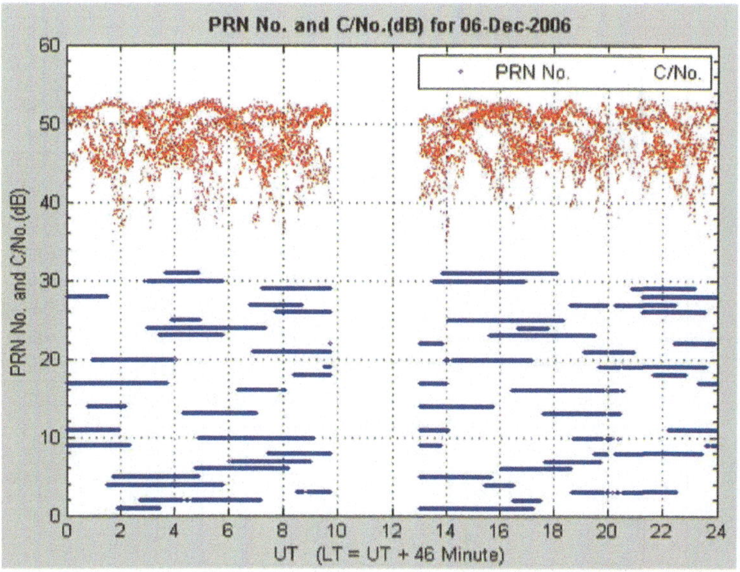

Fig. 14 The C/N0 ratio and lock time for different visible PRN's

moderate or intense geomagnetic storms were observed. The study's main findings are summarised below: During all the disturbed days, the amplitude and phase scintillations of moderate-intensity were observed. The occurrence of these scintillations usually corresponds to the decrement in the value of VTEC. The value of the C/No ratio describing the effect of the scintillations was found to fall below 40 dB during all the selected events.

Moreover, it was also found to fluctuate rapidly during the occurrence of both amplitude and phase scintillations. The loss of lock was significantly affected by the occurrence of scintillation during all the events. The amplitude and phase scintillation become frequent, and the visible PRNSs do not remain stable for more extended periods. The loss of lock frequently occurs whenever the GPS signals scintillate. In present studies, TEC's diurnal variation goes higher during the daytime, but it decreases gradually in the Antarctica region. Temporal variations were also observed in amplitude and phase scintillation indices. This study presents the daily and seasonal variation in GPS-measured TEC over the Antarctica region using simultaneous GPS receivers measurements. Observations revealed that the diurnal variation at high latitude station reached its maximum value between 12:00 and 14:00 LT.

Similarly, the daily minimum in GPS TEC occurs between 05:00 and 06:00 at the same station. The diurnal variation in GPS TEC shows a range of about 0–60 TECU. The latitudinal and longitudinal variations show a sharp, steep increase of about 12–16 TECU occurring between 01:00 and 02:00 LT. An intense and short-lived daytime minimum of about 0–2 TECU occurs between 04:00 and 06:00 LT. TEC increases with time across all the longitudes until noon-time. The speed and

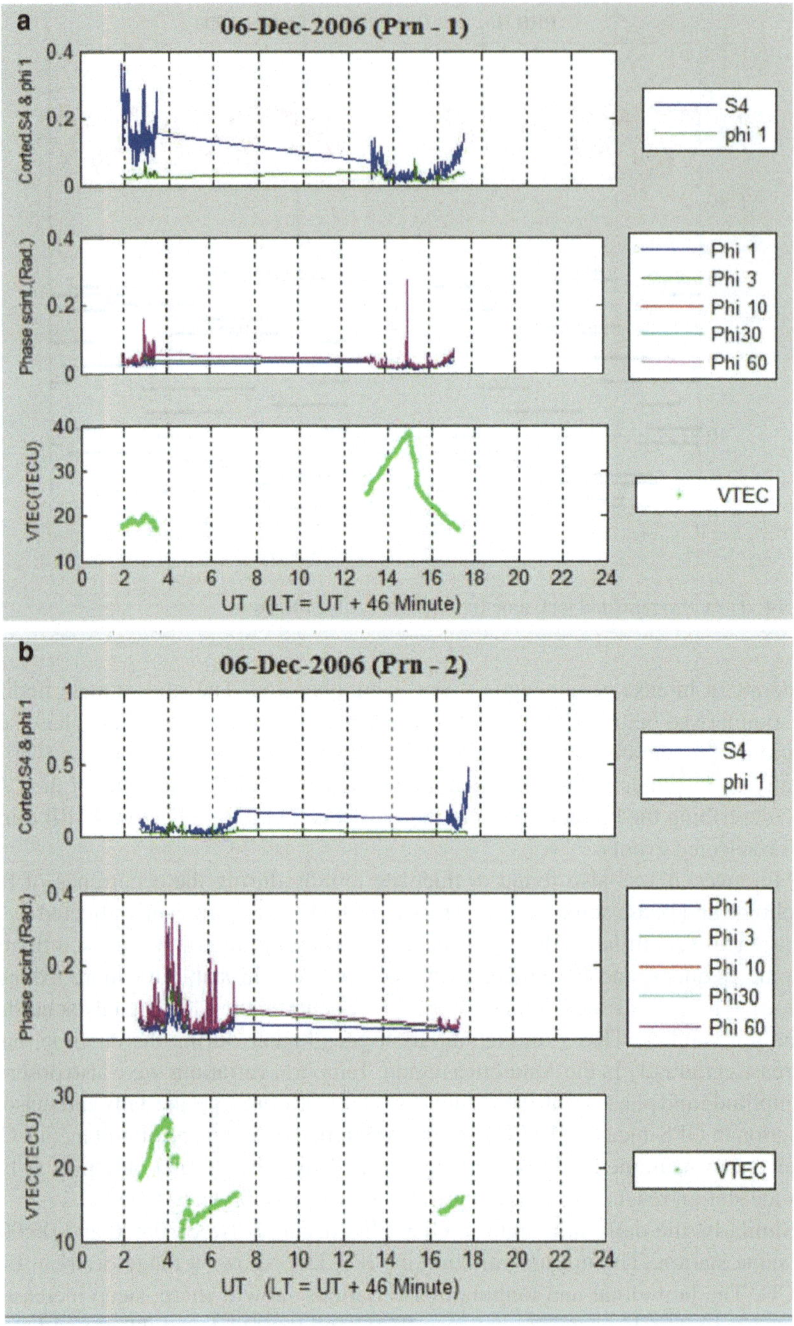

Fig. 15 **a**: PRNs showing loss of lock-in TEC curve with the rise scintillation. **b**: PRNs showing loss of lock-in TEC curve with the rise scintillation. **c** PRNs were showing loss of lock-in TEC curve with the rise scintillation. **d**: PRN's showing loss of lock-in TEC curve with the rise scintillation. **e**: PRNs showing loss of lock-in TEC curve with the rise scintillation

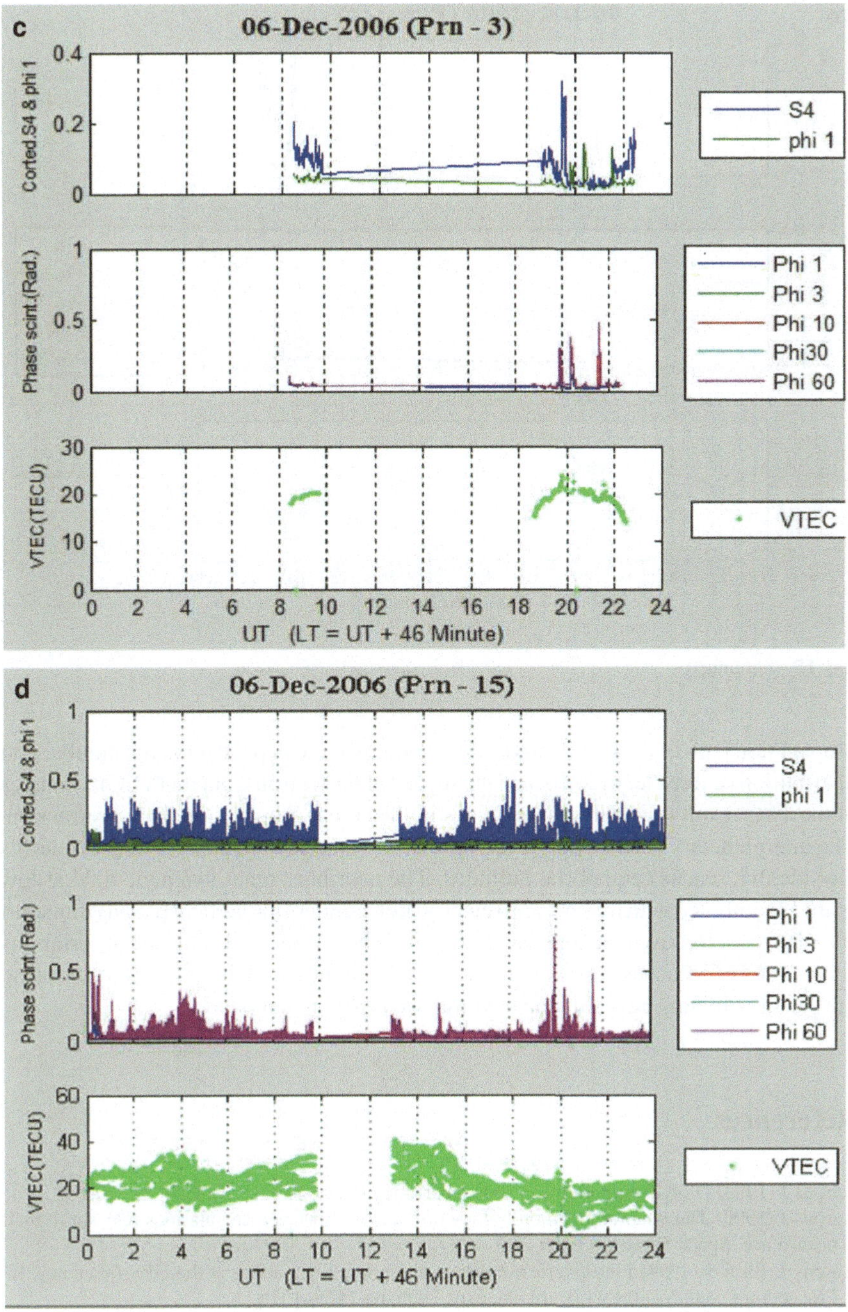

Fig. 15 (continued)

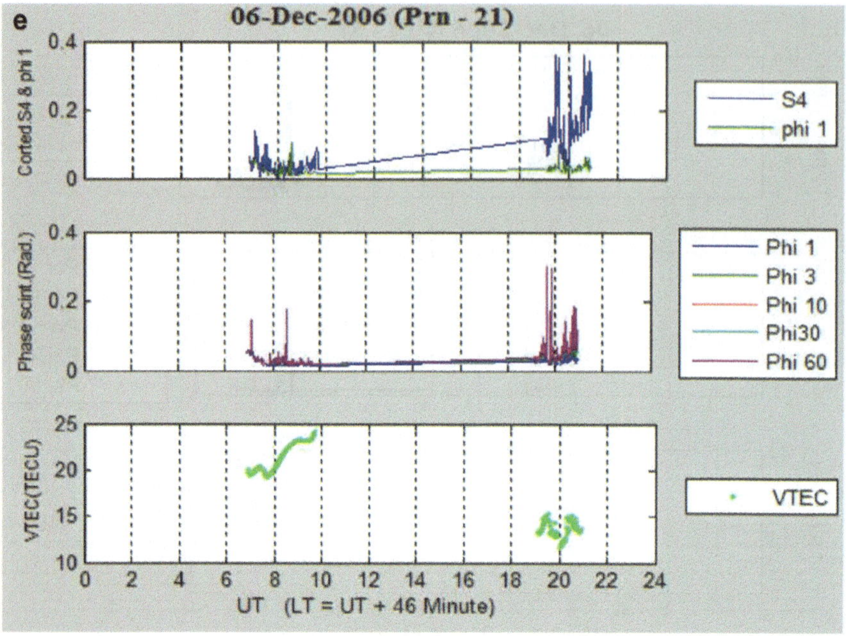

Fig. 15 (continued)

propagation path of the GNSS signals in the ionosphere depend upon the number and distribution of free electrons in the course, or Total Electron Content (TEC). Changes in the TEC result in variable time delays in the signal propagation due to refraction. The interplanetary electric field's increase drives an eastward electric field on earth's dayside that reaches equatorial latitudes. The near-horizontal magnetic field at low latitude combined with the eastward electric field renders the vertical plasma transport (E × B) very effective in lifting the plasma to higher altitudes. The seasonal variations show that TEC reaches a maximum during the equinox months and is lowest during the solstice months, revealing an asymmetric semi-annual pattern.

References

Aarons J (1982) Global morphology of ionospheric scintillations. Proc IEEE 70(4):360–378

Aarons J (1993) The longitudinal morphology of equatorial F-layer irregularities relevant to their occurrence. Space Sci Rev 63:209–243

Aarons J, Basu S (1994) Ionospheric amplitude and phase fluctuations at the GPS frequency. In: Proceedings of ION-GPS-94, Salt Lake City, Utah, pp 1569–1578

Aarons J, Mullen JP, Koster JR (1980) Seasonal and geomagnetic control of equatorial scintillations in two longitude sectors. J Atmos Terr Phys 42:861–866

Aquino M et al (2005) Implications of ionospheric scintillation for GNSS users in Northern Europe. J Navig 58(2):241–256

Basu S, MacKenzie E, Basu S (1988) Ionospheric constraints on VHF/UHF communications links during solar maximum and minimum periods. Radio Sci 23(3):363–378
Banola S, Pathan BM, Rao DRK (2001) Strength of the equatorial electrojet and geomagnetic activity control on VHF scintillations at the Indian longitudinal zone. Indian J Radio Space Phys 30:163–171
Biktash LZ (2004) The role of magnetospheric and ionospheric currents in generating equatorial scintillations during a geomagnetic storm. Ann Geophys 22:3195–3202
Bhattacharya S et al (2010) Study of GPS based ionospheric scintillation and its effects on the dual-frequency receiver. J Eng Sci Manag Educ 1:55–61
Bhattacharya S, Purohit PK, Gwal AK (2011) Influence of magnetospheric and ionospheric currents on radio wave scintillation storm time condition. Int J Environ Sci 1(5):924–937
De Oliveira Moraes A et al (2011) Analysis of the characteristics of low-latitude GPS amplitude scintillation measured during solar maximum conditions and implications for receiver performance. Surv Geophys 33(5):1107–1131
Doherty PH et al (2000) Ionospheric scintillation effects in the equatorial and auroral regions. In: Proceeding of ION-GPS 2000, Salt Lake City, Utah, pp 662–671
Dubey S, Wahi R, Gwal AK (2006) Ionospheric effects on GPS positioning. Adv Space Res 38(11):2478–2484
Forte B (2012) Analysed ionospheric solid scintillation events measured employing GPS signals at low latitudes during disturbed conditions. Radio Sci 47(4):RS4009. https://doi.org/10.1029/2011RS004789.
Groves KM et al (1997) Equatorial scintillation and systems support. Radio Sci 32:2047–2064
Gwal AK et al (2006) Amplitude and phase scintillation study at Chiang Rai, Thailand. Adv Space Res 38(11):2361–2365
Kintner PM, Ledvina BM, De Paula ER (2007) GPS and ionospheric scintillations. Space Weather 5:S09003. https://doi.org/10.1029/2006SW000260
Klobuchar JA (1991) Ionospheric Effects on GPS. GPS World, 48–51
Kumar S, Singh AK, Chauhan P (1993) Multi stations analysis of VHF radio wave scintillations at low latitudes. J Radio Space Phys 22:267–272
Kumar S, Gwal AK (2000) VHF ionospheric scintillations near the equatorial anomaly crest: solar and magnetic activity effects. J Atmos Solar-Terr Phys 62:157–169
Krankowski A, Baran LW, Shagimuratov II (2002) Influence of the northern ionosphere on positioning precision. Phys Chem Earth 27:391–395
Li G et al (2006) Observations of GPS ionospheric scintillations over Wuhan during geomagnetic storms. Ann Geophys 24:1581–1590
Moreno B et al (2010) On the effects of the ionospheric disturbances on precise point positioning at equatorial latitudes. GPS Solut 15(4):381–390
Phoomchusak P, Leelaruji N, Hemmakorn N (2003) The deterioration of GPS accuracy caused by ionospheric amplitude scintillation, 2–55
Pi X, Mannucci AJ, Lindqwister UJ, Ho CM et al (1997) Monitoring global ionospheric irregularities using the worldwide GPS network. J Geophys Res 24:2283–2286
Pathan BM et al (1991) Dynamics of ionospheric irregularities producing VHF radio wave scintillations at low latitude. Ann Geophys 9:126–132
Rastogi RG, Mullen JP, MacKenzie E (1981) The effect of geomagnetic activity on equatorial VHF scintillations and spread-F. J Geophys Res 86:3661–3664
Rastogi RG, Koparkar PV, Pathan BM (1990) Nighttime radio wave scintillation at equatorial stations in Indian and American zones. J Geomagn Geoelect 42:1–10
Secan JA, Bussey RM, Fremmouw EJ (1995) An improved model of equatorial scintillation. Radio Sci 30:607–617
Wanninger L (1993) Effects of the equatorial ionosphere on GPS. GPS World Adv Stat Commun 48–54

Study of Positional Error on Ionospheric Scintillation Over Antarctic Region and Loss due to Locking of GPS signal

A. K. Gwal, Suryanshu Choudhary, and Ritesh Yadav

Abstract The effect of high geomagnetic activity on the positional error in GPS signal by considering five geomagnetic storms events of different intensities during December 2006 has been studied. The main conclusions drawn from the study are that during all the five geomagnetically disturbed days the positional error increases significantly. During all the events increase in the latitudinal error is more than the longitudinal error. Study further records that in most of the events the increase in both the errors usually occurs during the main phase of the storm. The scatter of the 2D positional error around the actual position was also found to be large during all the events. However, scatter of error was found be larger along the longitude than the latitude. The scatter of 2D latitudinal positional error in meters was found to be larger than the longitudinal error. The error points are within the confidence level before the onset of geomagnetic storm event and move significantly out of the 95% error eclipse and error circle.

Keywords Geomagnetic activity · Positional error · Dilution of precision (DOP) · Ionosphere · Antarctica

1 Introduction

The ionosphere is the upper atmospheric region from altitudes 60 km up to 1000 km. It is part of the atmosphere where a fraction of the atmospheric particles is ionized, or particles are separated into positive ions free from electrons. When the ionosphere is in a refractive medium, radio signals resent through refraction, and the magnitude of the refraction is proportional to the local electron density. The presence of irregularities causes rapid fluctuation in the refractivity; therefore, radio signals

Present Address:
A. K. Gwal (✉) · S. Choudhary · R. Yadav
Department of Physics, Rabindranath Tagore University, Raisen 464993, India

Present Address:
Department of Physics, Dr A P J Abdul Kalam University, Indore 452016, India

N. Khare (ed.), *Assessing the Antarctic Environment from a Climate Change Perspective*, Earth and Environmental Sciences Library,
https://doi.org/10.1007/978-3-030-87078-2_12

are passes through it. Due to the signal's enhancements and fades, its amplitude from their mean value is manifested as amplitude scintillation. The reliability of the Global Positioning System (GPS) navigational system is affected by the scintillation and resulting from the positional error of above 10 m larger during high scintillation activities (Phoonchusak et al. 2003). Strong ionospheric disturbances have an impact on the performances of the GPS receivers, and the ionospheric effects on GPS receivers have been studied by many researchers (Doherty 2000; Groves 2000; Hegarty et al. 2001; Skone 2001; Conker et al. 2003; Bhattacharya et al. 2008; Shukla et al. 2009; Jain et al. 2010).

The quality of GPS-derived position estimate depends upon both the measurement geometry, which is represented by Dilution of Precision (DOP) values and range errors caused by signal strength, ionospheric effects, multipath, etc.

The presence of ionospheric irregularities can cause degradation in the GPS navigational accuracy and limitations in the GPS tracking performance (Bandyopadhayay et al. 1997; Skone 2001). Essentially, free electrons contained in the ionosphere affect the propagation of the signal as it passes through. Since the signals are travelling at the speed of light and GNSS is based on nanosecond timing, it does not take much interference to introduce error. To provide robust and reliable positioning, strict control of the causes of satellite positioning errors is demanded. GPS technology has broad applications, and a few of them require high precision, such as crustal deformation, geodesy, aviation, and emergency services. However, when the satellite signals from the satellite propagate through a disturbed ionospheric medium, their characteristics change according to disturbance level. Increased knowledge of the ionospheric structure and its variability is vital for precise positioning since the ionosphere impacts GPS L-band radio signals, especially during troubled geomagnetic conditions, due to its free electrons. When the ionosphere changes from its undisturbed state to more turbulent states, GPS applications are affected. Therefore, it is necessary to have a real-time analysis to provide the model to correct this error. Simultaneously, these perturbations in the GPS signals are taken as scientific information used to investigate ionospheric scenarios.

2 Position Accuracy and Precision

Accuracy and precision are often used to describe how good is the position acquired by the GPS receiver. A distinction should be made between accuracy and precision. Accuracy is the degree of closeness of an estimate to its true but unknown value, and precision is the degree of intimacy of observations to their means. Figure 1 illustrates various relationships between these two parameters. The actual value is located at the intersection of the crosshairs. The centre of the shaded area is the location of the mean estimate. The radius of the shaded area is a measure of the uncertainty contained in the calculation.

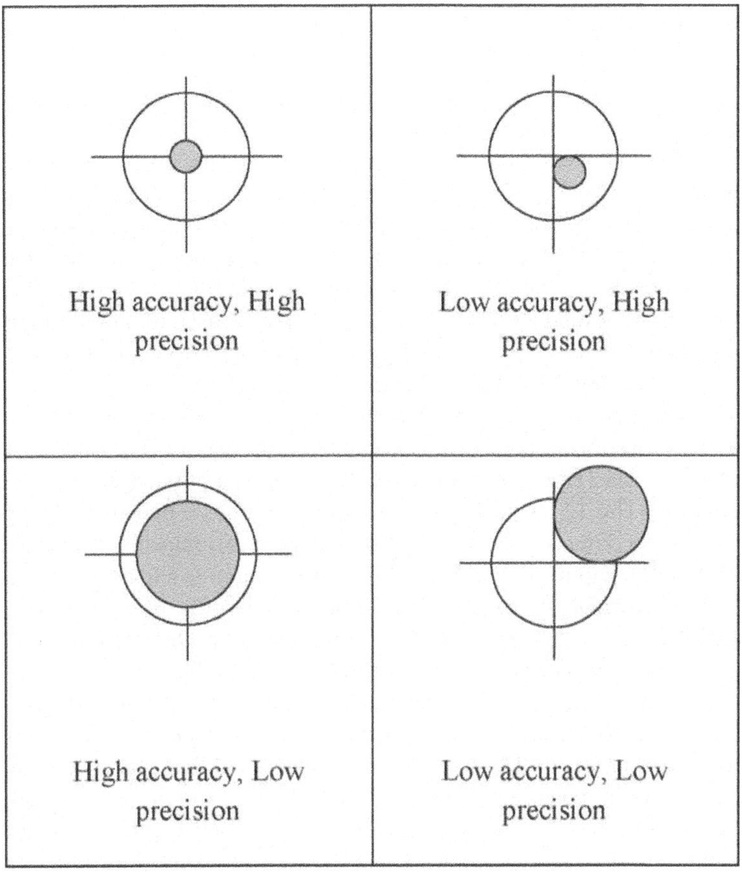

Fig. 1 Accuracy versus precision

3 Circular Error Probability (CEP)

Many different measures have been used for describing the accuracy obtainable from GPS. The most common two terms are CEP and 2DRMS. CEP refers to the radius of a circle in which contained 50% of the values occur. Plot the results from a large number of the position scatters and draw a circle centred on the mean of the GPS receiver. If a CEP of 5 m is quoted, then 50% of horizontal point positions should be within 5 m of the proper place. The radius of 95% is often quoted, and the term R95 used. R95 is CEP with the radius of the 95% probability circle.

4 Data Procurement and Methodology

In this chapter, we have studied the change in the positional error under changing ionospheric condition. To accomplish the study, we have considered the five most disturbing days of the year 2005. To investigate the correlation between the disturbed ionosphere and positional error, we used the Total Electron Content (TEC) parameter using GPS observations. The TEC data was collected at the high latitude Indian Station, Maitri (70.4N,11.4E), Antarctica, during low solar activity 2005. The TEC was recorded using the GISTM- (GPS Ionospheric Scintillation and TEC Monitor) based GPS receiver model GSV4004A. The receiver is used to monitor both ionospheric behaviour and GPS performance in equatorial regions and high latitude regions. The system is NovAtel's dual-frequency GPS receiver.

The receiver performs amplitude and phase measurements at the 50 Hz sampling rate and measures carrier-code divergence at the 1 Hz rate for each satellite tracked on L1. It computes TEC from combined L1 and L2 pseudo-range and carrier phase measurements. The 12 channel is used to measure a noise floor for C/N_0 as well as scintillations computations. The dual-frequency measurements were carried out with a 30 s data sampling interval to reduce processing time and are usually used to estimate phase fluctuations and cut off of elevation angles was set to 40°. BESTPOSA command has been taken out in ASCII format by using OEM Convert4 solution software to study the receiver's exact position in each minute.

In comparison, RANGB Command had been used for a set of visible PRN's and with their C/N_0 recorded every minute with all satellite geometry being tracked during the time of the experiment. The position performance of GPS is evaluated by defining positional errors in meter and studying their variation. The satellite geometry can significantly affect the position determination in GPS. Wrong satellite geometry can lead to significant errors in positioning. The satellite geometry is measured through a parameter commonly known as Dilution of Precision (DOP). The DOP is further classified into five categories, namely Horizontal Dilution of Precision (HDOP), Vertical Dilution of Precision (VDOP), Positional Dilution of Precision (PDOP), Geometric Dilution of Precision (GDOP) and Time Dilution of Precision (TDOP). As the first step in computing DOP, let x, y and z indicate the location of the receiver and x_i, y_i and z_i denote the position of the satellites.

5 Results and Discussion

We studied the effect on GPS signal positional error during the five most disturbing days of December 2006. During disturbed days, the irregularities of different scale are pretty standard, severely affecting GPS signal propagation. This can affect the causing of scintillation and hence degrade the positioning capability of the GPS. We have investigated these irregularities' effect on GPS's precise position by considering the geomagnetic storm events. We will show and discuss the results for each of these storms one by one.

5.1 15th December 2006 Storm

During December 2006, the geomagnetic activity remained high for many days. The 15th of December was one of such days when the geomagnetic activity was relatively high, and an intense geomagnetic storm was observed on this day. The Dst achieved a peak of minimum value of −162nT. Similarly, the Kp index and AE also underwent a considerable increase, as indicated by the peak values of 8.3 and 1372 nT, respectively. These indices' growth clearly shows that the geomagnetic activity was relatively high on this day; hence was designated as a disturbing day. The effect of the increased geomagnetic activity on positioning has been illustrated in the following figures. The temporal changes in the positional error during the 15th of December 2006 is shown in Fig. 2.

The positional error used in this figure has been calculated in terms of the error in the longitude and latitude from the actual point. The temporal variation of the positional error has been only shown during the 10:00 UT to 24:00 UT. The positional error has been calculated in meters. We find a significant variation of latitudinal and longitudinal positional error during the event from the plot. The error varies from −6 to 5 m. The interpretation of the 2DRMS or the absolute positional error in meters is shown against the no. of satellites locked is displayed in Fig. 3. We find that if the

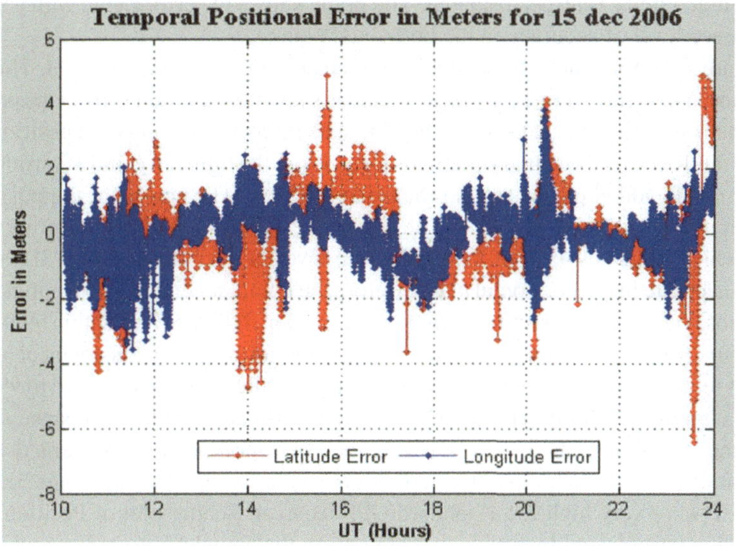

Fig. 2 Temporal variation of positional error in meters during the 15th of December 2006

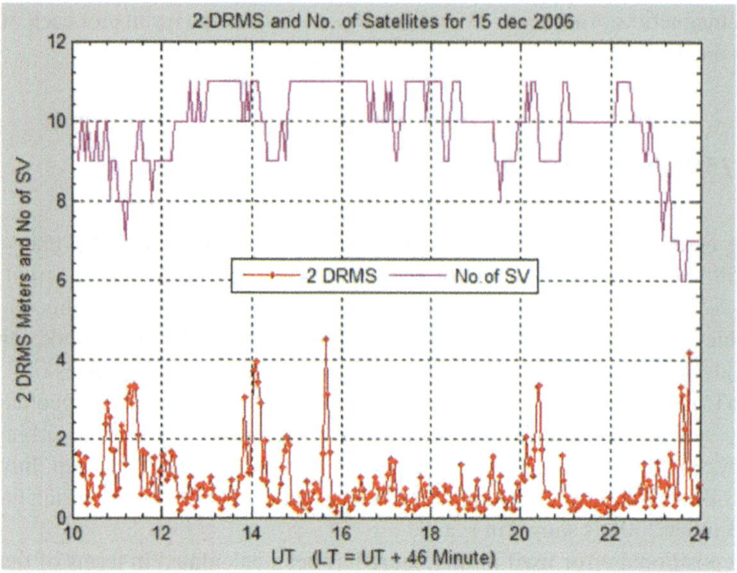

Fig. 3 Variation of 2DRMS and no. of satellites locked during the 15th of December 2006

more number satellites are closed, the absolute positional error is more minor, but as soon as the number of locked satellites decreases, the positional error increases. Therefore, the figure shows a perfect correlation between the 2DRMS value and the no. of locked satellites. Moreover, not only the number of closed satellites matters but their geometry is also substantial. The graph shows the error range with a 99.9% confidence level of accuracy in the horizontal plane.

Figure 4 demonstrates the scatter plot of error from the actual position. This plot was chosen to signify how far from the receiver's position solution is from the point of reference or origin, or exact status. The blue stars indicate the computed position, and the red circle shows the actual situation's level. From the figure, we find a considerable scatter of the added places around the exact position. The spread of the positional error in meters is shown in Fig. 4. The range of error was expressed in terms of the CEP, which indicates 49% of the confidence level of accuracies, while 3DRMS is 99.9% accurate in the horizontal plane. Before the commencement of the initial phase of storms, the errors are within the confidence level of 3DRMS, i.e., within 99.97% of the accuracy level. As soon as the main stage of the storm's commencement occurred, the Dst reached maximum negative value around, while the errors crossed to 99.97% of the confidential limit. The figure shows the ultimate positional mistake during that time, and during the recovery phase, the horizontal position converged within 99.97% of confidence levels. During this event, the recorded value of the 3DRMS and CEP were very high for a standard GPS receiver. Scatter plot of Position Error induced in latitude and longitude in meters from the actual position during the 15th of December 2006.

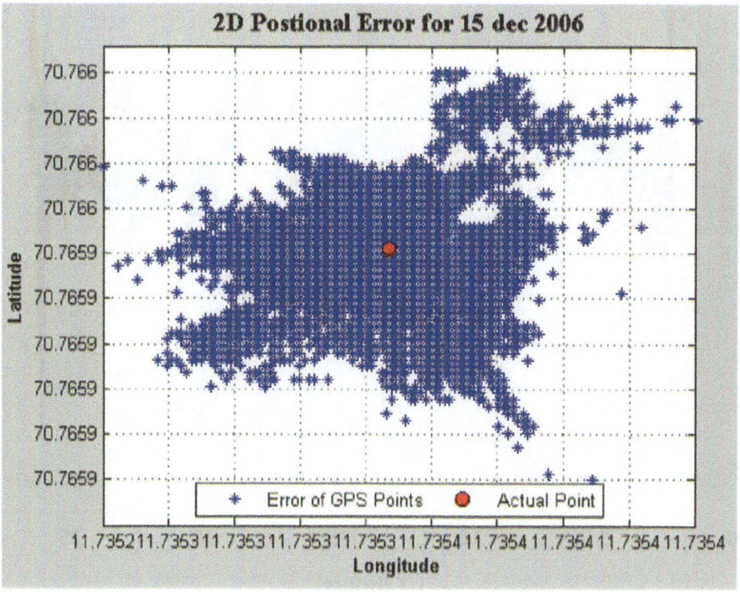

Fig. 4 Scatters 2D positional error plot

5.2 *The 14th of December 2006 Storm*

The 14th of December 2006 was also a disturbing day, as indicated by the increase in the geomagnetic indices. The temporal variation of position dilution of precision (PDOP) during this event is shown in Fig. 5. Although the time series of PDOP shows a fluctuating nature, it increases significantly during the storm's main phase. A moderate geomagnetic storm was observed on this day with a peak Dst value of −69 nT. The Kp index and AE index value also significantly increased with peak values of 5.3 and 1616 nT. Therefore, we can say that the global geomagnetic activity was relatively high on the 14th of December 2014. Thus, the 14th of December was designated as the disturbed day and was the second most disturbing day of December 2006. Errors occur in the computed position Due to fluctuations in the GPS signals while encountering irregularities. The temporal changes in the positional error, in meters, during different hours of the 14th of December, are shown in this result. The red line shows the latitude calculation error, while the blue line indicates the error in the longitude analysis. The error varies from −4 to 4 m. During the storm's main phase, the latitudinal error increases drastically and reaches a value of 8 m.

The detailed overview of 2D positional error can be seen in Fig. 6, where the scatter graph of 2D positional error has been shown as a function of latitude and latitude. The red dot in the centre indicates the place of the GPS Antenna, and the blue stars represent the positions computed from GPS measurement during the storm event of the 14th of December 2006. We find from the figure that the blue stars are

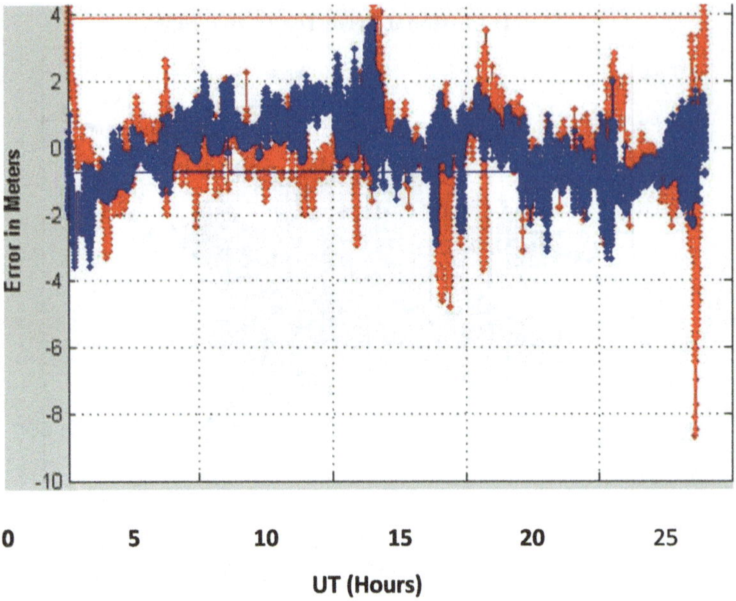

Fig. 5 Temporal variation of positional error in meters during the 14th of December 2006

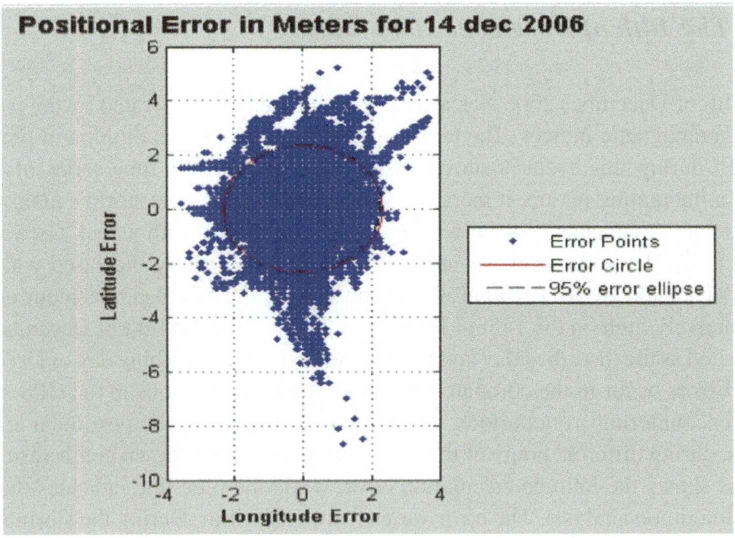

Fig. 6 Scatter plot of 2D position error in terms during the 14th of December 2006

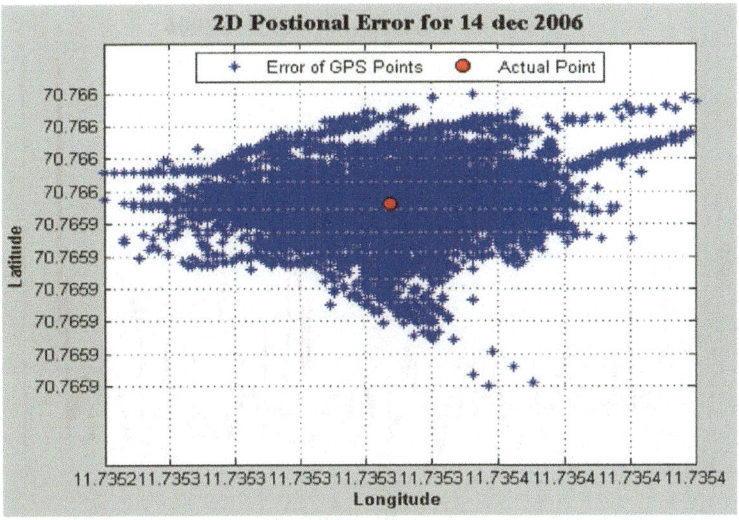

Fig. 7 Positional error in meters the 15th of December 2006

scattered around the red dot, and we notice that the longitudinal error is much more than the latitudinal error.

The scatter of the 2D positional error in meters has been shown in Fig. 7. The blue diamonds represent the error points, while the solid red line has also drawn the error circle. The black dotted line shows the 95% error eclipse. The latitudinal error varies between −6 to 6 m, while the longitudinal error varies between −2 to 2 m. Therefore latitudinal mistake is more than the longitudinal error. Moreover, the error increases significantly during the main phase of the storm.

The time profile of the positional dilution of precision has also been computed for this storm event and is shown in Fig. 8. A fluctuating behaviour of the PDOP can be seen from the figure. The PDOP increases dramatically during the main phase of the storm, reaching a maximum value of 5.5. The satellite locks are not stable for long during the storm's main stage, which brings the variations in the pseudorange in the form of absolute error, as shown in the above figure.

5.3 *The 12th of December 2006 Storm*

We now consider the 12th of December 2006 and show how the positional error behaved during this event. A moderate geomagnetic storm was observed, the peak value of storm intensity index (Dst index) was observed at 05:00 hrs UT, and the peak value of −55 nT was recorded. Moreover, during this storm, Dst's value showed another dip at 21:00 hrs with a −49 nT. The value of other geomagnetic indices

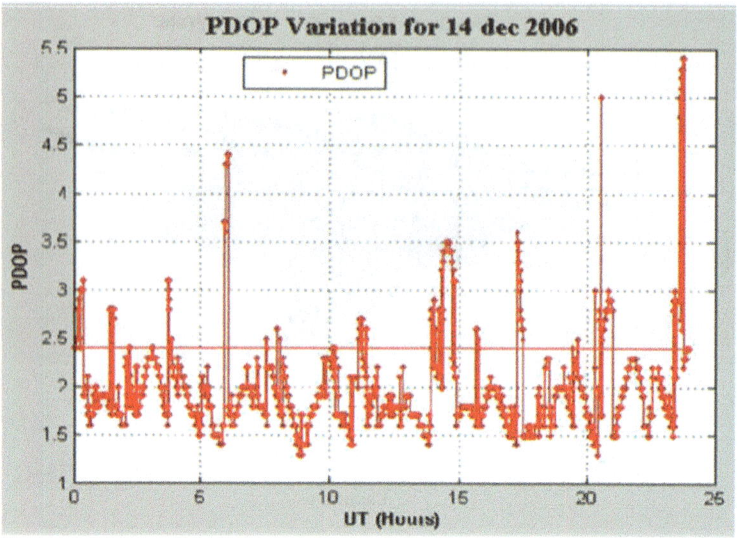

Fig. 8 Position dilution of precision (PDOP) during the 14th of December 2006

describing the state of geomagnetic activity like the Kp index and AE index also underwent a significant increase in achieving the maximum values of 5.3 and 824 nT.

From all these indices' values, we figure out that the 12th of December 2006 was a geomagnetically disturbed day. It was the third most disturbing day in December 2006. The temporal variation of the positional error in meters is plotted to show the positional error behaviour during this storm event, as shown in Fig. 9. The red line shows the latitude error, while the blue line represents the longitude error changes. The figure shows the error variation from 10:00 hrs UT to 24:00 hrs UT due to data's

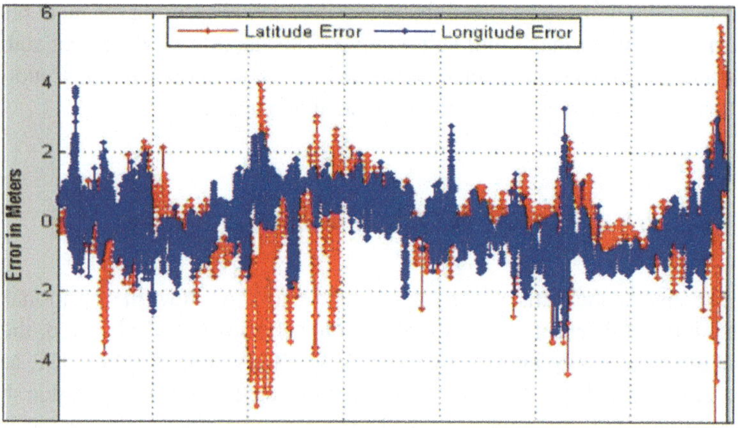

Fig. 9 Positional error on the 12th of December 2016

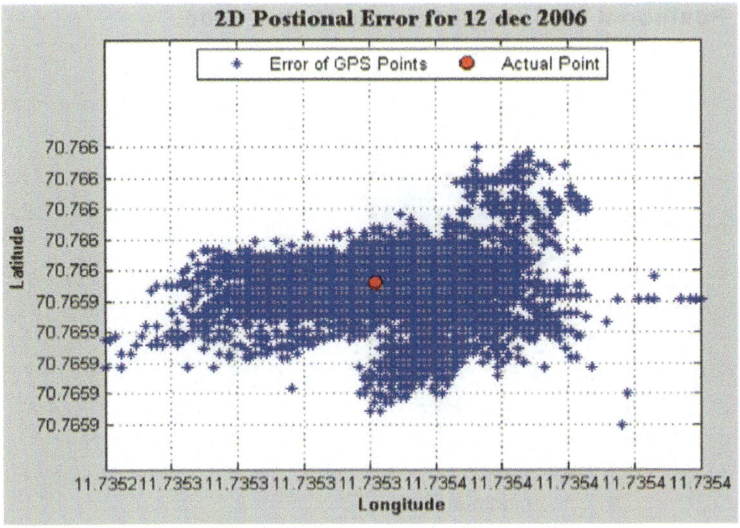

Fig. 10 Scatter plot of 2D position error in terms during the 12th of December 2006

non-availability. From the model, we find both the errors increase during the second dip of Dst. Here we also notice the increase in errors during the recovery phase of the storm. The maximum values achieved by the longitudinal and latitudinal error are 3 and 6 m, respectively.

The overview of 2D positional error from the actual position is shown in Fig. 10. The red dot indicates GPS Antenna's place, while the blue stars show the computed positions. From the figure, we find that the longitudinal scatter is comparatively much larger than the latitudinal error.

The range of 2D positional errors in meters is shown in Fig. 11; the X-axis represents the longitudinal error in meters while the Y-axis represents the latitudinal error in meters. The error points are represented by blue diamond's while the blue dotted eclipse represents the 95% error eclipse, and the red circle shows the error circle. Before the storm, the error points are in the confidence level, but during the storm, particularly during the second dip of Dst, we found that the error increases and the error points move out of the confidence level. The longitudinal error is in the range of −2 to 2 m while the latitudinal error is between −4 to 5 m, thereby showing that the latitudinal mistake is more than the longitudinal error. We also computed the position dilution of precision (PDOP) during the event. The time profile of the PDOP is shown in Fig. 12. We find that the PDOP does not undergo significant or large changes during the storm event from the figure. Only at the end of the day, when a second dip was observed in the Dst, the value of PDOP shoots to a maximum weight of about 8 units. Also, around 15:00 hrs UT, the value of PDOP shows an increase, and it was found to be consistent with the rise in the Dst index or geomagnetic activity around the same time.

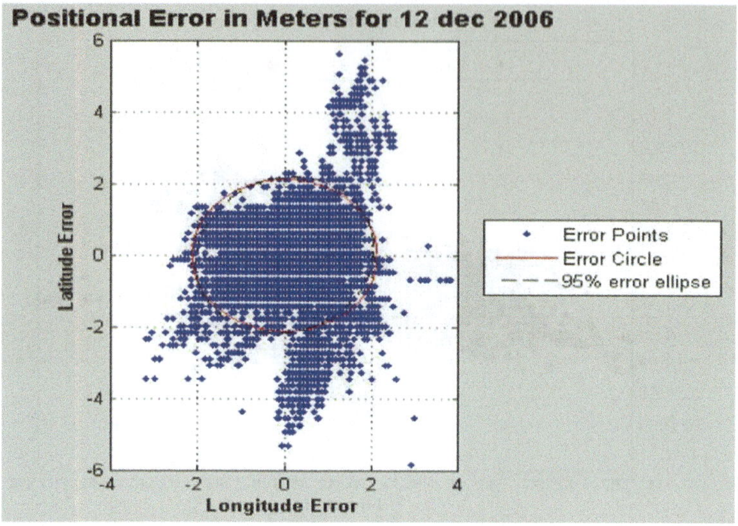

Fig. 11 Positional error in meters during the 12th of December 2006

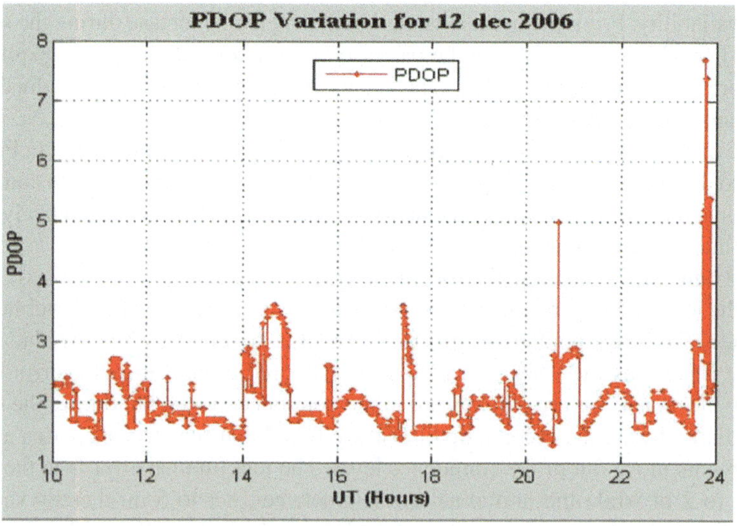

Fig. 12 Position dilution of precision (PDOP) during the 12th of December 2006

5.4 The 07th of December 2006 Storm

Consider the disturbing day of December 2006 and present the positional error changes due to the increased geomagnetic activity. The 07th of December 2006 was the fourth most problematic day. On this day, a weak geomagnetic storm was

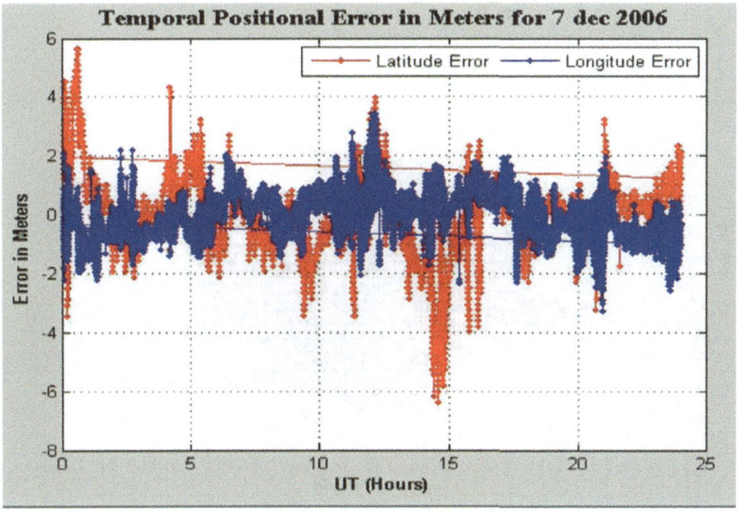

Fig. 13 Temporal variation of positional error in meters during the 07th of December 2006

observed as indicated the Dst index reached the peak value of −47 nT at 22:00 hrs UT. The other index, like the Kp and AE index describing the state of geomagnetic activity, also recorded a significant increase from their average values. The peak values of these indices were 5.0 and 697 nT, respectively. The increased importance of these indices shows that the 07th of December 2006 was a geomagnetically disturbed day. We now evaluate the effect of this increased geomagnetic activity on the positional error.

The temporal variation of the positional error is shown in Fig. 13. The figure shows the hourly changes in the latitudinal and longitudinal error in meters during the 07th of December 2006. The red line represents the latitude error, while the blue line represents the longitude error. From the figure, we find the positional error usually varies between −1.5 to 1.5 m. However, when the geomagnetic activity increases, both errors increase and reach a maximum value of 6 m.

Moreover, we also notice that during this geomagnetic storm, the increase in the latitude error is more than the rise in longitude error. The deviation of the computed positions from the actual work is shown in Fig. 14. It shows the scatter of the positional 2D mistake for the 07th of December 2006. The red dot indicates the receiver's actual position, while the blue stars show the functions computed by using the GPS observation during the high geomagnetic activity. We notice from the figure that the deviations of computed positions from the actual point are pretty significant. Here again, we see that the scatter is large along longitude as compared to latitude. The spread of 2D positional error in meters shown in the error circle and the 95% error eclipse are also drawn to show confidence. We see the error point are within the confidence level before the onset of the geomagnetic storm. Still, as the storm enters the primary phase and achieves its peak value, the error increases and the error point

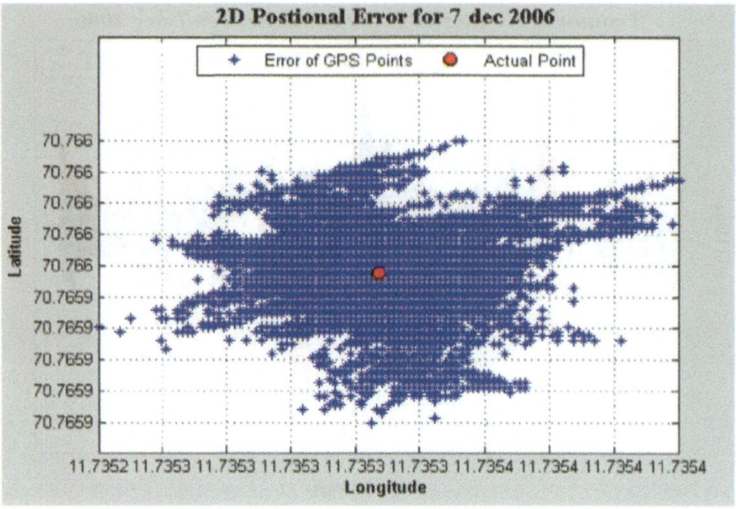

Fig. 14 Scatter plot of 2D position error in terms during the 07th of December 2006

out of the confidence level. Here we also notice that the error eclipse and the error circle do not overlap. The scatter of the latitude error in meters is much larger than the longitude error spread.

The last event considered for this study is the 06th of December 2006. It was also disturbed as indicated by the temporal variation of different geomagnetic indices like Dst, Kp and AE and was identified as the fifth most disturbing day of December 2006. The Dst reached its peak value of −55 nT at 12:00 hrs UT indicating a geomagnetic storm of moderate intensity. Kp and AE indices' value was also well above their typical or average values and recorded 4.7 and 810 nT's peak values. Therefore, from the variation of all these geomagnetic indices, we conclude that the 06th of December 2006 was disturbing. The temporal variation of the positional error during the 06th of December 2006 is shown in Fig. 15. It describes the changes in the latitude error and longitude error in meters. The red line shows the latitude error, and the blue line indicates the longitude error. From the figure, we find that significant changes in both the mistakes from their expected or average values take place at 14:00 hrs which corresponds to the main phase of the geomagnetic storm.

Figure 16 show the time profile of 2DRMS error with the number of visible or locked satellites. From the figure, we can easily see the value of 2DRMS increases when the number of satellites decreases. Thus absolute positional error critically depends upon the number of locked satellites. The scatter of the positional 2D mistake for the 06th of December 2006 is shown in Fig. 17. The red dot indicates the actual work position, while the blue stars show the functions computed from observations. From the figure, we can notice a significant scatter of the added places around the actual position. However, the spread along the longitude is more than along the latitude. The 2D positional error in meters is shown that the drawn the error circle

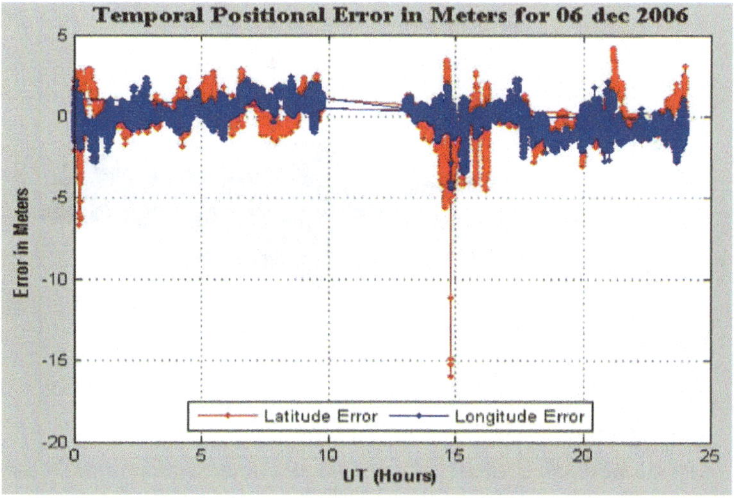

Fig. 15 Temporal positional error in meters on the 06th of December 2016

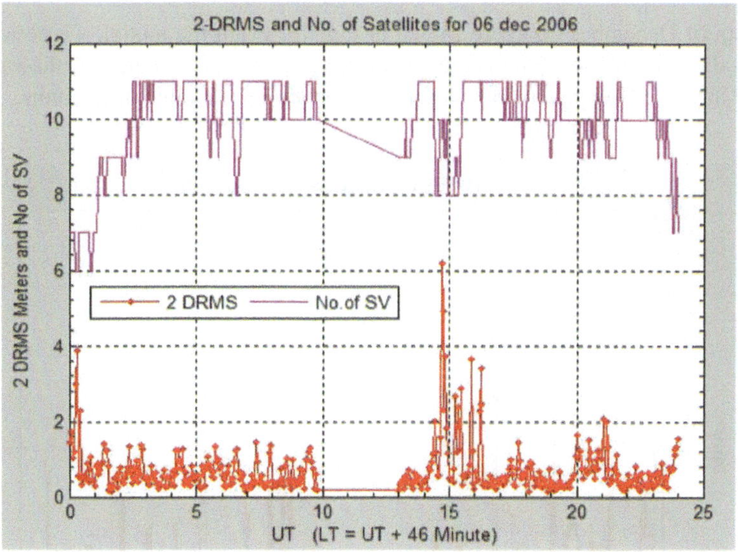

Fig. 16 Variation of 2DRMS and no. of satellites locked during the 06th of December 2006

and the 95% error eclipse. Here we find the latitudinal error in meters is more than the longitudinal error. The errors are within the confidence level before the onset of the geomagnetic storm, but as soon as the geomagnetic storm enters the primary phase, the error increases. The error point is shown in blue moves out the confidence level. The temporal changes in the Positional Dilution of Precision (PDOP) during

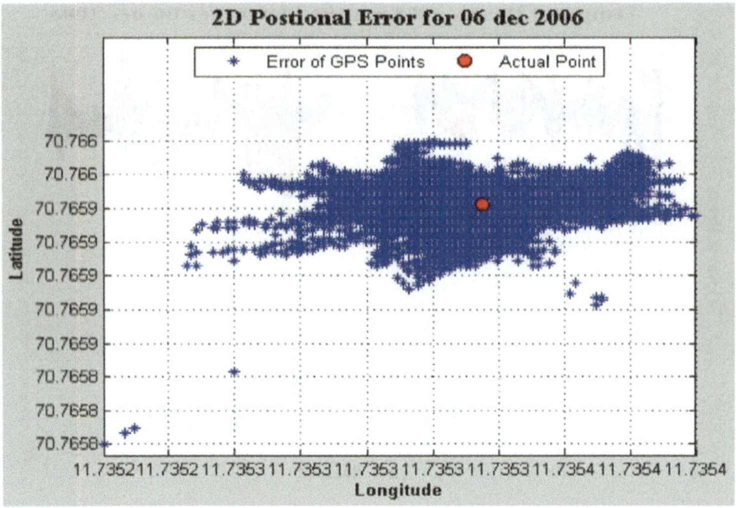

Fig. 17 Scatter plot of 2D position error in terms during the 06th of December 2006

the 06th of December 2006 are shown in Fig. 18. From the model, we notice that during the main phase of the geomagnetic storm around 14:00 hrs UT, the value of PDOP increase from the average level and reaches a peak value of 3.6 units.

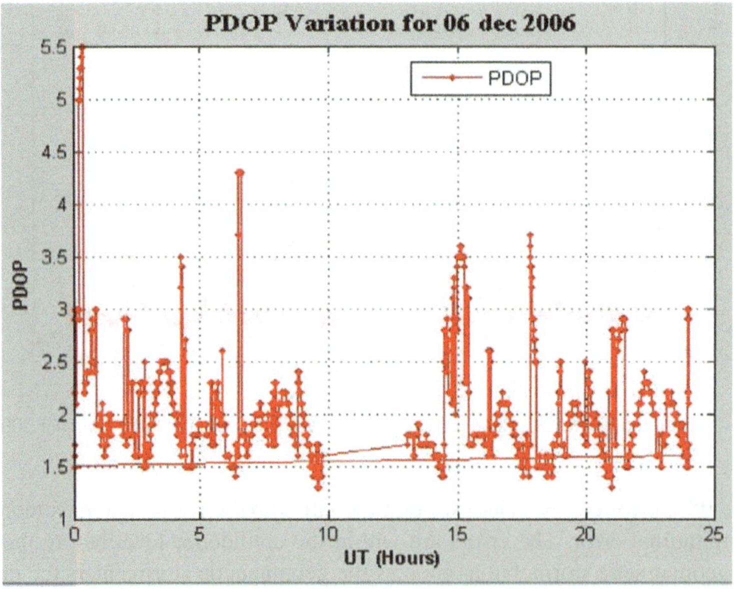

Fig. 18 Position dilution of precision (PDOP) during the 06th of December 2006

6 Conclusions

We studied the effect of high geomagnetic activity on the GPS signal's positional error by considering five geomagnetic storms of different intensities during December 2006. The main conclusions drawn from the study are enumerated below:

- We found that during all the five geomagnetically disturbed days, the positional error increases significantly. We found that increase in the latitudinal mistake is more than the longitudinal error during all the events.
- In most of the events, we found that the increase in both errors usually occurs during the storm's main phase.
- The scatter of the 2D positional error around the actual position was also prominent during all the events. However, the spread of error was found to be more pronounced along the longitude than the latitude.
- The scatter of 2D latitudinal positional error in meters was found to be larger than the longitudinal error. The error points are within the confidence level before the onset of a geomagnetic storm event and move significantly out of the 95% error eclipse and error circle.

The Position Dilution of Precision (PDOP) also increases during high geomagnetic activity. However, significant changes are observed during the main phase of the storm event and the recovery phase and the initial phase of the storm.

References

Bhattacharya S et al (2008) The effect of magnetic activity on ionospheric time delay at low latitude. J Astrophys Astron 29:269–274

Bandyopadhayay T, Guha A, Das A, Gupta P, Banerjee A, Bose A (1997) Degradation of avigation accuracy with global positioning system during the period of scintillation at equatorial latitudes. Electron Lett 33(12):1010–1011

Conker RS et al (2003) Modelling the effects of ionospheric scintillation on GPS/satellite-based augmentation system availability. Radio Sci 38(1): https://doi.org/10.1029/2000RS002604.

Doherty PH et al (2000) Ionospheric scintillation effects in the equatorial and auroral regions. In: Proceeding of ION-GPS 2000, Salt Lake City, Utah, pp 662–671

Groves K et al (2000) A comparison of GPS performance in scintillation environment at Ascension Island. In: Proceeding of ION-GPS 2000, Salt Lake City, Utah, pp 672–679

Hegarty C et al (2001) Scintillation modelling of GPS wide-area augmentation system receivers. Radio Sci 36(5):1221–1232

Jain A et al (2010) TEC response during severe geomagnetic storms near the crest of equatorial ionization anomaly. Indian J Radio Space Phys 39:11–24

Phoomchusak P, Leelaruji N, Hemmakorn N (2003) The deterioration of GPS accuracy caused by ionospheric amplitude scintillation, 2–55

Shukla AK et al (2009) Comparative analysis of ionospheric delay on user position accuracy using single and dual-frequency GPS receivers over Indian region. Indian J Radio Space Phys 38:57–61

Skone SH (2001) The impact of the magnetic storm on GPS receiver performance. J Geodyn 75(6):457–472

Climate Change and Seabirds: Insights from Ecological Monitoring of Snow Petrels in the Indian Antarctic Program

Anant Pande and Kuppusamy Sivakumar

Abstract Climate driven changes in the Southern Ocean impact biological communities and processes. Monitoring these changes requires systematic and periodic data collection on indicator taxa such as seabirds, which act as ecosystem sentinels. Understanding their breeding behaviour and phenology helps assess the impacts of anthropogenic pressure and environmental variations on seabird populations. Antarctic Wildlife Monitoring Program of Wildlife Institute of India is currently evaluating the population status, distribution and genetic structure of key seabird species (Adelie penguin, snow petrel, south polar skua, Wilson's storm petrel) breeding around Indian research stations. This chapter discusses the results of work being conducted on snow petrel, a climate-dependent seabird found in the ice-free coastal areas and inland mountains in Antarctica. Monitoring snow petrel populations in east Antarctica is critical to understanding their populations' response to climate change and predicting future impacts.

Keywords Climate change · Biological communities · Breeding behavior · Phenology · Antarctic Wildlife Monitoring Program

1 Introduction

Climate change is considered a significant driver of change across ecosystems as geographically diverse as tropical (Barlow et al. 2018), temperate (Schlaepfer et al. 2017) and polar (Hansen 2005; Du Pontavice et al. 2020). These changes amplify further in areas highly vulnerable to its impacts, exceptionally high mountain glaciers (Banerjee and Shankar 2013) and polar ice caps in the Arctic (Box et al. 2019) and Antarctica (Lee et al. 2017a, b). These regions comprise some of the planet's remotest parts, making scientific measurements challenging to undertake, limiting our understanding of physical, chemical, geological and biological processes driving or impacted by climate change.

A. Pande (✉) · K. Sivakumar
Department of Endangered Species Management, Wildlife Institute of India, Chandrabani, Dehradun, Uttarakhand 248001, India

N. Khare (ed.), *Assessing the Antarctic Environment from a Climate Change Perspective*, Earth and Environmental Sciences Library,
https://doi.org/10.1007/978-3-030-87078-2_13

Antarctic continent, the fifth-largest continent in the world (larger than Australia and Europe) recognised globally as the last wilderness (Sanderson et al. 2002), has been under severe pressure from the rapidly shifting global climate (Shaw et al. 2014; Lee et al. 2017a, b). Though highly remote, with year-round harsh weather patterns, the continent and its surrounding Southern Ocean regulate global ocean circulations and atmospheric processes. An internationally controlled Antarctic Treaty System prohibits commercial activities (except tourism) and restricts human activities in Antarctica; its polar ecosystem and biodiversity are considered to be under serious threat (Croxall et al. 2002; Barbraud and Weimerskirch 2006; Chown et al. 2012; Constable et al. 2014; Shaw et al. 2014; Cimino et al. 2016). Thus, despite a global treaty protecting Antarctica from globally mounting environmental change pressure, the white continent designated as a natural reserve, devoted to peace and science (NRC 1993), is at the receiving end of the Anthropocene. The rapid changes in the physical environment of the Southern Ocean affect marine life at all trophic levels, primary prey species (zooplankton including Antarctic Krill), to mesopredators (like squids) to top predators such as marine mammals and seabirds (Atkinson et al. 2004). It is expected that modifications in the Southern Ocean's cold climate will impact the community composition of primary producers, thereby affecting the higher trophic levels (Croxall et al. 2002; Agusti et al. 2010; Constable et al. 2014), which includes seabirds.

1.1 Seabirds as Indicators of Environmental Change

Seabird populations across the globe are threatened with human-induced changes. Long-term monitoring programs have highlighted these threats and the declining status of seabirds worldwide (Croxall et al. 2012; Thiebot et al. 2016; Pertierra et al. 2017); Antarctica is no exception. In the Southern Ocean, where seabird populations have declined substantially over the last few decades (Paleczny et al. 2015), interdisciplinary approaches are being utilised to aid their conservation and management (Friesen 2007; Croxall et al. 2012; Taylor and Friesen 2012). Current knowledge on seabird distribution in Antarctica has been known through the efforts of long-term datasets generated by National Antarctic Programs. With Antarctic researchers' efforts working in megafauna ecology, a Biogeographic Atlas of the Southern Ocean documents the distribution of seabirds and marine mammals in the Southern Ocean and Antarctica (Ropert-Coudert et al. 2014). The atlas gives large spatial scale distribution maps from the data on the species sighted at sea or using the tracking data available from multiple studies on seabirds and marine mammals. It came out as a product of the discussion during the International Polar Year 2007–2009 (www.ipy.org) and later from the Census of Marine Life 2000–2010 (www.com l.org), provided by the SCAR Marine Biodiversity Information Network (www.sca rmarbin.be) and Census of Antarctic Marine Life (www.caml.aq) (De Broyer et al. 2014). Seabird species distribution has also been extensively studied utilising data from ground surveys and through remotely sensed data along several portions of the Antarctic coast (Mehlum et al. 1988; Schwaller et al. 1989, 2018; Fretwell and Trathan 2009; Lynch et al. 2010; LaRue et al. 2014; Lynch and LaRue 2014).

Marine top predators such as seabirds serve as indicators of environmental changes in the Antarctic environment (Woehler 1990; Croxall et al. 2012). Seabirds, being top predators, maintain the structure of marine food webs, regulate island and marine ecosystem processes and act as indicators of aquatic ecosystem health (Lascelles et al. 2012; Paleczny et al. 2015). Monitoring their populations thus acts as a potent tool to assess the Antarctic environment's anthropogenic impact (Croxall et al. 2002; Micol and Jouventin 2001) and understand the variability of the climatic effect on the Antarctic biota. Scanty information exists from very few long-term population studies on pelagic seabird populations in Antarctica (Jouventin and Viot 1985; Chastel et al. 1993; Lorentsen 1996; van Franeker et al. 1999; Barbraud 1999, 2000; Barbraud and Weimerskirch 2001, 2006; Jenouvrier et al. 2005; Barbraud et al. 2015; Descamps et al. 2016a, b). Limited quantitative knowledge on dynamics of interactions between top predators, their prey, and their environment hinders the understanding of complex processes occurring in Antarctica (Croxall et al. 2002), including anthropogenic activities and climate change (Kennicutt et al. 2014; Rodríguez et al. 2019).

Several studies focusing on seabird population monitoring have highlighted the threatened status of seabirds across the globe (Croxall et al. 2012; Thiebot et al. 2016; Pertierra et al. 2017), especially in the Southern Ocean, where seabird populations have declined substantially over the last few decades (Paleczny et al. 2015). This has led to efforts focusing on understanding seabird population dynamics using interdisciplinary approaches to aid conservation and management across their distribution range (Friesen 2007; Croxall et al. 2012; Taylor and Friesen 2012). Focused studies on Antarctic seabird populations have been carried out on the Antarctic peninsula, especially on penguins (Lynch et al. 2012a, b; Clucas et al. 2014), skuas (Borghello et al. 2019; Phillips et al. 2019) etc. Besides, site-specific monitoring of pelagic species has been carried out at sub-Antarctic islands (Brown et al. 2015; Quillfeldt et al. 2017) and multiple sites along the Antarctic coast (Barbraud and Weimerskirch 2001, 2006; Techow et al. 2010; Brown et al. 2015).

Seabirds, a numerically significant group in the Southern Ocean fauna (Warham 1996; Olivier 2006), exert control over the marine trophic web and are affected by the same environmental variations as their prey. Preventing the further decline of their populations is essential as these species have broad ecological impacts (Welch et al. 2012) in the Southern Ocean. The baseline data necessary to study seabird population changes over time are scarce and are challenging to obtain given remoteness and inaccessibility of seabird habitats in Antarctica. Monitoring breeding success and temporal variations in seabird populations have been successfully validated as a potent tool to assess the anthropogenic impact and effects of environmental variations (Croxall et al. 2002; Micol and Jouventin 2001). However, seabirds' data collection is often hindered by the extreme climatic conditions in the Antarctic continent. Accurate forecasting of any environmental or anthropogenic impacts would require a thorough understanding of drivers of change in the seabird population demographics (Barbraud et al. 2011) or breeding phenology (Lynch et al. 2012a, b).

1.2 Long-Term Monitoring of Antarctic Seabirds in Indian Antarctic Program

Antarctic Wildlife Monitoring Program of the Indian Antarctic Program has been monitoring seabirds and marine mammals since the early 1990s (Sathyakumar 1995; Bhatnagar and Sathyakumar 1999; Hussain and Saxena 2008; Sivakumar and Sathyakumar 2012; Kumar and Johnson 2014; Pande et al. 2017, 2018, 2020). The Phase-I of this program was conducted to ascertain the feasibility of conducting long-term research on Antarctica wildlife species. Later, as the Phase-II of the program, in two successive expeditions, extensive spatial scale surveys were carried out in the Indian sector of operation in Antarctica and the Southern Ocean to assess penguins' distribution and abundance seals (Sivakumar and Sathyakumar 2012; Kumar and Johnson 2014). With knowledge of existing species around the Indian research station and logistical capabilities, the Phase-III of the program was launched in 2013–14 to undertake detailed long-term monitoring work on selected indicator species of the polar ecosystem. The Phase-III of the program was conducted during three successive expeditions (33rd, 34th and 35th Indian Scientific Expeditions to Antarctica) and resulted in a critical understanding of species' distribution and breeding biology as pelagic seabirds, penguins and seals (Pande et al. 2017, 2018). The program's Phase-IV is in progress, ascertaining the population status, distribution and genetic structure of select seabird species (Adelie penguin, snow petrel, south polar skua, Wilson's storm petrel) breeding around Indian research stations (Pande et al. 2020). It aims to understand two significant aspects of seabird biology in east Antarctica, viz. nesting ecology and population genetics, with a long-term objective of looking at changes concerning climatic variations in the environment. This chapter discusses the work being conducted on snow petrel, a climate-dependent seabird found in the ice-free coastal areas and inland mountains in Antarctica.

*1.3 Nesting Ecology of Snow Petrel (***Pagodroma nivea***)*

The snow petrel (*Pagodroma nivea*) is endemic to Antarctica, being the most southerly breeding bird species of the world (Loy 1962; Harrison 1983). It is the only member of the genus *Pagodroma* (Bonaparte 1856) in the family Procellariidae. It spends its entire life in the Southern Ocean waters surrounding the continent while breeding during the austral summer ice-free areas on the Antarctic coast, rarely found breeding inland (Ryan and Watkins 1989). Snow Petrels breed colonially at ice-free islands along the Antarctic coast and on exposed rocky mountain areas over 300–400 km inland from the open sea during the austral summers (Løvenskiold 1960). Some earlier reports on breeding locations of snow petrels include Maher (1962) near Cape Hallet, Pryor (1968) at Haswell Island, Ryan and Watkins (1989) in Dronning Maud Land and Chastel et al. (1993) at Terre Adélie. Several other authors have reported snow petrel breeding in locations in the vicinities of Australian stations,

Davis and Mawson (Brown 1966; Johnstone et al. 1973; Bonner and Lewis-Smith 1985; Woehler and Johnstone 1991). Detailed studies on the distribution and abundance of snow petrels on the Ardery and Odbert Islands were conducted in the 1980s–90s by Bonner and Lewis-Smith (1985) and van Franeker et al. (1990). A comprehensive review of the existing literature and unpublished records reported breeding at 298 localities in a circumpolar distribution (Croxall et al. 1995). However, several reports have documented this species' presence and breeding sites from different parts of the continent; the environmental factors influencing its breeding distribution are relatively unknown in Antarctica (Olivier and Wotherspoon 2008). Despite comprehensive knowledge of their breeding distribution, only two long-term studies have been conducted on snow petrels. These include a ~60 year long study on snow petrel breeding biology by French Antarctic researchers at Pointe Géologie Archipelago, Terre Adélie, Antarctica, site of the French station Dumont d'Urville (Jouventin and Viot 1985; Viot et al. 1993; Barbraud 1999; Barbraud 2000; Barbraud and Weimerskirch 2001; Barbraud et al. 2015) and; at the Australian Antarctic Territory encompassing Australian Antarctic stations of Mawson, Davis and Casey (Woehler 1990; Olivier 2006; Olivier and Wotherspoon 2008; Einoder et al. 2014).

Snow petrel nesting behaviour and reproductive success might differ with nest site location (Pierotti 1982; Gaston and Elliot 1996), where reproductive success is influenced by behavioural factors such as breeding synchronisation, incubation scheduling, etc., predator defence (Coulson 2002; Hamer et al. 2002). It has been positively demonstrated that nesting behaviour and reproductive success can differ with nest site location in a seabird colony (Pierotti 1982; Gaston and Elliot 1996), where reproductive success is influenced by behavioural factors such as breeding synchronisation, incubation scheduling and predator defence (Coulson 2002; Hamer et al. 2002). Apart from location, the physical characteristics of a nest site can influence intensities of disturbance from conspecifics (Kim and Monaghan 2005a), predation events (Gaston and Elliot 1996; Gilchrist and Gaston 1997) and nest microclimate (Kim and Monaghan 2005a, b) and thereby influence parental behaviour at the nest. Moreover, nest sites with high visibility may cause the parents to spend more time in vigilance or defending their nest against conspecific intruders or patrolling predators (Drent 1975; Hatch and Nettleship 1998) with drastic impacts on both energy stores and the time spent in parenting.

Snow petrels form colonies of variable sizes and suitable nesting locations in ice-free areas, cliffs and rock faces in Antarctica. They are highly mobile and have few apparent physical barriers to dispersal like other colonially nesting seabird species. Thus, they are capable of flying vast distances in search of epipelagic prey (Avise et al. 2000). However, many seabird species exhibit strong philopatry and can become genetically distinct over short geographical distances (Milot et al. 2008). The birds' movement between breeding sites influences their population dynamics, gene flow and individual fitness, with subsequent significant consequences for population persistence and viability (Hanski 2001; Bowler and Benton 2005). Though several studies have focused on investigating dispersal mechanisms in individual animals using capture-recapture (Lebreton et al. 2003) or bio-telemetry (Shaffer et al. 2006), these approaches are relatively difficult to implement in Antarctica due to extreme weather conditions and inaccessible terrain supplemented by limited logistics. Alternatively, molecular techniques have been effectively used for explaining colonisation patterns, population genetic structure, gene flow and individual immigrants (Rousset 2001) across a varied range of taxa (Knight et al. 1999; Jehle et al. 2005; Welch et al. 2012).

Nest site selection, protecting both adults and young from environmental conditions and predation, is a substantial factor in bird survival and reproduction, particularly for order Procellariiformes (Warham 1996; Thompson et al. 1993) species nest in cavities or burrows. Processes of nest site formation and selection in snow petrels were suggested in the early work of Brown (1966). Other studies also have proposed topography as the significant determining factors explaining snow petrel colonies' distribution (Ryan and Watkins 1989). More recently, detailed habitat selection models were established for the snow petrel at Casey in East Antarctica (Olivier and Wotherspoon 2006). However, the specific process of spatiotemporal selection of nest sites by snow petrel has not yet been entirely clarified (Olivier et al. 2004; Olivier 2006; Olivier and Wotherspoon 2008; Einoder et al. 2014).

Isotopic records of stomach oil spits deposited at the nest cavities of snow petrels suggest that they might be continuously occupied for over 14,000 years (Hiller et al. 1995). Philopatry has been demonstrated in snow petrels though there are studies that indicate otherwise, too (Chastel et al. 1993). Morphological studies in snow petrels have shown two forms of different size (*P. N. Nivea* and *P. N. Major*) which are sympatric at some breeding sites (Isenmann 1970; Cowan 1981; van Franeker et al. 1990; Marchant and Higgins 1990). The origin, status and significance of these two forms remains controversial (Jouventin and Viot 1985) and demands the need to clarify the issue using molecular techniques.

CCAMLR Ecosystem Monitoring Program (CEMP) has overlooked snow petrel in their program to understand population size changes, breeding success, body mass, and foraging behaviour in select indicator species (https://www.ccamlr.org). An in-depth understanding of these dynamics is vital as colonies of seabirds sometimes contain a disproportionately large population at a small number of sites. Seabirds are some of the most threatened groups of birds globally. It becomes imperative to understand gene flow between snow petrel colonies as an essential prerequisite for their successful conservation and management. Their wide distribution across the Antarctica coast makes them an important indicator species for monitoring the Antarctic marine ecosystem's health.

Snow Petrels are long-lived, upper-trophic-level predators greatly dependent on the Southern Ocean's seasonal ice system, which increases their vulnerability to climate change (Olivier 2006). They are specialist foragers near pack ice areas (Griffiths 1983; Ainley et al. 1986), usually occurring abundantly at latitudes south of 60°S (Griffiths 1983; Hunt and Veit 1983; Ainley et al. 1986; Bretagnolle and Thomas 1990). Dietary studies have reported Antarctic Krill *Euphausia superb* to be the significant component of the snow petrel diet during the breeding season, apart from fish and squids (Brown 1966; Griffiths 1983; Ainley et al. 1992). Commercial krill harvesting has raised concerns about its potential damage to the dependent predators, highlighting the need to generate accurate information on the distribution and abundance of Snow Petrels. Climate-dependent variables such as sea ice extent have been shown to affect their breeding success by affecting nesting rates and chick body condition (Barbraud and Weimerskirch 2001). Long-term datasets on these birds link climate change to variations in phenological events in the species' life history (Barbraud and Weimerskirch 2006).

2 Methods

2.1 *Study Areas*

This work was carried out around the Indian sector of operations at two significant sites in east Antarctica viz. Larsemann Hills and Schirmacher Oasis. Larsemann Hills (69° 20′S to 69° 30′S Latitude; 75° 55′E to 76° 30′E Longitude), is a group of islands at Prydz Bay (Fig. 1a). It is an ice-free oasis on the Ingrid Christensen Coast, Princess Elizabeth Land, located approximately midway between the eastern end of the Amery Ice Shelf and the Vestfold Hills (Kiernan et al. 2009). Together the islands form the second largest group of four major ice-free oases found along East Antarctica's 5000 km long coastline spread over an area of about 50 km^2 (Hodgson et al. 2005). The region is bordered on both sides by two large peninsulas, the western

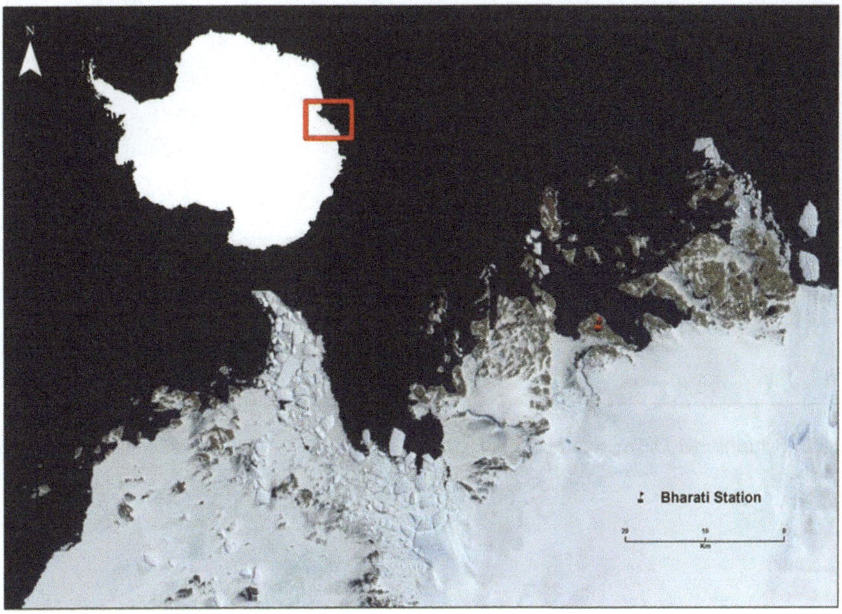

Fig. 1 Larsemann Hills, East Antarctica, site of Bharati, Indian Antarctic research station

Stornes and the eastern Broken, which enclose a group of variously sized islands and peninsulas. India's 2nd permanent research station in Antarctica, *BHARATI*, is located on the North Groves peninsula (Fig. 1). Four other Antarctic stations, viz. the progress I and Progress II (Russia), Law-Racovita (Australia-Romania) and Zhongshan (China), are located along the edge of the Broken peninsula.

Weather at Larsemann hills is influenced by persistent, intense katabatic winds that blow from east to south-east during austral summer. Daytime ambient average air temperatures range from a maximum of 4 °C (Dec–Feb) to a minimum of –40 °C (May–July) (Turner and Pendlebury 2004). Precipitation occurs as snow not exceeding 250 mm of water equivalent annually (Hogdson et al. 2001). Pack ice is extensive in the north-eastern side throughout the austral summer, and the fjords and bays are hardly ice-free even during peak summer. Snow cover is generally higher and persistent on Stornes Peninsula compared to Broken Peninsula. The sea ice grows slowly during March–September, reaching its peak in April–June (NCAOR 2006). Since this study also incorporated the phylogeographic assessment of snow petrel, part of the sampling was conducted at Schirmacher Oasis, Central Dronning Maudland. Schirmacher Oasis is situated on the Princess Astrid Coast of Dronning Maud Land, Antarctica, between the Fimbul ice shelf and continental icecap (Fig. 1). This ice-free land is spread across 34 km^2 between the coordinates 70° 44′–46′ S and 11° 26′–49′ E (Singh et al. 2014). Indian research station *MAITRI* is located on the south-eastern part of the oasis (Fig. 2).

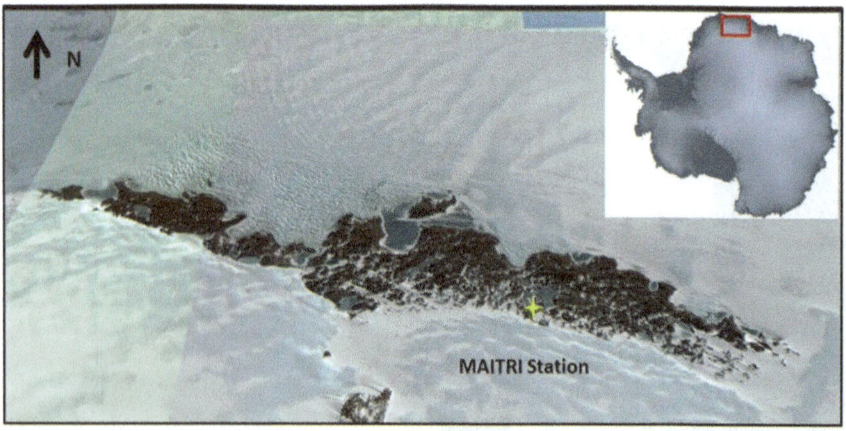

Fig. 2 Schirmacher Oasis, central Dronning Maudland, site of Maitri, Indian Antarctic research station

2.2 Field Sampling

Field surveys were conducted to locate snow petrel nests over several Larsemann hills' islands under the "Antarctic Wildlife Monitoring Program" of Indian Scientific Expeditions to Antarctica (ISEA). These surveys were spread over three austral summers (November–March) of 2013–14, 2014–15 and 2015–16, coinciding with the Antarctic seabird species' breeding season. At Larsemann hills, all the named islands/peninsulas and their adjoining rocky outcrops were surveyed for the presence of snow petrel nesting sites (Pande et al. 2020). Snow petrel nests were physically located using a hand-held flashlight (300 lumens) using the area search method. Snow petrels' nest in rock cavities or crevices formed within natural boulders in steep rocky slopes (Olivier and Wotherspoon 2008; Tveraa and Christensen 2002). Cavities large enough to hold snow petrel breeding pairs were marked as an occupied nest (OCN) based on the presence of one or more adult bird or an unhatched egg or a live chick or a potential unoccupied or potential nest (UPN) based on the fact of dead egg/s or broken eggshells or hatched eggshells or dead chick or quiet adult or guano marks or mumiyo deposits. Once a snow petrel nest was detected, an extensive search was conducted in a 50 × 50 m area around it to locate all occupied and unoccupied or potential nest cavities. Each OCN and UPN nest cavity was marked using non-toxic, odourless paint (red or yellow), and its geographic coordinates were recorded on a hand-held GPS unit (GARMIN eTrex 30xTM). The periphery of the colony was mapped on the GPS device by walking around the outermost detected nests. Once a rough estimate of the colony perimeter was ascertained, 3 × 3 m plots were placed at fixed intervals along lines running diagonally from the bottom to the colony's top (Mehlum et al. 1988). Random nests were then chosen from these intensive study plots (3 × 3 m) within snow petrel colonies to study nest cavity characteristics (Figs. 3 and 4).

The nest cavities' physical characteristics were manually profiled into rock type, nest bowl metrics, nest orientation, etc. Nest cavity metrics were obtained using a measuring tape, i.e. nest entrance measurements and nest bowl measurements. In cases where the access to nest bowl was not possible, an extension mechanical arm tool was used to reach deeper cavities. Nest orientation and aspect was measured using a hand-held clinometer and compass verified using a digital compass on the GPS unit. Each potential nest cavity within the study plot was marked using non-toxic paint, and its geographic coordinates were recorded on a hand-held GPS unit. The nest locations were also observed on Google Earth Pro v.7.1.8 and then later extracted as KML files for visualisation and planning for monitoring visits over the expedition duration (Figs. 5 and 6). The monitoring planned to cover all phases of nesting of the species, starting from November (egg-laying) and ending in February–March (fledging).

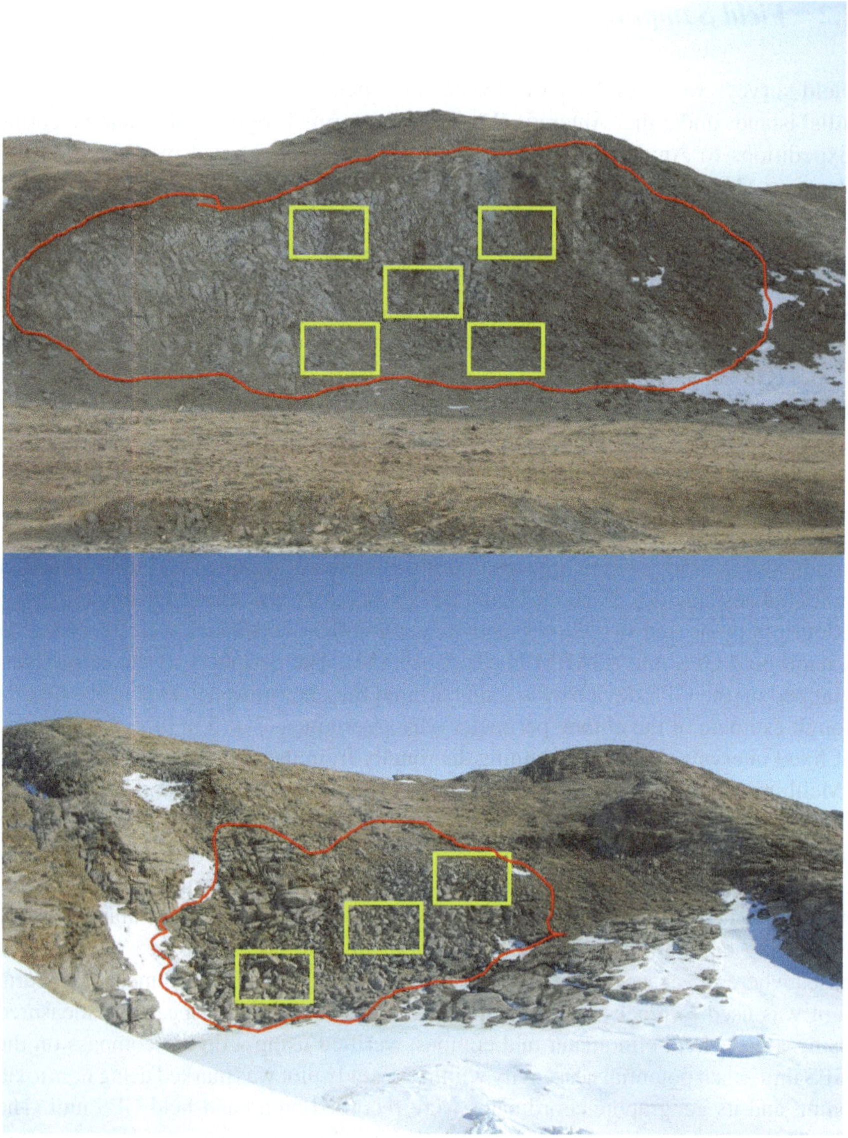

Fig. 3 Typical cliff habitats where snow petrels nest in Larsemann hills. Yellow boxes represent plots where nests were marked over these cliffs, while red polygon is the colony's approximate periphery

Fig. 4 Top: Slab type of snow petrel nest cavity; the snow petrels occupy narrow spaces under these flat rocks. Bottom (left): Boulder type of snow petrel nest cavity; the snow petrels occupy spaces created within two or more boulders. Bottom (right): Crack or Crevice type of snow petrel nest cavity; the snow petrels occupy spaces created within a crack in a rock

2.3 Nest Monitoring

Due to logistical constraints and limitations in visiting each marked nest site regularly or in a planned manner over the fieldwork period, it was decided to maximise the nest visits by visiting all substantial nests opportunistically. These visits were conducted in conjunction with sampling for genetic samples to look at the species' breeding success in the study area. Further, few nest sites were chosen for automatic camera

Fig. 5 Snow petrels occupy natural rock cavities at Grovnes peninsula, Larsemann Hills, East Antarctica

Fig. 6 Snow petrel nest cavities were measured physically, and nest occupancy was determined at Larsemann Hills, East Antarctica

operated monitoring to understand nest attendance and parental care strategies in snow petrels.

In association with Indraprastha Institute of Information Technology—Delhi, we designed motion-sensing cameras for monitoring breeding pairs of snow petrels. Being cryptic cavity-nesting species, manual observation methods do not work with snow petrel. Moreover, inclement weather conditions in the study area make it difficult to collect continuous observation data on the species. We decided to experiment with automated modes of image collection using motion-sensing cameras after due discussion with engineers and technicians working in wildlife biology. The camera design needed to be small enough to be fitted inside the nest cavities and sustain the study area's sub-zero temperatures (Fig. 7).

The main idea behind developing an automated camera system was to ensure:

(i) An inexpensive device that can be quickly recovered. The camera system's cost needed to be less as the chances of losing a device is inaccessible, extreme climate prone, remote areas is very high.
(ii) The camera system needed to be durable for the harsh climatic conditions of Antarctica. Larsemann hills change very frequently and receive precipitation from the katabatic winds blowing from the south-east.
(iii) The camera system needed to be insulated from external temperatures and snow precipitation to conserve battery power and provide output for a longer duration.

Fig. 7 Snow petrel individuals at nest cavities, Larsemann Hills, East Antarctica

With these constraints in mind, we designed a camera system with an optimised power consumption profile to prolong battery life. Therefore, the system consisted of two standard USB cameras connected to a Raspberry Pi A+, the smallest and most power-efficient board of the Raspberry family. Both cameras were programmed to take an image every second and compile them into short videos after every 10 min. If there is a change due to, e.g., a movement, the current incarnation was recorded along with a time stamp on a connected USB pen drive. The timestamp was provided by a custom PCB equipped with a DS3231 Real-Time-Clock (RTC). Since the system is self-sufficient, the PCB also provided a user interface to replace the USB pen drive and a depleted power supply on the fly without shutting down the system. The device itself recognised these events and acted accordingly by, e.g., turning off the cameras in case the pen drive has been removed. In case a fault is detected, the system shut down to a saved state, preserving the battery and preventing any further damages. Given the harsh environment of Larsemann Hills, where temperatures around −20 °C to −30 °C are recorded regularly, maintaining a power supply remained a challenge. We fitted one Lithium-ion battery with a rated capacity of 13,000 mAh to supply each operating system for approximately 24 h before it needed to be replaced. This was achieved by insulating both the Raspberry Pi and the Li-Ion battery against the sub-zero temperatures and using the electronic components' heat to warm the battery (Fig. 8).

A total of five camera systems were deployed, each nest camera system comprising of USB cameras (n = 2), Raspberry Pi microcomputers (n = 2), 16 GB USB storage

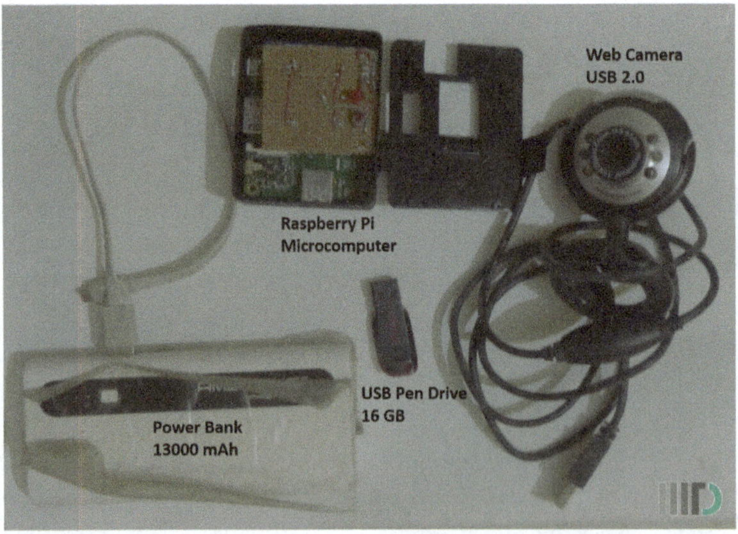

Fig. 8 Nest camera system designed with the support of Indraprastha Institute of Information Technology–Delhi. The images were recorded by a USB Web Camera connected to a pre-programmed Raspberry Pi microcomputer powered by a 13,000 mAh power bank battery. The images were recorded on a 16 G.B. data storage pen drive

devices (n = 2) powered by Li-ion battery (1) at nest sites adjacent to each other. The nest sites chosen to be monitored using the camera system were selected from pre-marked nest sites from field surveys conducted earlier (Pande et al. 2017; see Chap. 1 for details). The nest sites were selected based on their accessibility in inclement weather conditions and ease of replacement of batteries every 24 h. Nest sites on the north Grovnes peninsula, the Bharati research station site, were selected for undertaking this camera-operated continuous monitoring due to accessibility and feasibility in visiting throughout the study period.

3 Results

A total of 9 islands were covered to establish long-term plots for nest monitoring of snow petrels. Over 250 cavities/crevices large enough to contain snow petrel breeding pairs were paint-marked during the breeding season of 2014–15 and 2015–16 for long-term monitoring. Being alternate year breeders (Chastel et al. 1993; Olivier et al. 2005), it is imperative to monitor snow petrel nests for a minimum of two seasons to understand breeding success. In the first season of 2014–15, 95 nest sites (including OCN and UPN nests) were visited and marked for monitoring. In the second season (2015–16), 159 more nests were drawn and observed in the previous year. However, due to various logistical constraints during the fieldwork, an equal number of visits could not be made to all the nest sites in both seasons. Only 198 nests out of the total 254 (78%) could be visited in 2015–16. Moreover, only a subset of these marked nests (n = 66) could be seen more than twice in the next season to check for any nesting activity. A total of 238 nests were visited throughout two austral summer seasons (Table 1).

3.1 Nest Cavity Characteristics

Snow petrel nest cavities were classified into three types, viz. crack or crevice, boulder and slab, based on Einoder et al. (2014). A crack or crevice type of nest cavities are formed by glacial or cold weathering of rocks; boulder type of nest cavities are spaces available in between two or more stones, whereas; slab type of holes are rooms open between the ground and large flat boulders.

Out of the 192 cavities classified for rock type (Fig. 9), the boulder and slab type were equally occupied for nesting (40.6% and 40.2%, respectively). In contrast, lesser cavities of crack or crevice type were occupied (19.2%). The niches occupied by snow petrels had about 16% smaller entrance area (Fig. 10) and were about 18% lower in volume (Fig. 11) compared to the unoccupied ones. An independent-sample t-test of cavity volumes showed no significant difference between unoccupied and occupied nest cavities (t-stat = 0.94, $p > 0.05$).

Table 1 Nest occupancy, laying success and hatching success of snow petrel nests monitored over two breeding periods 2014–15 and 2015–16, Larsemann Hills, East Antarctica. Numbers in parentheses (n) indicate several nests, (–) represents sites not visited to ascertain the status of snow petrel nesting, (DNA—Data Not Available)—means places that could not be visited later in the season for confirming nest status

Island	2014–15			2015–16		
	Occupied nests	Laying success%	Hatching success%	Occupied nests	Laying success%	Hatching success%
Betts	–	–	–	6	63.6 (4)	DNA
Bharati	31	90.3 (28)	87.1 (27)	55	63.6 (35)	16.4 (9)
Breadloaf	–	–	–	5	100 (5)	100 (5)
Broknes	34	91.1 (31)	85.3 (29)	41	87.8 (36)	4.8 (2)
Cook	–	–	–	5	20 (1)	120 (6)
Easther	26	50.0 (13)	42.3 (11)	15	93.3 (14)	73.3 (11)
Fisher	–	–	–	24	100 (24)	54.2 (13)
Manning	–	–	–	1	100 (1)	100 (1)
McLeod	4	100 (4)	100 (4)	6	66.6 (4)	DNA
Total nest monitored	95	76	71	152	105	47

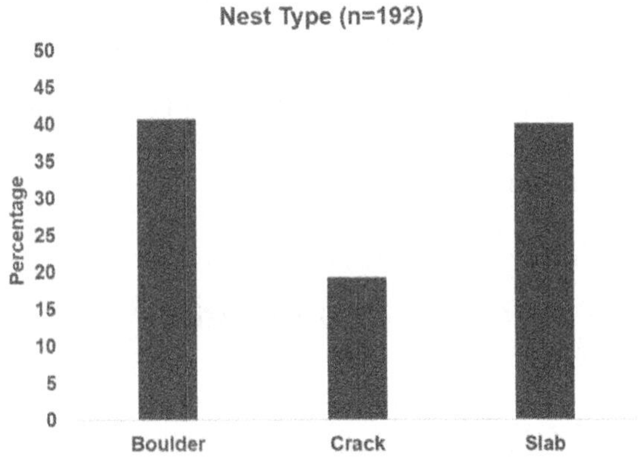

Fig. 9 Rock type of nest cavities occupied by snow petrels at Larsemann Hills, East Antarctica

Nest cavities were mostly oriented towards the east direction (~39%) in the study area (mean Vector 61.749°). Wind direction data (weekly average for December 2015, incubation period) was acquired from a weather station deployed by India Meteorological Department at *Bharati* station about 1.5 km away. The wind in the area was predominantly from the north direction (mean Vector 12.379°). Rose plots plotted using software Oriana v.4.0 for nest orientation and wind direction exhibit complete exclusion of north-facing cavities by the snow petrels (Fig. 12). Rayleigh's test performed on the circular data of nest orientation was significant (Rayleigh's $z_{7.298}$, $p < 0.001$; see Table 2 for circular statistics), which means that nest cavities

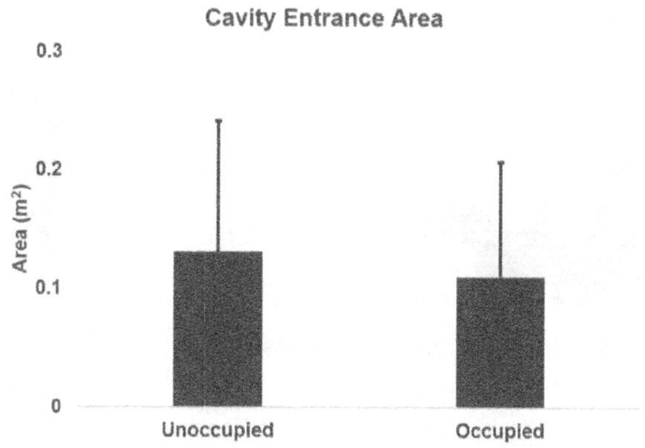

Fig. 10 Nest cavity entrance areas of occupied and unoccupied nests of snow petrels at Larsemann Hills, East Antarctica

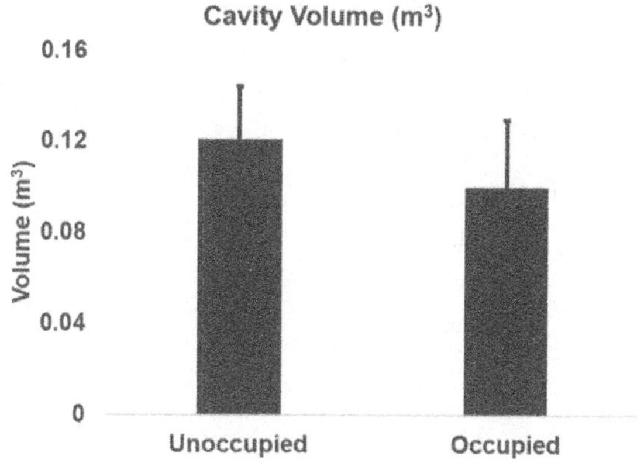

Fig. 11 Nest cavity volumes of occupied and unoccupied nests of snow petrels at Larsemann Hills, East Antarctica

were clustered towards specific directions, in this case towards east-northeast and north-northwest.

3.2 *Breeding Success*

Breeding success was classified into (i) laying success calculated as the number of eggs laid in the occupied nests and (ii) hatching success as the number of eggs hatched of eggs laid. The fledging success could not be determined as the fieldwork could not be done post-late-February (last dates for visiting nests in 2014–15 and 2015–16 was 19-Feb 2015 and 12-Feb 2016, respectively). For each year, the percentage of

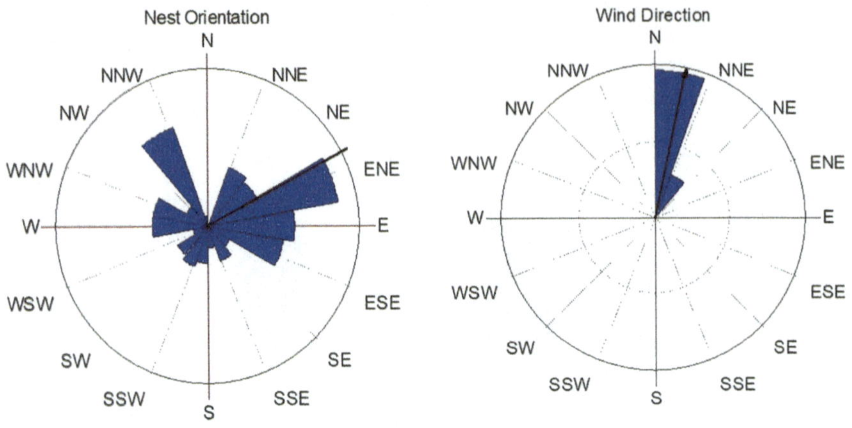

Fig. 12 Snow petrel nest orientation and wind direction rose plots of Larsemann Hills, East Antarctica. The blue triangles represent the wind direction and nest orientation clustering concerning the movement, and the black arrow represents the mean direction

Table 2 Summary circular statistics of nest orientation and wind direction measurements at Larsemann hills using Oriana v.4.0

Variable	Nest orientation	Wind direction
Data type	Angles	Angles
Number of observations	171	10,134
Mean vector (μ)	61.749°	12.379°
Length of mean vector (r)	0.207	0.995
Concentration	0.422	100.1
Circular variance	0.793	0.005
Circular standard deviation	101.755°	5.741°
One-sample tests		
Rayleigh test (Z)	7.298	10,032.76
Rayleigh test (p)	6.77E-04	<1E-12
Rao's spacing test (U)	275.211	348.028
Rao's spacing test (p)	<0.01	<0.01

occupied nests was calculated as the number of nests in which laying was observed divided by the total number of nests monitored. In 2014–15, the mean laying and hatching success for the first season 2014–15 was higher than that of the second season 2015–16 (82.8% for 2014–15 and 77.2% for 2015–16; 74.7% for 2014–15 and 58.6% for 2015–16 respectively; Fig. 13).

To account for non-parametric and unequal variances, Mann–Whitney U tests (two-tailed) were performed to test the effect of cavity characteristics concerning nest occupancy and laying success, respectively. Sample sizes were not sufficient to test for differences in the distribution of values in the nests with successful hatching. The rock cavity volume did not have any effect on occupancy or laying success. Nest

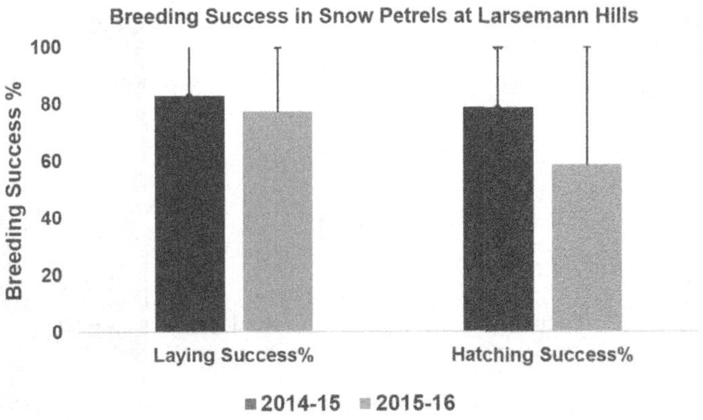

Fig. 13 Breeding snow petrel success derived from monitored nests during the austral summers of 2014–15 and 2015–16 at Larsemann Hills, East Antarctica

Table 3 Mann–Whitney U test (two-tailed) statistics for nesting success comparison between cavity volume and nest bowl volumes

	Nest occupancy		Nest with egg (successful laying)	
	Cavity volume	Nest volume	Cavity volume	Nest volume
Samples	$N_1 = 137, N_2 = 33$	$N_1 = 137, N_2 = 33$	$N_1 = 59, N_2 = 66$	$N_1 = 66, N_2 = 71$
U-value	1778.5	1670	1697	1820
z-score	−1.89702	−1.89702	−1.23389	2.2508
p-value	0.06	0.06	0.21	0.02*

*Indicate significant value, $p < 0.05$, 2-tailed

bowl volumes did not affect occupancy but significantly affected spreading success (see Table 3).

3.3 Breeding Phenology

A total of 12 nests were monitored using the automated motion-sensing camera systems deployed at North Grovnes peninsula. Cameras were moved if the breeding pair deserted the nest cavity or the laying did not happen. Out of 12 cavities occupied, 4 teams failed to lay an egg, while two out of those later could not hatch. Hatching was successful in only 6 of the nests monitored. Over 3 million images from the automated nest camera systems obtaining crucial insights into the breeding biology of snow petrels in the study area. Nest attendance patterns of snow petrels were ascertained during the breeding period using these images.

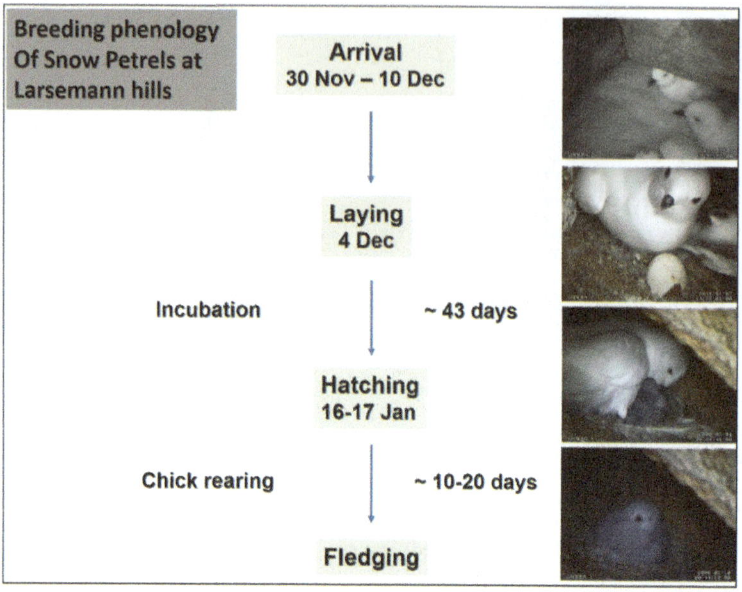

Fig. 14 Breeding phenology of snow petrels revealed through automated motion-sensing cameras deployed at North Grovnes peninsula, Larsemann Hills, East Antarctica

Snow petrels arrived in the last week of November in the area after the pre-laying exodus and occupied rock cavities for nesting. Nesting began by the first week of December, though individuals kept coming till Dec 10. Seven snow petrel breeding individuals were documented laying a single egg during the intervening night of 4th and 5th December 2015. One of the nests could not be assessed for the egg's presence as the bird moved very little to give away any hint. After an incubation period of 43 days, the eggs hatched on 16th or 17th January ($n = 6$). Later, the chick-rearing period varied between 10 and 20 days in different breeding pairs. The monitoring could not go 26–27 days beyond the hatching date due to the expedition vessel's departure from Larsemann hills on Feb 12 2016. This first-time automated monitoring of an Antarctic cavity-nesting seabird provided crucial insights into its breeding phenology and serves as the first baseline information on the species from Larsemann hills (see Fig. 14).

4 Discussion

Multiple environmental factors affect the nest-site selection and breeding success (Bourgeois and Vidal 2007; Catry et al. 2003; Drummond and Leonard 2010; Einoder et al. 2014). Various physical factors such as cavity depth, volume, rock type, substrate, slope, aspect and environmental parameters like wind, temperature, precipitation etc. are critical for selecting a nesting site to avoid inclement weather conditions and provide protection from predators (Ramos et al. 1997; Bourgeois and Vidal 2007). However, collecting data on these variables needs repeated systematic surveys over several years to assess each variable's role in nest-site selection studies significantly. This study forms the first attempt under the Indian Antarctic Program to gather systematic data on Antarctica's single species. After several years of field surveys under limited logistics (Sathyakumar 1995; Bhatnagar and Sathyakumar 1999; Hussain and Saxena 2008; Sivakumar and Sathyakumar 2012), snow petrel colonies were mapped and marked for long-term monitoring around Indian research station Bharati (Pande et al. 2017, 2018).

In the Indian Antarctic Expeditions, the field visits are limited by the availability of travel support to Larsemann Hills islands. When the sea ice is thick around the station area (roughly >1 m thickness), the travel to various nearby islands or peninsulas is supported by snowmobiles. However, due to the thinning of sea ice in the Quilty bay and Thala fjord area, the use of snowmobiles is curtailed for safety reasons, somewhere around mid-December every year. Later, after the arrival of the expedition vessel near the station area, the field visits are supported by single-engine light utility helicopters that facilitate the visit to far-flung islands. Considering these logistical limitations, the effort to visit and mark colonies of snow petrels varied over the two seasons of 2014–15 and 2015–16. In the 2014–15 season (34th Indian Antarctic Expedition), the support to visit field sites was significantly reduced due to the helicopter's technical glitch. Only 95 nest sites could be seen and marked in 10 days (Jan 29 and Feb 19 2015). On the other hand, the duration spent at Larsemann hills was more in the second season (2015–16; 35th Indian Antarctic Expedition). Thus, more field visits could be made (24 days between Nov 30, 2015, and Feb 11 2016). Consequently, 62.5% more nests were monitored in the second season.

The laying as well hatching success was higher for the second season compared to the first season. However, this difference in breeding success could be attributed to various factors, including differences in sample sizes and local weather pattern, as excessive snow accumulation is known to negatively impact snow petrel breeding success (Einoder et al. 2014). At Larsemann hills, snow petrels established nests between boulders and under flat slab-like rocks and preferred fewer crevices or cracks within stones. However, a detailed study on the availability of these spaces suitable for snow petrel nesting versus actual use could be done in the future to look at the nesting habitat preference of the species.

Snow petrels breed in naturally formed rock cavities (Einoder et al. 2014; Olivier et al. 2004). At Larsemann hills, snow petrels selected cavities with smaller entrances and lower overall volume for breeding. Narrower gates presumably reduce airflow (Einoder et al. 2014) and subsequently lead to lesser ice accumulation during precipitation or snowdrifts. This aspect of nest-site selection could be studied in detail in future monitoring studies. The wind direction in the study area is mainly from a north direction, and snow petrels select sites which are towards mostly east-northeast and north-northwest directions. However, the wind direction data was taken from the weekly average for December 2015. Long-term weather data is needed to look for prevailing wind conditions during late November and early December, the months when the snow petrels arrive for occupying nest cavities.

Research on breeding biology of seabirds has been conducted chiefly from direct observations or repeated site visits or with the use of hand-held camera devices (Bourgeois and Vidal 2007; Einoder et al. 2014; Lacey 2018; Mallory 2009; Mejías et al. 2017; Olivier and Wotherspoon 2008). However, in recent times and with the advent of advanced remotely operated camera technology, many cryptic cavity-nesting species have been monitored worldwide (Landers 2011; Prinz et al. 2016; Sabine et al. 2005). The camera monitoring system designed by us at Wildlife Institute of India and Indraprastha Institute of Information Technology- Delhi were the first attempt to study a cavity-nesting Antarctic seabird species. It gave helpful information on the breeding phenology of snow petrels at Larsemann hills, specifically about the dates of laying, hatching and nest attendance by parents. The image dataset obtained from nest monitoring requires quantification for detailed analysis of snow petrels' intra-season breeding behaviour pattern at Larsemann Hills. More data on the breeding of the only predator in the area, the south polar skua, on long-term weather patterns and sea ice conditions should also be collected simultaneously to investigate the impacts of climate change on the breeding success of snow petrel and other co-habiting seabird species (Barbraud et al. 2015; Barbraud and Weimerskirch 2001; Constable et al. 2014). Species-specific monitoring work with a systematic long-term approach would yield crucial data on species' biological responses to climate change in the continent.

Acknowledgements This work was funded by Wildlife Institute of India, Ministry of Environment, Forest, Climate Change, Government of India (No. WII/33rdISEA/2013-14) and National Centre for Polar and Ocean Research, Ministry of Earth Sciences, Government of India (No. NCAOR/14(159)/13/BES-3). The authors would like to thank the National Centre for Polar and Ocean Research, Goa, for the logistics support during the Indian Antarctic Expeditions. Special thanks to Dr. Alexander Fell, Indraprastha Institute of Information Technology-Delhi for helping design the Snow Petrel Nest Cameras and providing the detailed supporting text material on the design. Thanks to the Director, Wildlife Institute of India, Rahul Mohan (NCPOR) and M. Javed Beg (NCPOR) for their constant support. The authors are grateful for generous field support provided by voyage leaders (PK Manna, Shailendra Saini), stations leaders (Kailash Bhindwar, Raghavendra Kaila) and members of 33rd, 34th and 35th Indian Scientific Expeditions to Antarctica.

References

Ainley DG, Fraser WR, Ribic CA (1986) AMERIEZ 1986: oceanic factors affect seabirds in the Scotia and Weddell Seas. Antarct J U S 22(5):172–173

Atkinson A, Siegel V, Pakhomov E, Rothery P (2004) Long-term decline in krill stock and increase in salps within the Southern Ocean. Nature 432:100–103. https://doi.org/10.1038/nature02996

Agustí S, Sejr MK, Duarte CM (2010) Impacts of climate warming on polar marine and freshwater ecosystems.

Avise JC, Nelson WS, Bowen BW, Walker D (2000) Phylogeography of colonially nesting seabirds, with special reference to global matrilineal patterns in the sooty tern (*Sterna fuscata*). Mol Ecol 9(11):1783–1792

Banerjee A, Shankar R (2013) On the response of Himalayan glaciers to climate change. J Glaciol 59(215):480–490

Barbraud C (1999) Subspecies-selective predation of snow petrels by skuas. Oikos 86:275. https://doi.org/10.2307/3546445

Barbraud C (2000) Natural selection on body size traits in a long-lived bird, the snow petrel *Pagodroma Nivea*. J Evol Biol 13:81–88. https://doi.org/10.1046/j.1420-9101.2000.00151.x

Barbraud C, Delord K, Weimerskirch H (2015) Extreme ecological response of a seabird community to unprecedented sea ice cover. R Soc Open Sci 2:140456–140456. https://doi.org/10.1098/rsos.140456

Barbraud C, Weimerskirch H (2001) Contrasting effects of the extent of sea-ice on an Antarctic top predator's breeding performance, the Snow Petrel *Pagodroma Nivea*. J Avian Biol 32:297–302. https://doi.org/10.1111/j.0908-8857.2001.320402.x

Barbraud C, Weimerskirch H (2006) Antarctic birds breed later in response to climate change. Proc Natl Acad Sci 103:6248–6251. https://doi.org/10.1073/pnas.0510397103

Barbraud C, Rivalan P, Inchausti P, Nevoux M, Rolland V, Weimerskirch H (2011) Contrasted demographic responses facing future climate change in Southern Ocean seabirds. J Anim Ecol 80(1):89–100

Barlow J, França F, Gardner TA, Hicks CC, Lennox GD, Berenguer E, Castello L, Economo EP, Ferreira J, Guénard B, Leal CG, Isaac V, Lees AC, Parr CL, Wilson SK, Young PJ, Graham NA (2018) The future of hyperdiverse tropical ecosystems. Nature 559(7715):517–526

Bhatnagar YV, Sathyakumar, S (1999) Developing a long-term monitoring programme for birds and mammals in the Indian Ocean and Antarctica. Fifteenth Indian Expedition to Antarctica, Scientific Report, Department of Ocean Development, Technical Publication No. 13, pp 131–164

Bonaparte CL (1856) Especes nouvelles d'oiseaux d'Asie et d'Amedque et tableaux paralleliques des pdaglens ou Gaviae. C R Acad Sci 42:764–776

Bonner WN, Smith R (eds) (1985) Conservation areas in the Antarctic. Scientific Committee on Antarctic Research, International Council of Scientific Unions, Cambridge

Borghello P, Torres DS, Montalti D, Ibañez AE (2019) Diet of the Brown Skua (Stercorarius antarcticus lonnbergi) at Hope Bay, Antarctic Peninsula: differences between breeders and non-breeders. Polar Biol 42(2):385–394

Bourgeois K, Vidal É (2007) Yelkouan shearwater nest-cavity selection and breeding success. CR Biol 330(3):205–214. https://doi.org/10.1016/j.crvi.2006.12.007

Bowler DE, Benton TG (2005) Causes and consequences of animal dispersal strategies: relating individual behaviour to spatial dynamics. Biol Rev 80:205–225

Box JE, Colgan WT, Christensen TR, Schmidt NM, Lund M, Parmentier FJW, Brown R, Bhatt US, Euskirchen ES, Romanovsky VE, Walsh JE, Overland JE, Wang M, Corell RW, Meier WN, Wouters B, Mernild S, Mård J, Pawlak J, Olsen MS (2019) Key indicators of Arctic climate change: 1971–2017. Environ Res Lett 14(4):045010

Bretagnolle V, Thomas T (1990) Seabird distribution between Tasmania and Adélie Land (Antarctica) and other Antarctic sectors. Emu 90(2):97–107

Brown RM, Techow NMSM, Wood AG, Phillips RA (2015) Hybridisation and back-crossing in giant petrels (*Macronectes giganteus* and *M. halli*) at Bird Island, South Georgia, and a summary of hybridisation in seabirds. PLoS One 10:e0121688. https://doi.org/10.1371/journal.pone.0121688

Catry P, Campos A, Segurado P et al (2003) Population census and nesting habitat selection of thin-billed prion Pachyptila belcheri on New Island, Falkland Islands. Polar Biol 26:202–207. https://doi.org/10.1007/s00300-002-0465-9

Chastel O, Weimerskirch H, Jouventin P (1993) High annual variability in reproductive success and survival of an Antarctic seabird, the snow petrel *Pagodroma Nivea*: a 27-year study. Oecologia 94:278–285. https://doi.org/10.1007/BF00341328

Chown SL, Lee JE, Hughes KA et al (2012) Challenges to the future conservation of the Antarctic. Science 337:158–159

Cimino MA, Lynch HJ, Saba VS, Oliver MJ (2016) Projected asymmetric response of Adélie penguins to Antarctic climate change. Sci Rep 6. https://doi.org/10.1038/srep28785

Clucas GV, Dunn MJ, Dyke G, Emslie SD, Levy H, Naveen R, Polito MJ, Pybus OG, Rogers AD, Hart T (2014) A reversal of fortunes: climate change 'winners' and 'losers' in Antarctic Peninsula penguins. Sci Rep 4(1):1–7

Constable AJ, Melbourne-Thomas J, Corney SP et al (2014) Climate change and the Southern Ocean ecosystems I: how changes in physical habitats directly affect marine biota. Glob Change Biol 20:3004–3025. https://doi.org/10.1111/gcb.12623

Coulson, J.C. (2002) Colonial breeding in seabirds. In: Schreiber EA, Burger J (eds) Biology of marine birds. CRC, New York, pp 87–113

Cowan AN (1981) Size variation in the Snow Petrel (Pagodroma nivea). Notornis 28(3):169–188

Croxall JP, Trathan PN, Murphy EJ (2002) Environmental change and Antarctic seabird populations. Science (New York, NY) 297:1510–1514. https://doi.org/10.1126/science.1071987

De Broyer C, Koubbi P, Griffiths HJ, Raymond B, Udekem d'Acoz Cd', Van de Putte AP, Danis B, David B, Grant S, Gutt J, Held C, Hosie G, Huettmann F, Post A, Robert-Coudert Y (eds) (2014) Biogeographic atlas of the Southern Ocean. Scientific Committee on Antarctic Research, Cambridge, XII + 498 pp

Descamps S, Tarroux A, Cherel Y, et al. (2016a) At-sea distribution and prey selection of antarctic petrels and commercial krill fisheries. PLoS One 11:e0156968. https://doi.org/10.1371/journal.pone.0156968

Descamps S, Tarroux A, Lorentsen S-H, et al. (2016b) Large-scale oceanographic fluctuations drive Antarctic petrel survival and reproduction. Ecography 39:496–505. https://doi.org/10.1111/ecog.01659

Drent RH (1975) Incubation. In: Farner DS, King JR (eds) Avian biology, vol 5. Academic, New York, pp 333–420

Drummond BA, Leonard ML (2010) Reproductive consequences of nest site use in fork-tailed storm-petrels in the Aleutian Islands, Alaska: potential lasting effects of an introduced predator. Avian Conserv Ecol 5. https://doi.org/10.5751/ACE-00414-050204

Du Pontavice H, Gascuel D, Reygondeau G, Maureaud A, Cheung WW (2020) Climate change undermines the global functioning of marine food webs. Glob Change Biol 26(3):1306–1318

Einoder LD, Emmerson LM, Southwell DM, Southwell CJ (2014) Cavity characteristics and ice accumulation affect nest selection and breeding in snow petrels *Pagodroma Nivea*. Mar Ornithol 42:175–182

Fretwell PT, Trathan PN (2009) Penguins from space: faecal stains reveal the location of emperor penguin colonies. Glob Ecol Biogeogr 18:543–552. https://doi.org/10.1111/j.1466-8238.2009.00467.x

Friesen VL (2007) New roles for molecular genetics in understanding seabird evolution, ecology and conservation. Mar Ornithol 35:89–96

Gaston AJ, Elliot RD (1996) Predation by ravens *Corvus corax* on Bru¨nnich's guillemot *Uria lomvia* eggs and chicks and its possible impact on breeding site selection. Ibis 38:742–748

Gilchrist HG, Gaston AJ (1997) Effects of murre nest site characteristics and wind conditions on predation by glaucous gulls. Can J Zool 75:518–524

Griffiths AM (1983) Factors affecting the Snow Petrel distribution (*Pagodroma Nivea*) and Antarctic Petrel (*Thalassoica Antarctica*). Ardea 71:145–150

Hamer KC, Schreiber EA, Burger J (2002) Breeding biology, life histories, and life history-environment interactions in seabirds. In: Schreiber EA, Burger J (eds) Biology of marine birds. CRC, New York, pp 217–261

Hansen JE (2005) A slippery slope: how much global warming constitutes "dangerous anthropogenic interference"? Climatic Change 68(3):269–279

Hanski I (2001) Population dynamic consequences of dispersal in local populations and metapopulations. In: Clobert J, Danchin E, Dhondt AA, Nichols JD (eds) Dispersal. Oxford University Press, Oxford, U.K., pp 283–298

Harrison P (1983) Seabirds. An identification guide. Croom Helm, Beckenham

Hatch SA, Nettleship DN (1998) Northern fulmar (*Fulmarus glacialis*). In: Poole A, Gill F (eds) The birds of North America No. 361. The Birds of North America, Philadelphia

Hiller A, Hermichen WD, Wand U (1995) Radiocarbon-dated subfossil stomach oil deposits from petrel nesting sites: novel paleoenvironmental records from continental Antarctica. Radiocarbon 37(2):171–180

Hodgson DA, Verleyen E, Sabbe K, Squier AH, Keely BJ, Leng MJ, Saunders KM, Vyverman W (2005) Late Quaternary climate-driven environmental change in the Larsemann Hills, East Antarctica, multi-proxy evidence from a lake sediment core. Quat Res 64(1):83–99

Hunt GL, Veit RR (1983) Marine bird distribution in Antarctic waters. Antarct J US 18:167–169

Hussain SA, Saxena A (2008) Distribution and status of Antarctic seals and penguins along the Princess Astrid Coast, East Antarctica. Indian J Mar Sci 37:7

Isenmann P (1970) Contribution aÁ la biologie de reproduction du peÂtrel des neiges (Pagodroma nivea Forster); le probleÁme de la petite et de la grande forme. Oiseau 40: 99±134

Jehle R, Wilson GA, Arntzen JW, Burke T (2005) Contemporary gene flow and the Spatio-temporal genetic structure of subdivided newt populations (*Triturus cristatus*, *T. marmoratus*). J Evol Biol 18:619–628

Jenouvrier S, Barbraud C, Weimerskirch H (2005) Long-term contrasted responses to climate of two Antarctic seabird species. Ecology 86:2889–2903

Johnstone GW, Lugg DJ, Brown DA (1973) The biology of the Vestfold Hills, Antarctica. ANARE Sci Rep Ser B (Zool) 123:1–62

Jouventin P, Viot CR (1985) Morphological and genetic variability of Snow Petrels *Pagodroma Nivea*. Ibis 127:430–441. https://doi.org/10.1111/j.1474-919X.1985.tb04839.x

Kennicutt MC, Chown SL, Cassano JJ, Liggett D, Massom R, Peck LS, Rintoul SR, Storey JW, Vaughan DG, Wilson TJ, Sutherland WJ. (2014) Polar research: six priorities for Antarctic science. Nat News 512(7512):23

Kiernan K, Gore DB, Fink D et al (2009) Deglaciation and weathering of Larsemann Hills, East America. Antarct Sci 21:373. https://doi.org/10.1017/S0954102009002028

Kim SY, Monaghan P (2005a) Interacting effects of nest shelter and breeder quality on herring gulls' behaviour and breeding performance. Anim Behav 69:301–306

Kim SY, Monaghan P (2005b) Effects of vegetation on nest microclimate and breeding performance of lesser black-backed gulls (*Larus fuscus*). J Ornithol 146:176–183

Knight ME, Van Oppen MJH, Smith HL et al (1999) Evidence for male-biased dispersal in Lake Malawi cichlids from microsatellites. Mol Ecol 8:1521–1527

Kumar RS, Johnson JA (2014) Aerial surveys for pack-ice seals along with the Ingrid Christensen and Princess Astrid Coasts, East Antarctica. J Threatened Taxa 6:6230–6238. https://doi.org/10.11609/JoTT.o3817.6230-8

Lacey LM (2018) Role of nest site microclimate and food availability in chick development and reproductive success in blacklegged kittiwakes (Rissa tridactyla). Honors theses 442. Bucknell University.

Landers TJ, Bannock CA, Hauber ME (2011) Dynamics of behavioural rhythms in a colonial, nocturnal, burrowing seabird: a comparison across different temporal scales. Notornis 58(2):81–89

LaRue MA, Lynch HJ, Lyver POB et al (2014) A method for estimating colony sizes of Adélie penguins using remote sensing imagery. Polar Biol 37:507–517. https://doi.org/10.1007/s00300-014-1451-8

Lascelles BG, Langham GM, Ronconi RA, Reid JB (2012) From hotspots to site protection: identifying marine protected areas for seabirds around the globe. Biol Conserv 156:5–14. https://doi.org/10.1016/j.biocon.2011.12.008

Lebreton JD, Hines JE, Pradel R, Nichols JD, Spendelow JA (2003) Estimation by capture-recapture of recruitment and dispersal over several sites. Oikos 101(2):253–264
Lee JR, Raymond B, Bracegirdle TJ, et al. (2017a) Climate change drives the expansion of Antarctic ice-free habitat. Nature 547:49–54. https://doi.org/10.1038/nature22996
Lee JR, Raymond B, Bracegirdle TJ, Chades I, Fuller RA, Shaw JD, Terauds A (2017b) Climate change drives the expansion of Antarctic ice-free habitat. Nature 547(7661):49–54
Løvenskiold HL (1960) The Snow Petrel *Pagodroma nivea* nesting in Dronning Maud Land. Ibis 102:132–134
Loy W (1962) Ornithological profile from Iceland to Antarctica. Gerfaut 52:626–640
Lynch HJ, Fagan WF, Naveen R (2010) Population trends and reproductive success at a frequently visited penguin colony on the western Antarctic Peninsula. Polar Biol 33:493–503. https://doi.org/10.1007/s00300-009-0726-y
Lynch HJ, LaRue MA (2014) First global census of the Adélie Penguin. Auk 131:457–466. https://doi.org/10.1642/AUK-14-31.1
Lynch HJ, Fagan WF, Naveen R, Trivelpiece SG, Trivelpiece WZ (2012a) Differential advancement of breeding phenology in response to climate may alter staggered breeding among sympatric pygoscelid penguins. Mar Ecol Prog Ser 454:135–145
Lynch HJ, Naveen R, Trathan PN, Fagan WF (2012b) Spatially integrated assessment reveals widespread changes in penguin populations on the Antarctic Peninsula. Ecology 93(6):1367–1377
Maher WJ (1962) Breeding biology of the snow petrel near Cape Hallett, Antarctica. Condor 64:488–499. https://doi.org/10.2307/1365472
Mallory ML (2009) Incubation scheduling by northern fulmars (Fulmarus glacialis) in the Canadian High Arctic. J Ornithol 150(1):175–181. https://doi.org/10.1007/s10336-008-0332-8
Marchant S, Higgins PJ (1990) Handbook of Australian. New Zealand & Antarctic birds. Univ. Press, Melbourne
Mehlum F, Gjessing Y, Haftorn S, Bech C (1988) Census of breeding Antarctic Petrels Thalassoica Antarctica and physical features of the breeding colony at Svarthamaren, Dronning Maud Land, with notes on breeding Snow Petrels Pagodroma Nivea and South Polar Skuas Catharacta MacCormick. Polar Res 6(1):1–9
Mejías MA, Wingate DB, Madeiros JL, Wiersma YF, Robertson GJ (2017) Nest-cavity selection and nesting success of Bermudian white-tailed Tropicbirds (Phaethon lepturus catesbyi). Wilson J Ornithol 129(3):586–599. https://doi.org/10.1676/16-115.1
Micol T, Jouventin P (2001) Long-term population trends in seven Antarctic seabirds at Pointe Géologie (Terre Adélie) Human impact compared with environmental change. Polar Biol 24(3):175–185
Milot E, Weimerskirch H, Bernatchez L (2008) The seabird paradox: dispersal, genetic structure and population dynamics in a highly mobile, but philopatric albatross species. Mol Ecol 17:1658–1673. https://doi.org/10.1111/j.1365-294X.2008.03700.x
NCAOR (2006) Draft Comprehensive Environmental Evaluation of New Indian Research Base at Larsemann Hills, Antarctica. National Centre for Antarctic and Ocean Research, Ministry of Earth Sciences, Government of India, Goa, p 103
NRC (1993) Science and stewardship in the Antarctic. National Academies Press, National Research Council
Olivier F (2006) Nesting habitat preferences of Snow Petrels (*Pagodroma Nivea*) and Wilson's Storm petrels (*Oceanites oceanicus*) in East Antarctica. PhD thesis, University of Tasmania
Olivier F, Lee AV, Woehler EJ (2004). Distribution and abundance of snow petrels *Pagodroma Nivea* in the Windmill Islands, East Antarctica. Polar Biol 27:257–265. https://doi.org/10.1007/s00300-004-0595-3
Olivier F, Van Franeker JA, Creuwels JC, Woehler EJ (2005) Variations of snow petrel breeding success in relation to sea-ice extent: detecting local response to large-scale processes? Polar Biol 28(9):687–699. https://doi.org/10.1007/s00300-005-0734-5
Olivier F, Wotherspoon SJ (2006) Modelling habitat selection using presence-only data: a case study of a colonial hollow nesting bird, the snow petrel. Ecol Modell 195(3–4):187–204

Paleczny M, Hammill E, Karpouzi V, Pauly D (2015) population trend of the world's monitored seabirds, 1950-2010. PLoS One 10:e0129342. https://doi.org/10.1371/journal.pone.0129342

Pande A, Rawat N, Sivakumar K, et al. (2018) Cross-species screening of microsatellite markers for individual identification of snow petrel *Pagodroma Nivea* and Wilson's storm petrel Oceanites oceanicus in Antarctica. PeerJ 6:e5243. https://doi.org/10.7717/peerj.5243

Pande A, Sivakumar K, Sathyakumar S et al (2017) Monitoring wildlife and their habitats in the Southern Ocean and around Indian Research Stations in Antarctica. Proc Indian Natl Sci Acad 90:483–496. https://doi.org/10.16943/ptinsa/2017/48958

Pande A, Mondol S, Sathyakumar S, Mathur VB, Ray Y, Sivakumar K (2020) Records and current distribution of seabirds at Larsemann Hills and Schirmacher Oasis, East Antarctica. Polar Record 56

Pertierra LR, Hughes KA, Vega GC, Olalla-Tárraga MA (2017) High-resolution spatial mapping of human footprint across Antarctica and its implications for the strategic conservation of avifauna. PLoS One 12:1–20. https://doi.org/10.1371/journal.pone.0168280

Phillips RA, Silk JR, Massey A, Hughes KA (2019) Surveys reveal increasing and globally important populations of south polar skuas and Antarctic shags in Ryder Bay (Antarctic Peninsula). Polar Biol 42(2):423–432

Pierotti R (1982) Habitat selection and its effect on reproductive output in the herring gull in Newfoundland. Ecology 63:854–868

Prinz AC, Taank VK, Voegeli V, Walters EL (2016) A novel nest-monitoring camera system using a Raspberry Pi microcomputer. Journal of Field Ornithology. 87(4):427–435. https://doi.org/10.1111/jofo.12182

Pryor ME (1968). The avifauna of Haswell Island, Antarctica. In: Austin OL (ed) Antarctic Bird Studies. Antarct. Res. Ser. 12:57–82

Quillfeldt P, Moodley Y, Weimerskirch H, Cherel Y, Delord K, Phillips RA, Navarro J, Calderón L, Masello JF (2017). Does genetic structure reflect differences in non-breeding movements? A case study in small, highly mobile seabirds. BMC Evol Biol 17:1–11

Ramos JA, Monteiro LR, Sola E, Moniz Z (1997) Characteristics and competition for nest cavities in burrowing Procellariiformes. Condor 99(3):634–641

Rodríguez A, Arcos JM, Bretagnolle V, et al (2019) Future directions in conservation research on petrels and shearwaters. Front Mar Sci 6. https://doi.org/10.3389/fmars.2019.00094

Ropert-Coudert Y, Hindell MA, Phillips RA (2014) Biogeographic atlas of the Southern Ocean

Rousset F (2001) Genetic approaches to the estimation of dispersal rates. In: Clobert J, Danchin E, Dhondt AA, Nichols JD (eds) Dispersal. Oxford University Press, Oxford, U.K., pp 18–28

Ryan PG, Watkins BP (1989) Snow petrel breeding biology at an inland site in Continental Antarctica. Colon Waterbirds 12:176. https://doi.org/10.2307/1521338

Sabine JB, Meyers JM, Schweitzer SH (2005) A simple, inexpensive video camera setup for the study of avian nest activity. J Field Ornithol 76(3):293–297. https://doi.org/10.1648/0273-8570-76.3.293

Sanderson EW, Jaiteh M, Levy MA, et al (2002) The human footprint and the last of the wild. BioScience 52:891–904. https://doi.org/10.1641/0006-3568

Sathyakumar S (1995). Developing a long-term monitoring programme for birds and mammals in the Indian Ocean and Antarctica using GPS and GIS technologies. Technical Publication. Department of Ocean Development 12:207–219

Schlaepfer DR, Bradford JB, Lauenroth WK, Munson SM, Tietjen B, Hall SA, Wilson SD, Duniway MC, Jia G, Pyke DA, Lkhagva A, Jamiyansharav K.(2017) Climate change reduces the extent of temperate drylands and intensifies drought in deep soils. Nat Commun 8(1):1–9

Schwaller MR, Lynch HJ, Tarroux A, Prehn B (2018) A continent-wide search for Antarctic petrel breeding sites with satellite remote sensing. Remote Sens Environ 210:444–451. https://doi.org/10.1016/j.rse.2018.02.071

Schwaller MR, Olson CE Jr, Ma Z, Zhu Z, Dahmer P (1989) A remote sensing analysis of Adélie penguin rookeries. Remote Sens Environ 28:199–206

Shaffer SA, Tremblay Y, Weimerskirch H et al (2006) Migratory shearwaters integrate oceanic resources across the Pacific Ocean in an endless summer. Proc Natl Acad Sci USA 103:12799–12802

Shaw JD, Terauds A, Riddle MJ, et al (2014) Antarctica's protected areas are inadequate, unrepresentative, and at risk. PLoS Biol 12:e1001888. https://doi.org/10.1371/journal.pbio.1001888

Singh SM, Singh PN, Singh SK, Sharma PK (2014) Pigment, fatty acid, and extracellular enzyme analysis of a fungal strain *Thelebolus microsporus* from Larsemann Hills, Antarctica. Polar Rec 50(1):31–36

Sivakumar K, Sathyakumar S (2012) Climate change and its potential impacts on the distribution of birds in Southern Indian Ocean and Antarctica. Daya Publishers, New Delhi

Taylor SA, Friesen VL (2012) Use of molecular genetics for understanding seabird evolution, ecology and conservation. Mar Ecol Prog Ser 451:285–304. https://doi.org/10.3354/meps09694

Techow NMSM, O'Ryan C, Phillips RA et al (2010) Speciation and phylogeography of giant petrels *Macronectes*. Mol Phylogenet Evol 54:472–487. https://doi.org/10.1016/j.ympev.2009.09.005

Thiebot JB, Delord K, Barbraud C et al (2016) 167 individuals versus millions of hooks: bycatch mitigation in longline fisheries underlies conservation of Amsterdam albatrosses. Aquat Conserv Mar Freshwat Ecosyst 26:674–688. https://doi.org/10.1002/aqc.2578

Thompson DR, Furness RW, Lewis SA (1993) Temporal and spatial variation in mercury concentrations in some albatrosses and petrels from the sub-Antarctic. Polar Biol 13(4):239–244

Turner J, Pendlebury S (2004) The international Antarctic weather forecasting handbook. British Antarctic Survey

Van Franeker JA, Bell PJ, Montague TL (1990) Birds of Ardery and Odbert Islands, Windmill Islands, Antarctica. Emu 90:74–80

Van Franeker JA, Gavrilo M, Mehlum F, et al (1999) Distribution and Abundance of the Antarctic Petrel. Waterbirds: Int J Waterbird Biol 22:14. https://doi.org/10.2307/1521989

Viot CR, Jouventin P, Bried J (1993) Population genetics of southern seabirds. Mar Ornithol 21:1–25

Warham J (1996) The behaviour, population biology and physiology of the Petrels. Academic Press

Welch AJ, Fleischer RC, James HF, Wile AE, Ostrom PH, Adams J, Duvall F, Holmes N, Hu D, Penniman J, Swindle KA (2012) Population divergence and gene flow in an endangered and highly mobile seabird. Heredity 109:19–28. https://doi.org/10.1038/hdy.2012.7

Woehler EJ (1990) The distribution of seabird biomass in the Australian Antarctic Territory: implications for conservation. Environ Conserv 17:256. https://doi.org/10.1017/S0376892900032409

Antarctic Lichen Response to Climate Change: Evidence from Natural Gradients and Temperature Enchantment Experiments

Sanjeeva Nayaka and Himanshu Rai

Abstract Antarctica is one of the very few ecosystems in the world with minimum anthropogenic interventions and pollution load. The extreme climatic conditions such as temperature, precipitation, and smaller ice-free regions allow only cryptogams such as bryophytes and lichens to grow dominantly. Although lichens are well-known biomonitors and bioindicators of climate change, environmental pollution and anthropogenic perturbations, their potential has been explored very recently. In this chapter, various climate-change studies in Antarctica employing lichens as an integrated bioindicator system are reviewed. The studies utilized either natural gradients of climate across the continent or passive or active air temperature enhancement experiments. The lichen communities in Antarctica has been found sensitive to both climatic clines and temperature manipulations. The lichens' response was species-specific, the species with wider distribution were more adaptive to climate change than those with restricted distribution. The studies also indicated that climate warming would cause the extinction of sensitive species. Simultaneously, some will increase their geographical extension due to the increased water availability and nutrients in changed ecosystems.

Keywords Cryptogam · Lichenized-fungi · Ecophysiology · Global warming · Biodiversity loss · Bioindicator

1 Introduction

Earth's environment has been unusually stable for the past 10,000 years. Anthropogenic actions since the industrial revolution have become the primary driver of global environmental change. Climate change, biodiversity loss, nitrogen and phosphorus cycles have crossed the critical levels of the threshold for the sustenance of humanity on earth (Rockström et al. 2009). The mitigation of these factors primarily requires their early detection and thorough understanding of each component (Cash

S. Nayaka (✉) · H. Rai
Lichenology Laboratory, Plant Diversity, Systematics and Herbarium Division, CSIR-National Botanical Research Institute, Lucknow, Uttar Pradesh 226001, India

N. Khare (ed.), *Assessing the Antarctic Environment from a Climate Change Perspective*, Earth and Environmental Sciences Library,
https://doi.org/10.1007/978-3-030-87078-2_14

et al. 2003). Traditionally, climate change monitoring involves satellite data, ground studies, and modelling of the predications (Appenzeller et al. 2008). The higher maintenance cost of the traditional climate change monitoring systems and the inconsistency of monitoring data to actual events have necessitated the researchers to utilize bioindicators such as fish, insects, mussels, lichens, algae, plants and birds (Guralnick 2002; Caza et al. 2016).

Amid the various bioindicators, lichens, the symbiotic association of a fungus and green or blue-green algae (sometimes both) has been utilized to monitor air pollution as early as 1866 (Nylander 1866). Apart from being excellent accumulators of pollutants (heavy metals, aromatic hydrocarbons and nitrogen), lichens also indicate land-use change and various other anthropogenic activities world-wide (Garty 2001; Wolterbeek et al. 2003; Blasco et al. 2008; Rai et al. 2012; Loppi 2019; Landis et al. 2019; Nascimbene et al. 2019; Serrano et al. 2019; Zhao et al. 2019; Capozzi et al. 2020; Wang et al. 2020). Lichens are also a proven indicator of climate change (Sancho et al. 2007; Ellis 2019). Unlike pollution monitoring studies, climate change studies are influenced by other ecological factors, such as phorophyte types, forest structure and pollution loads. Therefore, climate changes studies employing experimental, and modelling systems must incorporate different aspects of the environment (Ellis et al. 2007; Aptroot 2009; Nascimbene et al. 2016; Alatalo et al., 2017; Ellis 2019; Giordani et al. 2019). In Antarctica these interfering factors are minimized, especially in lichen-based climate changes studies as the continent has minimal air pollution (Upreti and Pandey 1994; Mietelski et al. 2000; Bargagli et al., 2004) and anthropogenic influence but has diverse lichen communities and environmental gradients (Øvstedal and Smith 2001).

Situated in the southern hemisphere, Antarctica is the fifth-largest continent with an area of 14,200,000 km^2. The average elevation of Antarctica is 2500 m, the highest for any continent. The ice sheet covers about 98% of the continent with about 1.9 km of average thickness (Fretwell et al. 2013). The low-elevational areas are either in proximity to the seacoast or at the coast. The ice-free area includes only 0.18% (21,745 km^2), the majority of which lies in the Antarctic peninsula, Trans-Atlantic mountains, along with Dronning Maud land, various coastal islands and nunataks (Rai et al. 2011; Burton-Johnson et al. 2016). Antarctica is the coldest continent with an average minimum temperature ranging from −63 °C in coastal areas to −89.2 °C in continental Antarctica. It is a cold desert with an annual precipitation maximum of up to 200 mm in the coastal habitats and much lesser in continental Antarctica. Two biogeographic zones can be recognized within the continent—the continental and maritime Antarctica (Smith 1984; Convey 2001), which are further categorized into sixteen eco-regions (Terauds and Lee 2016) (Fig. 1). Antarctica vegetation is primarily dominated by cryptogams, i.e. bryophytes and lichens, except for two vascular plants *Deschampsia antarctica* Desv. and *Colobanthus quitensis* (Kunth) Bartl. are limited to the peninsular region (Smith 1994; Convey 2006).

The lichen community of Antarctica is represented by 427–500 species accounting for about 2.5% of the total estimated lichens globally (i.e. ~20,000 species, Lücking et al. 2016), of which 40% are endemic to the continent (Smith and Øvstedal 1991; Upreti and Pant 1995; Upreti 1996, 1997; Gupta et al. 1999; Pandey and Upreti 2000;

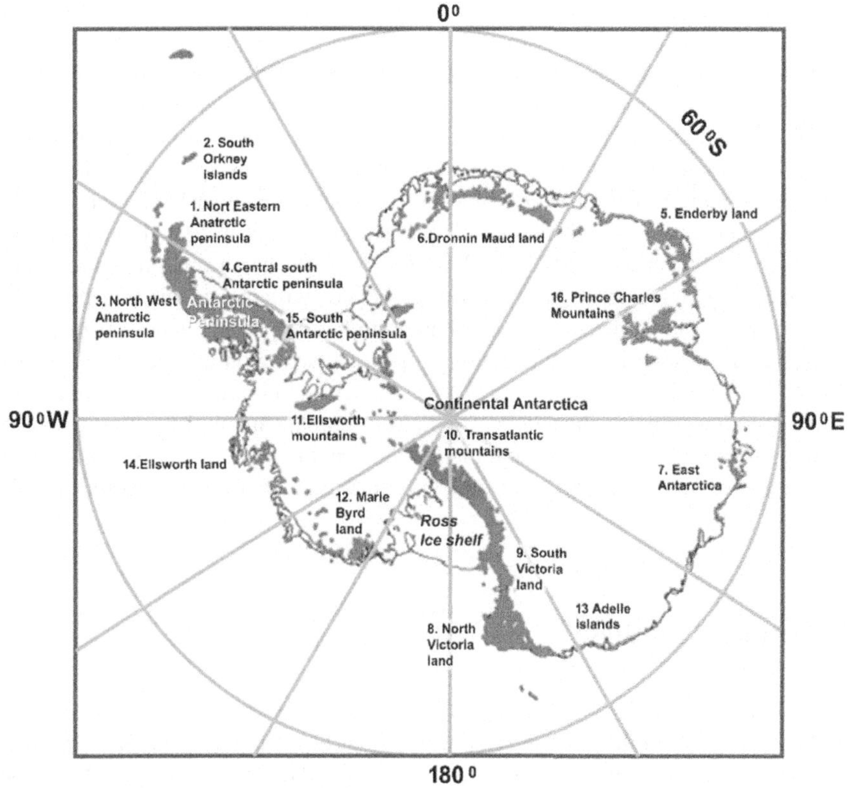

Fig. 1 The two biogeographic zones and sixteen eco-regions of Antarctica (after Smith 1984; Convey 2001; Terauds and Lee 2016)

Øvstedal and Smith 2001; Sancho et al. 2004; Krzewicka and Smykla 2004; Nayaka and Upreti 2005; Kim et al. 2006; Singh et al. 2007; Krzewicka and Maciejowski 2008; Olech and Czarnota 2009; Osyczka et al. 2010; Ruprecht et al. 2010; Upreti and Nayaka 2011; Green et al. 2015; Park et al. 2018). The lichens with cyanobacteria as primary (i.e. bi-partite) or secondary photobionts (i.e. tripartite) are specifically restricted to the wetter and warmer peninsular region (Green et al. 2011a, b). The green algae containing chlorolichens are distributed throughout the Antarctic with the dominance of crustose forms and few endemic fruticose species (e.g. *Usnea Antarctica* Du Rietz) (Green et al. 2011a, b). Lichens play a crucial role in the ecosystem functioning of the Antarctic. They are the pioneer colonizer on denuded surfaces which initiate ecological succession and soil formation (Walton 1985; Ascaso et al. 1990; Sancho and Valladares 1993). Lichen also contributes to nutrient cycling and a food source to the invertebrates in the Antarctic habitats (Lindsay 1978; Greenfield 2004; Kennedy 2004; Cannone et al. 2006; Bokhorst et al. 2007a, b; Leishman et al. 2020).

Antarctic lichen research was initially focused on the taxonomy and floristics from the peninsular and accessible regions of the continent (Øvstedal and Smith 2001; Krzewicka and Maciejowski 2008; Park et al. 2018). Studies then gradually focused on ecophysiology and generating baseline data for environmental gradients (i.e. climatic and species). In situ, in vitro temperature and nutrient manipulation experiments were carried to understand the effects of a warming climate in contracting habitats and climatic conditions (Lange and Kappen1972; Barták et al. 2003; Schroeter et al. 2011; Nayaka et al. 2011; Balarinová et al. 2014; Laguna-Defior et al. 2016; Raggio et al. 2016; Cho et al. 2020). Due to the harsh environmental conditions in continental Antarctica, limitations associated with movement, transportation and unforeseen situations, the climate change studies employing lichens as response organisms are primarily reported from the peninsular region and maritime islands (Table 1). The lichen-based climate change studies can broadly categorize into two types (i) studies utilizing the natural gradients of climate and lichen communities as a proxy to climate change, (ii) manipulating the ambient air temperature using either passive and/or active experimental setups to imitate the warming conditions (Table 1).

2 Gradient Studies

The analysis of lichen response along climatic, species, growth form and time series gradients are some of the most appropriate methods to accurately monitor and predict the climate change in stressed habitats of Antarctica (Huiskes et al. 2004; Anderson et al. 2012; Rodriguez et al. 2018; Folgar-Cameán and Barták 2019). The gradients forces species to change their adaptability to optimize their physiology, ecological distribution and reproductive strategies (Chen et al. 2011; Tomiolo and Ward 2018; Determeyer-Wiedmann et al. 2019; Pérez-Ramos et al. 2020; Rolshausen et al. 2020). In Antarctica, the environmental gradient effect is more pronounced on the lichen communities due to minimum interference from the phytosociological competition and other abiotic factors such as pollution.

The growth of lichen thallus is one of the most reliable parameters for studying the response of lichens to microclimatic (e.g. water availability and nutrient deposition) as well as macroclimatic (e.g. temperature and precipitation) variables (Kershaw 1985). Sancho and Pintado (2004) studied the growth rates of six lichens *Acarospora macrocyclos* Vain., *Bellemerea* sp. *Buellia latemarginata* Darb., *Caloplaca sublobulata* (Nyl.) Zahlbr., *Rhizocarpon geographicum* (L.) DC. and *Usnea antarctica* in Livingston Island, South Shetland Islands, maritime Antarctica. The study compared the growth rates of the lichens between 1991 and 2002. All the studied lichens showed a higher increased growth rate ever reported for crustose lichens of the region. In contrast, two lichens, *R. geographicum* and *Bellemerea* sp. showed significantly higher growth rates. Such a tendency was attributed to the Livingston Island ice cap and glaciers' rapid retreat during that decade, which was speculated to global warming (Sancho and Pintado 2004). Sancho et al.

Table 1 Climate change studies carried out in Antarctica using various natural gradients and temperature enhancement experiment

S. No	Study type	Experimental setup used	Lichen species/communities used	Location of study	Major findings	References
1	Natural gradient	Species growth rate gradient	*Acarospora macrocyclos, Bellemerea* sp., *Buellia latemarginata, Caloplaca sublobulata, Rhizocarpon geographicum, Usnea antarctica*	Livingston Island, Maritime Antarctica	Higher growth rate due to global warming	Sancho and Pintado (2004)
2	Natural gradient	Climate gradient	*Buellia latemarginata, Buellia frigida*	Western Antarctic Peninsula and Dry Valleys, near Ross Island	The growth rate difference following climatic gradient	Sancho et al. (2007)
3	Natural gradient	Latitude linked climatic gradient	Lichen communities	Peninsular and continental sites	Two separate zones were identified based on the influence of macroclimatic and microclimatic factors influencing the lichen communities	Green et al. (2011a, b)
4	Natural gradient	Latitude linked microclimatic climatic gradient	Lichen communities	Sites along the western Ross sea coast	Lichen communities were influenced by microclimatic conditions, e.g., availability of water	Colesie et al. (2014)

(continued)

Table 1 (continued)

S. No	Study type	Experimental setup used	Lichen species/communities used	Location of study	Major findings	References
5	Natural gradient	The long-term atmospheric temperature gradient	*Bellemerea* sp., *Buellia latemarginata*, *Caloplaca sublobulata*, *Rhizocarpon geographicum*, *Acarospora macrocyclos* and *Usnea antarctica*	Livingston Island, Maritime Antarctica	Species-specific response corresponding to temperature variations. "Snow kill" affects negatively on the growth of some crustose lichens	Sancho et al. (2017)
6	PTES	Temperature enhancement using OTCs, comparative assessment	Lichen communities	Livingstone Island, Maritime Antarctica, Falkland island, sub-Antarctic	Open plant communities (lichens and mosses) are negatively affected by warming	Bokhorst et al. (2007a, b)
7	PTES	Temperature enhancement using OTCs	*Usnea antarctica*	Signy Island, maritime Antarctic	Temperature stress negatively affect the lichen growth	Bokhorst et al. (2016)
8	PTES	Temperature enhancement using OTCs	*Placopsis antarctica*	Fildes Peninsula, maritime Antarctica	Temperature decreases the photosynthetic efficiency of lichens	Casanova-Katny et al. (2019)
9	Field manipulations	Water and nutrient addition as a proxy to global warming	*Usnea* sp.	Windmill Islands in East continental Antarctica	Water and nutrient availability influences lichens growth	Wasley et al. (2006)
10	ATES	Growth chamber temperature enhancement	*Stereocaulon alpinum*, *Placopsis contortuplicata*, *Usnea aurantiaco-atra*	Livingston Island, maritime Antarctica	Antarctic lichens can't tolerate the temperature increase	Colesie et al. (2018)

(continued)

Table 1 (continued)

S. No	Study type	Experimental setup used	Lichen species/communities used	Location of study	Major findings	References
11	Natural gradient, PTES and ATES	Temperature enhancement using polystyrene glasshouse, geothermal heat, laboratory incubation	Lichen communities	Islands in maritime Antarctica	Temperature increases the lichen cover and diversification	Kennedy (1996)

(2007) estimated the growth rates of two taxonomically related crustose lichen species *Buellia latemarginata* and *Buellia frigida* Darb., from sites located in maritime Antarctica and continental Antarctica, respectively, for nine to 25 years. *B. latemarginata* (87 mm 100 y^{-1}) recorded the fastest growth rates ever recorded from Antarctica, whereas *B. frigida* recorded minimal, barely detectable growth (1 mm 100 y^{-1}). The extreme difference in growth rate was attributed to temperature linked precipitation at two habitats. The study concluded that extreme cline in the Antarctic lichen species' growth rate could be used to indicate climate change-induced temperature variation when the growth rate is recorded at a specified time interval for long period monitoring (Sancho et al. 2007).

Sancho et al. (2017) monitored the effect of regional temperature variations on the growth of six lichen species (*Acarospora macrocyclos, Bellemerea* sp., *Buellia latemarginata, Caloplaca sublobulata, Rhizocarpon geographicum* and *Usnea antarctica*) in Livingston Island, South Shetland Islands, maritime Antarctica for 24 years. The regional mean summer temperature (MST) of the study location between 1991 and 2002 increased to 0.42 °C, whereas a decline of 0.58 °C was observed from 2002 to 2015. The temperature fluctuation response on the lichens' growth was species-specific, where some species (*Buellia latemarginatai*) showed no reaction. In contrast, some species (*Bellemerea* sp., *Rhizocarpon geographicum* and *Usnea antarctica*) responded positively with an increase in growth rate from 1991 to 2002 followed by a decrease accordingly to MST in 2015. The lichen species *Caloplaca sublobulata* and *Acarospora macrocyclos* reacted differently with a reduction and increase respectively, from 1991 to 2002, followed by an abrupt decline in thallus growth from 2002 to 2015. Such response was attributed to the thallus' snow kill due to the longer spans of snow cover (Sancho et al. 2017). Among the six species, *Usnea antarctica,* due to its robust upright fruticose thallus and water utilization capacities, emerged as an appropriate proxy of Antarctica's temperature variations (Sancho et al. 2017). Here, the "snow kill" represents a threshold of lower temperature in Antarctica, which can be a driver for the extinction of specific lichen communities and extension for others (Sancho et al. 2017). The studies revealed that Antarctic lichens' growth indeed responds to climate warming, where the pioneer colonizer utilizes the resources made available by retreating glaciers. The steep difference in temperature of maritime and continental Antarctica affects lichen growth through precipitation variability. The long-term temperature variation affects the lichen growth according to the increase and decreases and physical damage caused by a more extended snow deposition period.

Antarctica's latitudinal gradient influences the solar insolation on the continent, which affects the climate creating a cline that can act as a proxy to climate change or global warming conditions. Green et al. (2011a, b) studied Antarctica's terrestrial vegetation as a predictor of climate change, considering the latitudinal incline and associated climatic variation as proxies. The study involved regression analysis of the number of lichen species on latitude and meant annual temperature gradient. The study determined two zones of terrestrial lichen vegetation—the micro-environmental and macro-environmental zone. The micro-environmental area lay south of 72° S. The vegetation is predominately influenced by the micro-climatic parameters such as the occasional occurrence of warm ambient environmental temperature, water availability, sunlight, and shelter. Further in this zone, the lichen physiology is essentially modulated to survival mode by curtailing the thallus growth (Green et al. 2011a, b). The macro-environmental area, which lays north of 72° S in maritime Antarctica, is characterized by lichen species richness, higher cover and growth. Due to the increase in water availability, moderate atmospheric temperature, and longer active periods, the macro-environmental zone allows great net productivity switching lichen vegetation from surviving mode to growth mode (Green et al. 2011a, b). Cyanolichens incapable of performing physiological activities in sub-zero temperatures were found distributed to maritime Antarctica but with only four species, i.e. *Leptogium puberulum* Hue, *Massalongia* aff. *carnosa* (Dicks.) Körb., *Pannaria hookeri* (Borrer) Nyl. and *Pyrenopsis* sp. These species were distributed on the main continent in proximity to the maritime Antarctic Peninsula. The study further concluded that temperature increase due to global warming would have some predictable effects, such as—the lichen diversity will increase in the macro-environment zone, and there will be a southward extension of this zone. There will be a significant change in the micro-climatic zone's species composition as a local microclimate guide. The cyanolichens will fan out in the coastal habitats limited by the availability of appropriate substratum. The study further highlights Antarctica's unique latitudinal cline, which can be imitated to predict the present and future lichen community diversity and changes triggered by global or regional warming.

Continental Antarctica is characterized by a harsher climatic regime where the microclimatic resources such as water availability are limited due to occasional thawing events that decide lichen species composition and community structure. Colesie et al. (2014) studied variations in lichen communities in six sites (Cape Hallett, Terra Nova Bay, Botany Bay, Taylor Hill, Diamond Hill, and Queen Maud mountains) latitudinal gradient at the western Ross Sea coastline in north Antarctica. The study found a decrease in lichen diversity from Cape Hallett to Diamond Hill except Queen Maud mountains, which harbours similar lichen diversity as Cape Hallett. The study concluded that lichen diversity was potentially guided by micro-climatic conditions (i.e. water availability) rather than macroclimatic parameters in continental Antarctica.

The long-term studies in Antarctica on the effect of climate on the lichen communities and their dynamics have also exposed the direct impact of climate warming and the glacial retreat. Olech and Słaby (2016) studied the change in lichen communities 1988–1990 and 2007–2008 in King George Island, maritime Antarctica, to the regional climate change. The comparative study recorded significant differences, especially in the glacial forcefield recently exposed due to glacier retreat (White Eagle Glacier between 1988 and 2008) due to global warming. The study reported the extinction of some species (*Polyblastia gothica* Th. Fr., *Thelenella kerguelena* (Nyl.) H. Mayrhofer, *Thelocarpon cyaneum* Olech & Alstrup), reduction in the distribution of species (*Leptogium puberulum*, *Staurothele gelida* (Hook. f. & Taylor) I. M. Lamb and increase in extant of pioneering species (*Bacidia chrysocolla* Olech,

Czarnota & Llop, *Caloplaca johnstonii* (C.W. Dodge) Søchting & Olech, *Candelariella aurella* (Hoffm.) Zahlbr., *Lecanora dispersa* (Pers.) Röhl.). The study emphasized the lichen communities' change in the glacial moraine forefield exposed due to glaciers' retreat due to climate warming during the last 5–6 decades. The study points towards the biodiversity loss in the form of lichen species extinction and proliferation of early successional lichens, both triggered by the significant glacial retreat in the past 50 years. Thus, Olech and Słaby (2016) study provide a strong reason for including lichens in Antarctica climate change studies.

3 Experimental Manipulation Studies

The experimental temperature enhancement has been used extensively to study the effect of global warming in the past two decades (Marion et al. 1997). Manipulative warming or temperature enhancement can be achieved by two methods, (i) Passive temperature enhancement systems (PTES) using open-top chambers (OTC). Here, the study surface temperature is enhanced in situ by concentrating and retaining solar insolation for a more extended period diurnally and seasonally (Rai et al. 2010). The PTES have continuity of experimental microenvironment with the ambient natural environment. (ii) Active temperature enhancement systems (ATES) using closed experimental equipment such as growth chambers or incubator. Here the predetermined temperature is enhanced along with manipulating light–dark hours, relative humidity operating heating assemblies and fluorescence lights.

PTES using OTC have been successfully used to elevate the temperature triggering the warming effect on the ground lichens (Barták et al. 2019). Bokhorst et al. (2007a, b) studied the effect of experimental warming using OTC on the cryptogam communities (i.e. lichen and bryophytes) of coastal ecosystems in two Maritime islands (i.e. Signy Island and Anchorage Island) and one sub-Antarctic Island (i.e. Falkland Island) for three summers. The study recorded a significant decrease in the moss vegetation cover in Falkland Island and lichens in Signy and Anchorage Island. The results were linked to drought stress induced by the elevated temperatures within the OTCs. The study concluded that the open plant communities (lichens and mosses) are compassionate to even the slightest change in the climatic variables in the sub-Antarctic to Antarctic habitats (Bokhorst et al. 2007a, b). Bokhorst et al. (2016) studied the effect of simulated warming using OTCs on the cover of the dominant lichen *Usnea antarctica* in Signy Island (northern maritime Antarctic) for 10 years (2003–2013). They recorded about 71% loss in the lichen cover. The study concluded that the lichen cover decrease was due to lichen's inability to compensate for the increased carbon loss in summer. The higher net respiration rate was driven by the elevated ambient temperature in the OTCs.

Casanova-Katny et al. (2019) studied climate change on a tripartite cyanolichen *Placopsis antarctica* D.J. Galloway using OTCs, which enhanced the ambient atmospheric temperature up to 2.2 °C than the control sites at Fildes Peninsula, King George Island (maritime Antarctica). The effective quantum yield of photosystem

II (Φ_{PSII}), photosynthetic electron transport rate (ETR), photosynthetically active radiation (PAR) and hydration state of the thallus at 10 min interval were measured for 12 days (Fig. 2). The study concluded that elevated temperature within OTCs and dehydration of the thallus limit the photosynthetic processes in *P. antarctica*. That was further reflected in decreased ETR and chlorophyll fluorescence of samples. The effects of manipulated warming were also confirmed by laboratory hydration experiments where the chlorophyll fluorescence (FM) and Φ_{PSII} corroborated the field studies. The photosynthetic parameters reacted according to the hydration level of the thallus (Fig. 3). OTCs study reflected the physiological stress lichens will encounter global warming, which can be used as an early indicator for long-term studies and further understand the lichen physiology under high-temperature conditions.

Fig. 2 Experimental set up of the MONI-PAM with measuring probes installed over *Placopsis antarctica* thalli. **A** An overview of the OTCs located on La Cruz Plateau; **B–D** The OTC after snowfall (Note the absence of snow in the OTC in D due to elevated temperature), **E–F** The measuring probe on the control plot (E- after a rainfall, F- after a snowfall), **G** The saturation pulse spot on *Placosis antarctica* thallus, H. The measuring probe and Cu-Co thermocouple measuring air temperature at the measuring location inside OTC (Photographs © M. Barták, after Casanova-Katny et al. 2019)

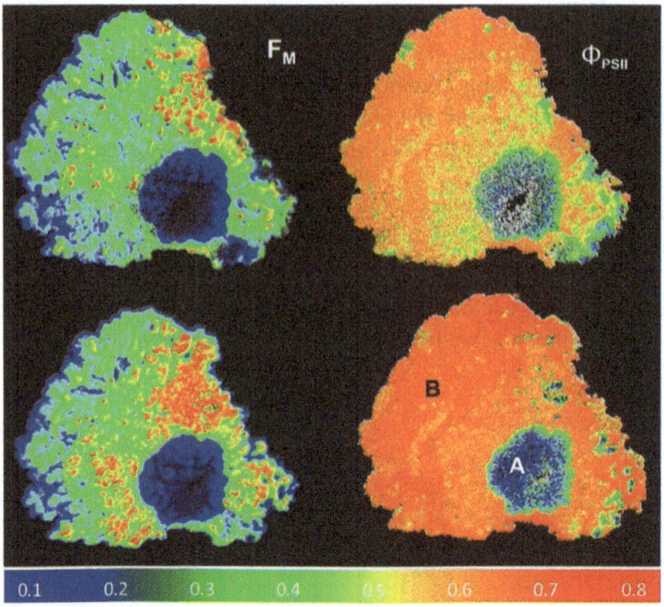

Fig. 3 Chlorophyll fluorescence imaging of *Placosis antarctica* thallus hydrated for 1 h (upper panels) and 48 h (lower panels). Φ PSII—effective quantum yield of photosystem II, A-Cephalodium possesses *Nostoc* commune, B—marginal part of the thallus possessing green microalga (after Casanova-Katny et al. 2019)

The ground lichen vegetation can be manipulated with water and nutrients as a proxy to climatic warming. Wasley et al. (2006) studied the effect of increased water and nutrient availability on lichen (*Usnea*) and bryophyte communities in the Windmill Islands in East Antarctica. The primary productivity (chlorophyll content, fluorescence, and nutrient content) was used indicator. The study concluded that though water availability plays a vital role in developing and maintaining lichen communities, the nutrients act as a determinative factor (Wasley et al. 2006).

ATES using growth chambers or incubators has been rarely used to study the effect of simulated warming but can successfully demonstrate some species-specific results and insight into the species reaction to climate change. Colesie et al. (2018) studied the effect of temperature elevation on three lichens (*Placopsis contortuplicata* I. M. Lamb, *Stereocaulon alpinum* Laurer, *Usnea aurantiaco-atra* (Jacq.) Bory) collected from Livingston Island in peninsular maritime Antarctica using a growth chamber. The activated lichen samples were subjected to temperature gradients of 5° (control), 15° and 23 °C, and hydration-desiccation cycles (3–4 days) for six weeks (apparently to mimic the natural conditions). The respiration, net photosynthesis, and chlorophyll fluorescence of the samples were analyzed. The study found that 15 °C was the upper limit of temperature for photobionts viability in *P. contortuplicata* and *U. aurantiaco-atra* whereas widely distributed *S. alpinum* was able to retain its photosynthetic vitality. The study indicated species-specific response to lichens growing in Antarctica where specialized species (*P. contortuplicata* and *U. aurantiaco-atra*) mostly restricted to the continent negatively affected climate warming widely distributed species which have wide acclimatization range to temperature variability. The study highlights Antarctic lichen's extreme sensitivity, which is incapable of coping with extreme atmospheric temperature fluctuation induced by climate change.

Studies employing PTES, ATES and natural gradients decipher the intricate correlation of lichen response to the wild and experimental setups. Kennedy (1996), by analyzing the proxies of climate warming such as passive polystyrene glasshouses, geothermal heated ground and laboratory incubation of soil samples from maritime Antarctica (i.e. Signy island, Candlemas island, Deception island), concluded that climate or substratum warming increases the lichen cover and supports initiation and diversification of lichens in the otherwise harsh climate of Antarctic. The study indicated the more significant expansion of cryptogamic communities is limited by physiological constraints under global warming in Antarctica.

4 Discussion and Conclusions

Antarctica offers several advantages for climate change studies using lichens as integrative bioindicators due to its unique global ecology positioning. The minimal human interference, lowest pollution levels, dominant ground vegetation of lichens with minimal to no vegetative competition, a wide range of climatic with gradients of temperature and precipitation are some of these advantages. However, studies carried out so far are scarce in comparison to other habitats of the planet. The lichen communities of Antarctica show diverse behaviour in the pattern of distribution and physiology as an indicator to natural climatic gradients and experimental warming, which can be summarised as follows

1. The lichen communities in maritime Antarctica show high diversity due to milder temperature regime and high precipitation, which tend to extend southwards with climate warming.
2. The cyanolichens are limited to maritime Antarctica and peninsular regions. Climate change-induced warming will increase their extension deep into the continental area with limitations of physiological parameters.
3. Climate change-induced warming will increase the pace of glacier retreat, increasing the current extent of lichen communities in all the habitats of Antarctica.
4. The increased temperature will tend to alter the net photosynthesis of lichens. It will have a detrimental effect on the endemic species as they are less adaptive than lichens having circumpolar distribution. This will lead to the extinction of some species during extension and invasion of other species.

The studies carried out so far on Antarctica's lichens indicate that climate change is harmful, leading to biodiversity loss.

Acknowledgements Authors are thankful to Director, CSIR-NBRI, Lucknow, for providing infrastructure facilities and financial assistance through OLP 101, to the Director, National Centre for Polar and Ocean Research (formerly NCAOR), Goa, for facilitating the participation of SN in the 22nd and 28th Indian Antarctic expedition, and to Prof. Miloš Barták, Section of Experimental Plant Biology, Faculty of Science, Kotlarska 2, Brno, the Czech Republic for providing necessary permission to use specific photos.

References

Alatalo JM, Jägerbrand AK, Chen S, Molau U (2017) Responses of lichen communities to 18 years of natural and experimental warming. Ann Bot 120:159–170. https://doi.org/10.1093/aob/mcx053

Anderson JT, Panetta AM, Mitchell-Olds T (2012) Evolutionary and ecological responses to anthropogenic climate change. Update on Anthropogenic Climate Change 160:1728–1740. https://doi.org/10.1104/pp.112.206219

Appenzeller C, Begert M, Zenklusen E, Scherrer SC (2008) Monitoring climate at Jungfraujoch in the high Swiss Alpine region. Sci Total Environ 391:262–268. https://doi.org/10.1016/j.scitotenv.2007.10.005

Aptroot A (2009) Lichens as an indicator of climate and global change. In: Letcher TM (ed) Climate change. Elsevier, Amsterdam, pp 401–408. https://doi.org/10.1016/B978-0-444-53301-2.00023-3

Ascaso C, Sancho LG, Rodriguez-Pascual C (1990) The weathering action of saxicolous lichens in maritime Antarctica. Polar Biol 11:33–39. https://doi.org/10.1007/BF00236519

Balarinová K, Barták M, Hazdrová J, Hájek J, Jílková J (2014) Changes in photosynthesis, pigment composition and glutathione contents in two Antarctic lichens during light stress and recovery. Photosynthetica 52:538–547. https://doi.org/10.1007/s11099-014-0060-7

Bargagli R, Battisti E, Focardi S, Formichi P (2004) Preliminary data on the environmental distribution of mercury in northern Victoria Land, Antarctica. Antarct Sci 5:3–8. https://doi.org/10.1017/S0954102093000021

Barták M, Láska K, Hájek J, Váczi P (2019) Microclimate variability of Antarctic terrestrial ecosystems manipulated by open-top chambers: Comparison of selected austral summer seasons within a decade. Czech Polar Rep 9:88–106. https://doi.org/10.5817/CPR2019-1-8

Barták M, Vráblíková H, Hájek J (2003) Sensitivity of photosystem II of Antarctic lichens to high irradiance stress: Fluorometric study of fruticose (*Usnea antarctica*) and foliose (*Umbilicaria decussata*) species. Photosynthetica 41:497–504. https://doi.org/10.1023/B:PHOT.0000027513.90599.ad

Blasco M, Domeño C, Nerín C (2008) Lichens biomonitoring as a feasible methodology to assess air pollution in natural ecosystems: combined study of quantitative PAHs analyses and lichen biodiversity the Pyrenees Mountains. Anal Bioanal Chem 391:759–771. https://doi.org/10.1007/s00216-008-1890-6

Bokhorst S, Convey P, Huiskes A, Aerts R (2016) *Usnea antarctica*, an essential Antarctic lichen, is vulnerable to aspects of regional environmental change. Polar Biol 39:511–521. https://doi.org/10.1007/s00300-015-1803-z

Bokhorst S, Huiskes A, Convey P, Aerts R (2007a) The effect of environmental change on vascular plant and cryptogam communities from the Falkland Islands and the Maritime Antarctic. BMC Ecol 7:15. https://doi.org/10.1186/1472-6785-7-15

Bokhorst S, Ronfort C, Huiskes A, Convey P, Aerts R (2007b) Food choice of Antarctic soil arthropods clarified by stable isotope signatures. Polar Biol 30:983–990. https://doi.org/10.1007/s00300-007-0256-4

Burton-Johnson A, Black M, Fretwell PT, Kaluza-Gilbert J (2016) An automated methodology for differentiating rock from snow, clouds and sea in Antarctica from Landsat 8 imagery: a new rock outcrop map and area estimation for the entire Antarctic continent. Cryosphere 10:1665–1677. https://doi.org/10.5194/tc-10-1665-2016

Cannone N, Ellis Evans JC, Strachan R, Guglielmin M (2006) Interactions between climate, vegetation and the active layer in soils at two Maritime Antarctic sites. Antarct Sci 18:323–333. https://doi.org/10.1017/S095410200600037X

Capozzi F, Sorrentino MC, Di Palma A, Mele F, Arena C, Adamo P, Spagnuolo V, Giordano S (2020) Implication of vitality, seasonality and specific leaf area on PAH uptake in moss and lichen transplanted in bags. Ecol Indic 108:105727. https://doi.org/10.1016/j.ecolind.2019.105727

Casanova-Katny A, Barták M, Gutierrez C (2019) Open-top chamber microclimate may limit photosynthetic processes in Antarctic lichen: a case study from King George Island, Antarctica. Czech Polar Rep 9:61–77. https://doi.org/10.5817/CPR2019-1-6

Cash DW, Clark WC, Alcock F, Dickson NM, Eckley N, Guston DH, Jäger J, Mitchell RB (2003) Knowledge systems for sustainable development. Proc Natl Acad Sci 100:8086–8091. https://doi.org/10.1073/pnas.1231332100

Caza F, Cledon M, St-Pierre Y (2016) Biomonitoring climate change and pollution in marine ecosystems: a review on *Aulacomya ater*. J Mar Biol 7183813. https://doi.org/10.1155/2016/7183813

Chen IC, Hill JK, Ohlemüller R, Roy DB, Thomas CD (2011) Rapid range shifts of species associated with high levels of climate warming. Science 333:1024-1026. https://doi.org/10.1126/science.1206432

Cho SM, Lee H, Hong SG, Lee J (2020) Study of ecophysiological responses of the Antarctic fruticose lichen *Cladonia borealis* using the PAM fluorescence system under natural and laboratory conditions. Plants 9:85. https://doi.org/10.3390/plants9010085

Colesie C, Büdel B, Hurry V, Green TGA (2018) Can Antarctic lichens acclimatize to changes in temperature? Glob Change Biol 24:1123–1135. https://doi.org/10.1111/gcb.13984

Colesie C, Green TGA, Türk R, Hogg ID, Sancho LG, Büdel B (2014) Terrestrial biodiversity along the Ross Sea coastline, Antarctica: lack a latitudinal gradient and potential limits of bioclimatic modelling. Polar Biol 37:1197–1208. https://doi.org/10.1007/s00300-014-1513-y

Convey P (2001) Antarctic ecosystems. In: Levin S (ed) Encyclopedia of biodiversity, vol 1. Academic Press, San Diego, pp 171–184

Convey P (2006) Antarctic climate change and its influences on terrestrial ecosystems. In: Bergstrom DM, Convey P, A. Huiskes HL (eds) Trends in Antarctic Terrestrial and Limnetic Ecosystems: Antarctica as a Global Indicator. Dordrecht, Springer, Netherlands, pp 253–272. https://doi.org/10.1007/1-4020-5277-4_12

Determeyer-Wiedmann N, Sadowsky A, Convey P, Ott S (2019) Physiological life-history strategies of photobionts of lichen species from the Antarctic and moderate European habitats in response to stressful conditions. Polar Biol 42:395–405. https://doi.org/10.1007/s00300-018-2430-2

Ellis CJ (2019) Climate change, bioclimatic models and the risk to lichen diversity. Diversity 11:54. https://doi.org/10.3390/d11040054

Ellis CJ, Coppins BJ, Dawson TP (2007) Predicted response of the lichen epiphyte *Lecanora populicola* to climate change scenarios in a clean-air region of Northern Britain. Biol Cons 135:396–404. https://doi.org/10.1016/j.biocon.2006.10.036

Folgar-Cameán Y, Barták M (2019) Evaluation of photosynthetic processes in Antarctic mosses and lichens exposed to controlled rate cooling: species-specific responses. Czech Polar Rep 9:114–124. https://doi.org/10.5817/CPR2019-1-10

Fretwell P, Pritchard HD, Vaughan DG, Bamber JL, Barrand NE, Bell R, Bianchi C, Bingham RG, Blankenship DD, Casassa G, Catania G (2013) Bedmap2: improved ice bed, surface and thickness datasets for Antarctica. Cryosphere 7:375–393. https://doi.org/10.5194/tc-7-375-2013

Garty J (2001) Biomonitoring atmospheric heavy metals with lichens: theory and application. Crit Rev Plant Sci 20:309–371. https://doi.org/10.1080/20013591099254

Giordani P, Malaspina P, Benesperi R, Incerti G, Nascimbene J (2019) Functional over-redundancy and vulnerability of lichen communities decouple across spatial scales environmental severity. Sci Total Environ 666:22–30. https://doi.org/10.1016/j.scitotenv.2019.02.187

Green TGA, Sancho LG, Pintado A, Schroeter B (2011a) Functional and spatial pressures on terrestrial vegetation in Antarctica forced by global warming. Polar Biol 34:1643. https://doi.org/10.1007/s00300-011-1058-2

Green TGA, Sancho LG, Pintado A, Schroeter B (2011b) Functional and spatial pressures on terrestrial vegetation in Antarctica forced by global warming. Polar Biol 34:1643. https://doi.org/10.1007/s00300-011-1058-2

Green TGA, Seppelt RD, Brabyn LR, Beard C, Türk R, Lange OL (2015) Flora and vegetation of Cape Hallett and vicinity, northern Victoria Land, Antarctica. Polar Biol 38:1825–1845. https://doi.org/10.1007/s00300-015-1744-6

Greenfield LG (2004) Retention of precipitation nitrogen by Antarctic mosses, lichens and fellfield soils. Antarct Sci 4:205–206. https://doi.org/10.1017/S0954102092000312

Gupta RK, Sinha GP, Singh DK (1999) A note on lichens of Shirmacher Oasis, East Antarctica. Indian J For 22:292–294

Guralnick J (2002) Biological indicators as an early warning of ESNO events. Regional Disaster Information Centre (CRID)

Huiskes AHL, Gremmen NJM, Francke JW (2004) Morphological effects on the water balance of Antarctic foliose and fruticose lichens. Antarct Sci 9:36–42. https://doi.org/10.1017/S0954102097000059

Kennedy AD (1996) Antarctic fellfield response to climate change: a tripartite synthesis of experimental data. Oecologia 107:141–150. https://doi.org/10.1007/BF00327897

Kennedy AD (2004) Microhabitats occupied by terrestrial arthropods in the Stillwell Hills, Kemp Land, East Antarctica. Antarct Sci 11:27–37. https://doi.org/10.1017/S095410209900005X

Kershaw KA (1985) Physiological ecology of lichens. Cambridge University Press, UK

Kim JH, Ahn IY, Hong SG, Andreev M, Lim KM, Oh MJ, Koh YJ, Hur JS (2006) Lichen flora around the Korean Antarctic Scientific Station, King George Island, Antarctic. J Microbiol 44:480–491

Krzewicka B, Maciejowski W (2008) Lichen species from the northeastern shore of Sørkapp Land (Svalbard). Polar Biol 31:1319–1324. https://doi.org/10.1007/s00300-008-0469-1

Krzewicka B, Smykla J (2004) The lichen genus *Umbilicaria* from the neighbourhood of Admiralty Bay (King George Island, maritime Antarctic), with a proposed new key to all Antarctic taxa. Polar Biol 28:15–25. https://doi.org/10.1007/s00300-004-0638-9

Laguna-Defior C, Pintado A, Green TGA, Blanquer JM, Sancho LG (2016) Distributional and ecophysiological study on the Antarctic lichens species pair *Usnea antarctica*/*Usnea aurantiaco-atra*. Polar Biol 39:1183–1195. https://doi.org/10.1007/s00300-015-1832-7

Landis MS, Studabaker WB, Patrick Pancras J, Graney JR, Puckett K, White EM, Edgerton ES (2019) Source apportionment of an epiphytic lichen biomonitor to elucidate the sources and spatial distribution of polycyclic aromatic hydrocarbons in the Athabasca Oil Sands Region, Alberta, Canada. Sci Total Environ 654:1241–1257. https://doi.org/10.1016/j.scitotenv.2018.11.131

Lange OL, Kappen, L (1972) Photosynthesis of lichens from Antarctica. In: Llano GA (ed) Antarctic Terrestrial biology, vol 20, American Geophysical Union, pp 83–95. https://doi.org/10.1002/9781118664667.ch4

Leishman MR, Gibson JAE, Gore DB (2020) Spatial distribution of birds and terrestrial plants in Bunger Hills. Antarct Sci 32:153–166. https://doi.org/10.1017/S0954102020000012

Lindsay DC (1978) The role of lichens in Antarctic ecosystems. Bryologist 81:268–276. https://doi.org/10.2307/3242188

Loppi S (2019) May the diversity of epiphytic lichens be used in environmental forensics? Diversity 11:36. https://doi.org/10.3390/d11030036

Lücking R, Hodkinson BP, Leavitt SD (2016) The 2016 classification of lichenized fungi in the Ascomycota and Basidiomycota - Approaching one thousand genera. Bryologist 119:361–416. https://doi.org/10.1639/0007-2745-119.4.361

Marion GM, Henry GHR, Freckman DW, Johnstone J, Jones G, Jones MH, Levesque E, Molau U, Mølgaard P, Parsons AN, Svoboda J (1997) Open-top designs for manipulating field temperature in high-latitude ecosystems. Glob Change Biol 3:20–32. https://doi.org/10.1111/j.1365-2486.1997.gcb136.x

Mietelski JW, Gaca P, Olech MA (2000) Radioactive contamination of lichens and mosses collected in South Shetlands and Antarctic Peninsula. J Radioanal Nucl Chem 245:527–537. https://doi.org/10.1023/A:1006748924639

Nascimbene J, Benesperi R, Giordani P, Grube M, Marini L, Vallese C, Mayrhofer H (2019) Could hair-lichens of high-elevation forests help detect the impact of global change in the Alps? Diversity 11:45. https://doi.org/10.3390/d11030045

Nascimbene J, Casazza G, Benesperi R, Catalano I, Cataldo D, Grillo M, Isocrono D, Matteucci E, Ongaro S, Potenza G, Puntillo D (2016) Climate change fosters the decline of epiphytic *Lobaria* species in Italy. Biol Cons 201:377–384. https://doi.org/10.1016/j.biocon.2016.08.003

Nayaka S, Upreti DK, Singh R (2011) Water relations of some common lichens occurring in Schirmacher Oasis, East Antarctica. In: Singh J, Dutta HN (eds) Antarctica: The most Interactive Ice-Air-Ocean Environment. Nova Science Publishers, Inc, pp 163–172

Nayaka S, Upreti DK (2005) Schirmacher Oasis, East Antarctic, a lichenologically interesting region. Curr Sci 89:1059–1060

Nylander MW (1866) Les lichens du jardin du Luxembourg. Bulletin De La Société Botanique De France 13:364–371. https://doi.org/10.1080/00378941.1866.10827433

Olech M, Słaby A (2016) Changes in the lichen biota of the Lions Rump area, King George Island, Antarctica, over the last 20 years. Polar Biol 39:1499–1503. https://doi.org/10.1007/s00300-015-1863-0

Olech M, Czarnota P (2009) Two new *Bacidia* (Ramalinaceae, lichenized Ascomycota) from Antarctica. Polish Polar Res 30:339–346

Osyczka P, Kukwa M, Olech M (2010) Notes on the lichen genus *Lepraria* from maritime (South Shetlands) and continental (Schirmacher and Bunger Oases) Antarctica. Polar Biol 33:627–634. https://doi.org/10.1007/s00300-009-0738-7

Øvstedal DO, Smith RL (2001) Lichens of Antarctica and South Georgia: a guide to their identification and ecology. Cambridge University Press, Cambridge, UK

Pandey V, Upreti DK (2000) Lichen flora of Schirmacher Oasis and Vettiya Nunatak. In: Scientific report: Eleventh Indian Expedition to Antarctica, Department of Ocean Development, Technical Publication No. 15, pp 185–201

Park CH, Hong SG, Elvebakk A (2018) *Psoroma antarcticum*, a new lichen species from Antarctica and neighbouring areas. Polar Biol 41:1083–1090. https://doi.org/10.1007/s00300-018-2265-x

Pérez-Ramos IM, Cambrollé J, Hidalgo-Galvez MD, Matías L, Montero-Ramírez A, Santolaya S, Godoy Ó (2020) Phenological responses to climate change in communities of plants species with contrasting functional strategies. Environ Exp Bot 170:103852. https://doi.org/10.1016/j.envexpbot.2019.103852

Raggio J, Green TGA, Sancho LG (2016) *In situ* monitoring of microclimate and metabolic activity in lichens from Antarctic extremes: a comparison between the South Shetland Islands and the McMurdo Dry Valleys. Polar Biol 39:113–122. https://doi.org/10.1007/s00300-015-1676-1

Rai H, Nag P, Upreti DK, Gupta RK (2010) Climate warming studies in alpine habitats of Indian Himalaya, using lichen based passive temperature-enhancing system. Nat Sci 8:104–106. https://doi.org/10.6084/m9.figshare.12199652.v1

Rai H, Khare R, Nayaka S, Upreti DK, Gupta RK (2011) Lichen synusiae in East Antarctica (Schirmacher Oasis and Larsemann Hills): substratum and morphological preferences. Czech Polar Rep 1:65–77. https://doi.org/10.5817/CPR2011-2-6

Rai H, Upreti DK, Gupta RK (2012) Diversity and distribution of terricolous lichens as an indicator of habitat heterogeneity and grazing induced trampling in a temperate-alpine shrub and meadow. Biodivers Conserv 21:97–113. https://doi.org/10.1007/s10531-011-0168-z

Rockström J, Steffen W, Noone K, Persson Å, Chapin FS, Lambin EF, Lenton TM, Scheffer M, Folke C, Schellnhuber HJ, Nykvist B (2009) A safe operating space for humanity. Nature 461:472–475. https://doi.org/10.1038/461472a

Rodriguez JM, Passo A, Chiapella JO (2018) Lichen species assemblage gradient in South Shetlands Islands, Antarctica: relationship to deglaciation and microsite conditions. Polar Biology 41:2523–2531. https://doi.org/10.1007/s00300-018-2388-0

Rolshausen G, Hallman U, Grande FD, Otte J, Knudsen K, Schmitt I (2020) Expanding the mutualistic niche: parallel symbiont turnover along climatic gradients. Proc R Soc b: Biol Sci 287:20192311. https://doi.org/10.1098/rspb.2019.2311

Ruprecht U, Lumbsch HT, Brunauer G, Green TGA, Türk R (2010) Diversity of *Lecidea* (Lecideaceae, Ascomycota) species revealed by molecular data and morphological characters. Antarct Sci 22:727–741. https://doi.org/10.1017/S0954102010000477

Sancho LG, Pintado A (2004) Evidence of high annual growth rate for lichens in the maritime Antarctic. Polar Biology 27:312–319. https://doi.org/10.1007/s00300-004-0594-4

Sancho LG, Valladares F (1993) Lichen colonization of recent moraines on Livingston Island (South Shetland I., Antarctica). Polar Biol 13:227–233. https://doi.org/10.1007/BF00238757

Sancho LG, Allan Green TG, Pintado A (2007) Slowest to fastest: extreme range in lichen growth rates supports their use as an indicator of climate change in Antarctica. Flora - Morphol Distrib Funct Ecol Plants 202:667–673. https://doi.org/10.1016/j.flora.2007.05.005

Sancho LG, Kappen L, Schroeter B (2004) The lichen genus *Umbilicaria* on Livingston Island, South Shetland Islands, Antarctica. Antarct Sci 4:189–196. https://doi.org/10.1017/S0954102092000294

Sancho LG, Pintado A, Navarro F, Ramos M, De Pablo MA, Blanquer JM, Raggio J, Valladares F, Green TGA (2017) Recent warming and cooling in the Antarctic Peninsula region has rapid and large effects on lichen vegetation. Sci Rep 7:1–8. https://doi.org/10.1038/s41598-017-05989-4

Schroeter B, Green TGA, Pannewitz S, Schlensog M, Sancho LG (2011) Summer variability, winter dormancy: lichen activity over 3 years at Botany Bay, 77°S latitude, continental Antarctica. Polar Biol 34:13–22. https://doi.org/10.1007/s00300-010-0851-7

Serrano HC, Oliveira MA, Barros C, Augusto AS, Pereira MJ, Pinho P, Branquinho C (2019) Measuring and mapping the effectiveness of the European Air Quality Directive in reducing N and S deposition at the ecosystem level. Sci Total Environ 647:1531–1538. https://doi.org/10.1016/j.scitotenv.2018.08.059

Singh SM, Nayaka S, Upreti DK (2007) Lichen communities in Larsemann Hills, East Antarctica. Curr Sci 93:1670–1672

Smith RI (1984) Terrestrial plant biology of the sub-Antarctic and Antarctic. In: Laws RM (ed) Antarctic ecology 1. Academic Press, London, pp 61–162

Smith RIL (1994) Vascular plants as bioindicators of regional warming in Antarctica. Oecologia 99:322–328. https://doi.org/10.1007/BF00627745

Smith RIL, Øvstedal DO (1991) The lichen genus *Stereocaulon* in Antarctica and South Georgia. Polar Biol 11:91–102. https://doi.org/10.1007/BF00234271

Terauds A, Lee JR (2016) Antarctic biogeography revisited: updating the Antarctic conservation biogeographic regions. Divers Distrib 22:836–840. https://doi.org/10.1111/ddi.12453

Tomiolo S, Ward D (2018) Species migrations and range shifts: A synthesis of causes and consequences. Perspect i Plant Ecol Evol Syst 33:62–77. https://doi.org/10.1016/j.ppees.2018.06.001

Upreti DK, (1996) Lecideoid lichens from Schirmacher Oasis, East Antarctica. Willdenowia 25:681–686

Upreti DK (1997) Notes on some crustose lichens from Schirmacher Oasis, East Antarctica. Feddes Repertorium 25:681–686

Upreti DK, Nayaka S (2011) Affinities of the lichen flora of Indian subcontinent vis-à-vis the Antarctic and Schirmacher Oasis. In: Singh J, Dutta HN (eds) Antarctica: The most Interactive Ice-Air-Ocean Environment. Nova Science Publishers, Inc, pp 149–161

Upreti DK, Pant G (1995) Lichen flora in and around Maitri Region, Schirmacher Oasis, East Antarctica. In: Scientific report: Eleventh Indian Expedition to Antarctica, Department of Ocean Development, Technical Publication No. 9, pp 229–241

Upreti DK, Pandey V (1994) Heavy metals of Antarctic lichens 1. *Umbilicaria* Feddes Repertorium 105:197–199. https://doi.org/10.1017/10.1002/fedr.19941050312

Walton DWH (1985) A preliminary study of the action of crustose lichens on rock surfaces in Antarctica, Berlin, Heidelberg. https://doi.org/10.1007/978-3-642-82275-9_25

Wang CH, Hou R, Wang M, He G, Li BG, Pan RL (2020) Effects of wet atmospheric nitrogen deposition on epiphytic lichens in the subtropical forests of Central China: evaluation of the lichen food supply and quality of two endangered primates. Ecotoxicol Environ Saf 190:110128. https://doi.org/10.1016/j.ecoenv.2019.110128

Wasley J, Robinson SA, Lovelock CE, Popp M (2006) Climate change manipulations show Antarctic flora is more strongly affected by elevated nutrients than water. Glob Change Biol 12:1800–1812. https://doi.org/10.1111/j.1365-2486.2006.01209.x

Wolterbeek HT, Garty J, Reis MA, Freitas MC (2003) Biomonitors in use: lichens and metal-air pollution. In: Markert BA, Breure AM, Zechmeister HG (eds) Trace Metals and other Contaminants in the Environment, vol 6, Elsevier, pp 377–419. https://doi.org/10.1016/S0927-5215(03)80141-8

Zhao L, Zhang C, Jia S, Liu Q, Chen Q, Li X, Liu X, Wu Q, Zhao L, Liu H (2019) Element bioaccumulation in lichens transplanted along two roads: the source and integration time of elements. Ecol Ind 99:101–107. https://doi.org/10.1016/j.ecolind.2018.12.020

Higher Rate of Pigment Synthesis in Antarctic Plants: A Strategy of Survival Under UV Radiations

Jaswant Singh, Rudra P. Singh, and Rajni Khare

Abstract Antarctica is known for the adverse climatic conditions and high UV-B radiations; thus, it reveals only climatically adapted flora and fauna. Cryptogams are the main flora of Antarctica and are dominated by lichens, followed by mosses and algae. Antarctic floral diversity reveals that pigments' synthesis plays an essential role in their survival, growth, development, and diversity during the annual spring. Antarctic cryptogams are growing in the photosynthetically active radiation (PAR) and ultraviolet radiations (UV-R), closely associated with photosynthetic and photoprotective synthesise pigments. Antarctic cryptogams cope with high UV radiation stress by synthesising UV-absorbing compounds; UV-B absorbs pigments and other compounds; the pigment synthesis protects cryptogamic flora. In lichens, usnic acid, perlatolic acid, and fumar photometric acid, mainly induced by, UV-B radiation, provide protection. In other lichens, secondary metabolites such as phenolics, atranorin, parietin and melanin also enhance the plant defence against UV radiation. In mosses, neoxanthin, violaxanthin, lutein, epoxide, anteraxanthin, zeaxanthin, UV absorbing, and phenolics are the important pigments synthesised by the plants under stress conditions. In Antarctic aquatic algae, algal pigments such as mycosporine-like amino acids, violaxanthin and β-carotene are present for protection. In a comparative study of pigments of plants growing in different regions of Antarctica, it was observed that these plants have a well-developed mechanism of synthesis of a wide variety of pigments in higher concentrations to cope with the UV radiation and other adverse environmental conditions.

Keywords Antarctica · Biodiversity · Cryptogams · Pigments · UV radiation

J. Singh (✉) · R. P. Singh
Department of Environmental Sciences, Dr R M. L. Avadh University, Ayodhya, Uttar Pradesh 224001, India

R. Khare
Delhi Public School, Sector Gama-2, Greater Noida, Gautam Buddh Nagar, Uttar Pradesh 201308, India

N. Khare (ed.), *Assessing the Antarctic Environment from a Climate Change Perspective*, Earth and Environmental Sciences Library,
https://doi.org/10.1007/978-3-030-87078-2_15

1 Introduction

The antarctic ecosystem exhibits one of the extremely adverse environmental conditions across the Earth planet. Antarctica is the coldest, driest and windiest continent, and it has exposed to high UV radiations. Antarctica is considered a desert, with the lowest annual precipitation (200 mm) along the coast and far less inland. Under harsh environmental conditions, floral diversity is shallow. A limited number of Lichens, Mosses, Algae, Fungi and few vascular plants can grow in different parts of the Antarctic continent.

However, lichens are the dominant floral component in most continental Antarctica parts and are the most visually recognised organisms in Antarctica terrestrial ecosystems; lichens have high resistance to freezing (Kappen 2000) and other conditions. The Antarctic soil has less humus but rich in nitrogen and phosphorus, which results in the luxuriant growth of lichens and mosses on rocks and dry and moist habitats (Crittenden 1998). Cold adapted algal community also grows at floating or melted glaciers, lakes and wherever waters are present. In favourable conditions, hard adopted algae form visible surface crusts and mats. The algal flora of Antarctica is dominated by blue-green algae, green algae, yellow-green algae and diatoms. Antarctic cryptogams possess the ability to enter a dormant state of physiological inactivity (Robinson et al. 2003).

Antarctic fungi have remarkable physiological plasticity, and some species produce several types of pigments (Duran et al. 2002; Zhou and Liu 2010; Akilandeswari and Pradeep 2016). Fungal pigments demonstrate different functions such as protection against lethal photo-oxidation and protection to environmental stress and can act as cofactors in enzymatic catalysis (Mapari et al. 2005). The production of photoprotective compounds (melanin, carotenoids, and mycosporines) by yeasts and fungi could be a strategy to survive in extreme environments. This behaviour can be evidenced by the high number of pigments (Rosa et al. 2009).

Antarctic plants have become a promising source of pigments and are supposed to be the basis for plants' survival under extreme environmental conditions. Wynn-Williams et al. (2002) highlighted pigments synthesis as an essential strategy in Antarctica's photosynthetic plants' survival. Our earlier studies (Singh et al. 2011, 2012; Singh and Singh, 2014) on the synthesis of pigments under harsh environmental conditions also indicate a higher rate of synthesis of pigments plants exposed high UV-B radiation. Therefore, in the present communication, we have summarised the synthesis of plants pigments and their comparison with other Antarctica regions and the role of pigment synthesis in plants' survival under harsh environmental conditions.

2 Bio-Geographic Zones of Antarctica and Biodiversity

Antarctic biome can be divided into several zones related to their different environmental factors (Longton 1988). "Antarctic botanical zone" divided into six longitudinal sectors Greene (1968), i.e., Maud (30°W–30°E), Enderby (30–90°E), Wilkes (90–150°E), Ross (150°E–150°W), Byrd (150–90°W) Scotia (90–30°W) (Greene 1968). Numerous bio-geographical schemes were applied to classify the Antarctic terrestrial environment, and most of the researchers have accepted recognition of three bio-geographical zones of Antarctica (Fig. 1), i.e., the Sub-, Maritime and Continental Antarctic zone (Smith 1984; Longton 1988; Ovstedal and Smith 2001; Convey 2005).

Later on, Peat et al. (2007) studied the Antarctic plant biogeography and established the boundaries and components of maritime Antarctica; in the same year, Chown and Convey (2007) described the biogeographic potential of the division between the southern Antarctic Peninsula and their remainder of the continent. Recently, Terauds et al. (2012) have identified greater biogeographic regionalisation than previously thought, defining 15 distinct "Antarctic Biogeographic Conservation Regions" (ACBRs) within the continent and Antarctic Peninsula. Recently, Terauds and Lee (2016) establish a revised version of the ACBRs, highlighting underlying spatial layers, together with the results of new analysis and reported 16th bioregion.

The sub-Antarctic zone includes isolated Islands and archipelagos at high latitudes in the Southern Ocean. With South Georgia's exclusion, Heard and McDonald Islands are nearby to or north Polar Frontal Zone (PFZ) of the Ocean. These islands are

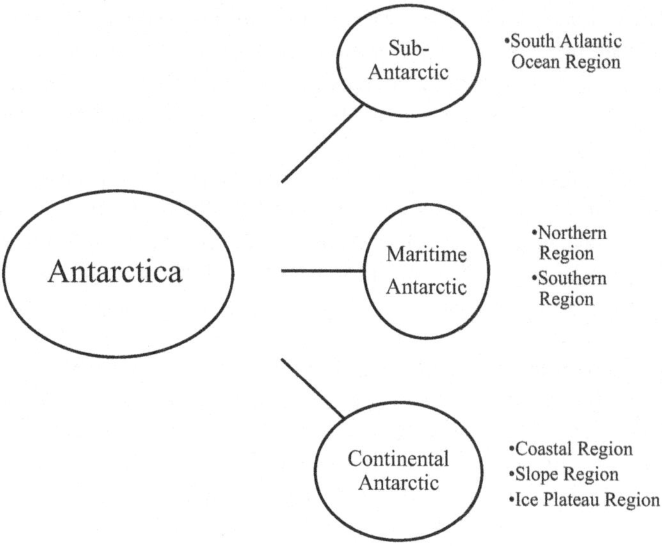

Fig. 1 Radial representation of Bio-geographic regions of Antarctica (Reference: Ovstedaland Smith 2001)

situated under the substantial maritime influence, which controls and adjusts the temperature differences year-round, as they are not usually affected by the pack or fast ice. Some researchers reported different groups (Falkland Islands, Gough Island, Iles Amsterdam and St. Paul, New Zealand's outlying groups of Snares, Campbell, Bounty, and Chatham, and parts of Tierra del Fuego) either within the sub-Antarctic or in a separate "mild Antarctic" classification.

Maritime Antarctic zone is also under strong maritime influences of the Southern Ocean, in this situation effects being highly seasonal and limited to the short Antarctic summer period. This zone includes Antarctic Peninsula (from the western coast) to 72°S, South Shetland, South Orkney and the South Sandwich Islands, and the isolated Bouvetoya and Peter Ioya.

Continental Antarctic covers the largest bio-geographical zone in terms of area; it includes the whole East and West Antarctica south of 70°S, the Balleny Islands, and those parts of the Antarctic Peninsula not covered in the maritime zone. In contrast to sub and marine zones, continental Antarctica's terrestrial zone habitats are restricted and isolated. However, this zone includes rocky coastal regions and is similar to some maritime Antarctic zone (Ovstedal and Smith 2001). Later on, Peat et al. (2007); Chown and Convey (2007) raise doubt on the latter element's inclusion, except climatic conditions.

The plant diversity of sub- Antarctic is tundra-like non-arborescent vegetation; vascular plants dominate at low altitude, along with different pteridophytes, bryophytes and lichens. Maritime Antarctica consists of the northern and southern maritime Antarctic zone. Northern maritime Antarctica dominated by cryptogams, along with other locally diverse flora in nearby coastal areas.

Growth of mosses restricted to nearby wetter habitats, peats and liverworts diversity are also frequent in this part. The lichens are predominant in exposed drier and inland conditions; snow algae and macrofungi quickly occur in summer. In southern Maritime Antarctic, biotic features are similar to Northern region, but cryptogamic diversity comparatively less and frequent in closed stands; two phanerogams (*Deschampsia antarctica* and *Colobanthu squitensis*) relatively common and one *Diptera* occasional at west coast sites; moss peat accumulation is somewhat less (Ovstedal and Smith 2001, 2004; Convey and Stevens 2007). Continental Antarctica consists of coastal, slope and ice plateau parts. In the coastal regions, biotic features are semi-desert with the growth of moss and algae. Bryophyte diversity is low in comparison to lichens. The snow algae are rarely growing in these localities. The slope continental Antarctic zone in the desert with very occasional lichen growth stands while mosses are rarely growing. In the ice plateau part, macroscopic life is limited to periodic growth of lichens, algae, cyanobacteria and some microorganisms (Ovstedal and Smith 2001, 2004; Peat et al. 2007).

Despite severe climatic conditions for plant growth, cryptogamic plants are found on the Antarctic continent, including the two Angiosperms, restricted to the relatively mild maritime zone (Smith 1984). In an investigation, Longton (1985) reported 2 Angiosperms, 85 Mosses, 25 Liverworts, 200^{+}, Lichens and 28 Macrofungi at Antarctic continent (Continental and maritime Antarctic zones). In the maritime zone, the cryptogamic vegetation is initially limited to small rocky outcrops, coast parts, dry valleys, and inland nunatak (Hansom and Gordon 1998).

The mosses have been reported from as far south as 84° S, and although bryophyte fruiting events are rare in the Antarctic continental zone (Filson and Willis 1975; Wise and Gressitt 1965). Seppelt et al. (1992) have reported that the moss sporophytes growth-restricted from as 77° 55′ S. In addition to the relatively conspicuous mosses and lichens, the continental Antarctic terrestrial vegetation includes groups that are often overlooked, including the chasmendolithic algae, which occur only within rock fissures. The lichens, moss, and algae are widespread in Antarctica's coastal regions and are believed to underlie 20% of the rock surface in few Antarctica parts (Hansom and Gordon 1998; Longton 1985). Later on, Convey (2005) reported Antarctica's biodiversity under three bio-geographical zones; Sub-Antarctic's floral diversity was dominant than Maritime and continental Antarctic (Table 1). The Sub- Antarctic zone's total plant diversity records the maximum 655 species, and the maritime includes 405 species. In comparison, the lowest plant diversity of 176 species were reported from continental Antarctic zones**.** Most recently, Lize-Marie van der Watt said, about 800 plant species form Antarctica, out of which about 350 are lichens and 100 species of Bryophytes (mosses and liverworts) predominate in maritime regions, rest all 400 species contributes with a diversity of moulds, yeasts, and other fungi, as well as freshwater algae and bacteria (https://www.britannica.com/place/Antarctica/Plant-life).

3 Antarctic Terrestrial Plants and Their Pigments

Antarctica reveals only climatically well-adopted plants (Singh et al. 2018), and the changing environmental conditions are attracting researchers' particular interest. Antarctic harsh ecological conditions, annual ozone depletion, and high UV radiation exposure to plants attract world researchers' attention. In continental Antarctica, environmental conditions are highly hostile to terrestrial vegetation, icy conditions, UV irradiance exposure, intense wind speed, and the lowest precipitation (Green and Schroeter 1999). In an Antarctic ecosystem, snow cover offers protection to PAR and high UV-B radiation exposure (Marchand 1984). Antarctic Peninsula's vegetation is poikilohydric, and the plants' occurrence depends upon the availability of water. In the spring and summertime period, plants' growth is high in Antarctica (Hovenden et al. 1994). Several pigments like chlorophyll, carotenoid in lichens, algae, bryophytes, and green alga are the advance parameter rather than morphology and anatomy to detect ultraviolet radiation's effects on Antarctic flora (Table 2).

Lichens

Lichens produce many secondary chemicals such a atronorin, usnic acid, zeorin, triterpins are widely distributed in lichen thallus and protect against climatic stress. The colour of the lichen thallus, such as dark black, orange, red, yellow and grey colours also helps in survival under harsh conditions. Elix (1996) reported that these

Table 1 Distribution of Antarctic plants in bio-geographic zones of Antarctica

Bio-geographic zones	Lichens	Mosses	Liverworts	Macro-fungi and other plant species	Angiosperms	Total plant diversity	Reference
Sub- Antarctic	250+	250	85	70	–	655	Convey (2005)
Maritime Antarctic	300	100		400	–	800	Van der Watt (2020)
	250	100	25	30	–	405	Convey (2005)
	150	75	25	22+	2	274	Smith (1984)
Continental Antarctic	150	25	1	–	–	176	Convey (2005)
	125	30	1	2	–	158	Smith (1984)

Table 2 Photosynthetic and other important pigments synthesised Antarctic plants

S. No	Lichens	Photosynthetic pigments	Other pigments	References
1	*Umbilicaria decussate*		β-Carotene, Lutein, Canthaxanthin, Astaxanthin	Czeczuga et al. (1996)
2	*Umbilicaria antarctica*		α-tocopherol, γ-tocopherol, plastochromanol	Strzalka et al. (2011)
3	*Umbilicaria aprina*	Chlorophyll, Carotenoids	UV absorbing, Phenolics	Singh and Singh (2014)
4	*Acarosporag wynnii*		β-Carotene, lutein, Hydroxyechinenone	Czeczuga et al. (1996)
5	*Buellia pallid*		β-Carotene, lutein, Zeaxanthin	Czeczuga et al. (1996)
6	*Lecidea cacriformis*		β-Carotene, lutein	Czeczuga et al. (1996)
7	*Lecanora fuscobrunnea*		α-Carotene, β-Carotene, Lutein	Czeczuga et al. (1996)
8	*Physciacaesia*		β-Carotene, β-Cryptoxanthin, lutein, Astaxanyhin	Czeczuga et al. (1996)
9	*Rhizocarpon flavum*		β-Cryptoxanthin, Canthaxanthin, Lutein,	Czeczuga et al. (1996)
10	*Rinodina olivaceobrumea*		β-Carotene, lutein, Zeaxanthin, Canthaxanthin	Czeczuga et al. (1996)
11	*Umbilicaria antarctica*		β-Carotene, β-Cryptoxanthin, lutein,	Czeczuga et al. (1996)
12	*Xanthoria elegans*		β-Carotene, β-Cryptoxanthin, Lutein, Asthaxanthin	Czeczuga et al. (1996)
13	*Usnea antarctica*		β -Carotene, β-Cryptoxanthin, Lutein, Hydroxyechinenone	Czeczuga et al. (1996)
14	*Usnea aurantiaco-atra*		α-tocopherol, γ-tocopherol, plastochromanol	Strzalka et al., (2011)
15	*Sphaerophorus globosus*		Astaxanthin, Adonixanthin	Czeczuga et al. (1996)

(continued)

Table 2 (continued)

S. No	Lichens	Photosynthetic pigments	Other pigments	References
16	*Parmelia saxatilis*		β-Carotene, β-Cryptoxanthin, Lutein	Czeczuga et al. (1996)
17	*Turgidosumc oplicatulum*	Chl. a, Chl. b	β-Carotene, lutein, Neoxanthin, Violaxanthin	Lud et al. (2001)
18	*Caloplaca regalis*		α-tocopherol, γ-tocopherol, plastochromanol	Strzalka et al. (2011)
19	*Stereocaulon alpinum*		α-tocopherol, γ-tocopherol, plastochromanol	Strzalka et al. (2011)
20	*Himantornia lugubris*		α-tocopherol, γ-tocopherol, plastochromanol	Strzalka et al. (2011)
21	*Ochrolechia frigid*			Strzalka et al. (2011)
	Bryophytes			
22	*Bryum argenteum*	Chl. a, Chl. b, Chlorophyll, Carotenoids	Violaxanthin, β-Carotein; UV absorbing; Phenolics	Markham et al. 1990; Singh and Singh 2014
23	*Brymp seudotriquetrum*	Chl. a, Chl. b	Violaxanthin, β-Carotein	Robinson et al. (2006)
24	*Ceratodon perpureus*	Chl. a, Chl. b	Violaxanthin, β-Carotein	Dunn (2000)
25	*Grimmia antarctici*	Chl. a, Chl. b	Violaxanthin, β-Carotein	Dunn 2000; Robinson et al. 2005
26	*Sanioniauncinata*	Chl. a, Chl. b	Neoxanthin, lutein, Zeaxanthin, β, β-Carotene	Newsham et al. (2002)
27	*Andrea regularis*	Chl. a, Chl. b	Neoxanthin, violaxanthin, Lutein, Zeaxanthin	Newsham (2003)
28	*Cephaloziella varians*	Chl. a, Chl. b	Neoxanthin, lutein, Zeaxanthin, β,β-Carotene	Newsham et al. 2002, 2005
29	*Schistidium antarctici*	Chl. a, Chl. b	Anthocyanin	Robinson et al. (2006)

(continued)

Table 2 (continued)

S. No	Lichens	Photosynthetic pigments	Other pigments	References
30	*Syntrichiam agellanica*	Chl. a, Chl. b and Carotene	Neoxanthin, Violaxanthin, Lutein, Epoxide, Anteraxanthin, Lutein, Zeaxanthin	Strzalka et al. (2011)
31	*Polytrichastrum alpinum*	Chl. a, Chl. b and Carotene	Neoxanthin, Violaxanthin, Lutein, Epoxide, Anteraxanthin, Lutein, Zeaxanthin	Strzalka et al. (2011)
32	*Sanioni ageorgicouncinata*	Chl. a, Chl. b and Carotene	Neoxanthin, Violaxanthin, Lutein, Epoxide, Anteraxanthin, Lutein, Zeaxanthin	Strzalka et al. (2011)
33	*Warnstro fiaarmentosa*	Chl. a, Chl. b and Carotene	Neoxanthin, Violaxanthin, Lutein, Epoxide, Anteraxanthin, Lutein, Zeaxanthin	Strzalka et al. (2011)
	Algae			
34	*Entomeneisk jellmannii*		Mycosporine like amino acids	Ryan et al. (2002)
35	*Berkeleya adeliense*		Mycosporine like amino acids	Ryan et al. (2002)
36	*Nitzschia stellata*		Mycosporine like amino acids	Ryan et al. (2002)
37	*Prasiol acrispa*	Chl. a, Chl. b	Violaxanthin	Post and Larkum (1993)
38	*Choricystis minor*	Chl. a, Chl. b	β-carotene	Zidarova and Pouneva (2006)
39	*Phormidium sp.*	Chl. a	MAAs (porphyra-334) β-Carotene	Quesada et al. (1998)
40	*Lyngbya sp.*	Chl. a	MAAs (porphyra-334) β-Carotene	Quesada et al. (1998)
41	*Oscillatoria sp.*	Chl. a	MAAs (porphyra-334) β-Carotene	Quesada et al. (1998)
42	*Nodularia sp.*	Chl. a	MAAs (porphyra-334) β-Carotene	Quesada et al. (1998)

(continued)

Table 2 (continued)

S. No	Lichens	Photosynthetic pigments	Other pigments	References
43	*Anabaena sp.*	Chl. a	MAAs (porphyra-334) β-Carotene	Quesada et al. (1998)
44	*Palmari adecipiens*	Chl. a		Dhargalkar (2004)
45	*Phyllophora antarctica*	Chl. a		Dhargalkar (2004)
46	*Porphyra endiviifolium*	Chl. a		Dhargalkar (2004)
47	*Iridaea cordata*	Chl. a		Dhargalkar (2004)
	Fungal Species			
48	*Sporobolomyces Salmonicolor*		β-Carotene, torularhodin, and torulene	Dimitrova et al. (2010)
49	*Sporobolomyces Metaroseus*		β-Carotene and 4-ketotorulene, furthermore, β-cryptoxanthin and spirilloxanthin	Barahona et al. (2016)
50	*Dioszegia sp.*		OHK torulene	Villarreal et al. (2016)
51	*Rhodotorula larynges*		Torulene and Lycopene	Villarreal et al. (2016)
52	*Cryptococ cusgastricus*		Torulene, γ-carotene, and lycopene	Villarreal et al. (2016)
53	*Rhodotorula mucilaginosa*		2-γ-Carotene	Villarreal et al. (2016)
54	*Cryptococ cusalbidus*		β-Carotene	Dimitrova et al. (2010)
55	*Cryptococ cuslaurentii*		β-Carotene	Dimitrova et al. (2010)
56	*Dioszegi apatagonica*	Total Carotenoid		Trochine et al. (2017)
57	*Nadsoniella nigra*		Melanine	Chyizhanska and Beregova (2009)
58	*Thelebolus microspores*		β-Carotene	Singh et al. (2014)
59	*Arthrobotrys ferox*	Carotenoid	Mycosporines	Arcangeli et al. 2000; Arcangeli and Cannistraro(2000)
60	*Exophiala xenobiotica*		Melanine	Vasileva-Tonkova et al. (2014)
	Higher Plants			
61	*Deschampsia antarctica*	Chl. a, Chl. b	Orientin, Luteolin Hydroxycinnamic acids, Flavonoids Ferulic acid, p-coumaric acid	Ruhland et al. 2005; Ruhland and Day 2000

(continued)

Table 2 (continued)

S. No	Lichens	Photosynthetic pigments	Other pigments	References
62	*Colobanthus quitensis*	Chl. a, Chl. b	Orientin, Luteolin Hydroxycinnamic acids, flavonoids ferulic acid, p-coumaric acid	Lud et al. 2001; Ruhland and Day 2000

secondary chemicals reach the upper cortex of lichens thallus to protect against adverse conditions. Huneck (1999) observed that these secondary chemicals sometimes react with rock substratum and secrete a powdery substance on the upper surface called "Pruna". These pruna shade the thallus to protect the thallus against temperature as well as UV radiations. The darkness of melanin pigments of *Lobaria-pulmonaria* thallus reflects the exposure of solar light gradient. The brownest thalli are observed in the most sun-exposed position, and increasing pigmentation causes a reduced cortical transmittance, mainly UV radiation and short wave radiation. Riley (1997) analysed the melanin pigments in lichens and reported that melanin is a predominantly indolic polymer that occurs on the surface of lichen; these are black and generally set up in Antarctic lichens. A photoprotective role for melanin compound should imply a higher resistance against excessive light in thalli where these pigments are abundant (Riley 1997). The melanin exhibit the most robust protection at energy-rich wavelength; the percentage reduction in transmittance caused by melanin appears to be prominent in the UV range. McEvoy et al. (2007) reported that melanin and photosynthetic pigments increase acclimation under high light intensity exposure; besides this, melanic thalli absorbs higher solar energy under increased temperatures relatively to pale colour thalli. Gauslaa and Ustvedt (2003) explored the bright orange colour of an anthraquinone, i.e., parietin found in lichens and is a UV screening compound. Parietin pigment of lichens may reduce the adverse impact of ultraviolet radiations. UV-B absorbing pigments, i.e., flavonoids are wavelength selective UV-B screening pigment, is synthesised rapidly under high UV-B radiation exposure levels (Caldwell et al. 1983). The secondary medullary compounds, stictic and nastictic acid, accumulated in the highest quantities in the photobiont layer and may help UV radiation protection (Bachereau and Asta 1997).

Bryophytes

The bryophytes can be found in moist areas and are supporting Antarctic plant biodiversity. Along with lichens, bryophytes share the distinction of being the largest group of plants in the Antarctic region. Despite general assumption, Joseph D. Hooker made the first collections of mosses in Antarctica during 1829–30. The bryophytes are also poikilohydric; the plant growths are limited to free water during the summer month for photosynthetic carbon gain. Antarctic bryophyte communities are primarily confined to the margins of melt lake at Casey (the Australian base in the Windmill Island region, 660 17′S, 1100 32′E), the three leading moss species *Ceratodon purpureus*,

Grimmia antarctici and *Bryumps eudotriqu etrum* are establish in both pure and mixed communities (Selkirk and Seppelt 1987) and stream areas subject to snow deposition. In addition to their role in energy storage compounds, sugars and polyols perform multiple plant functions. They are thought to be active as a cryoprotectant, as an osmotic regulator in the drought and salt stress plants (Popp and Smirnoff 1995). Various soluble carbohydrates can interact with the polar head groups of phospholipids, taking water molecules and maintaining membrane integrity during desiccation (Crowe and Crowe 1986); polyols accumulate in higher plants in response to water stress. These molecules most likely have numerous roles such as compatible solutes, scavengers of active oxygen species, and macromolecules' stabilisers (Loescher 1987; Smirnoff and Cumbes 1989; Popp and Smirnoff 1995). Differences in the response of the three moss species to desiccation and the phenotypic plasticity were assessed at two sites and were different in water availability; (1) the WC (water content) at complete hydration, (2) the rates of drying over time, and (3) the relationship between the decline in chlorophyll fluorescence and relative WC. Initial water loss rates were faster in moss obtained from the wet sites compared to dry areas. Some moss species of high latitudes have been studied to understand the impact of UV radiation on growth and morphology (Sonesson et al. 1996; Searles et al. 2002; Robson et al. 2003; Robinson et al. 2005). Short moss turf and cushion mosses are frequently reported from sandy and gravelly soils. These moss communities' extent is limited to meltwater availability in the summer period and ranges from the extensive mosses bounded to around melt lakes and streams (Selkirk and Seppelt 1987). Mosses are growing in colonies, which make them possible to store and retain more water. They also lose less water by evaporation and show a marked ability to use water rapidly whenever it remains available. Mosses have also become well adapted to almost continuous light exposure during a long day of polar summer.

Over a few decades, Antarctic plants have been exposed to the most significant relative increase in UV-B radiation due to ozone depletion. Although photosynthetic rates were not affected, there was evidence of UV effects on Antarctic bryophytes' morphology. These findings suggested that *G.antarctici* may be disadvantaged in some settings under a climate with continuing high springtime UV-B radiation levels (Robinson et al. 2005). Newsham et al. (2002) conducted a similar UV related onsite study with two Antarctic plants (*Cephaloziell avarians* and *Sanioni auncinata*) and reported no change in photosynthetic pigments except an increase in carotenoid content. In another study, Newsham et al. (2005) said that the chlorophyll concentrations of *C. variants* were reduced in sunlight exposure. A similar study conducted by Robinson et al. (2005) reported a lower concentration of chlorophyll in *G.antarctici* under near-ambient UV radiation exposure.

In the same way, a high concentration of carotenoids was found under reduced UV radiation exposure. Lud et al. (2001) did not see any changes in chlorophyll, carotenoid, UV-B absorbing compounds and photosystem II efficiency in *Turgidosculum complicatulum* under different combinations of UV radiation exposure and temperature. Day et al. (1999), Searles et al. (2001), Lud et al. (2002), and Newsham (2003) were reported no effects of UV-B radiation on chlorophyll concentration of the selected plants. Chlorophyll content was significantly lower in plants grown under near-ambient UV, while the relative proportions of photoprotective pigments carotenoids, β-carotene and zeaxanthin were higher (Newsham et al. 2005). According to Paul (2001), the carotenoid concentration plays an essential role in an extended freezing period. The *Bryum argenteum* produce more energy by photosynthesis in low light at 5 °C than it does at 15 °C or higher. Photosynthesis can start within a few hours of thawing after an extended time of freezing and almost immediately following in a short time duration; for these reasons, it may survive in Antarctica.

Algae

Antarctic algae grow in semi-permanent to permanent snow or ice in the world's alpine or polar region. Their optimum growth temperature is generally below 10 °C. According to Green et al. (1999), more than 300 species of non-marine alga have been found in Antarctica and are successfully adapted to harsh environmental conditions through the development of some adaptive features, which includes the formation of pigments, polyols (sugar, alcohols, e.g. glycerin), sugar and lipid (oils), mucilage sheaths, motile stages and spore development. The floral vegetation of Antarctica is characterised by a high degree of adaptation under harsh environmental conditions. The pigment found in photoautotrophic organisms has a great interest in the ecologist and biologist for many reasons, e.g., chlorophyll and their derivatives have been used to measure productivity. The photosynthetic pigment contents in lichenised alga are chlorophyll a, chlorophyll b, carotenoid, phycocyanin, lutein and b-carotene lipid-soluble antioxidant alpha-tocopherol (Table 2).

Cyanobacteria have a cosmopolitan distribution ranging from hot springs to the coldest continent (the Arctic and Antarctic regions). They are characterised by high variability to adapt to various environmental conditions (Rozema et al. 2002). Blue-green and other algae are growing under damp sand and gravel around lakes, pools, melted water streams, or low-lying areas. The blue-green algae from fabulous red, yellow or green patches over permanent snow area. These red, yellow or green pigments protect the cells from ultraviolet radiation and high light damage during the summer. The pigments may be in the form of iron tannin compounds, as *Mesotaenium berggrenii*, or orange to red-pigmented lipids, as in the majority of the snow algae. Some species' cells also secrete copious amounts of mucilage, making them unable to adhere to one another and snow crystal and prevent the cell from being washed away by meltwater. The adhesive role is to form a protective cover of UV radiation and delayed the water loss in algae. According to Sinha et al. (2005), Ultraviolet-B radiation reaching the earth surface destroys phycobiliproteins like phycoerythrin, phycocyanin, allophycocyanin in blue-green algae. In general, cyanobacteria are protected by mycosporine-like amino acids (MAAs) and scytonemins (Table 2), while terrestrial plants contain flavonoids.

Fungi

Like lichen, moss and algae, Antarctic fungi are also an essential source of pigments due to their physiological plasticity. Some species produce several kinds of pigments with diverse characteristics (Duran et al. 2002; Zhou and Liu 2010;

Duarte et al. 2019). When the colonies are established, fungi produce pigments when the vital supplies become partial (Isaac 1994). Fungal pigments can reveal diverse functions such as protection against lethal photo-oxidation (like carotenoids) and environmental stress (melanin) and act as cofactors in enzymatic catalysis (Mapari et al. 2005). Among several studies, limited findings addressed the production of pigments by filamentous fungi. Among these, production of β-carotene by *Thelebolus microspores* (Singh et al. 2014) and carotenoid and mycosporine by *Arthrobotrys ferox* were reported by Arcangeli et al. (2000), Arcangeli and Cannistraro(2000).

In the laboratory, the first screening for pigment production by Antarctic fungi is performed in a solid culture medium (yeast malt agar) through direct visualisation (Vaz et al. 2011). These pigments are either intracellular or extracellular. The extracellular pigments are screening sun rays, and other pigments have quenching properties against UV-B radiation. Simultaneously, some absorb UV-B radiation inside the cell before metabolically essential molecules can be damaged (Wynn-Williams et al. 2001). Antarctic fungal pigments such as carotenoids have been identified with photoprotective functions against radiation exposure (Moline et al. 2009; Dimitrova et al. 2010).

In situ studies conducted by Vendruscolo et al. (2010) reported that fungal pigments could be produced under controlled conditions through the submerged, solid, or semi-solid fermentation by using a wide variety of eco-friendly substrates. The environmental conditions, i.e., temperature, aeration, agitation, pH, and culture medium, directly influence the pigments' production and quantity (Medentsev et al. 2005). Fungi produce pigments in different colours and are recognised as black or dark brown (for melanin), orange (for β-carotene), orange-red (for γ-carotene and xanthophyll), dark red (for lycopene) (Vasileva-Tonkova et al. 2014; Barahona et al. 2016; Villarreal et al.2016).

Higher Plants

In Antarctica, only two endemic vascular plants, *Deschampsia antarctica* (Antarctic hair grass) and *Colobanthu squitensis* (cushion forming pearlwort) survive south of 56°S. They are distributed under small clumps near the shore of the west coast of the Antarctic Peninsula. Both plants can tolerate very harsh environmental conditions. The accumulation of UV-B absorbing pigments are advantageous in Antarctic plants because such passive screens could protect them from UV-B damage (Lovelock et al. 1995; Cockell and Knowland 1999). Xiong and Day (2001) assessed UV-B radiation's influence on biomass production and photosynthesis of *C.quitensis* and *D.antarctica* at Antarctic Peninsula region. Plant leaves under exposed UV-B radiation were denser, probably thicker, and had higher concentrations of photosynthetic and UV-B absorbing pigments (Xiong and Day 2001).

Ruhland and Day (2000) examined insoluble phenylpropanoids in *C.quitensis* and *D.antarctica,* and the HPLC analysis revealed that ferulic and p-coumaric acid were the main components of both insoluble and soluble phenylpropanoids. In another study, Ruhland et al. (2005) analysed the ultraviolet-B radiation effect on phenylpropanoid concentrations of *D. antarctica.* They reported that p-coumaric, caffeic and ferulic acids were the primary hydroxycinnamic acids, and luteolin derivatives were the major flavonoids in both insoluble and soluble leaf extracts. In *D.antarctica* and *C.quitensis*, the synthesis of photoprotective pigments depends on UV radiation exposure and protects against UV radiation's adverse impact.

4 Pigments in Cryptogamic Plants Growing in Other Regions Than Antarctica

Energy flow is one of the most vital functions of any ecosystem; therefore, plant pigments play an essential role in the flourishing of vegetation. The number of photosynthetic pigments may indicate the health of flora and the productivity of the ecosystem. The climatic factors such as variations in temperature, moisture and radiations are essential in maintaining the plants' pigment concentrations (Garty et al. 1985; Lauge et al. 1999). The quantification of pigments in cryptogams has been done for different purposes, e.g. pigment distribution within the thallus (Boonpragob 2000; Krenelamp 1971), observations of the effect of pollutants (Puckett et al. 1973) or environmental conditions and as a quotient for composition rate (Farrar 1976). Chlorophyll content and chlorophyll degradation are prevalent parameters used to assess the impact of stress on cryptogams (Singh et al. 2008). Stratospheric ozone depletion increased UVB radiations (280–320 nm) exposure on the earth surface therefore; plant life and ecosystem functions are affected by changes in plants pigment concentration (Mc Peters et al. 1996; Herman et al. 1999). Singh et al. (2008) studied the chlorophylls, carotenoids concerning altitudinal changes, the effect on the pigment concentration, and reported that cryptogams contain higher chlorophyll content at the lower altitude *Umbilicaria* and *Spirogyra* while Chlorophyll b concentration was maximum in *Spirogyra* followed by *Umbilicaria*, *Xanthoria*, and *Candilari*. McEvoy et al. (2007) studied the resynthesis of usnic acid in *Xanthoparmelia stenophylla* thalli under the natural UV radiation exposure levels.

Prasad et al. (2005) noticed a significant reduction in chlorophyll content with increasing doses of UVR. The suppression was more prominent on chlorophyll a than chlorophyll b, showing decreasing trends in Chl a/b ratio. This reduction could be due to chlorophyll destruction, as reported in most UV radiation exposed plants (Teramura and Sullivan 1994). In contrast to chlorophyll concentrations and protein content, carotenoid showed increasing altitudes, i.e., higher UV radiation (Buffoni et al. 2002). The carotenoid concentration was maximum in cryptogams growing at a higher altitude as compared to a lower height.

Interestingly, the carotenoid contents are higher in those species with low chlorophyll content and were surviving well. Carotenoids protect the chlorophyll pigments against the excess radiation energy, which might otherwise photobleach the chlorophyll (Nybakken and Solhaug 2004). According to Prasad et al. (2005), carotenoids are the scavenger of singlet oxygen molecules formed during intense light and protect chlorophyll from photooxidative damage. Therefore, increased carotenoid contents at a higher altitude due to UV radiations could be a protective mechanism for chlorophyll pigments. In another study, Singh et al. (2012) measured the effect of ultraviolet-B (UV-B) radiation on two cryptogamic plants (*Xanthoria elegans* and *B. argenteum*) growing at a high altitude of the central Himalayan region of India. In the field experiments, both of these plants contain higher amounts of UV-B absorbing compounds and phenolics. Simultaneously, no significant changes were recorded in total chlorophyll and carotenoid under UV-B exposed conditions. A short-term study observed that B. argenteum contains higher amounts of total chlorophyll, carotenoids, UV-B absorbing compounds, and phenolics than the *X. elegans*. Larsson et al. (2009) observed no significant reduction in chlorophyll a and b in *Lobaria pulmonaria* and *Xanthoria aureola* at different UV-B levels under the laboratory conditions.

5 Conclusions

The synthesis of pigments in cryptogams growing under Antarctica's adverse environmental conditions is of great physiological significance. Extreme cold conditions, low temperature, lesser availability of nutrients, and higher UV-B radiations affect the synthesis of photosynthetic pigments and other important pigments, protecting against UV radiation and different Antarctica's adverse condition. The amount and synthesis of photosynthetic and other pigments indicate the growth and development of flora and the productivity of one of the world's most fragile ecosystem. Therefore, recently the scientific community has started quantifying these pigments, growth, development, morphology, physiology, and plants' biochemical changes growing in Antarctica. The synthesis of UV-absorbing pigments, UV screening and secondary plant pigments protecting the Antarctic flora against harsh environmental conditions is of particular significance. Our research finding reveals that plants growing at high altitude regions of the central Himalayas also synthesise more elevated levels of carotenoids, phenolics and other protective pigments to cope with the adverse environmental conditions, including the UV radiation stress. The synthesis of protective pigments depends on the levels of adverse environmental conditions or high-stress factors. During the period of ozone depletion, a higher amount of UV radiation reaches Antarctica. That's why during this period, the plant synthesises higher amounts of UV-absorbing pigments and other protective compounds to cope with the adverse effects. This is the basis of plants' survival in Antarctica's most adverse conditions and other polar regions of the world.

References

Akilandeswari P, Pradeep BV (2016) Exploration of industrially important pigments from soil fungi. Appl Microbiol Biotechnol 100:1631–1643

Arcangeli C, Cannistraro S (2000) *In situ* Raman microspectroscopic identification and localisation of carotenoids: approach to monitoring UV-B irradiation stress on the Antarctic fungus. Biopolymers 57:179–186

Arcangeli C, Yu W, Cannistraro S, Gratton E (2000) Two-photon autofluorescence microscopy and spectroscopy of an Antarctic fungus: a new approach for studying the effects of UV-B irradiation.Biospectroscopy 57:218–225

Bachereau F, Asta J (1997) Effects of solar UVR at high altitude on the physiology and the initials biochemistry of terrestrial lichen *Cetraria islandica* (L.) Ach. Symbiosis 23:197–217

Barahona S, Yuivar Y, Socias G, Alcaino J, Cifuentes V, Baeza M (2016) Identification and characterisation of yeasts isolated from sedimentary rocks of Union Glacier at Antarctica. Extremophiles 20:479–491

Boonpragob, K (2000) Monitoring physiological changes in lichens.Total chlorophyll content and chlorophyll degradation In Nimis PL, Schedegger C and Wolseley PA (eds) Monitoring with lichena-lichen monitoring. Kluwer Academic Publishers, pp 323–326

Buffoni-Hall RS, Bornman JF, Bjorn LO (2002) UV-induced changes in pigment content and light penetration in the fruticose lichen Cladonia arbuscula sp. Mitis J Photochem Photobiol b: Biol 66:13–20

Caldwell M, Robberecht R, Flint S (1983) Internal filters: prospects for UV-acclimation in higher plants. Physiol Plant 58:445–450

Chown SL, Convey P (2007) Spatial and temporal variability across life's hierarchies in the terrestrial Antarctic. Philos Trans R Soc Ser B 362:2307–2331

Chyizhanska N, Beregova T (2009) Effect of melanin isolated from Antarctic yeasts on preserving pig livestock after ablactation. UAZH 8:382–385

Cockell CS, Knowland J (1999) Ultraviolet radiation screening compounds. Biol Rev 74:311–345

Convey P (2005) Antarctic terrestrial ecosystems: responses to environmental change. Polarforschung 75:101–111

Convey P, Stevens MI (2007) Antarctic biodiversity. Science 317:1877–1878

Crittenden PD (1998) Nutrient exchange in an Antarctic macro-lichen during summer snowfall-snowmelt events. New Phytology 139:697–707

Crowe JH, Crowe LM (1986) Stabilisation of membranes in anhydrobiotic organisms. In: Leopold C (ed) Membranes, metabolism and dry organisms. Cornell University Press, Ithaca, USA pp 188–230

Czeczuga B, Inoue I, Upreti DK (1996) Carotenoids in lichens from the Antarctic. Rep Nankyoku Shiryo 40:247–254

Day TA, Ruhland CT, Grobe CW, Xiong F (1999) Growth and reduction of Antarctic vascular plants in response to warming and UV radiation reductions in the field. Oecologia 199:24–35

Dhargalkar VK (2004) Effect of different temperature regimes on the chlorophyll-a concentration in four Antarctica macroalgae species. Seaweed Res Util 26:237–243

Dimitrova S, Pavlova K, Lukanov L, Zagorchev P (2010) Synthesis of coenzyme q10 and β-carotene by yeasts isolated from Antarctic soil and lichen in response to ultraviolet and visible radiations. Appl Biochem Biotechnol 162:795–804

Duarte AWF, de Menezes GCA, Silva TR, Bicas, JL, Oliveira VM, Rosa LH (2019) Antarctic fungi as producers of pigments. In: Rosa LH (ed) Fungi of Antarctica. Springer Nature Switzerland, pp 305–318. https://doi.org/10.1007/978-3-030-18367-7_14

Dunn JL (2000) Seasonal variation in pigment content of three species of Antarctic bryophytes. BSc. Honours. University of Wollongong, Wollongong

Duran N, Teixeira MF, de Conti R, Esposito E (2002) Ecological-friendly pigments from fungi. Crit Rev Food Sci Nutr 42:53–66

Elix JA (1996) Biochemistry and secondary metabolites. In: Nash TH (ed) Lichen biology. Cambridge University Press, UK, pp 154–180

Farrar JF (1976) The lichen as an ecosystem. Observation and experiment. In: Brown DH, Hawksworth DL, Bailey RH (eds) lichenology: progress and problem. Academic Press, London, pp 385–406

Filson RB, Willis JH (1975) A fruiting occurrence of *Bryum algens* card. In East Antarctica. Muelleria 3:112–116

Garty J, Ronen R, Galun M (1985) Correlation between chlorophyll degradation and some elements in the lichen Ramalina duiaei. J Env Exp Bot 25:67–74

Gauslaa Y, Ustvedt EM (2003) Is parietin a UV-B or a blue light screening pigment in lichens Xanthoria parietina. Photochem Photobiol Sci 2:423–424

Green TGA, Schroeter BLG (1999) Plant life in Antarctica. In: Valladares F (ed) Handbook of functional plant ecology. Macel Dekker. Inc., Press, New York, pp 496–543

Greene SW(1968) Studies in Antarctic bryology. I. A basic checklist for mosses. Rev Bryol Lichenol 36:132–138

Hansom JD, Gordon JE, (1998) Antarctic Environments and Resources: Geographical perspective. Addison Wesley Longman Ltd, Harlow, Essex

Herman JR, Mc Kenzie RL, Daiz S, Ker J, Madroinch S, Seckmever G (1999) UV radiation at the earth's surface. In: Albettion DP, Aucam P, Megie G, Walson R (eds) Scientific assessment of ozone depletion: 1998. World Meteorological Organization, Geneva, pp 9.1–9.46

Hovenden MJ, Jackson AE, Seppelt RD (1994) Field photosynthetic activity of lichens in the Windmill Islands oasis, Wilkes Land, continental Antarctica. Physiol Plant 90:567–576

Huneck S (1999) The significance of lichens and their metabolites. Naturwissenschaften 86:559–570

Isaac S (1994) Mycology answers. Mycologist 8:178–179

Kappen L (2000) Some aspects of the great success of lichens in Antarctica. Antarct Sci 72:314–324

Krenelamp I (1971) Distribution of chlorophyll in the lichen *Cladonia alpestis.* Ann Univ Turku Ser A II Biol Geogr 46:1–6

Larsson P, Vecerova K, Cempirkova H, Solhaug KA, Gauslaa Y (2009) Does UV-B influence biomass growth in lichen deficient in sun-screening pigment? Environ Exp Bot 67:215–221

Lauge OL, Lesner JMR, Bilger W (1999) Chlorophyll fluorescence characteristics of the cyanobacterial lichen *Peltigera rufescens*under field conditions. II. Diel and annual distribution of metabolic activity and possible mechanisms to avoid photoinhibition. Flora 144:413–430

Loescher WH (1987) Physiology and metabolism of sugar alcohols in higher plants. Physiol Plant 70:553–557

Longton RE (1985) Terrestrial habitats vegetation. In: Walton (ed) DWH key environments Antarctica, Pergamon Press, Oxford, pp 73–105

Longton RE (1988) Biology of polar bryophytes and lichens. Cambridge University Press, Cambridge

Lovelock CE, Osmond CB, Seppelt RD (1995) Photoinhibition in the Antarctic moss *Grimmia antarctica* Card. When exposed to cycles of freezing and thawing. Plant Cell Environ 18:1395–1402

Lud D, Huiskes A, Moerdijk T, Rozema J (2001) The effects of altered levels of UV-B radiation on an Antarctic grass and lichen. Plant Ecol 154:89–99

Lud D, Moerdijk T, Van de Poll W, Buma AGJ, Huiskes AHL (2002) DNA damage and photosynthesis in the Antarctic and Arctic *Sanionia uncinata*(Hedw.)Loeske under ambient and enhanced levels of UV-B radiation. Plant Cell Environ 25:1579–1589

Mapari SA, Nielsen KF, Larsen TO, Frisvad JC, Meyer AS, Thrane U (2005) Exploring fungal biodiversity for the production of water-soluble pigments as potential natural food colourants. Curr Opin Biotechnol 16:231–238

Marchand PJ (1984) Light extinction under a changing snow cover. In: Merritt JF (ed) Winter ecology of small mammals, Pittsburg, pp 33–37

Markham KR, Franke A, Given DR et al (1990) Historical Antarctic ozone level trends from herbarium specimen flavonoids. Bull De Liaison Groupe Du Polyphenols 15:230–235

McEvoy M, GauslaaY, Solhaug KA, (2007) Changes in a pool of depsidones and melanins, and their function, during growth and acclimation under contrasting natural light in the lichen *Laboria pulmonaria*. New Phytol 157:271–282
Medentsev AG, Arinbasarova AY, Akimenko VK (2005) Biosynthesis of naphthoquinone pigments by fungi of the genus *Fusarium*. Appl Biochem Microbiol 41:503–5076
Moline M, Libkind D, Dieguez MC, van Broock M (2009) Photoprotective role of carotenoid pigments in yeasts: experimental study contrasting naturally occurring pigmented and albino strains. J Photochem Photobiol B 95:156–161
Newsham K (2003) UV-B radiation arising from stratospheric ozone depletion influences the pigmentation of the moss *Andreaea regularis*. Oecologia 135:327–331
Newsham K, Geissler P, Nicolson M et al (2005) Sequential reduction of UV-B radiation in the field alters an Antarctic leafy liverwort's pigmentation. Environ Exp Bot 54:22–32
Newsham K, Hodgson D, Murray A et al (2002) Response of two Antarctic bryophytes to stratospheric ozone depletion. Global Change Biol 8:972–983
Nybakken L, Solhaug KA (2004) The lichens *Xanthoria elegans* and *Cetraria islandica* maintain high protection against UV-B radiation in Arctic habitats. Oecologia 140:211–216
Ovstedal DO, Smith RIL (2001) Lichens of Antarctica and South Georgia. A guide to their identification and ecology. Cambridge University Press, Cambridge
Ovstedal DO, Smith RIL (2004) Additions and corrections to the lichens of Antarctica and South Georgia. Cryptogam Mycol 25:323–331
Paul N (2001) Plant response to UV-B; time to look beyond stratospheric ozone depletion. New Phytol 150:1–8
Peat HJ, Clarke A, Convey P (2007) Diversity and biogeography of the Antarctic flora. J Biogeogr 34:132–146
Popp M, Smirnoff N (1995) Polyol accumulation and metabolism during water deficit. In: Smirnoff N (ed) Environment and plant metabolism-flexibility and acclimation. Bios Scientific Publishers, Oxford, UK, pp 199–215
Post A, Larkum AWD (1993) UV-absorbing pigments, photosynthesis, and UV exposure in Antarctica, comparing terrestrial and marine algae. Aquat Bot 45:231–243
Prasad SM, Srivastava G, Mishra V, Dwivedi R, Zeeshan M (2005) Active oxygen species generation, oxidative damage and antioxidant defence system in *Possum sativa* exposed to UV-B irradiation. Physiol Mol Bio Plant 11:303–311
Puckett KJ, Nieboer E, Flora WP, Richardson DHS (1973) Sulpher dioxide: its effect on photosynthetic 14C fixation in lichen and suggested mechanism of phytotoxicity. New Phytol 72:141–154
Quesada A, Goff L, Karentz D (1998) Effects of natural UV radiation on Antarctic cyanobacterial mats. Polar Biol 11:98–111
Riley, PA (1997) Melanin: molecules in focus. Int. J. Biochem Cell Biol. 29(11):1235–1239
Robinson SA, Lovelock CE, Wasley J (2006) Climate change manipulations show Antarctic flora is more strongly affected by high nutrients than water. Global Change Biol 12:1800–1812
Robinson SA, Turnbull JD, Lovelock CE (2005) Impact of changes in natural ultraviolet radiation on pigment composition, physiological and morphological characteristics of the Antarctic moss, *Grimmia antarctici*. Global Change Biol 11:476–489
Robinson SA, Wasley J, Tobin AK (2003) Living on the edge plants and global change in continental and maritime Antarctica. Global Change Biol 9:1681–1717
Robson TM, Panocotto VA, Flint SD, Ballare CL, Sala OE, Scopel AL, Cadwell MM (2003) Six years of solar UV-B manipulations affect the growth of *Sphagnum* and vascular plants in a Tierra del Fuego peatland. New Phytol 160:379–389
Rosa LH, Vaz ABM, Caligiorne RB, Campolina S, Rosa CA (2009) Endophytic fungi associated with the Antarctic grass *Deschampsia antarctica* Desv. (Poaceae). Polar Biol 32:161–167
Rozema J, Bjorn L, Bornman J et al (2002) The role of UV-B radiation in aquatic and terrestrial ecosystems- an experimental and functional analysis of the evolution of UV absorbing compounds. J Photochem Photobiol B 66:2–12

Ruhland CT, Day TA (2000) Effect of ultraviolet-B radiation on leaf elongation, production and phenylpropanoid concentrations of *Deschampsia antarctica* and *Colobanthus quitensis* in Antarctica. Physiol Plant 109:244–251

Ruhland CT, Xiong FS, Clark WD, Day TA (2005) The influence of ultraviolet-B radiation on growth hydroxycinnamic acids and flavonoids of *D. antarctica* during spring time ozone depletion in Antarctica. Photochem Photobiol 81:1086–1093

Ryan KG, McMinn A, Mitchell KA, Trenerry L (2002) Mycosporine-like amino acids in Antarctica sea algae, and their response to UVB radiation. Z. Naturforsch 57c:471–477

Searles PS, Flint SD, Caldwell MM (2001) A meta-analysis of plant field studies simulating stratospheric ozone depletion. Oecol 127:1–10

Searles PS, Flint SD, Diaz SB et al (2002) Plant response to solar ultraviolet-B radiation in a southern South American *Sphagnum* peatland. J Ecol 90:704–713

Selkirk PM, Seppelt RD (1987) Species distribution within a moss bed in Greate Antarctica. Symposia Biol Hung 35:279–284

Seppelt RD, Green TGA, Schwarz AMJ, Frost A (1992) Extreme southern locations for moss sporophytes in Antarctica. Antarct Sci 4:37–39

Singh J, Dubey AK, Singh RP (2011) Antarctic terrestrial ecosystem and role of pigments in enhanced UV-B radiations. Rev Environ Sci Biotechnol 10:63–67

Singh J, Singh RP, Dubey A (2012) Effects of ultraviolet-B (UV-B) radiation on two cryptogamic plants pigments growing at higher altitudes of the central Himalayan region, India. Afr J Environ Sci Technol 6(1):9–16

Singh J, Singh RP, Khare R (2018) Influences of climate change on Antarctic Flora. Polar Sci 18:94–101

Singh J, Singh RP (2014) Adverse effects of UV-B radiation on plants growing at schirmacher oasis, East Antarctica. Toxicol Int 21:101–106

Singh J, Upreti DK, Bajpai R, Singh RP, Dubey A (2008) Effect of altitudinal changes in photosynthetic pigment concentration in some cryptogam. J Ecophysiol Occup Health 8:107–110

Singh SM, Singh PN, Singh SK, Sharma PK (2014) Pigment, fatty acid and extracellular enzyme analysis of a fungal strain *Thelebolus microsporus* from Larsemann Hills, Antarctica. Polar Rec 50:31–36

Sinha RP, Kumar A, Tyagi MB, Hader DP (2005) Ultraviolet-B induced destruction of phycobiliproteins in cyanobacteria. Physiol Mol Biol Plant 11(2):313–319

Smirnoff N, Cumbes QJ (1989) Hydroxyl radical scavenging activities of compatible solutes. Phytochemistry 28:1057–1060

Smith RIL (1984). In: Laws RM (ed) Terrestrial plant biology of the Sub-antarctic and Antarctic. Antarct Ecol, vol 1. Academic Press, London, pp 61–162

Sonesson M, Callaghan TV, Carlsson BA (1996) Effects of enhanced ultraviolet radiation and carbon dioxide concentration on the moss *Hylocomium splendens*. Global Change Biol 2:67–73

Strzalka K, Szymanska R, Suwalsky M (2011) Prenyllipids and pigments content in selected Antarctic lichens and mosses. J Chil Chem Soc 56(3):808–811

Teramura AH, Sullivan JH (1994) Effect of UV-B radiation on photosynthesis and terrestrial plants' growth. Photosynth Res 39:463–47

Terauds A, Lee JR (2016) Antarctic biogeography revisited: updating the Antarctic conservation biogeographic regions. Diversity Distrib 22:836–840. https://doi.org/10.1111/ddi.12453

Terauds A, Chown SL, Morgan F et al (2012) Conservation biogeography of the Antarctic. Diversity Distrib 18:726–741

Trochine A, Turchetti B, Vaz ABM, Brandao L, Rosa LH, Buzzini P, Rosa C, Libkind D (2017) Description of *Dioszegia patagonica* sp. Nov., a novel carotenogenic yeast isolated from cold environments. Int J Syst Evol Microbiol 67:4332–4339

Van der Watt LM (2020) Antarctica. https://www.britannica.com/place/Antarctica/Plant-life

Vasileva-Tonkova E, Romanovskaya V, Gladka G, Gouliamova D, Tomova I, Stoilova-Disheva M, Tashyrev O (2014) Ecophysiological properties of cultivable heterotrophic bacteria and yeasts dominating in phytocenoses of Galindez Island, Maritime Antarctica. World J Microbiol Biotechnol 30:1387–1398

Vaz ABM, Rosa LH, Vieira ML, Garcia V, Brandao LR, Teixeira LCRS, Moline M, Libkind D, Van BMG, Rosa CA (2011) The diversity, extracellular enzymatic activities and photoprotective compounds of yeasts isolated in Antarctica. Braz J Microbiol 43:937–947

Vendruscolo F, Pitol LO, Carciofi BAM, Moritz DE, Laurindo JB, Schmidell W, Ninow JL (2010) Construction and application of a vane system a rotational rheometer for determination of the rheological properties of Monascusruber CCT 3802. J Biorheol 24:29–35

Villarreal P, Carrasco M, Barahona S, Alcaíno J, Cifuentes V, Baeza M (2016) Tolerance to ultraviolet radiation of psychrotolerant yeasts and analysis of their carotenoid, mycosporine, and ergosterol content. Curr Microbiol 72:94–101

Wise KAJ, Gressitt JL (1965) Far southern animals and plants. Nature 207:101–102

Wynn-Williams DD, Edwards HGM, Newton EM, Holder JM (2002) Pigmentation as a survival strategy for ancient and modern photosynthetic microbes under high ultraviolet stress planetary surfaces. Int J Astrobiol 1:39–49

Wynn-Williams DD, Newton EM, Edwards HGM (2001) The role of habitat structure for biomolecule integrity and microbial survival under extreme environmental stress in Antarctica (and Mars?): ecology and technology. European Space Agency, (Special Publication) ESA SP, pp 225–237

Xiong FS, Day TA (2001) Effect of solar ultraviolet-B radiation during springtime ozone depletion on photosynthesis and biomass production of Antarctic vascular plants. Plant Physiol 125:738–751

Zhou ZY, Liu JK (2010) Pigments of fungi (macromycetes). Nat Prod Rep 27:1531

Zidarova R, Pouneva I (2006) Physiological and biochemical characterisation of Antarctic Isolate Choricystis minor during oxidative stress at different temperatures and light intensities. Gen Appl Plant Physiol, Special Issue:109–115

Geoscience Studies in Antarctica by CSIR-National Geophysical Research Institute, Hyderabad

Joshi K. Catherine, Ch Mohana Lakshmi, Amit Kumar, Saroj K. Mondal, V Rajesheswar Rao, and Vineet K. Gahalaut

Abstract Antarctica, the southern polar icy continent with its exceptional geodynamic scenery, has been a part of India's scientific research since 1981. Currently, the activities are being facilitated by the National Centre for Polar and Ocean Research (N.C.P.O.R.), Goa, through scientific expeditions to Antarctica. Since its inception, CSIR-NGRI has been participating in these expeditions and has established seismological, G.P.S. observatories for monitoring seismicity and to understand the tectonics of Antarctica plate and has carried out some geological and geophysical studies, too. We summarise some of the essential contributions of CSIR-NGRI.

Keywords Geodynamics · Seismology · GPS · Plate tectonics · Gravity measurements

1 Introduction

Antarctica, which is the fifth-largest of the seven continents on Earth, was part of the supercontinent Gondwana more than 170 million years ago. Over time Gondwana broke apart, and Antarctica formed around 35 million years ago. The Antarctic plate has a unique geodynamic setting since it is almost entirely (87%) surrounded by the mid-oceanic ridges. These divergent or conservative margins are formed due to the Antarctic plate's interaction with the South America plate, Africa plate, Australia plate, Pacific plate and Nazca plate. Only a tiny part of it abuts the subduction zone formed due to the South America plate, Scotia plate and Antarctic plate. Seismicity in the Antarctica plate interiors is generally low. Most of the seismicity is concentrated on the plate boundary. A few earthquakes occur in the volcanic region in western Antarctica and the peninsular region (Reading 2007). The continental interior shows suppression of crustal failure due to ice loading causing low seismicity. Ice quakes occur in the continental shelf region due to ice shelf break off. The Antarctic plate comprising the southern polar continent is considered to be stable and seismically

J. K. Catherine · C. Mohana Lakshmi · A. Kumar · S. K. Mondal · V. R. Rao · V. K. Gahalaut (✉)
CSIR-National Geophysical Research Institute, Uppal Raod, Hyderabad 500007, India

N. Khare (ed.), *Assessing the Antarctic Environment from a Climate Change Perspective*, Earth and Environmental Sciences Library,
https://doi.org/10.1007/978-3-030-87078-2_16

quiet. The Antarctica plate's geology varies from fossiliferous sedimentary rocks, lava and deep magmatic rocks, to a wide range of metamorphic rocks, and active volcanoes and glacial deposits.

India started scientific expeditions in Antarctica in 1981 through the Department of Ocean Development, Government of India and now Ministry of Earth Sciences. The National Centre for Polar And Ocean Research (N.C.P.O.R.), Goa, organises scientific expeditions. Since its first expedition, the CSIR-National Geophysical Research Institute (NGRI), Hyderabad, has been participating in these expeditions quite regularly. CSIR-NGRI has taken up several geological, geophysical, seismological and geodetic studies. CSIR-NGRI has also established a permanent seismological and geodetic observatory at Maitri. Due to logistic hardship, most of the time has been spent in establishing the infrastructure at Maitri. We attempt to summarise the outcome of these studies briefly.

2 Geological Studies

The east–west trending Schirmacher Oasis in the Central Droning Maud Land of East Antarctica mainly comprises of garnet-biotite gneiss, pyroxene granulite, calc-gneiss, khondalite along with migmatite and augen gneiss (Singh 1986; Sengupta 1986, 1988). During the 13th Indian Antarctic scientific expedition, several lamprophyre dykes were noticed. During the 17th expedition, a basaltic dyke, characterised by several quench morphologies of olivine textures, was reported from Schirmacher Oasis. The occurrences of these dykes, their petrography and geochemistry were recorded. The poly-metamorphosed rocks of Schirmacher Oasis are intruded by dolerite, basalt, pegmatite, lamprophyre dykes and an aplite dyke (Bormann et al. 1986). The available age data (K–Ar ages) on a few of the basalt dykes of Schirmacher Oasis range from 290 m.y. to 302 m.y. (Kaiser and Wand 1985). Wand et al. (1991) have suggested that the lamprophyre dykes are younger than the associated pegmatite dykes and older than the basalt dykes.

Lamprophyre dykes vary in length from less than 10 m to about 500 m and in width from a few centimetres to about one meter (Jafri 1997). Lamprophyre dykes, which occur in the central part of the Oasis, are dark brown, lustrous rocks with dark brown coloured mica (shining porphyries), which show chilled margins and are characterised by the occurrence of "ocelli" ranging from a few mm to one cm in diameter. The basaltic dyke is about 1 m across and is traceable over a length of about 100 m. It is fine-grained, aphyric and is characterised by chilled margins (Jafri 2000). The mineral compositions were detemined by Electron Probe Micro Analyser (EPMA) on polished and carbon-coated thin sections, using Henoc and Maurice's (1978) correction procedure (1978). Major and minor elements were analysed by using an X-ray Fluorescence Spectrophotometer. Trace and rare earth elements (REE) were determined using an I.C.P.- M.S. (Balaram et al. 1992). Representative rock sample analysis of lamprophyre dykes for major and minor elements from the central and the eastern parts of the Schirmacher Oasis suggested that these ultra-potassic dykes have been derived from partial melting thickened lithosphere (Jafri 1997). Its elemental composition indicates that it is an alkalic basalt and that the source magma was fractionated and contaminated with crustal rocks during its ascent through the continental crust (Jafri 2000). This dyke is also characterised by the occurrence of quench olivine textures that were formed by the rapid cooling of the magma at 15–40° c/h and are similar to those reported from submarine basalts lunar basalts and experimentally produced quench olivine textures. Further, the delicate nature of the quench olivine, which is mainly confined to the glassy margins of the dyke, suggests in situ crystallisation.

3 Geophysical Studies

3.1 Gravity Measurements

Gravity measurements in Antarctica started more than 65 years back during the Indian Scientific Expedition to Antarctic (I.S.E.A.), using a relative gravity meter tied to Goa's reference gravity station. These measurements with an accuracy of few mGal could neither be merged with another data set nor could be used for any temporal studies. The Antarctic landmass has a few absolute reference gravity stations but none near the Indian station, Maitri. Therefore During the 23rd Indian Antarctic expedition (the year 2004), the first absolute gravity (A.G.) measurements with an accuracy of a few microGal, were carried out by the CSIR-National Geophysical Research Institute (NGRI) in collaboration with the Department of Ocean Development (DoD) by installing FG-5 Absolute Gravimeter (A.G.) at Maitri, Indian station in Antarctica with two primary objectives of (i) establishing reference gravity station for future gravimetric surveys, (ii) for studying temporal gravity changes related to Antarctic ice mass loss. A concrete platform for A.G. measurements was constructed, and the instruments were installed on 28 January 2004. Measurements were taken from 4 to 27 February 2004 (except on bad weather days), connecting earlier gravity observations with A.G. reference point. The absolute gravity at the reference pillar was estimated as 982,578,797 ± 4 micro Gal, Fig. 1, (Tiwari et al. 2006) from the

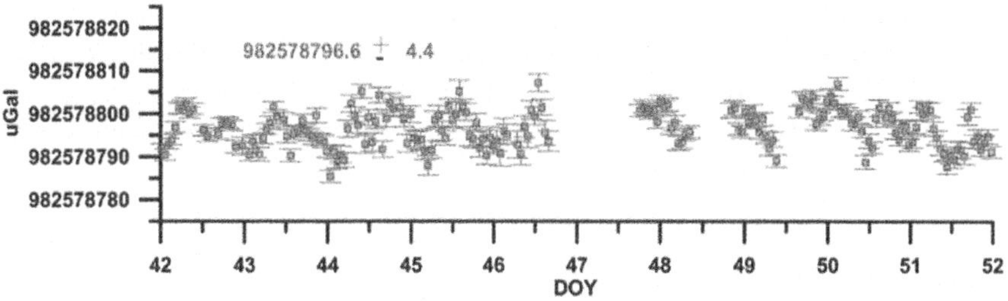

Fig. 1 The estimated absolute gravity at the reference pillar (982,578,797 ± 4 micro Gal)

recorded set of observations. This reference gravity station at Maitri could be used for any future gravity surveys in Antarctica for geodynamic studies and for monitoring temporal gravity changes at the exact location from time to time, which provides constraints on the elastic rebound due to deglaciation. Besides, a magnetic traverse was made, cutting all the exposed litho units of the Schirmacher area.

3.2 Magnetotelluric Studies

The first Magnetotelluric data by CSIR-NGRI were acquired during the 24th Indian Scientific Expedition to Antarctica (2004) to delineate the deep electrical conductivity structure in the north-central part of the Dronning Maud Land in East Antarctica, along an east–west transect in the Schirmacher Oasis region. A total of nine tensor soundings in the period range 0.001–1000 s were taken in a profile of length 16 km, along with the nearly east–west oriented Schirmacher Oasis. M.T. data have been recorded at only one site on ice cover, situated 4 km south of Schirmacher Oasis, for experimentation. Use of titanium electrodes along with bentonite and salt has helped in achieving a good contact resistance of 300 Ω (100–500 Ω) with the ice cover (Murthy et al. 2013). M.T. data were collected for 3–4 days at each site with a station spacing of ~2–3 km. Despite the odd challenges due to Antarctica's logistical requirements, good quality M.T. data could be acquired (Murthy et al., 2012). The most striking feature is the variation of apparent resistivity in transverse electric (T.E.) and transverse magnetic (T.M.) modes, which are the two principal impedances parallel and perpendicular to the two-dimensional structure. These results suggest a highly resistive layer (at short periods) followed by a conductive layer at more extended periods. It is evident from the magnetotelluric response computed from three-dimensional forward modelling at six different locations in Schirmacher Oasis that the apparent resistivities and the phase data do not show much variation due to the coast effect. These studies revealed a well-defined layer with an average thickness of 15 km, a highly resistive (8000–10,000-Ω m) upper crust, and a less resistive (500–600 Ω m) lower crust all along with the profile.

Thus, the lower resistivity of the lower crustal layer observed along the M.T. traverse is consistent with a cratonic nature and solid-phase conductors (Wannamaker et al. 1996). This layer's thickness varies over the profile's length, with 20 km on the eastern side of Maitri and a thinning (ca. 10 km) towards the western side. There is reasonably good fit between the observed and computed data for this profile in the form of apparent resistivity and phase pseudo-sections. The crustal resistivity section beneath Schirmacher Oasis is interpreted to represent that of a craton (Murthy et al. 2013). Due to accessibility problems, M.T. data could not be collected in a north–south oriented profile (i.e., in the strike perpendicular direction). To better understand the upper mantle's resistivity structure, M.T. sites over a more extended transect are required.

4 Seismological Studies

The permanent Seismological Observatory at Maitri—the Indian base station in Central Dronning Maud Land (D.M.L.), East Antarctica was established by CSIR-National Geophysical Research Institute (CSIR-NGRI) with the support from National Centre for Antarctic and Ocean Research (N.C.A.O.R.), with analogue and short period digital systems became operational during the 17th Indian Scientific Expedition to Antarctica (I.S.E.A.) on 26 January 1998. It was established to monitor the seismicity in and around Antarctica, crustal structure, earthquake source mechanism. This observatory was later upgraded with the installation of a Broadband seismic system during the 20th I.S.E.A. The temperature is maintained between 20 and 30 degrees Celsius in the recording room and in the vault where the seismometer is kept for better performance. This seismological station was operational continuously up to November 2009. Later on due to the system's technical issues, the station was not in operation. Cracks developed on the pier in a seismic vault that was constructed in the year 1997. The seismic vault's reconstruction works were carried out in the 33rd Indian Scientific Expedition to Antarctica (I.S.E.A.) during 2013–2014 (Fig. 2). The newly constructed seismic vault is suitably covered by adiabatic walls and is temperature-controlled (thermostat) electronically to maintain a constant temperature inside the vault at $15° \pm 0.5$ °C.

Other than earthquake monitoring, the Maitri observatory waveform data have been used to infer crustal structure beneath the Maitri seismic station using receiver function analysis (Gupta et al. 2017). The analysis and modelling of the teleseismic receiver functions calculated at the seismic station and the Vs structure suggest that the uppermost 2.5 km ice and sediments (Vs = 1.5–2.0 km/s) are underlain by 12.5 km of extrusive igneous rock (rhyolite, Vs = 2.25–2.6 km/s). Between 16 and 28 km depth, Vs gradually increases from 2.9 to 3.4 km/s. In the lower crust, a ca. 7 km thick layer of Vs = 3.9 km/s is followed by a 6 km thick underplated

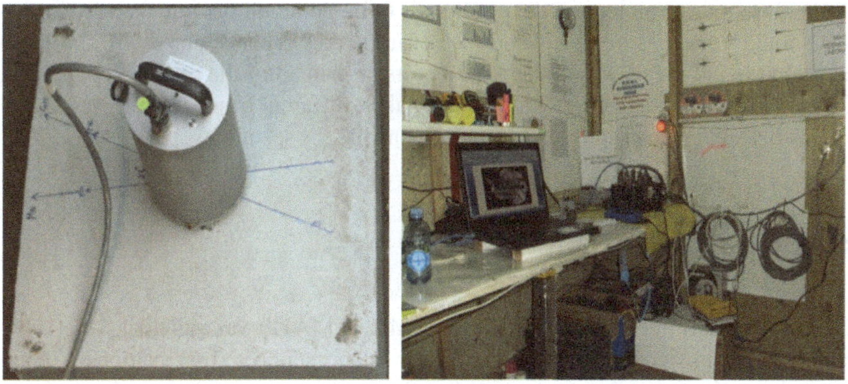

Fig. 2 Seismological Observatory at Maitri. The left panel shows the sensor in the vault, while the right panel shows the recorder

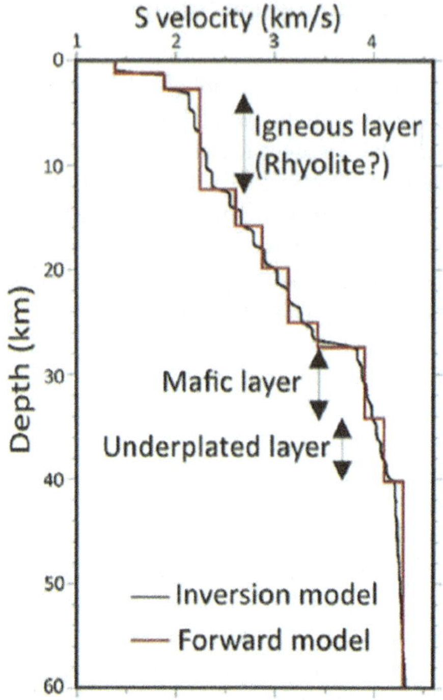

Fig. 3 Crustal shear velocity structure (red line) beneath the Maitri station in central D.M.L., East Antarctica through the receiver. A black line shows the final inversion model after Gupta et al. (2017)

layer (Vs = 4.1 km/s) at the crust-mantle boundary. The Moho is at ca. 40 km, and the uppermost mantle's Vs is ca. 4.3 km/s (Fig. 3). The presence of underplated material in the lowermost crust, extrusive volcanic rocks (Rhyolite) in the upper crust, seaward dipping reflectors in the surrounding area and the general paucity of seismicity suggest that the crust beneath the Maitri station is a volcanic passive continental margin and not a typical cratonic continental crust.

Analysis of seismological data indicates frequent small earthquakes in Antarctica's interior, suggesting the occurrence of icequakes. The icequakes are similar to earthquakes but occur within the ice sheet itself instead of the crust below the ice. Careful analysis of data for picking up icequakes, different from tectonic events, will give better understanding of processes responsible for seismicity.

5 Geodetic Studies

The continuously operating G.P.S. station at Maitri (MAIT) was established by CSIR-National Geophysical Research Institute with the support from National Centre for Polar and Ocean Research (N.C.P.O.R.) in the year 1997 to understand the Antarctic plate motion and its internal deformation. Observations from the MAIT GPS site until 2002 consisted of only a few epochs without any change in receiver and antenna pair. After the year 2002, a frequent change in the instrument set up, led to shifts in coordinates' time series. Further, the monument appears to have become unstable after 2002. Hence in February 2013, during the 33rd Indian Scientific Expedition to Antarctic, a new GPS site was established, and it was integrated with a met sensor (Fig. 3).

After establishing a new and stable GPS site at MAIT in 2013, GPS observations for more than 50 locations from 2008 to 2017 were processed in daily session using GAMIT/GLOBK. The estimated plate velocity for the Indian base station Maitri is now better constrained at 6 mm/yr predominantly towards the north. It is consistent with the plate motion estimated by the global plate models. The Euler pole evaluated for the Antarctic plate is Latitude = 58.6985 ± 0.41182, Longitude = –130.1819 ± 0.41613, Angular Velocity = 0.21534 ± 0.0021293°/Ma (Ghavri et al. 2017). Corresponding to the estimated pole, the residual motion for most of the sites lie within 2 mm/yr, which show that the Antarctic plate is quite a rigid plate. The slight relative movement of 0.9 ± 0.17 mm/yr towards N60° across Trans-Antarctic Mountains separating east and west Antarctica suggests deformation across it but needs longer time series to ascertain this. The sites exhibiting substantial seasonal variations have motion of more than 2 mm/yr. The significant seasonal variation in the horizontal components compared to the tectonic deformation at the sites close to the pole indicate that this variation is nontectonic. The variation in Antarctic plate motion from ~4 to 20 mm/year and its direction is quite spectacular (Fig. 4). Such a significant variation in the velocity across this primary plate is primarily because it is surrounded by the divergent plate margins, namely, the mid-oceanic ridges. There is no significant convergent plate margin for this plate.

Analysis of the spatial and temporal variation in the cyclic and negative trend in the GPS vertical displacement reveals the associated changes in Ice sheet melting and consequent mass loss using advanced methods like Singular Spectrum Analysis (S.S.A.), Wavelet spectrum and Multi Taper Method (M.T.M.) spectrum. The annual Seasonality rate at some of the sites shows an increase from January-May followed by a decrease from June-December, consistent with the region's summer and winter seasons. The annual mode is undergoing modulation by the decadal-scale climatic cycles, with the periodic variations in the vertical displacement being non-stationary. The results also demonstrate that the yearly mode variability is increasing from east to west Antarctica (Fig. 5), which may be related to the tectonic stability of eastern Antarctica, and the role of mass variation and tectonic dynamics needs to be investigated further.

6 Future Work

To better understand the crustal dynamics in and around the Antarctica plate and to monitor the spatial and temporal distribution of seismicity, long-term operation of the seismological observatory and permanent GPS station at Antarctica is required.

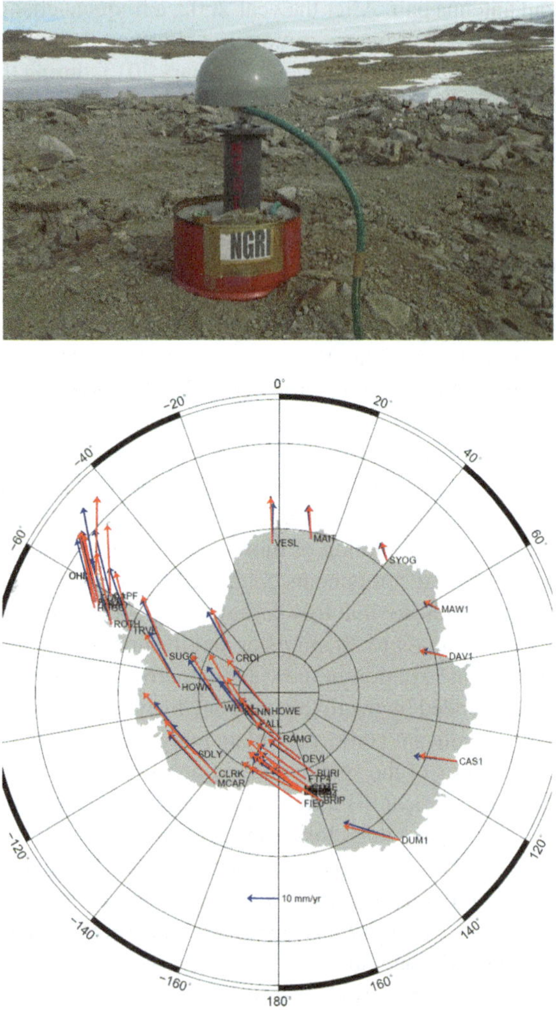

Fig. 4 The top panel shows the permanent G.P.S. antenna installed at Maitri. The lower panel shows the site velocity (observed) estimated from GPS measurements at Maitri and other Antarctica sites. Red vectors are the predicted site motion at these sites from the estimated pole location

CSIR-NGRI has regularly participated in the Indian Scientific Expedition to Antarctica to monitor these observatories' health and collect data. GPS data will be utilised to understand various seasonal, tectonic, hydrological and climatic processes in the vicinity of Indian station, Maitri, Antarctica. Our stations can contribute to understand the anatomy of earthquake slip for significant earthquake sources and pick up ice quakes characteristic of the ice continent to better understand the processes responsible for seismicity in Antarctica.

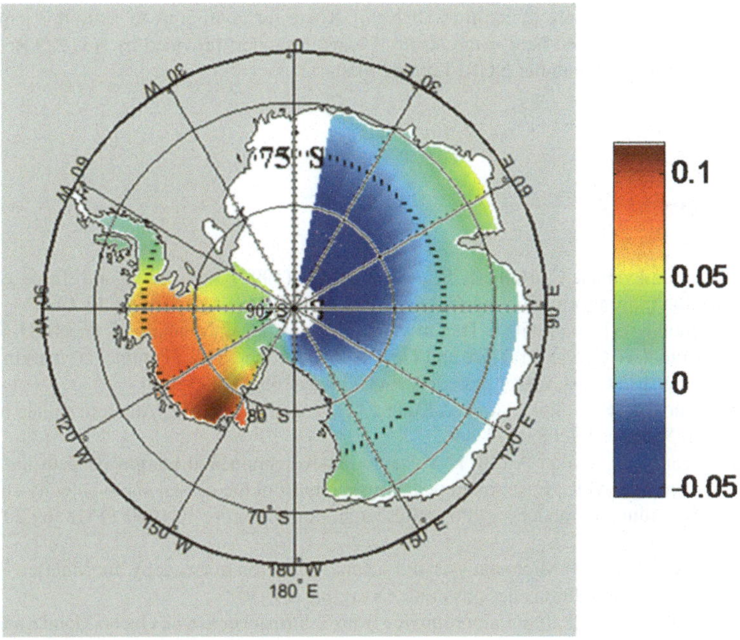

Fig. 5 Spectral power of annual mode variability of S.S.A. reconstructed G.P.S. vertical displacement estimated using M.T.M. spectrum

7 Conclusions

CSIR-NGRI has been partaking year after year in Antarctica's scientific expeditions for carrying out geological and geophysical research. CSIR-NGRI has established permanent seismological and GPS observatories for earthquake monitoring and understanding Antarctic plate motion. Modelling of teleseismic RFs calculated at the seismic station suggests that the Maitri station's crust is a volcanic passive continental margin and not a typical cratonic crust.

The Indian permanent GPS station at Maitri moves at a 6 mm/yr velocity predominantly towards the north, consistent with the Antarctic plate's expected plate motion at this site. The Euler pole is estimated for the Antarctic plate. The plate is largely aseismic and there is no significant internal deformation. Singular spectrum analysis and wavelet spectrum of GPS displacement vertical component demonstrate that annual mode variability is increasing from east to West Antarctica. The observed positive gradient of the vertical displacement probably indicates mass loss due to the melting of ice. The analysis further reveals that the annual mode is undergoing modulation by the decadal-scale climatic cycles.

Acknowledgements We are thankful to Dr Neloy Khare for inviting us to write this paper. All the research work presented here is the result of logistic support provided by N.C.P.O.R. This is CSIR-NGRI publication number NGRI/Lib/2020/Pub-52.

References

Balaram V, Manikyamba C, Ramesh SL, Anjaiah KV (1992) Rare Earth and Trace element determination in iron formation reference samples by ICP-MS. At Spectrosc 13:19–25

Bormann P, Bankwitz P, Bankwitz E, Damm V, Hurtig E, Kampf H, Menning M, Paech HJ, Schafer U, Staekebrandt W (1986) Structure and Development of the passive continental margin across the Princess Astrid Coast, East Antarctica, Jour. Geodyn 6:347–373

Ghavri S, Catherine JK, Ambikapathy A, Kumar A, Gahalaut VK (2017) Antarctica plate motion. Proc Indian Natl Acad Sci 83:437–440

Gupta S, Nagaraju K, Akilan A (2017) Volcanic passive continental margin beneath the Maitri station in central D.M.L., East Antarctica: constraints from the crustal shear velocity structure from receiver function modelling. Polar Research. https://doi.org/10.1080/17518369.2017.1332947

Henoc J, Maurice K (1978) Microanalysis and scanning electron microscopy. In: Maurice F, Mery L, Tixler R (eds) Les editions des physique, Orsay, pp 281–307

Jafri, S H., (1997). Calc-alkaline Lamprophyres from Schirmacher Oasis, Queen Maud Land, East Antarctica, Thirteenth Indian Expedition to Antarctica, Scientific Report Technical Publication No 11, pp 227–236, Department of Ocean Development, New Delhi

Jafri SH (2000) Quench textures in Basaltic dykes from Schirmacher Oasis, Queen Maud Land, East Antarctica, Seventeenth Indian Expedition to Antarctica, Scientific Report Technical Publication No 15, pp 129–144, Department of Ocean Development, New Delhi

Kaiser G, Wand U (1985) K-Ar dating of basalt dykes in the Schirmacher Oasis area, Drowning Maud Land, East Antarctica, Zeit. Geol Wits Berlin 13:299–307

Murthy YVVBSN (2001) Geophysical Activities at Maitri, Antarctica, Technical Publication No: 17, 19th I.A.E.

Murthy DN, Veeraswamy K, Harinarayana T, Singh UK, Santhosh M (2013) The electrical structure beneath Schirmacher Oasis, East Antarctica: a magnetotelluric study. Polar Res 32:17309. https://doi.org/10.3402/polar.v32i0.17309

Reading AM (2007) The Antarctic plate's seismicity. InStein S and Mazzotti S Ed, Continental IntraplateEarthquakes: Science, Hazard, and Policy Issues GeologicalSociety of America Special Paper 425:285–298. https://doi.org/10.1130/2007.2425(18)

Sengupta S (1986) Geology of Schirmacher Range (Dakshin Gangotri), East Antarctica. Third Indian Expedition to Antarctica, Scientific Report, Technical Publication No. 3, Department of Ocean Development, New Delhi

Sengupta S (1988) History of successive deformation in relation to metamorphism-migmatitic events in the Schirmacher Hills, Queen Maud Land, East Antarctica. J Geol Soc India 32:295–319

Singh RK (1986) Geology of Dakshin Gangotri hill range, Antarctica. Third IndianExpedition to Antarctica, Scientific Report, Technical Publication No. 3, Department of Ocean Development, New Delhi

Tiwari VM, Singh B, Vyaghreswara Rao MBS, Mishra DC, Absolute gravity measurements in India and Antarctica. Current Sci 91(5), 686–689

Wand U, Geisler M, Korich D (1991) Petrography and Geochemistry of lamprophyres from Schirmacher Oasis. East Antarctica. Z. Geol. Wiss, Berlin 19(2):199

Wannamaker PE, Stodt JA, Olsen SL (1996) Dormant state of rifting in central west Antarctica implied by magnetotelluric profiling. Geophys Res Lett 23:2983–2987

Seismo-Geophysical Studies in the Antarctic Region: Geodynamical Implications

O. P. Mishra

Abstract Conduction of integrated seismo-geophysical studies in the Antarctic region is a challenge as well as very much warranted to explore the region for its better geo-scientific understanding. Seismogenesis and seismic potential of the Antarctic region have not yet been well understood because of lack of common consensus on various issues, besides its unique and complex geotectonic settings associated with intriguing landscape evolution of the Antarctic plate since the breakup of Gondwana, West Antarctic Rift System (WARS), different patterns of exhumation events that occurred between the Early Cretaceous and Cenozoic. The hostile climatic situation and inaccessibility of the region due to the huge spatial distribution of thicker ice sheets hindered the mission of conducting comprehensive seismo-geophysical studies for the Antarctic Peninsula due to severe constraints of installations of ground-based sophisticated seismo-geophysical equipments in the region. Several causative factors associated with natural and anthropogenic are found still enigmatic in the sense to unravel the fact how the genesis of earthquakes are related to the glacial-dynamics and glacial mass change-induced earthquakes (GMCIE). It has become important to decipher the role and contribution of the East and the West Antarctic microplates and West Antarctic rift systems (WARS) in seismogenesis using advanced methodologies of geosciences. Seismicity of the Antarctic continent region is confined to different tectonic blocks, distributed into the southern ocean, continental margin, Lutzow-Holm Bay, Antarctic Peninsula, and in the volcanic regions in and around Deception Island, which helped estimate the seismic structure of Antarctica. In this chapter, a comprehensive overview of seismo-geophysical studies has been made to understand seismo-geodynamical implications for the Antarctic region in light of the Plate Reconstruction and seismo-geophysical structures of Antarctica.

Keywords The Antarctic · Seismogenesis · Seismic potential · GMCIE · WARS · Glacial dynamics · Plate reconstruction · Geodynamics · Seismo-geophysical structures

---

O. P. Mishra (✉)
National Centre for Seismology, Ministry of Earth Sciences, New Delhi, India
e-mail: omp.mishra@nic.in

N. Khare (ed.), *Assessing the Antarctic Environment from a Climate Change Perspective*, Earth and Environmental Sciences Library,
https://doi.org/10.1007/978-3-030-87078-2_17

1 Introduction

Antarctica is regarded as the hotspot for conducting multi-disciplinary scientific research since the advent of modern technology because of its unique geographical location and intricate tectonic structures in the South Pole region. The 1959 Antarctic Treaty under the Antarctic treaty system (ATS) that entered into force on 23 June 1961 defines Antarctica as all lands and ice shelves south of 60°S latitude, which advocates that the continent is an international scientific laboratory for peaceful purposes and allows freedom of scientific investigations as well as dissemination of scientific data through the exchange among global communities by prohibiting any attempts of countries in conducting nuclear explosions or the disposal of radioactive waste there in the Antarctic.

It is observed that the landscape of the Antarctic region begins since the breakup of the Gondwana that was controlled by the tectonic evolution of the Antarctic Plate when the Weddell Sea formed by rotation and translation of the West Antarctic Microplates (Fig. 1). It is reported that the rifting of the continents away from Antarctica, commencing with Africa and further movement in a clockwise direction around Antarctica compelled the Antarctic Plate to get isolated from the other surrounding plates and shows the present polar position of Antarctica. During the processes, development of seaways and the circum-Antarctic current, thermal isolation, climate change, and the present-day cold polar environment occurred (Lawver et al. 1992; Fitzgerald 2002). Fitzgerald (2002) reported that the landscape evolution of the Antarctic continent since the breakup of Gondwana has been controlled in the first order by changes in the location and configuration of plates. During the breakup, Antarctica moved towards its present polar location and then became geographically and thermally isolated that contributed to the development of the present-day hyper-arid, cold polar climate. Three stages of exhumation occurred between the Early Cretaceous to Early Cenozoic contributed to regional tectonic events, comprising: (a) the initial breakup between Australia and Antarctica in the Early Cretaceous; (b) the main phase of extension between East and West Antarctica in the Late Cretaceous accommodated on low-angle extensional faults; (c) propagation of southward of seafloor spreading from the Adare Trough into continental crust underlying the western Ross Sea in the early Cenozoic, which might have acted as a trigger for the flexural uplift of the East Antarctic lithosphere to form the Transantarctic Mountains (TAM). These geotectonic processes of three-stage exhumation may have a strong bearing on the seismogenesis and seismic potential of the Antarctic region.

Lawver et al. (1998) demonstrated a model with the fit of the large continents within Gondwana in which three continents, namely Australia, India, and Africa are found to fit well against the present-day rifted margins of Antarctica as shown in Fig. 1. It is further argued that the Gondwana reconstruction is less well constrained along the Trans2002antarctic margin of East Antarctica where there is considerable uncertainty over the number and position of some of the microplates within the West Antarctic region (Daziel and Elliot 1982; Storey 1996; Storey et al. 1998; Fitzgerald). The main West Antarctic microplates contain some pertinent tectonic

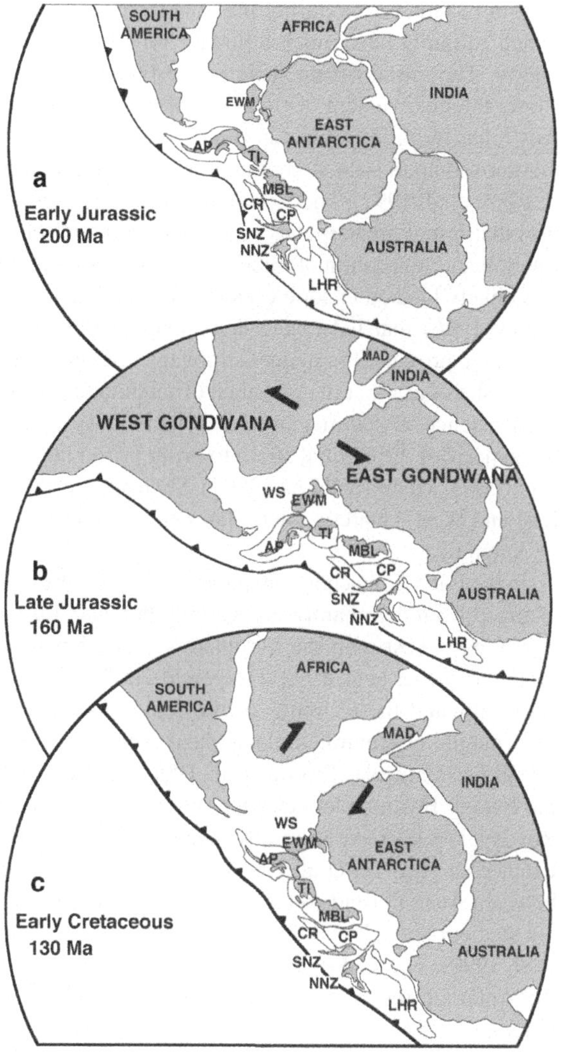

Fig. 1 Map showing the Gondwana tight fit reconstruction and breakup model assimilated by using the information on continent and microplate positions along with other inputs (Lawver et al. 1992, 1998; Storey 1996). Abbreviations, AP: Antarctic Peninsula; TI: Thurston Island; MBL: Marie Byrd Land; CR: Chatham Rise; CP: Campbell Plateau; SNZ: Southern Newzealand; NNZ: Northern New Zealand; LHR: Lord Howe Rise; WS: Weddell Sea (Adapted from Fitzgerald 2002)

units, such as the Ellsworth-Whitmore Mountains crustal block (EWM), Antarctic Peninsula, Thurston Island, and Marie Byrd Land. Several researchers used the Haag Nunatak, Berkner, and Filchner microplates in reconstructions whilst Marie Byrd Land can be thought of as comprising both west and east portion of the Antarctic (De Wit et al. 1988; Divenere et al. 1996; Fitzgerald 2002). These observations are supported by independent geophysical studies made using Paleomagnetic data with integration of structural modeling based on geological information to constrain the fit of these microplates. It is also found that Antarctic Peninsula, Thurston Island, and Marie Byrd Land contain a large proportion of subduction-related rocks, these blocks are retained in approximately their present location along the paleo-Pacific margin of Gondwana during the majority of reconstructions (Lawver et al. 1992), which certainly have much influence on nature and extent of seismogenesis in the region. The initial rifting stage in the breakup of Gondwana as shown in Fig. 1b shows the involvement of right-lateral transmission as East Gondwana (Antarctica, Australia, India, New Zealand) and West Gondwana (South America and Africa) that moved apart with stretching beginning in the north and propagating southward (Lawver et al. 1992). There is a common consensus among different group of scientists that initial breakup was also associated with plume-generated magmatism and rotation and translation of microplates, showing Plume-related magmatism having a head likely located in southeastern Africa (Cox 1988; White and MacKenzie 1989; Storey 1996) that led to the emplacement of huge within-plate mafic and felsic magmatic provinces in many Gondwana continents as well as Antarctica. It is well-established facts that the presence of magmatic, mafic, and felsic materials, besides the existence of sedimentary basin and rift system make the region more apt for the genesis of differential strain to bring brittle failure into earthquakes (Mishra and Zhao 2003; Mishra and Zhao 2004). Such appreciable compositional scenarios in the East and West Antarctic microplates can have contributed to more seismic potential of Antarctica.

In the Early Cretaceous, as shown in Fig. 1c, the Gondwana breakup stress regime changed from dominantly north–south between East and West Gondwana to dominantly east–west with the two-plate system that was replaced by a multiple plate system (Lawver et al. 1992; Fitzgerald 2002). This change in stress regime dramatically affected the tectonics of the region as evident from the large-scale ductile deformation concentrated along shear zones in gneissic and magmatic rocks in the Antarctic Peninsula (Storey et al. 1996), and existing sedimentary basins such as the Latady Basin underwent thin-skinned deformation and inversion (Kellogg and Rowley 1989). Separation also began between India and Antarctica in the Early Cretaceous (Lawver et al. 1991). It has also been noted that initial stretching began between Australia and Antarctica in the Early Cretaceous (Stagg and Willcox 1992), but seafloor spreading did not begin until 95 Ma (Candle and Mutter 1982; Veevers et al. 1990). These observations suggest the tectonic intricacies and complexities of the Antarctic region be studies in detail in light of its seismogenesis and seismic potential.

It is estimated that more than 98% of its total area comprising approximately 5.5 million square miles is blanketed by the Antarctic Ice sheet while the vividly exposed parts of the contents correspond to the coastal regions of the Antarctic Peninsula, the high ranges of the coastal hinterland of South Victoria Land, the isolated mountains and nunataks of the interior which pierce the ice carapace, and the marginal regions of East Antarctica (Adie 1962; Fitzgerald). The geological exploration has been carried out in West Antarctica since before and after the International Geophysical year of 1957–1958. Those studies, however, found purely of reconnaissance category, and the mission of exploring Antarctica has been extended to address several unknown tectonic intricacies of the interior of East and West Antarctica, which yielded significant information for the furtherance of extensive seismo-geophysical studies to explore the Antarctica region using the state-of-the-art techniques by several groups of geo-scientists to address a diverse set of geological problems that remained unexplored and completely unknown to us. Some of the pertinent problems like stratigraphical, structural, seismotectonical, and geodynamical remain to be resolved by generating high-quality data from the field on the deployment of advanced technology by rendering plausible explanation and interpretation to those unresolved problems. None-the-less, the state of the Antarctic geology is not static rather very progressive and advancing continuously and provides enough opportunity to enunciate the reasonable facts and figures that can usher ample avenues to adopt new ideas to conduct theme-based research for sake of outstanding outcomes through fascinating researches for the Antarctica continent of the Earth.

As mentioned above, the Antarctic plate is having unique tectonic settings associated with diverse sets of structural regimes, surrounded by divergent margins, convergent or transformed margins in which divergent margins are well characterized by the circum Antarctic seismic zones and covers nearly 92% of the surroundings of the Arctic plate whilst the convergent margins locate in the northwestern part of the South Shetland Islands and are less than 2% of the plate boundary with transformed margins cover less than 7% along the boundary of the Scotia micro-Plate (Kanao 2014) as shown in Fig. 2.

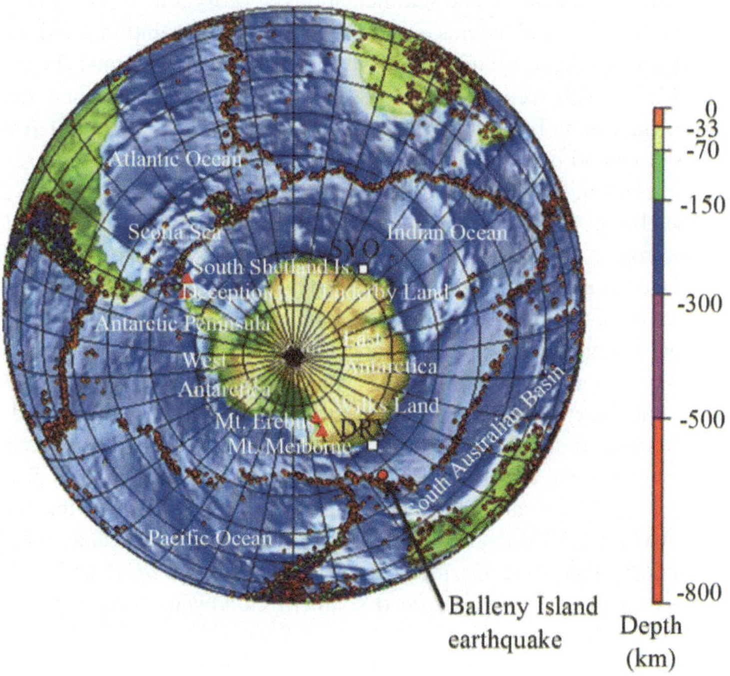

Fig. 2 Seismotectonic settings of the Antarctic along with the distribution of earthquakes occurred at different depths represented by solid circles in different colors (*source* NEIC of USGS). Volcanoes are denoted by solid red triangles (adapted from Kanao 2014)

2 Geo-Environmental Setup and Tectonics of Antarctica

The unique geology and seismotectonic settings of Antarctica make it possible for a diverse group of researchers to research diverse topics to understand implications of seismo-geophysical studies, modeling of geodynamical and mantle thermal states, crustal and sub-crustal seismogenesis, and seismic potential of the region. Applications of other geophysical tools to explore natural resources underneath Antarctica as well as the study on various indicators of climate dynamics for the South pole having bearing on the rest of the continental regions have become one of the leading-edge research topics in recent years.

Rock exposures in Antarctica are very limited in extent, occurring mainly as isolated mountain ranges, nunataks, ice-free fringing coastal areas, and offshore islands, but the geological record is found to be very much informative and most revealing about the composition, and structural evolutionary history of Antarctica (Adie 1962). Based on stratigraphical, petrological, structural, and paleogeographical studies, Antarctica comprises two important geological provinces: (1) Andean Province or West Antarctica; (2) Gondwana Province or East Antarctica as shown in Fig. 3.

The continent of Antarctica found suffered the vicissitudes of mountain building processes, severe denudation, and sedimentations in geosynclinals troughs. There is evidence of subsidence and incursion of the sea, large-scale crumpling, and fracturing of its sedimentary layers. Inundation of vast areas by lava flows and the spread of volcanic ashes were also reported in the past geological mapping of Antarctica (Adie 1962). Flora and fauna-based interpretation revealed that the climate of Antarctica has witnessed several reversals in the geological past.

In Permo-Carboniferous times glacial conditions prevailed, but in the Middle Jurassic, the climate record corresponds to tropical/temperate to sub-tropical conditions, supporting a luxuriant flora because at that time climate got deteriorated to final refrigeration in the Pleistocene to recent. The present glaciation has also evolved through several phases of advancement and retreat of glaciers in Antarctica. The main continental shield of Antarctica is analogous to those of the other southern continents associated with rocks 'which have undergone at least three phases of regional metamorphism, it is now clear that the continent is composed of two distinct and contrasting geological provinces though they evolved separately under different environmental and tectonic conditions since early Paleozoic. Geological studies of Antarctica revealed that East Antarctica, the older part of the continent, is the true Antarctic continental shield whilst the rest relatively undisturbed associated with early Paleozoic marine sediments and late Paleozoic-early Mesozoic terrestrial sediments of the widespread southern Gondwana System (Adie 1962; Fitzgerald 2002). The subsequent geological succession is found heavily intruded by thick, often differentiated, dolerite sheets of the Lower Jurassic age. Interestingly, the main tectonic disturbances in East Antarctica appear to be block faulting of a late Tertiary-Quaternary age with which wide-scale volcanism was reported to occur.

Fig. 3 A Map showing the two geological sub-division of Antarctica into the Andean Province and the Gondwana Province (adapted from Adie 1962)

Stratigraphic comparisons with other southern continents depict the similarity of East Antarctica with Southern Africa as shown in flowing Tables 1 and 2 based on six selected areas of Antarctica as shown in Fig. 4.

West Antarctica, including the Antarctic Peninsula, is found stratigraphically, structurally, and tectonically very akin to the west Patagonian Cordillera, which is a southern extension of the Andean Mountain chain through the Scotia Arc. In comparison with East Antarctica, it is much younger than having evolved mainly as a result of searching of the Andean geosynclinal sediments in the late Paleozoic. A huge and widespread lava and ash eruptions in the Upper Jurassic, marginal to the Andean geosyncline were associated with a period of minor folding. Further sedimentation and the formation of the mid- to late Cretaceous Magellan geosyncline and the large-scale intrusions of late Cretaceous-early occurred. It has been observed that sedimentation got abruptly terminated by mid-Miocene volcanism which has continued intermittently until Recent times. Block faulting and folding have played an important role in the formation of West Antarctica.

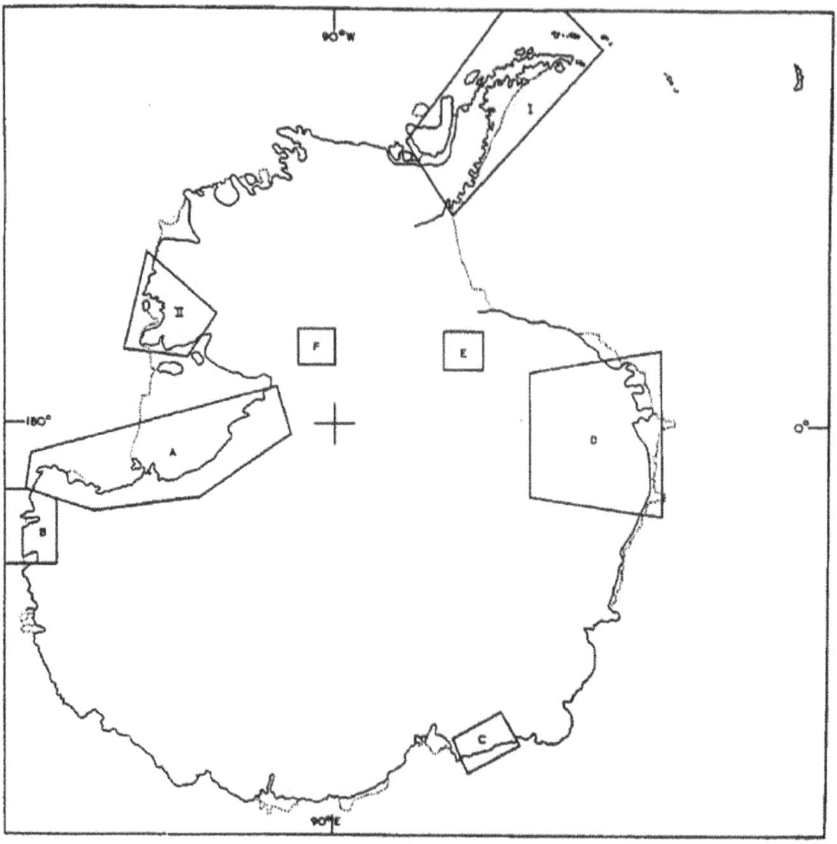

Fig. 4 A simple sketch map of Antarctica showing sic types of areas of East Antarctica with abbreviations, A: South Victoria Land; B: Oates Coast; C: MacRobertson Coast; D: Dronning Maud Land; E: Theron Mountains, WhichawayNunataks, and Shackleton Range; F: Central Horlick Mountains; West *Ant*arctica: I: Antarctic Peninsula; II: Edsel Ford Ranges and King Edward VII Peninsula (after Adie 1962)

It is found that Antarctica has two geographical regions as shown in Fig. 3. The more the geology of Antarctica is studied the more it becomes apparent that Antarctica is composed of two geological provinces, each of which has evolved independently but in adjacent positions in similar climatic environments. The geological history of the old stable platform of East Antarctica is closely similar to that of the other southern continents (Tables 1 and 2) where the great Gondwana System dominates the stratigraphy and for this reason the name 'Gondwana Province' (Fig. 3). West Antarctica is essentially considerably younger than East Antarctica but it has a more complex tectonic. There remain multitudes of problems related to the geology of the Antarctica region that needs to be solved and many thousands of square miles

Table 1 Stratigraphical comparison of South Africa, the Falkland Islands, Brazil, Uruguay, and Argentina

	South Africa	Falkland Islands (after *Adie* [1952])	Brazil, Uruguay, and Argentina
(?) Rhaetic to L. Jurassic	Karroo Dolerites Drakensberg Volcanics	Dolerite dykes	Serra Geral Eruptives Botucatú Sandstone
U. Triassic	Cave Sandstone Red Beds Molteno Beds U. Beaufort Series	West Lafonian Beds	
L-M. Triassic	M. Beaufort Series L. Beaufort Series	Bay of Harbours Beds	Rio do Rasto Beds Upper Estrada Nova Beds
U. Permian	U. Ecca Series	Choiseul Sound and Brenton Loch Beds	(?)
L. Permian	M. Ecca Series (*coal measures*)	Lafonian Sandstone	Lower Estrada Nova Beds
Lowest Permian	L. Ecca Series U. Dwyka Shales	Black Rock Slates	Iratý Shales Rio Bonito Beds
Permo-Carboniferous U. Carboniferous	Dwyka Tillite (*glacial*) L. Dwyka Shales	Lafonian Tillite (*glacial*) Bluff Cove (Fitzroy Basin) Beds	Itararé Beds (*glacial*)
L. Carboniferous M. Devonian	Witteberg Series U. Bokkeveld Series (terrestrial sediments)	Port Stanley Beds Port Philomel Beds (terrestrial sediments)	Barreiro sandstone (Brazil)
L. Devonian	L. Bokkeveld Series Table Mountain Series (marine sediments)	Fox Bay Beds Port Stephens Beds (marine sediments)	Ponta Grossa (Brazil) Rincon de Alonsa (Uruguay) Furnas Sandstone (Brazil) Carmen Sandstone (Uruguay) Sierra de la Ventana Quartzites (Argentina)
Precambrian	Basement Complex	Cape Meredith Complex	Basement Complex

of ice and snow to be covered in the search for new information that will eventually lead to the complete solution of Antarctic geology.

The geodynamical model of the West Antarctic Rift System is found to be very important for understanding the mantle thermal state (Harry et al. 2018). They assimilated two-dimensional finite element models with the extension of the West Antarctic Rift System (WARS), which exhibits three classes of behavior, which are dependent upon the pre-rift thermal state of the upper mantle. All of the models begin with relatively cool East Antarctica lithosphere juxtaposed against warmer West Antarctica. The models all undergo an initial period of extension that is broadly distributed across the WARS. Harry et al. (2018) enunciated that the West Antarctic Rift System (WARS) is a 750–1000-km-wide continental extensional province lying beneath the Ross Sea and Ross Ice Shelf between Marie Byrd Land on the east and the Transantarctic Mountains (TAM) at the edge of the East Antarctic craton on the west (Fig. 5). The timing and distribution of extension in the WARS are not tightly constrained, particularly beneath the West Antarctic Ice Sheet. However, it is clear that widespread extension began by Late Cretaceous time and continued at least into the Pleistocene Epoch. From Late Paleozoic time to ca. 105 Ma, the region now occupied by the WARS lay on the overriding plate of a convergent margin between East Gondwana and the Phoenix Plate (Lawver et al. 1992; Lawver and Gahagan 1994) as shown in Fig. 6a.

Table 2 Stratigraphical Comparison of six selected type areas (Fig. 4) in East Antarctica

	A, South Victoria Land (after *Harrington* [1958]; *Plumstead* [1960])	B, Oates Coast (after *Soloviev* [1960])	C, MacRobertson Coast (after *McLeod* [1960])
Quaternary	Moraines, raised beaches, etc.	(*Glaciation*)	(*Glaciation*)
Tertiary	McMurdo Volcanics		
L. Jurassic	Ferrar Dolerites		
Rhaetic			
Triassic U. Permian L. Permian	Beacon System (terrestrial sediments)	Beacon System (terrestrial sediments)	Sediments with coal seams (terrestrial sediments) (*coal measures*)
Permo-Carboniferous			
L. Devonian	Sediments with plant remains (terrestrial sediments?)		
U. Silurian		Schist-greywacke Series (405×10^6 yr)	
U. Cambrian		Berg Series (500×10^6 yr)	
L. Cambrian	*Archaeocyathus* Limestones (known only from erratics)		
Lower Paleozoic	Admiralty Intrusions		
Precambrian	Ross System (Moubray Group)	Basement Complex	Basement Complex (2 episodes of metamorphism)

(continued)

Table 2 (continued)

D, Western Dronning Maud Land (after *Roots* [1953]; *Reece* [1958])	E, Theron Mountains, Whichaway Nunataks, and Shackleton Range (after *Stephenson* [1960])	F, Buckeye Range, Central Horlick Mountains (after *Long* [1962])
(*Glaciation*)	(*Glaciation*)	(*Glaciation*)
Basalt dykes {Andesitic volcanics (pillow lavas) Gabbroic and dioritic sills and dykes}	Dolerite sills and dykes	Ferrar Dolerites (diabase)
∿∿∿∿∿∿	∿∿∿∿∿∿	∿∿∿∿∿∿
Beacon System {Clastic sediments (terrestrial sediments) ? Tillite (*glacial*)}	Beacon System {Sediments with plant remains (*coal measures*) and coal seams (terrestrial sediments) Sedimentary group (terrestrial sediments)}	Beacon System {Mount Glossopteris Formation (terrestrial sediments) (*coal measures*) (arkoses, sandstones and shales with plants and woods) Discovery Ridge Formation (carbonaceous shale, cone-in-cone limestone, platy shale) ∿∿∿∿∿∿ Buckeye Tillite (*glacial*) (with intercalated sandstones)} ∿∿∿∿∿∿ Horlick Formation (marine sediments) (sandstone-shale with marine fossils)
	∿∿∿∿∿∿	
	Slate-quartzite Group	
	∿∿∿∿∿∿	
	Archaeocyathus limestones (known only from erratics)	
	∿∿∿∿∿∿	∿∿∿∿∿∿
Metamorphic complex	Basement Complex	Basement (porphyritic biotite-quartz-monzonite)

3 Seismic Potential of the Antarctic and Adjoining Regions

Rigorous compilation of seismicity of the Antarctic and the surrounding region by the International Seismological Centre (ISC) resulted in very significant data as shown in Figure 2. Earlier studies highlighted that the Antarctic continent and surrounding

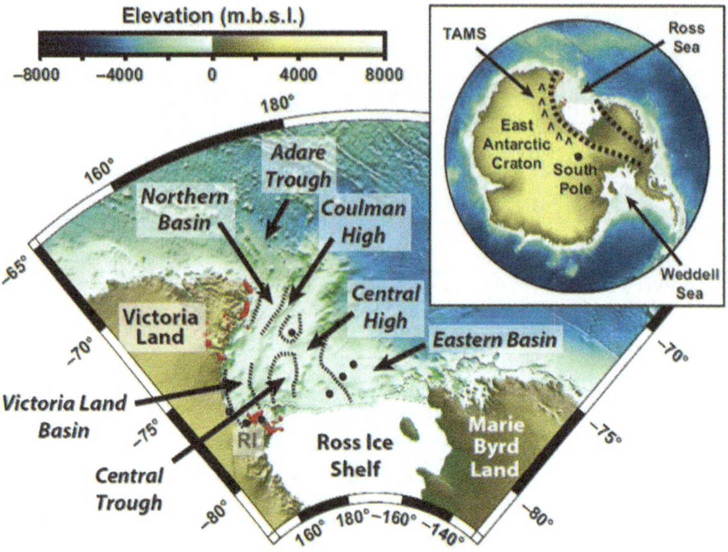

Fig. 5 Location map and major tectonic features of the Ross Sea and West Antarctic Rift System (WARS) (adapted from Harry et al. 2018). The Location of the Transantarctic Mountains (TAMS) and boundaries of the WARS (dashed lines) are shown in the inset. Dotted line—Late Cretaceous through Cenozoic basins and uplifts (Davey et al. 2006); black dots—drill holes penetrating Late Cretaceous through Cenozoic stratigraphy; red shading—Cenozoic alkaline magmatic rocks; RI—Ross Island. Relief from ETOPO2 global relief data set (National Geodetic Data Center 2006; Harry et al. 2018)

ocean region are one of the aseismic regions of the Earth for many decades. On installations of the Global Seismic Networks and local seismic arrays opened avenues for conducting an extensive seismic study that dispelled the doubt about the region not being aseismic. Several tectonic earthquakes were detected in and around the Antarctic continent, and earthquake recording was found to record several micro-earthquakes. Kanao (2014) showed that the entire Antarctic region has a total of 13 seismicity areas classified into the Antarctic continent (3 areas) and oceanic regions within the Antarctic Plate (10 areas). Studies demonstrated that seismic activity in the continental areas is associated with very low activity in Antarctica. Several smaller earthquakes are found to occur beneath Wilkes Land in East Antarctica through the area is a tectonically very active area in the continent. In the oceanic region, in contrast, seismic activity in the area between 120°W and 60° W sector is found much higher than those of other oceanic areas. Most interestingly, the tectonic stress concentration toward the Easter Island Triple Junction between the Antarctic Plate, the Pacific Plate, and the Nazuca micro-Plate showed appreciable seismicity. It is important to note that three volcanic areas, such as the Deception Island, the Mts. Erebus, and Melbourn found associated with very high seismic activities in comparison to the vicinity of the surrounding areas.

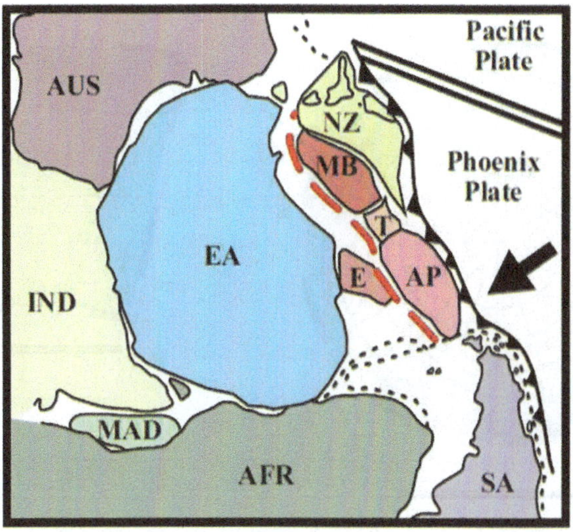

Fig. 6a Late Cretaceous reconstruction of West Antarctic Gondwana margin ca. 110 Ma showing the early collision of Phoenix/Pacific spreading system with New Zealand (adapted from Harry et al. 2018). The Red dashed line shows the trend of incipient West Antarctic Rift. SA—South America; AFR—Africa; IND—India; AUS—Australia; EA—East Antarctica; NZ—New Zealand. Major tectonic blocks comprising West Antarctica are AP—Antarctic Peninsula; E—Ellsworth Mountains; T—Thurston Island block; MAD—Madagascar; MB—Marie Byrd Land. Modified after Fitzgerald (2002) and Torsvik et al. (2008), Harry et al. (2018)

3.1 Seismic Network and Seismic Noise

The history of the Antarctic seismic monitoring starts from the deployment of six stations across West Antarctica between 1998 and 2001 under the first temporary broadband seismic network in Antarctica under the project of Antarctic Network of Unattended Broadband Seismometers (ANUBIS) [Anandakrishnan et al., 2000]. The Global Seismographic Station Network (GSSN), the Worldwide Standardized Seismograph Network (WWSSN) as well as the Integrated Research Institutions of Seismology (IRIS) have provided synoptic coverage since the mid-1960 (Butler et al., Hansen et al. 2015). Despite being Antarctica demonstrated a large gap in the distribution of seismic stations where till mid-1990 only eight permanent GSN-type stations were installed around the periphery of the continent. The mission of expanding the ANUBIS allows for the enhancement of equipment made of Gurlap CMG-3T and 24-bit REF TEK 72A-08 dataloggers by 46 in number (Pyle et al. 2010) as shown in Fig. 6b

The quality of data recording in Antarctica is dictated by the advancement in data recording and technological development related to array/network setup with large data storage as witnessed by steady developments for GAMSEIS array, containing 30-station network work since the year 2007 in the extremely hostile climatic condition in the East Antarctic plateau as shown in Fig. 6b. Different types of instruments

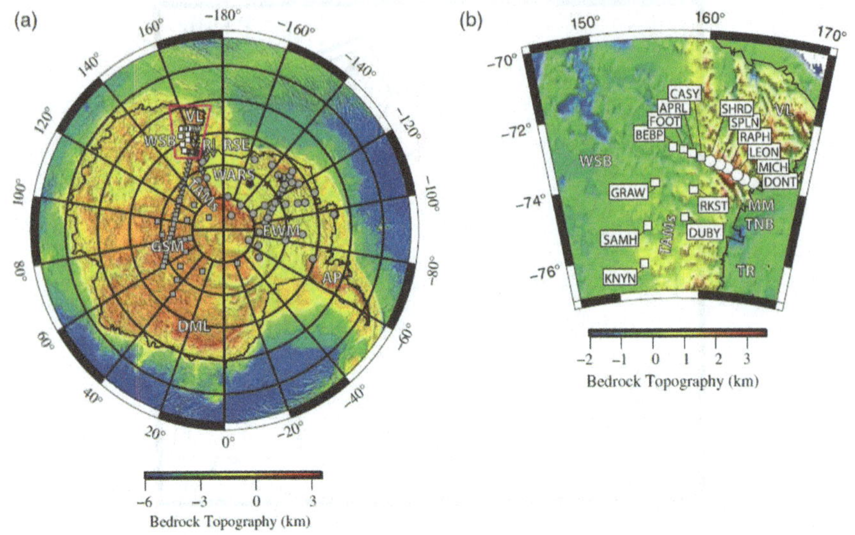

Fig. 6b **a** Subglacial bedrock topography from the BEDMAP2 model (Fretwell et al. 2013). Shapes indicate stations from various seismic deployments across the continent (black stars, Antarctic Network of Unattended Broadband Seismometers [ANUBIS]; gray triangles, Transantarctic Mountains Seismic Experiment [TAMSEIS]; gray squares, Gamburtsev Antarctic Mountains Seismic Experiment [GAMSEIS]; gray circles, Polar Earth Observing Network [POLENET]; white circles and squares, Transantarctic Mountains Northern Network [TAMNNET]). The polygon highlights the region shown in (**b**). Key geographic features are labeled: TAMs, Transantarctic Mountains; WSB, Wilkes subglacial basin; VL, Victoria Land; RI, Ross Island; RSE, Ross Sea embayment; WARS, West Antarctic rift system; GSM, Gamburtsev Subglacial Mountains; DML, Dronning Maud Land; AP, Antarctic Peninsula; and EWM, Ellsworth–Whitmore Mountains. **b** Regional map focused on the TAMNNET deployment, with station names indicated. Additional geographic features not labeled in (**a**) include TR, Terror Rift; TNB, Terra Nova Bay; and MM, Mt. Melbourne. Stations denoted by circles are those powered with absorbed glass mat (AGM) batteries and two lampshade-style solar panel systems, whereas stations denoted by squares have one lampshade-style solar panel, one AGM battery, and 15 lithium battery packs (after Hansen et al. 2015)

as mentioned above along with Quanterra Q 330 digital acquisition systems (DASs) with solid-state memory (Heeszel et al. 2013). Because of hostile and extremely low-temperature, special design has been adopted for installing seismo-geodetic equipment in which the seismometers/sensors were kept inside the insulated piers buried about 1m below the snow surface, and insulated boxes were used to house the electronics and batteries in line of earlier deployments of solar panels/wind generators were used to augment power and to recharge the lead-acid batteries during the austral summer (Johns et al. 2006; Hansen et al. 2014, 2015). The greatest challenge is to maintain power supply during the Antarctic winter, which is generally achieved by setting a system that automatically gets switched to non-rechargeable lithium battery packs with the provision of a heating pad for heating the system in such a way to maintain the temperatures inside the insulated box at 20 °C or greater than the ambient temperature, which gave enough scope of recording seismological data in Antarctica for round the year with impressive recovery of data by 95% (Heeszel et al. 2013; Hansen et al. 2015). To improve the installations of seismographs in Antarctica, snow vaults were also applied when instruments were installed on isolated rock outcrops throughout West Antarctica and within the Transantarctic mountains (Anthony et al. 2015). It is also documented that relatively warmer environment in West Antarctica in comparison to that of East Antarctica, the power consumption for running the equipment is comparatively as less as 1.5W, which are operational with lead-acid batteries and solar panel systems. Recent advancement as mentioned above is the availability of greater data storage (16–28 GB) with iridium telemetry to monitor station performance (Parker et al. 2008; Hansen et al. 2014, 2015). All of these developments in seismological networking have contributed to the successful deployment of several sets of temporary broadband arrays in the Antarctic region for a longer period to record huge data by different networks ascribed to different participating countries to investigate the 3-D seismic structure of deeper layers of Antarctica to understand its implications to crustal and mantle dynamics.

The quality of seismological recorded signals (desirable part of the recording) is of paramount importance because the level of noise (undesirable part of the record) needs to be kept as minimal as possible. Detection and identification of the source of noise is again a very daunting task as it needs filtering at various stages of recording and processing recorded data. Prior knowledge about Seismic noise will help us to optimize our approach towards future seismological deployments in Antarctica associated with very hostile climatic and intricate geological conditions. In addition, microseism arising from ocean wave activity contains useful climate proxy information on the state and variability of the relatively poorly sensed southern oceans (Aster et al. 2010; Stutzmann et al. 2009). It is suggested that microseism is very sensitive to sea ice concentration and areal coverage in the Polar Regions (Grob et al. 2011; Tsai and McNamara 2011; Anthony et al. 2014). Characterization of the seismic noise environment of Antarctica based on documentations of instrument performance, and comparisons of installations of equipment in different geo-environmental conditions with ice vaults/rock sites, such as volcanic and heat flow, ocean interference and wind velocity, glacial movements, and other sub-surface conditions that may have strong bearing to interfere in seismic recoding and can make the entire data spurious in Antarctica. Recent studies based on analyses of a range of data for Antarctica revealed the fact that the seismic noise environment of Antarctica is free from anthropogenic noise contamination (Anthony et al. 2014; Hansen et al. 2015).

Spatial distribution of the seismic noise state at various stations can be shown for Antarctica in Fig. 6c for which the median PSD of each station component has been separated into six-period bands (Anthony et al. 2014). They examined the median power in each station for Antarctica (Anthony et al. 2014). It is documented that the short-period band, 0.15–1.0 s, captures common sources of anthropogenic noise with seismic coupling due to wind (Peterson 1993; Galperin et al. 1986; Anthony et al. 2014). Signals ranging from local glaciological movements to teleseismic earthquakes also contribute to anthropogenic noise. In search of seismic signals, we are generally interested in the 1.0–5.0 s teleseismic body wave from exciting local,

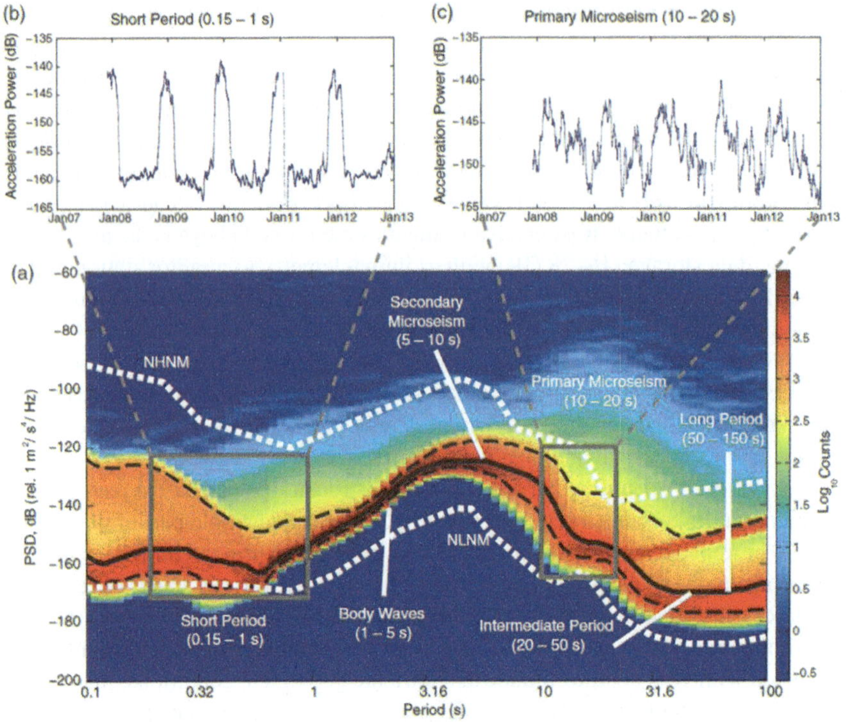

Fig. 6c **a** The probability density function of power spectral density (PSD) for the vertical-component of South Pole station QSPA (146 m borehole) for December 2007–December 2012 plotted on a logarithmic color scale to show transient high-amplitude signals (e.g., teleseismic earthquakes) and other probabilistically secondary features. The median PSD is plotted (solid black line) as well as 5th and 95th percentile statistics (dashed lines) and are compared to the global high- and low-noise models of Peterson (1993). In addition, the six-period bands referred to in the text are labeled. **b** Temporal evolutions in power in the short period and **c** primary microseism band are shown to illustrate the influence of seasonal anthropogenic noise at nearby (7.8 km) Amundsen–Scott South Pole Station and the unique seasonality (phase-shifted ~90° from the rest of the southern hemisphere; Aster et al. 2008) of the Antarctic microseism signal due to the annual growth and decay of sea ice (after Anthony et al. 2014)

regional, and teleseismic earthquake-generated body waves for conducting structural and source-related studies. In addition, several recent studies have attributed noise in this band at near-coastal and near-lake stations to local or regional swell activity (Bromirski et al. 2005; Tasi and McNamara 2011), which constitutes the shorter period portion of the double-frequency (secondary) microseism (Anthony et al. 2014). The secondary (5.0–10.0 s) and the primary (10.0–20.0 s) microseism bands are dominated in the absence of earthquake or other transient source excitation or by ocean-generated Rayleigh waves. Hasselmann (1963) showed that the primary microseism originates when deep-ocean waves break or shoal on a shallow seafloor and are primarily converted into Rayleigh waves. The typically much more powerful secondary microseism is usually generated by standing-wave components of the oceanic wavefield (e.g., coastal reflections, storm, or intrastorm wave interactions as discussed by Ardhuin et al. (2011), and they demonstrated that seafloor forcing at half the period of the constituent traveling ocean waves may the source of the noise (Tanimoto 2007; Anthony et al. 2014). Anthony et al. (2014) showed that variations in microseism power at specific stations in Antarctica are known to be strongly sensitive to both near coastal storms and to wave state and also highlighted that amplitude gets modulated by the annual formation and breakup of sea ice (Aster et al. 2008; Grob et al. 2011). It has been shown that the 20–50 s intermediate period band contains power from the longest period microseisms and is strongly excited by intermediate-period surface waves from global earthquakes whilst the 50–150 s long period band is controlled by low amplitude (e.g., ~300 time smaller in power than double frequency microseism excitation) oceanic excitation of long-period waves generated through infragravity wave excitation and difference-frequency interaction of opposing ocean wave trains (Traer et al. 2012; Anthony et al. 2014). A common source of instrumentally generated noise in this band is diurnal or another seismometer tilting that strongly couples into the horizontal components (Peterson 1993). This period band is also intermittently excited by long-period teleseismic surface waves from large earthquakes.

Seismic noise in Antarctica varies from surrounding formation on which stations are being installed and variability in noise is due to different processes, occurring due to interactions of wind/ocean wave/ storm with earth surfaces and ice sheets. We have to choose the ice borehole seismic station and rock borehole station very carefully. The quieter sensors in Antarctica at high frequencies (>~ 2 Hz) by placing instrument at depth of as more as 200 m that reduces 5 dB baseline natural high-frequency noise and a 10 dB reduction in anthropogenic noise compared to a sensor located within the wave guide, which is the quietest station on the earth. The rock bore hole station in the windy dry valleys at the foot of the seismic stations generally 10 dB quieter than that of purely rock sites because of lower power levels of microseism of 1–2 dB. Once sensor will be deployed typically 1–2 m below the surface and atop of thick ice sheets then 5–7 dB quieter in the short period band than comparable rock and near rock/shallow snow faults. Rock and shallow snow stations showed that long-period horizontal noise is attributable to tilt, and in the near-elimination of shorter period tilt may arise from direct wind forcing on the outcrop and station. Recently, seismic deployments in the remote interior of Antarctica have dramatically increased the quality and quantity of broadband data from large unsampled areas of the continent (Anthony et al. 2014).

It is well documented by Kanao (2014) in Fig. 2 that seismicity gets recorded by a series of seismograph stations installed in the Antarctic region that described seismotectonic settings with reference to its seismic potential. Some of the different participating countries conducted an experimental study to collect seismological using GNSS equipment to estimate velocity structure for better understanding of complex seismological and geodetic research in Antarctica; especially by Bulgaria from its Bulgarian Antarctic base (Dimitrova et al. 2015). The used McNamara method was used to study ambient seismic noise despite the effects of harsh weather conditions and the absence of man-made noise on the distribution of the noise power. Data available from various seismograph stations in a different network, namely, the Global Seismographic Station Network (GSSN), the Worldwide Standardized Seismograph Network (WWSSN) as well as from the Incorporated Research Institutions for Seismology (IRIS) are found not yet adequate for the comprehensive seismological study of Antarctic region. With the advent of technology and advancement in instrumentation, an attempt has been made to study Antarctica using a huge amount of seismological data recorded by the different seismographic network, such as, Transantarctic Mountains Seismic Experiment (TAMSEIS); the Gamburtsev Antarctic Mountains Seismic Experiment (GAMSEIS); and the Polar Earth Observing Network (POLENET) (Hansen 2015) that has significantly improved our understanding of the Antarctic's tectonic evolution even though more number of seismic stations are required for a denser network that may help us to unravel the mystery beneath Antarctica.

3.2 Seismicity Beneath Antarctic Continent

A general seismicity map in and around the Antarctic Plate is shown in Fig. 2. The earthquake locations are compiled by the National Earthquake Information Center (NEIC) of USGS. The Antarctic continent is roughly divided into two large tectonic provinces; East Antarctica (eastern hemisphere part from the Transantarctic Mountains) and the other younger province; West Antarctica. East Antarctica is characterized by a fragment of the Gondwana super-continent, one of the members of adjacent Pre-Cambrian terrains in the southern hemisphere (South America, Africa, India, and Australia; Fig. 2). West Antarctica, on the contrary, is attributed to a chain of many islands beneath the ice sheet of the Cenozoic era, including several active volcanoes. Four earthquakes inside the continental area of Antarctica were reported to the ISC catalog in the early stage of Antarctic scientific research. However, except for only one event out of the four, no significant earthquakes were located in the continent before the IGY, because neither their locations nor their magnitudes were accurately determined (Adams et al. 1985). One event which occurred on June 26, 1968, was located in Coats Land, 20°W, 80°S using the initial phase readings of five seismic stations on the Antarctic continent (Kaminuma and Ishida 1971). The magnitude of the event was 4.3 determined by the waveform amplitude of a three-component seismograph at South Pole Station (SPA). This was the first earthquake located instrumentally in the continent using the data of only seismic stations in Antarctica. However, this event was not listed in the ISC catalog used the data of five stations for hypocenter determination and two of the five stations were Byrd Seismic Array. As the focal depth was determined 1 km beneath the sea level, the event was considered to be an earthquake in the crust, not an ice quake. On the other hand, it has been reported that an earthquake occurred near the coast of Oates Land at 70.5°S and 161.3°E on October 15, 1974 (Adams 1982). This event was the only shock to be located on the Antarctic continent by international agency by that time. The magnitude of the event was estimated to be 5 and he concluded that the event might be an ice quake associated with ice movement or cracking. Identification of very shallow earthquakes near the surface has not been correctly made based on hypocentral information to understand whether they are tectonic events or ice-related signals. Regarding the wide inland area of East Antarctica, seismicity in Wilkes Land ("G" in Figs. 7 and 8) is found to be one order higher than that in the vicinity of Syowa Station (69.0°S, 39.0°E), the Lüzow-Holm Bay (LHB), East Antarctica (Kamimura 2000; Kanao and Kamimura 2006). Seven earthquakes with Mb 4.0–4.9 were located in the coastal area of 100°–170°E and 66°–82°S in Wilkes Land in 1964–1996. Another nine earthquakes were located in the inland area and two were located offshore. The magnitudes of these eleven earthquakes were not accurately determined. The formation, distribution, and stability of these sub-glacial lakes might affect tectonic processes involving relatively high seismicity in this area.

Not only micro-seismic activity but also small earthquake activity in Wilkes Land and surrounding coasts are higher than that in another area on the East Antarctic continent. The sub-glacial topography in Wilkes Land is characterized by a sub-glacial basin with 1000 m below sea level in minimum elevation of the bedrock (Drewry 1983). The maximum thickness of the ice sheet in the area is over 4000 m and the surface elevation of the ice sheet is mostly over 2000 m.

3.3 *Seismicity of Southern Ocean*

Seismic activity inside the Antarctic Plate has been known as very few distributions of their epicenters both in ocean and continent areas. As the spatial distribution and time variations in seismicity around the Antarctic Plate, in particular for the Indian Ocean sector (0°–160°E, 20°–80°S), was previously evaluated and intraplate seismicity was discussed associated with far-field tectonic stress in the oceanic lithosphere (Kanao et al. 2006). We need to understand the detailed distribution of hypocentral parameters within the whole Antarctic Plate in the area of 20°–80°S and every 60° of longitude has been investigated using the compiled data set by ISC for the recorded years during 1964–2002 as shown in Fig. 7. It is worth mentioning that Kanao (2014) demonstrated very intriguing results based on earthquake distribution among the individual 60° longitude sectors and South Pole area in 80°–90°S and he found that the earthquake activities are divided into 13 regions (from "A" to "M"). Since the seismic activities are extremely high along the plate boundaries around the Antarctic Plate between the surrounding plates, a criterion of the area for intra-plate was selected very carefully not to include the events associated with the plate boundaries.

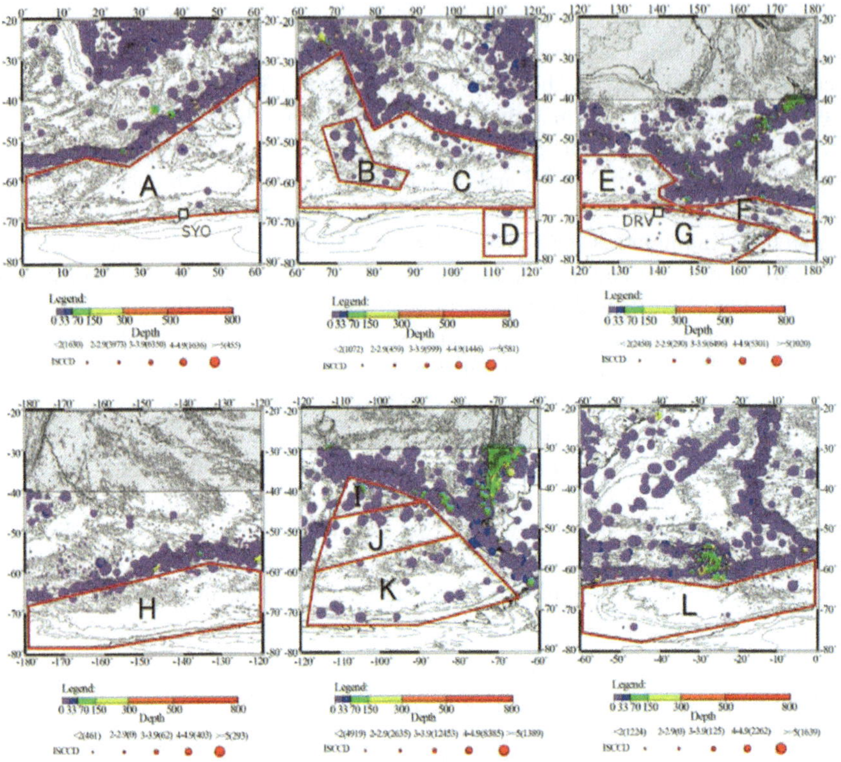

Fig. 7 Earthquake locations are determined by ISC in 20°–80°S and every 60° of longitude. Surrounded areas by solid redlines indicate individual blocks ("A"–"L") discussing seismicity in this paper. These areas were classified into the Antarctic continent ("D" and "G") and oceanic region within Antarctic Plate (other regions). Numerous denoted in the brackets correspond to magnitude ranges representing the number of included events for each 60° longitude sector (after Kanao 2014)

3.4 Seismicity of Continental Margin

Reading (2002) pointed out the considerable number of intra plate earthquakes in the 90°–180°E quadrant and divided the earthquakes into two groups as poorly located earthquakes and well-located ones. Inside the continental area in Antarctica, seismicity is almost very few in the "D", "G", and "M" areas of Fig. 7. However, the Wilkes Land ("M" in Figs. 7 and 8a) had been identified as the most active within the Antarctic continent during the IGY period. Poorly located earthquakes were lined from north to south along the 140°E longitude (Kaminuma 2000). In the earthquake locating area, the Resolution sub-glacial highland, the Adventure sub-glacial trench, and the Belgica sub-glacial highland are existed along the longitudinal direction from east to west (Drewry 1983). There is a possibility that the poorly located earthquakes were ice-quakes because the thick ice sheet and complicated sub-glacial topography must cause ice-shocks. Generally, the edge of the continent, the coast

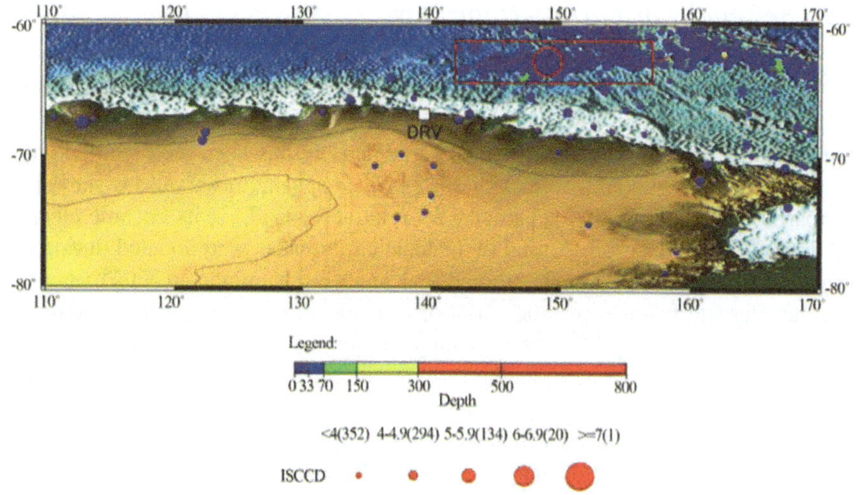

Fig. 8a Earthquake locations in 60°–80°S and 110°–170°E, including both the Wilkes Land and aftershock area (red square) of March 25, 1998, Balleny Island earthquake (Mw = 8.1, red circle). Numerous denoted in the brackets correspond magnitude ranges representing the number of included events (after Kanao 2014)

area, is an aseismic region. Over the past few decades, more seismic observations in the polar region have detected local seismicity by both temporary seismic networks and permanent stations. It is found that the majority of seismicity near the Scott Base (SBA, 167°E, 78°S) and Wright Valley area (VNDA,162°E, 78°S) located along the coast, particularly near large glaciers (Bannister and Kennett, 2002). They suggested a few generation mechanisms for these events, distinguishable by their focal mechanism and depth: basal sliding of the continental ice sheet, movement of ice streams associated with several scales of glaciers, movement of sea-ice, and tectonic earthquakes. The area has been studied by deploying a local seismic network around the Neumayer Station (08°W; 71°S), and determined hypocenters of local tectonic events, located along the coast and the mid area of the surrounding bay (Muller and Eckstaller 2003). A seismic array has been operated for more than one decade at the Neumayer Station. Since the deployment of the seismic network/array, several local events could be detected. Two seismic active regions were figured out at inland area and offshore of the continent. In addition, a broadband seismic network had been developed in the large region between Mawson and Casey stations and inland as far as 75°S by Australia (Reading, 2006). The aim to establish the seismic network is to discover the seismic structure of the continent under Antarctica (SSCUA). Moreover, India has also been carried out seismic observation at the Maitri Station (12°E, 71°S) since 1997. The seismic data have already been contributed to earthquake locations by ISC. India also has published the 'Seismological Bulletin of Maitri Station, Antarctica' every year (Chander et al. 2003).

3.5 *Seismicity of Lützow-Holm Bay*

Once a denser seismic network was established, small/micro earthquake activities became gradually clear. The LHB area around the Japanese Syowa Station, East Antarctica, is one of the areas where seismic activity has been well studied since the 1980s (Kaminuma and Akamatsu 1992). Since seismic observations started by a tripartite network in 1987, seismicity for relatively small events became clear in and around Syowa Station. A total of 18 local earthquakes were located during the 15 years in 1987–2003 (Fig. 8b) studied by Kanao and Kaminuma (2006).

Characteristic features in time variations of seismic activity are summarized as follows; Seismicity in 1987–1989: A three-station seismic network was operated around the Syowa Station. Epicenters of ten local earthquakes were determined during these three years. Many different types of earthquakes, such as a main-shock-aftershock, twin earthquake, earthquake swarms, were detected and identified at that time. The seismic activity during this period was higher than that of the following

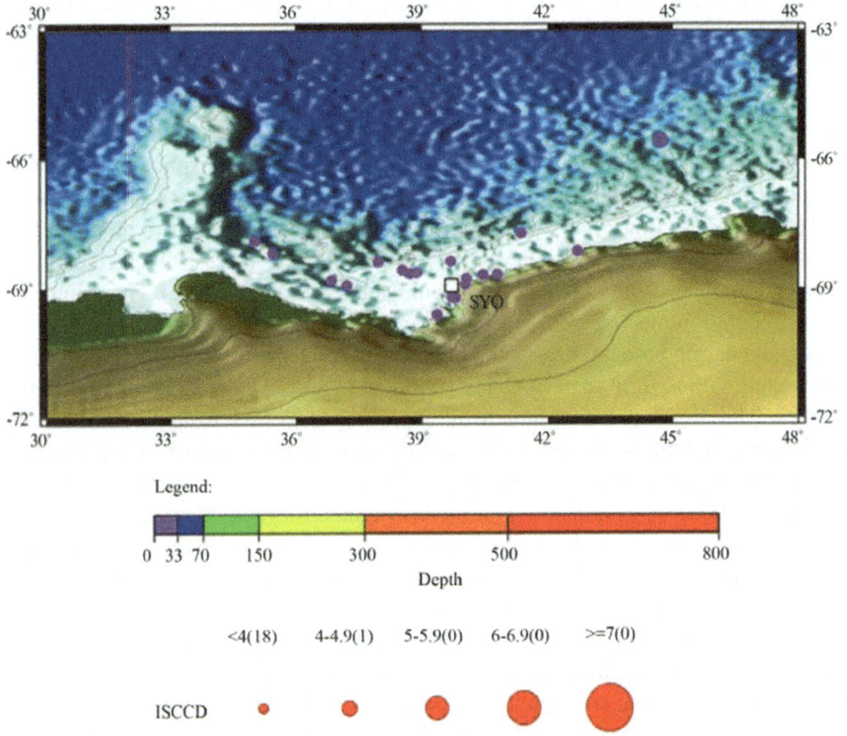

Fig. 8b Earthquake locations in and around the Lutzow-Holm Bay (LHB), East Antarctica detected at Syowa Station (Japan). A total number of 19 events were determined by their hypocenters. Numerous denoted in the brackets correspond to magnitude ranges representing the number of included events (after Kanao 2014)

decade. In 1990–1996, nine local earthquakes were recorded with many different types of events. The seismic activity during this period was very low and the magnitudes of the earthquakes ranged from − 0.5 to 1.4. One local event was detected in 1997, two events in 1998, and one event in 2001 and 2003, respectively. The low seismic activity continues to the present day in 2004. The seventeen events were only detected by the local seismic network deployed around the LHB, except for the September 1996 Mb = 4.6 earthquake in the southern Indian Ocean. Almost all the hypocenters were located along the coast, apart from a few on the northern edge of the continental shelf. Local earthquakes in and around Syowa Station were presumably caused by tectonic stress accumulated with crustal uplift after deglaciation (Kamimura and Kanao 1992). The effect of ice sheet changes may have caused phenomena such as crustal deformation, earthquake occurrence, faulting systems in the shallow part of the lithosphere.

4 Seismological Studies in the Antarctic Region and Geodynamical Implications

Seismological study in the Antarctic region has so far conducted before and after the International Geophysical Year, (IGY-1957–1958) using seismic data recorded by the different global seismographic network as well as in-situ seismographic permanent stations installed in the region by different countries under the International Antarctic treaty produced outstanding results for better understanding of seismological processes, crustal and sub-crustal seismic and thermal structures along with the deep insight into the mantle dynamics of the region.

Several sets of seismo-geophysical investigations have been deployed in the polar region, which includes the areas of both the Antarctic and the Arctic regions at various depth ranges from the surface layers to the deep interiors of the Earth. The polar region has an advantage for investigating deeper layers with the physical condition of the Earth as a "window" viewed from high latitudes (Kanao 2018a). Seismological tools have the advantage to investigate the inner structure and dynamics of the solid Earth by using time-space variations and changes of physical parameters determined from seismological methods (Kanao 2018a, 2018b; Mishra et al. 2021).

Seismic wave propagation is characterized by heterogeneous behavior that propagates through different strata of the Earth with varying speed with different attenuation characteristics that helped many international scientific groups to assimilate structures, and dynamics of the Earth's interiors in both areas of the Arctic and the Antarctic regions using seismic waves and travel-time data: inner structure and dynamics of the crust and mantle in high latitudes, seismicity and focal mechanism, seismic wave propagation within the global point of view. There are several Antarctic permanent stations ascribed to different nations, for example, Maitri and Bharti of India; Soywa of Japan, and several other countries as shown in Fig. 9 The extent of global seismicity has been recorded by these stations in the Polar region.

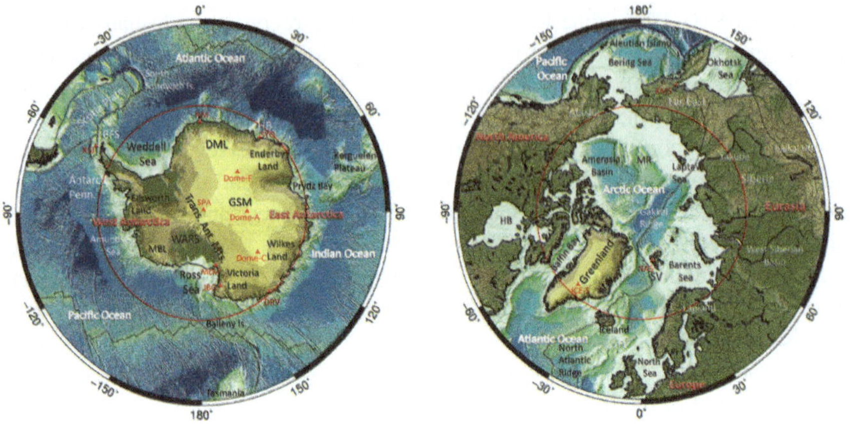

(a) (b)

Fig. 9 Surface topography and bathymetry in the Antarctic (**a**) and the Arctic (**b**) (adapted from Kanao 2018) (ETOPO1) with major geographic location Names of different Antarctic Research Stations. The red solid circle represents the "Antarctic circle" (66.6 °S). Abbreviations are as follows: LHB, Lützow-Holm Bay; DML, Dronning Maud Land; GSM, Gamburtsev Subglacial Mountains; Trans. Ant. Mts., Trans-antarctic Mountains; WARS, West Antarctic Rift System; MBL, Marie Byrd Land; and BFS, Bransfield Strait. Red solid triangles are the permanent stations. SYO, Syowa Station; NM, Neumayer Station; Dome-F (Fuji); Dome-A (Argus); Dome-C (Charlie); DRV, Dumont D'urville; SPA, South Pole Station; McM, McMurdo Station; JBG, Jang Bogo Station; and KSJ, King Sejong Station (original figure prepared for this InTech book). **b** Surface topography and bathymetry in the Arctic (ETOPO1) with major geographic location names treated in this review paper. The red solid circle represents the "Arctic circle" (66.6°N). Abbreviations are as follows: SV, Svalbard; MR, Mendeleev Ridge; and HB, Hudson Bay. Red solid triangles are the permanent stations.KMS, Kamenskoye Station; NAS, Ny-Alesund Station, ICE-S (South) (Original figure prepared for the book by InTech book)

Kanao (2018b) reported that a well defined comprehensive three-dimensional seismic velocity structure of the Antarctic Plate had been investigated by surface wave tomography using the shallow earthquakes that occurred at the plate boundaries and the surrounding plates (Ritwoller et al. 2001; Danesi and Morelli 2001; Kobayashi and Zhao 2004). The utilized seismic data have been compiled in the Data Managing System (DMS) of the Incorporated Research Institutions for Seismology (IRIS) as the stations belong to the digital Seismographic Network (FDSN). For instance, the seismic travel-time tomography beneath the Erebus Volcano of Ross Sea (near McM; Fig. 9) indicated the existence of a remarkably low-velocity anomaly associated with hot spots, which originate from the volcano (Kobayashi and Zhao 2004). The average thickness of the continental crust of East Antarctica was 10–20 km larger than that of West Antarctica; the corresponding lithosphere of East Antarctica posed high-velocity layers down to a depth of 150 km. Seismic body wave propagation within the upper mantle of the East Antarctica-based study reported that the presence of a strikingly low-velocity anomaly in the 200-km depths underneath the lithosphere (Kuge et al. 2005). The velocity models derived from both the observed and theoretical waveforms in corroboration with the presence of the unique chemical composition and thermal gradient revealed the fact that there exists a "depleted mantle", which characterizes the Archean age in the Earth's history. In this context, a three-dimensional seismic velocity model of the upper mantle beneath the Erebus Volcano of Ross Island was derived by travel-time tomography (Gupta et al. 2009) using the data from the Transantarctic Mountains Seismic Experiment (TAMSEIS; 2000–2002) (Wiens et al. 2008). Kanao (2012, 2018a) argued that the low-velocity region corresponds to a hot plume of the volcano that continued from the surface of the Earth to the 410-km seismic discontinuity. During the IPY, the total number of seismic stations was remarkably increased within the Antarctic continent by conducting several geophysical projects. The recording of earthquakes made by Sowa of Japan and Maitri of India (Figs. 9 and 10), which were used to determined seismotectonics and geodynamics of the Antarctic and the region covered by the seismic rays connecting the earthquake sources and recording permanent stations.

Rao et al. (2007) attempted to study Seismotectonics and geodynamical processes between Antarctica and India by monitoring the seismic activity in and around Antarctica and the Indian Ocean, a reconnaissance survey for site selection and the feasibility of operation of the seismological observatory in Antarctica was initiated during 16th Antarctic Expedition. A permanent Digital Seismological Station has been installed during the 17th expedition. The station was fully commissioned on 26 January 1998 with analog and short period digital systems. This observatory was further upgraded with the installation of the Broad Band seismic system in the 20th IAE. During the 21st IAE, only analog and Broad Band systems were used for daily recording and it is continuously operational and the seismic station is also a part of AnSWeR (Antarctic Seismic Web Resource). Then ongoing research activities at Antarctica were successfully carried out during 21st IAE by participating and contributing data to the Global data centers.

Both GPS and Seismic Observatory at Maitri, Antarctica has gone global and working in tandem aid mutually, the studies on tectonic processes, analyzing the seismic activities in and around Antarctica, yielded a comprehensive picture on Indian Plate Kinematics.

Acquisition of uninterrupted good quality Broad Band digital seismic data as well as GPS data continued. The seismic data was processed and analyzed at NGRI using SEISAN software, up to September 2002, and reported to International Seismological Centre, United Kingdom (Fig. 10). This data is quite useful in global epicentral determination, particularly about the earthquakes of the South of Africa, the Indian Ocean, and the South Sandwich Islands. About 314 events have been reported to ISC out of which 3 earthquakes of above 7 magnitudes, 30 earthquakes of above 6 magnitudes, and the rest are of magnitudes 5 and 4. The nearest region of the South Sandwich Islands region also experienced 27 major earthquakes and this data will be useful for further research activities (Fig. 10).

The GPS and the Seismic results show the strain accumulation and deformation processes towards the Indian Plate. Therefore these studies would continue for a longer period to precisely estimate the seismicity and tectonic activity in and

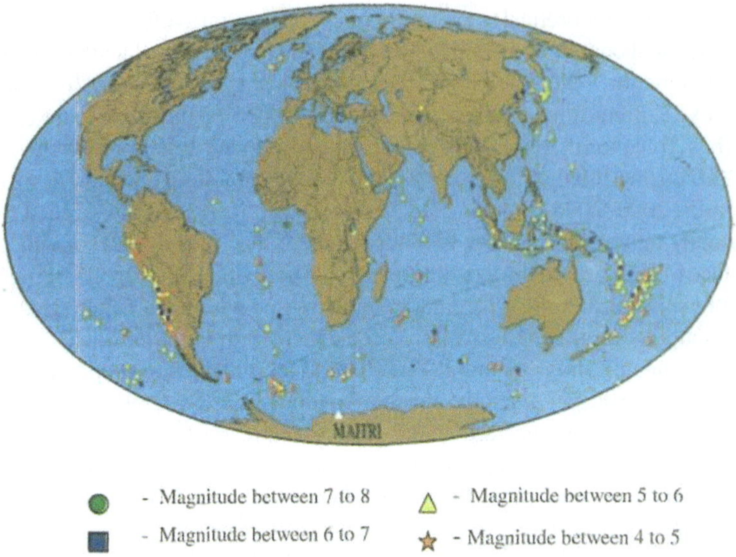

Fig. 10 A map showing global seismicity in the year 2002 recorded by Maitri (India), Antarctica (adopted from Rao et al. 2007)

around Antarctica and in the Indian Ocean. There is a need for extensive and integrated research in the Antarctic region by conducting collocation of a Permanent GPS Station and Seismic Observatory, which may also result in monitoring the space and time distribution of earthquake occurrences. This integrated approach may help estimate hypo- central parameters, magnitudes of earthquakes, the release of energy, strain accumulation, and stress drop and characterize the region resolvedly by involving velocity inversion to assimilate sub-surface structures at deeper depths with better insight into physical properties, earthquake source mechanism, receiver function analysis, attenuation of seismic waves and anisotropic behavior of the entire Antarctic region.

Advancement of seismological research has been reflected by studying the deep structure of the polar region when the boundary layers between the outer core and the lowermost mantle (i.e., the D- layer with a few hundred km of thickness above the core-mantle boundary) as well as the structure and dynamics of the inner core (central core region) were made before IGY. Seismic tomography uses the P-wave travel times of all the POLENET data; moreover, the heterogeneous structure of the upper mantle with wavelengths more than 1000 km was imaged with a high spatial resolution for the Antarctic continent, rather than those obtained from surface wave tomography (Hansen et al. 2014). In West Antarctica, particularly under MBL, hot plumes were recognized down to a depth of 800 km. The POLENET data were efficiently utilized in addition to the existing FDSN data. For instance, a very high-resolution three-dimensional shear velocity model was achieved for the upper mantle (both lithosphere and asthenosphere) of the Antarctic Plate, by applying a multi-filter technique to the surface waves generated by earthquakes that occurred at the surrounding plate boundaries (Kanao 2012b) (Fig. 10). The tomography study imaged a lithospheric root in East Antarctica almost reaching down to a depth of 200 km as well as clear boundaries to separate each tectonic terrain (geological fragments) within the continent. Moreover, low-velocity regions were found to spread out surrounding GSM, which might reflect the existence of a deep crustal root beneath the mountains.

Based on Shear-wave-splitting analysis of SKS phases, the major results obtained for the deep interiors of the Earth by using seismological data from the Federation of Digital Seismographic Network (FDSN) including the Japanese Syowa Station of East Antarctica, it is observed that heterogeneous structure of the D- layer exist beneath the Antarctic Plate where shear wave isotropy of 2.0% for the maximum within the D- layer beneath the Antarctic continent and the surrounding ocean has been determined by different researchers (Usui et al. 2005; Usui et al. 2008). The depths of the velocity discontinuity above the D- layer were determined as 50–100 km shallower than those of the Alaska region and the Caribbean Sea. Moreover, the heterogeneous structure and "super-rotation" of the inner core were observed by using the data from the Amundsen-Scott South Pole Station (ASSPS) and the Syowa Station including long-term analog records for more than 30 years (Song and Richard 1996; Isse and Nakanishi 2001). Thus, these long-term records obtained from the polar region have been efficiently utilized to study the Earth's interior.

5 Structures of Lithosphere and Near the Inner Core Boundary

A series of studies are made to assimilate lithospheric structure up to deeper layers extending to near the inner core boundary beneath Antarctica (Knopoff 1969; Knopoff and Vane 1973; Kanao et al. 2004; Gupta et al. 2006; Ohtaki et al. 2012; Kanao 2014; Hansen et al. 2015; Harry et al. 2018). Seismological results were further supported and corroborated by other geophysical methods consisted of electromagnetics of earth imaging, Magnetic and gravimetric studies of Antarctica (Hill 2020; Shimizu 2015), besides estimating ice thickness using controlled source (vibroseis) seismic method (Eisen et al. 2015).

(a) Structure of the Lithosphere and Upper Mantle

Seismic structures of various provinces of the Antarctic region have already been discussed above. A recent study by Hansen et al. (2014) showed lithospheric structure beneath the Transantarctic Mountain (TAMa) by estimating shallow structure at a depth of $\sim$3 km, which is associated with fluvial and shallow marine sediment deposition (Barrett 1981), and suggested the extent of rock uplift. The lack of compressional structures in the TAMs has led to considerable debate regarding their origin, and a variety of uplift mechanisms have been proposed. The seismic velocity structure and the crustal thickness beneath the TAMs are key components to distinguish between competing uplift models. Regional-scale tomography results from TAMSEIS reveal a low-velocity, high-attenuation upper mantle concentrated beneath the Ross Island region but extending 50–100 km beneath the TAMs (Lawrence et al. 2006; Watson et al. 2006). However, these models lose resolution away from Ross Island due to the limited aperture of the TAMSEIS deployment, making the lateral and depth extent of the upper mantle anomaly difficult to constrain. Crustal thickness beneath the TAMs has been primarily investigated using P- and S-wave receiver functions (e.g., Bannister et al. 2003). Beneath the central TAMs, the crust appears to thicken from 18–20 km beneath the Ross Sea coastline, immediately adjacent to the mountain range, to 40–45 km some 100–150 km inland. Beneath the more southerly front is necessary to more accurately assess competing origin models. The expanded seismic coverage provided by TAMNNET allows for such additional investigation. TAMs, constraints from receiver functions, and ambient noise tomography-computed with POLENET data indicate Moho depths up to ~45 km beneath the TAMs, with thinner crust underlying the adjacent West Antarctic rift system (Fig. 6b). However, none of the previous studies provided coverage of the northern TAMs, and additional characterization of the crustal and upper-mantle structure along this portion of the mountain.

(b) **The Inner Core Boundary regarding Seismic Velocity and Attenuation**

It is well-documented fact that the disposition of seismographic stations and scanty data poses a very crucial task to obtain good spatial coverage of seismic data points for better understanding the Earth's core. The more challenging issue is that the core beneath the Polar Regions remains largely unexplored. In the recent decade, Ohtaki et al. (2012) analyzed differential travel times and amplitude ratios of core phases whose ray paths run beneath Antarctica (Fig. 11) for determining the P-wave velocity (Vp) and P-wave Attenuation (Qp) structure near the inner core boundary in the south polar region.

Their south polar region model, SPR is revealed various interesting results for the preliminary reference Earth model (PREM) as follows: a 0.05 km/s lower Vp value at the top of the inner core, 1.5 times steeper Vp gradient in the upper 300 km of the inner core, a smaller Qp (300) in the upper 300 km of the inner core, and a 0.04 km/s lower Vp at the bottom of the outer core. The Vp values of SPR in the lowermost outer core lie between those of PREM and AK135, being closer to those of AK135. The lowermost outer core Vp inside the tangent cylinder is thus close to the global average. In the upper inner core, SPR has a lower Vp than AK135 and PREM. The SPR Vp profile is close to that of previous models for the Western Hemisphere, although most of the data sample used by Ohtaki et al. (2012) corresponds to the Eastern Hemisphere of the inner core. The spectacular findings suggest that the inner core does not have a simple hemispherical variation as usually supposed based on the support of data to an eyeball-shaped high-Vp anomaly with a compressional velocity higher than in 1-D reference Earth models, concentrated to a smaller region beneath eastern Asia (Fig. 12).

The study revealed that the average Qp in the upper 300 km of the inner core of SPR is 300, which is in the range (130–400) of the values reported by previous

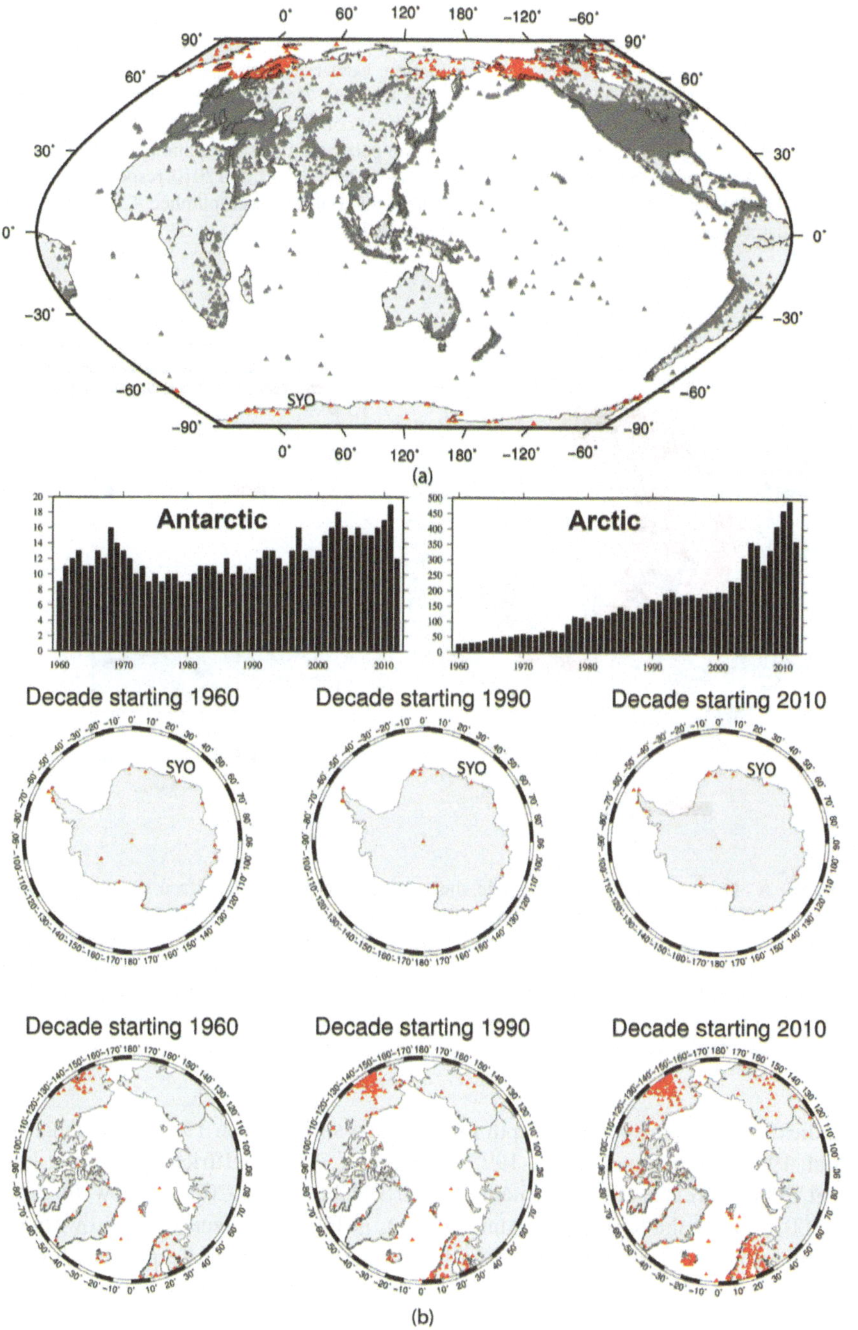
SYO
(a)
Antarctic
Arctic
Decade starting 1960
Decade starting 1990
Decade starting 2010
SYO
Decade starting 1960
Decade starting 1990
Decade starting 2010
(b)

◂**Fig. 11** **a** Upper: global distribution of the seismic stations (gray triangles) including those in polar regions (red triangles). SYO indicates Syowa Station. Lower: variation in the number of seismic stations reporting bulletin data to the ISC from the Antarctic (left) and the Arctic (right) regions. Copyright Clearance Center (CCC, http://www.copyright.com/). License number: 4278040314413, license date: January 29, 2018. **b** Distribution of the permanent seismic stations in polar regions (upper: the Antarctic; lower: the Arctic) for each decade in 1960, 1990, and 2010, respectively (after). Copyright Clearance Center (CCC, http://www.copyright.com/). License number: 4278040314413, license date: January 29, 2018 (adapted from Kanao 2018)

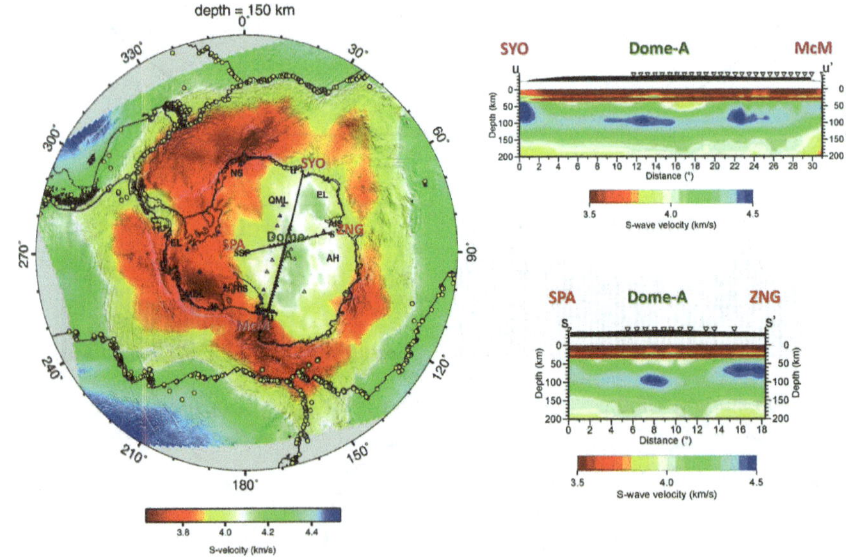

Fig. 12 A 3D image of the upper mantle shear velocity structure by surface wave tomography from the AGAP/POLENET data (modified after by Kanao 2018b). (Left) The 150-km depth slice for S-wave velocity distribution. (Upper right) Cross-section down to a depth of 200 km depth for the profile SYO (u)-Dome-A-McM (u′). (Lower right) Cross-section down to a depth of 200 km for the profile SPA (S)-Dome-A-ZHG (S′). Copyright Clearance Center (CCC, http://www.copyright.com/).License number: 4278580124881, license date: January 30, 2018 (adapted from Kanao 2018b, Intech Open Book, Chapter)

studies for the corresponding depth range [Niazi and Johnson 1992; Bhattacharyya et al. 1993; Souriau and Roudil 1995; Tseng et al. 2001; Helffrich et al. 2002; Ivan et al. 2006; Garcia et al. 2006; Iritani et al. 2010; Ohtaki et al. 2012]. Below the depth of 300 km from the ICB, they assumed that Q_p is 300. The effects on the Lowermost outer core region (F-region) are seen as different Q_p values for the deeper core through A(df/bc), two Q_p models with the same V_p profile as PREM are used in the study by Ohtaki et al. (2012). For the model whose Q_p value is 440 (which is the same as the PREM's value), the predicted amplitude ratios are higher than the observations and for a 100 km, a thick constant-Vp layer is found in the F region to fit the observations. On the other hand, for the model that has the Q_p value of 200 for the deeper inner core, the thickness of the constant-V_p layer in the F region which gives adequate fits to the observed amplitude ratios is about 75 km. This is the same as the thickness in SPR, which means the Q_p of the deeper inner core (below 300 km from the ICB) is not tightly constrained by the data set, recently used by Ohtaki et al. The Q_p structure of the lower half of the inner core has little effect on the observed amplitudes because our ray paths do not sample that region.

6 Geodynamics of the West Antarctic Rift System (Past and Present)

Harry et al. (2018) enunciated two-dimensional finite element models simulating extension of the West Antarctic Rift System (WARS) that exhibit three classes of behavior, which are dependent upon the pre-rift thermal state of the upper mantle. All of the models begin with relatively cool East Antarctica lithosphere juxtaposed against warmer West Antarctica. The models all undergo an initial period of extension that is broadly distributed across the WARS. Class 1 models (Fig. 13a–b) continue to extend in this way for more than 80 m.y. before abruptly developing a lithospheric neck at the edge of the model furthest from East Antarctica. The behavior of Class 1 models is dominated by a horizontal temperature gradient caused by the juxtaposition of the warm WARS lithosphere against the cooler East Antarctica lithosphere. This produces a corresponding strength gradient, which causes a neck to eventually develop at the warm, weak edge of the model. Class 1 models have relatively high pre-rift temperatures at the base of the crust (>800 °C), which inhibits the focusing of strain during the first 80 m.y. of extension. In Class 2 models the rift axis develops within the interior of the WARS.

Class 2 models (Fig. 14a–c) differ from Class 1 models in that the net heat production in the crust plays a larger role in determining the temperature at the top of the mantle before and during rifting. Necking at the edge of these models is inhibited because crustal thinning leads to cooling and strengthening of the lithosphere at the edge of the model. This causes the locus of extension to shift toward the weaker interior of the WARS (Harry et al. 2018).

In Class 3 models (Fig. 15a–c), the rift axis forms where the pre-rift lithosphere transitions between relatively cool and thick East Antarctica and warmer and thinner West Antarctica.

In these models, syn-extensional cooling and strengthening of the lithosphere cause the locus of strain to shift into the transitional region rather than the interior of the WARS. Class 3 models resemble the evolution of the WARS, which underwent a period of broad extension during the Late Cretaceous through the late Paleogene Periods and a more focused extension near the West Antarctica/East Antarctica boundary during the Neogene Period. All Class 3 models require the mantle potential temperature during the Late Cretaceous through Paleogene phase of the broad

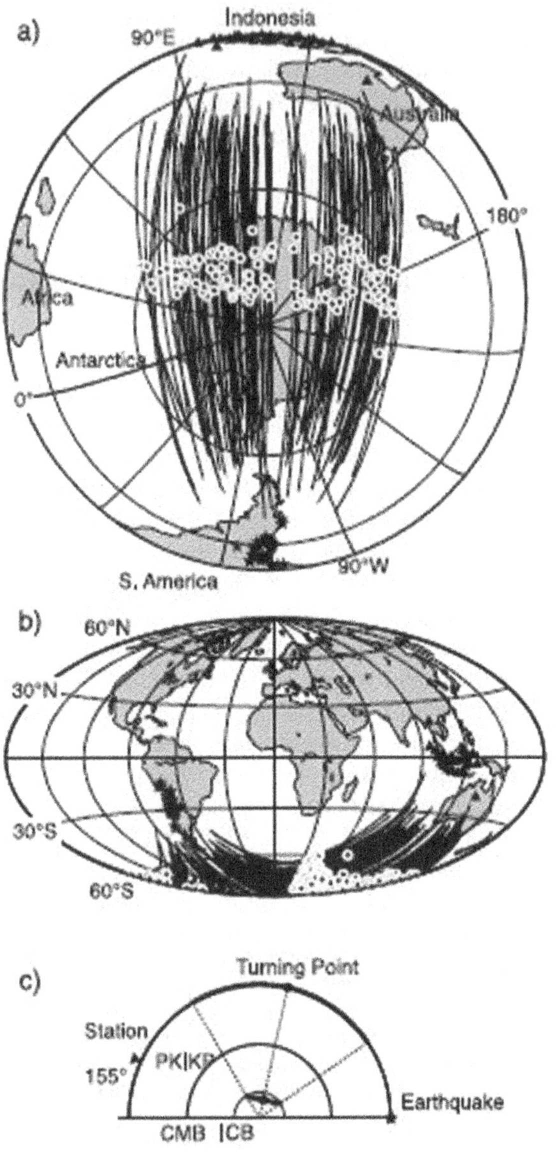

Fig. 13 **a** An orthographic projection map centered on (85 °S, 110 °E) showing the ray paths we used in this study. The rays of the core phases from the earthquakes beneath South America (stars) to the stations around Indonesia (triangles)pass through in the inner core beneath Antarctica. Solid lines show the portions of the ray paths in the inner core. Solid circles show the turning point of the rays of PKIKP and PKPbc and the midpoints of the diffracted segments for PKPc-diff. **b** The same ray paths as in Fig. 11a, but on the Hammer projection centered on 0 °E for comparison with other studies. **c** The ray paths and turning points are projected to the surface. The ray paths were computed using the TauP toolkit [Crotwell et al. 1999] (after Ohtaki et al. 2012)

extension to be no greater than 1270 °C, suggesting that an active mantle plume was not present beneath the WARS during the early stages of extension.

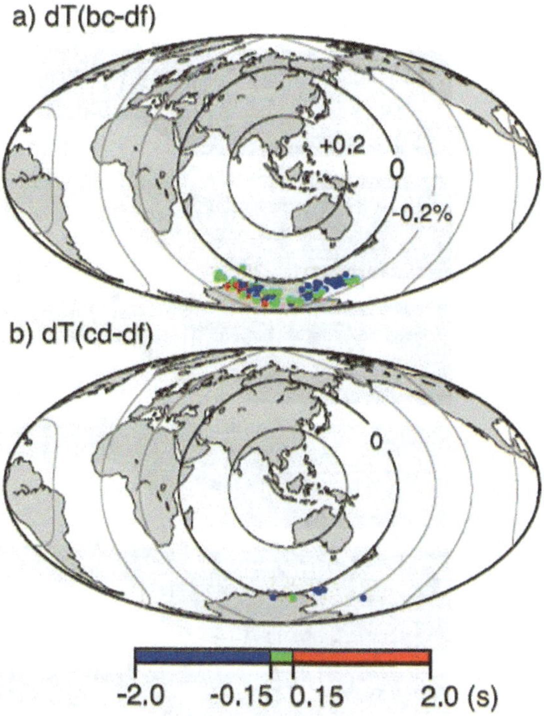

Fig. 14 The observed differential travel times relative to PREM are plotted using the Hammer projection and centered on 100 °E. All differential traveltime measurements are in seconds. Contours show the V_p perturbation model for the upper inner core by Tanaka and Hamaguchi (1997) expanded in spherical harmonics up to degree 1. Plots show **a** dT(bc–df) and **b** dT(cd–df). Black and gray lines represent the positive and negative model values, respectively. The contour interval is 0.2%. Note that rather large negative values are observed in this study where very small anomalies are expected from the model by Tanaka and Hamaguchi (1997) (after Ohtaki et al. 2012)

7 Geophysical Studies in Antarctic Region

The antarctic region is intensively studied using different tools of geophysics by different researchers (e.g., Behrendt et al. 1991; Lisker and Laufer 2011; Shimizu et al.2015; Murayama et al. 2015; Hill 2020; Dirscherl et al. 2020; Chunxia et al. 2008) to understand the crustal and sub-crustal behaviors beneath the Antarctic region.

(a) **Seismic-Gravimetric Magnetic studies**

Systematic gravity and magnetic methods were deployed along with aeromagnetic data interpretation for the region (Behrendt et al. 1991). A gravimetric and magnetic study conducted in Antarctica also required studying the topography of the region. It is observed that the West Antarctic rift system extends over 3000 × 750 km, the largely ice-covered area from the Ross Sea to the base of the Antarctic Peninsula comparable in area to the Basin and Range and the East African rift system. The rift system is characterized by bimodal alkaline volcanic rocks ranging from at least Oligocene to the present. A spectacular rift shoulders scarp along which peaks reach 4–5 km maximum elevation marks one flank and extends from northern Victoria Land-Queen Maud Mountains to the Ellsworth-Whitmore-Horlick Mountains. The

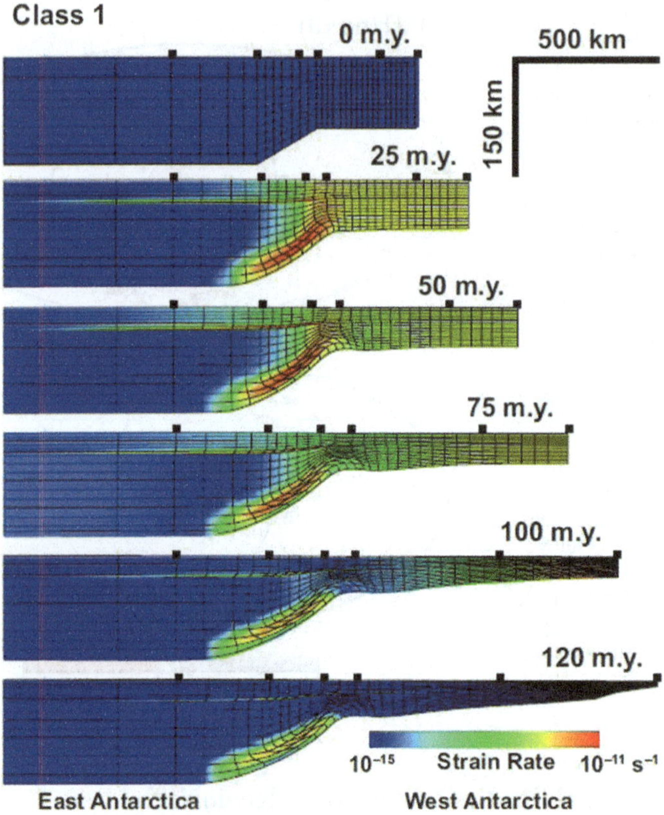

Fig. 15a Typical Class 1 model showing the second invariant of the strain rate. Black lines indicate finite element mesh. Squares at top of the mesh indicate reference positions shown in Fig. 13b–c. Moderately high strain rate is broadly distributed throughout West Antarctica for the first 100 m.y. of extension. By 120 m.y. an extension has become strongly focused within a narrow rift at the edge of the model furthest from East Antarctica. Minor extensional strain extending into the East Antarctica crust during the first 75 m.y. in this model is decoupled from the underly in guns trained mantle by a detachment at the base of the crust (narrow layer of high strain rates). High strain rates at the base of the lithosphere near the center of the model result from the flow of the lowermost lithosphere mantle, away from the transitional lithosphere separating East and West Antarctica and into the extending West Antarctic Rift System region. The simulation was terminated at 120 m.y. when the crust in the necking region had thinned to less than 5 km (after Harry et al. 2018)

rift shoulder has maximum present physiographic relief of 5 km in the Ross Embayment and 7 km in the Ellsworth Mountains-Byrd Subglacial Basin area (Fig. 16a). The Transantarctic Mountains part of the rift shoulder (and probably the entire shoulder) has been interpreted as rising since about 60 Ma, at episodic rates of 1 km/m.y., most recently since mid-Pliocene time, rather than continue at the mean rate of 100 m/m.y.

The gravimetric study demonstrated that a Bouguer anomaly range of approximately 200 (+ 50 to −150) mGal having 4–7 mGal/km gradients where measured in places marks the rift shoulder from northern Victoria Land possibly to the Ellsworth

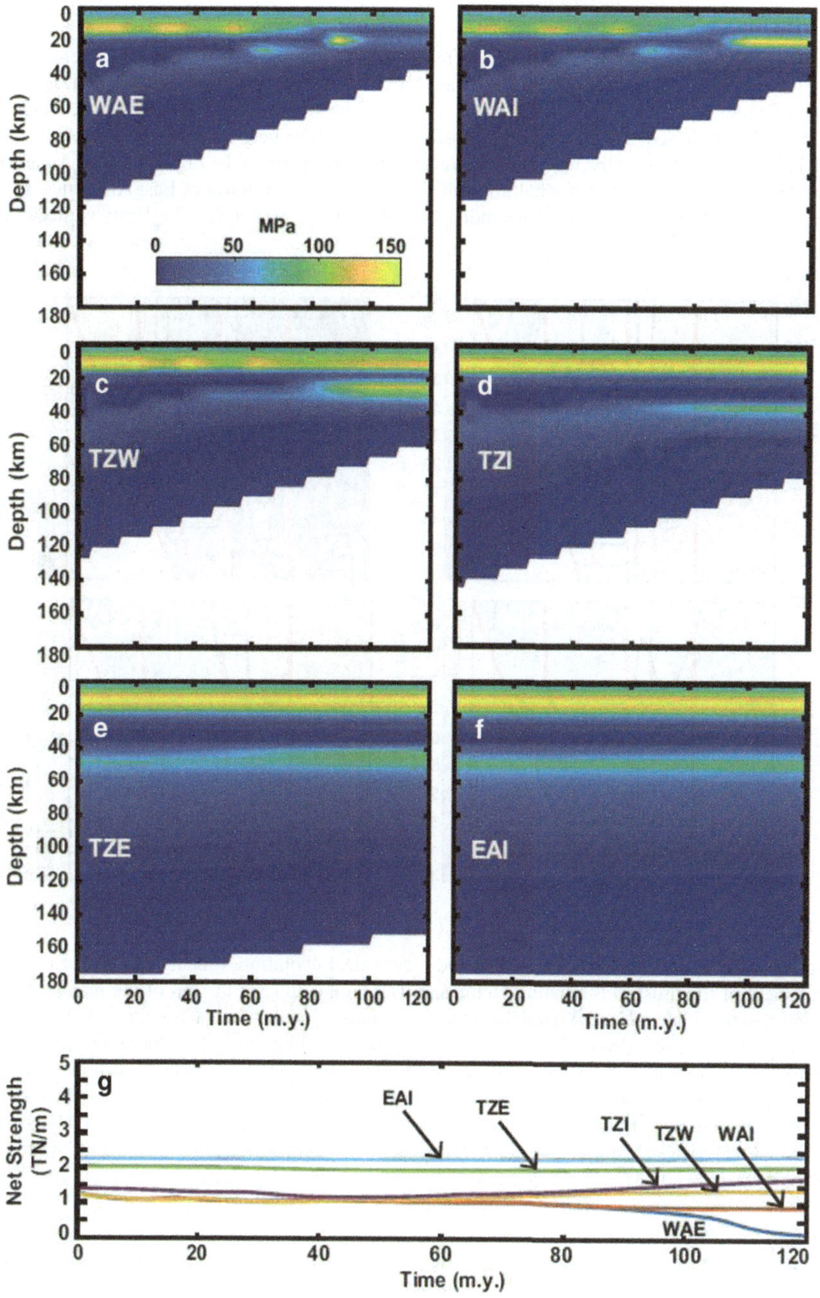
a
WAE
b
WAI
c
TZW
d
TZI
e
TZE
f
EAI
MPa
0 50 100 150
Depth (km)
Time (m.y.)
g
Net Strength (TN/m)
EAI
TZE
TZI
TZW
WAI
WAE

◂**Fig. 15b** Yield stress for Class 1 model as a function of depth and time at selected horizontal positions. The model is shown in Fig. 6. Yield stress is computed using the model temperature and rheological structure and a constant strain rate of 10–15 s–1. **a** Position PWAE, at the West Antarctica edge of the model. **b** PWAI, in the interior of the West Antarctic Rift System. **c** PTZW, at the West Antarctica edge of the region of transitional lithosphere between East and West Antarctica. **D** PTZI, in the interior of the region of the transitional lithosphere. **e** PTZE, at the East Antarctica edge of the region of the transitional lithosphere. **f** PEAI, in the interior of East Antarctica. **g** Net strength of the lithosphere for each position (the vertical integral of a–f) (after Harry et al. 2018)

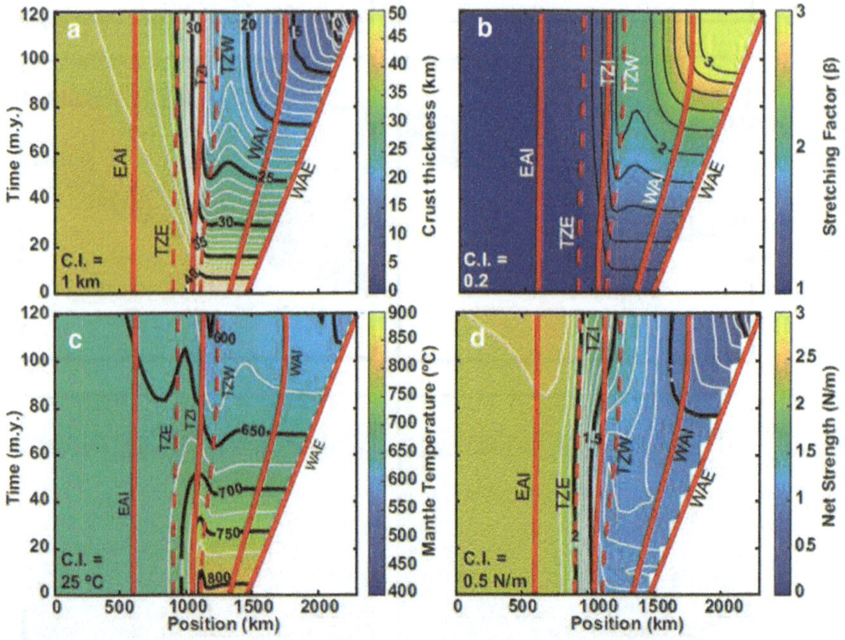

Fig. 15c Typical Class 1 model thermal and mechanical evolution. Model is shown in Fig. 6. **a** Thickness of the crust. **b** Stretching factor β. **c** The temperature at the top of the mantle. **D** Net strength of the lithosphere. Solid red lines show positions through a time of reference points located at the West Antarctica edge of the model (PWAE), within the interior of West Antarctica (PWAI), in the middle of the region where the pre-rift crust and lithosphere thicknesses are transitional between East and West Antarctica (PTZI), and within the interior of East Antarctica (PEAI). Dashed red lines mark the boundaries of the region with the transitional pre-rift lithosphere (PTZW on the eastern, or West Antarctica side, and PTZE on the western, or East Antarctica side). C.I.—contour interval (after Harry et al. 2018)

Mountains where data were too sparse to determine the maximum amplitude and gradient). The steepest gravity gradients across the rift shoulder require a high density of mafic to ultramafic rocks within the crust as well as at least 12 km of the thinner crust beneath the West Antarctic rift system in contrast to East Antarctica. Sparse land seismic data reported along the rift shoulder where velocities are greater than

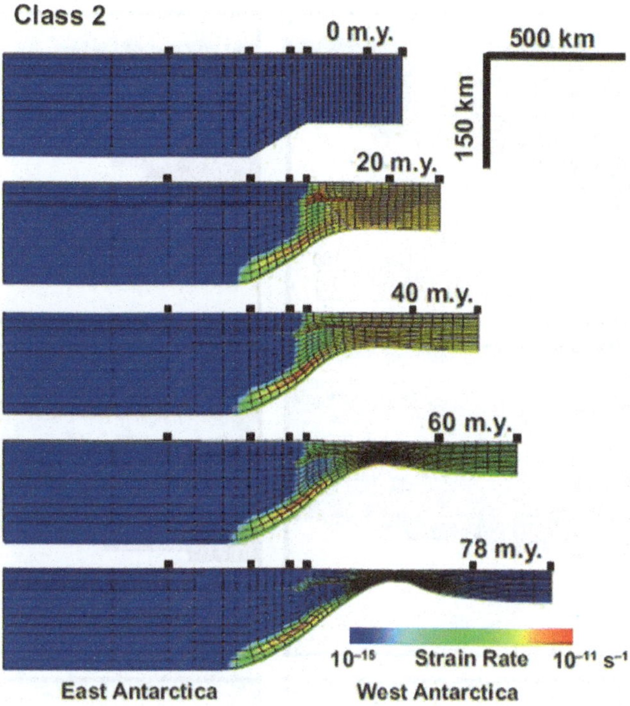

Fig. 16a Typical Class 2 model showing the second invariant of the strain rate. Black lines indicate finite element mesh. Squares at top of the mesh indicate reference positions shown in Fig. 14b, c. Moderately high strain rate is initially broadly distributed throughout West Antarctica. Extension becomes more focused and a narrow rift develops within the interior of West Antarctica after 60 m.y. High strain rates at the base of the lithosphere near the center of the model result from the flow of the lowermost lithosphere mantle into the extending West Antarctic Rift System region and away from the region of transitional lithosphere separating East and West Antarctica. The simulation was terminated at 78 m.y. when the crust in the necking region had thinned to less than 5 km (after Harry et al. 2018)

7 km/s, and marine data indicating velocities above 7 km/s beneath the Ross Sea continental elf support th isi interpretation. The maximum Bouguer gravity range in the Pensacola Mountains area of the Transantarctic Mountains is only about 130 mgal with a maximum 2 mGal/km gradient, which can be explained solely by 8 km of crustal thickening (Fig. 16b, c). Large offset seismic profiles over the Ross Sea shelf combined with earlier USGS and other results indicate 17–21 km thickness for the crust beneath the Ross Sea shelf which we interpret as evidence of extended rifted continental crust. A regional positive Bouguer anomaly (0 to + 50mGal), the width of the rift, extends from the Ross Sea continental shelf throughout the Ross Embayment and Byrd Subglacial Basin area of the West Antarctic rift system and indicates that the Moho is approximately 20 km deep tied to the seismic results, probably coincident with the top of the asthenosphere rather than the earlier studies reported the Moho was at 30 km.

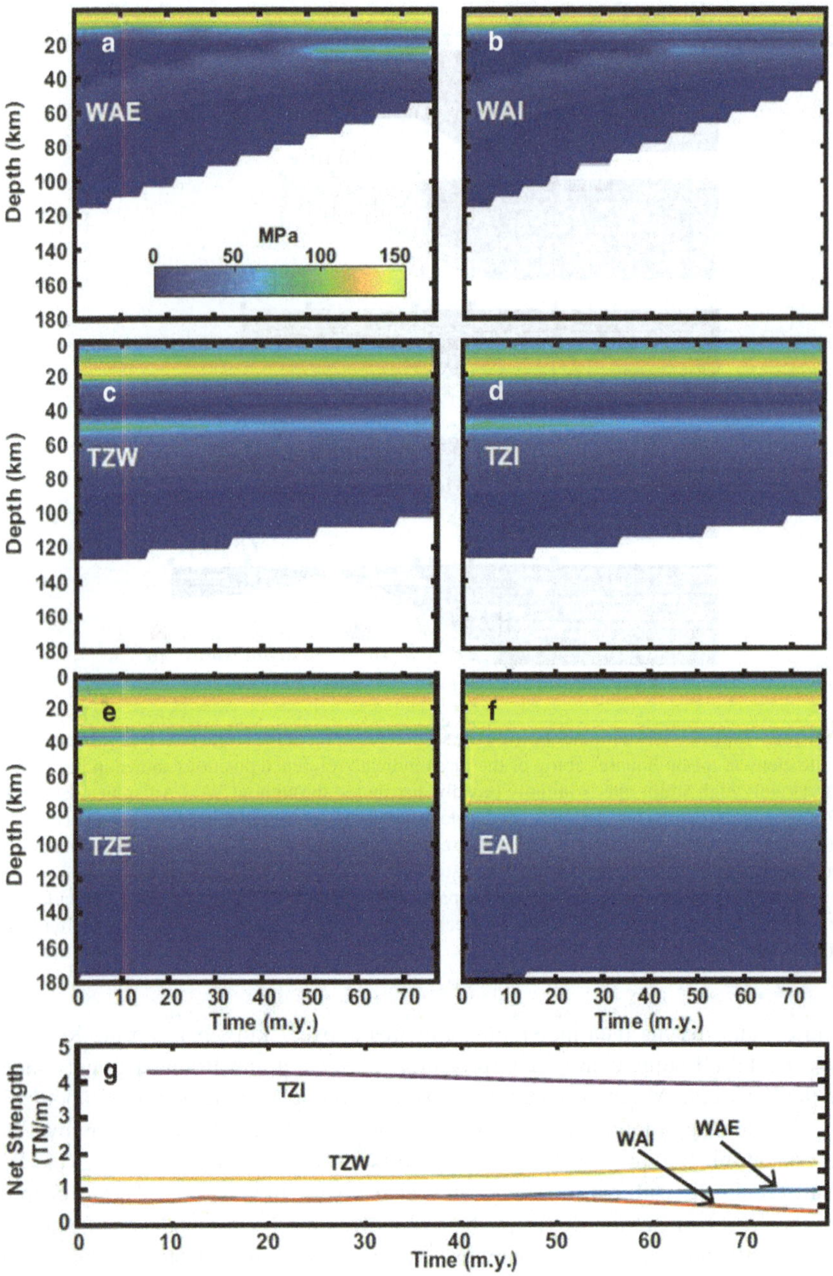
a
b
c
d
e
f
g
WAE
WAI
TZW
TZI
TZE
EAI
MPa
0
50
100
150
Depth (km)
Time (m.y.)
Net Strength (TN/m)
TZI
TZW
WAI
WAE

◂**Fig. 16b** Yield stress for Class 2 model as a function of depth and time at selected horizontal positions. The model is shown in Fig. 9. Yield stress is computed using the model temperature and rheological structure and a constant strain rate of 10–15 s–1. **a** Position PWAE, at the West Antarctica edge of the model. **b** PWAI, in the interior of the West Antarctic Rift System. **c** PTZW, at the West Antarctica edge of the region of transitional lithosphere between East and West Antarctica. **d** PTZI, in the interior of the region of the transitional lithosphere. **e** PTZE, at the East Antarctica edge of the region of the transitional lithosphere. **f** PEAI, in the interior of East Antarctica. **g** Net strength of the lithosphere for each position (the vertical integral of a–f). The net strength of East Antarctica and the eastern transitional region (reference positions EAI and TZE) are greater than 12 TN/m, and are not shown on this figure. (after Harry et al. 2018)

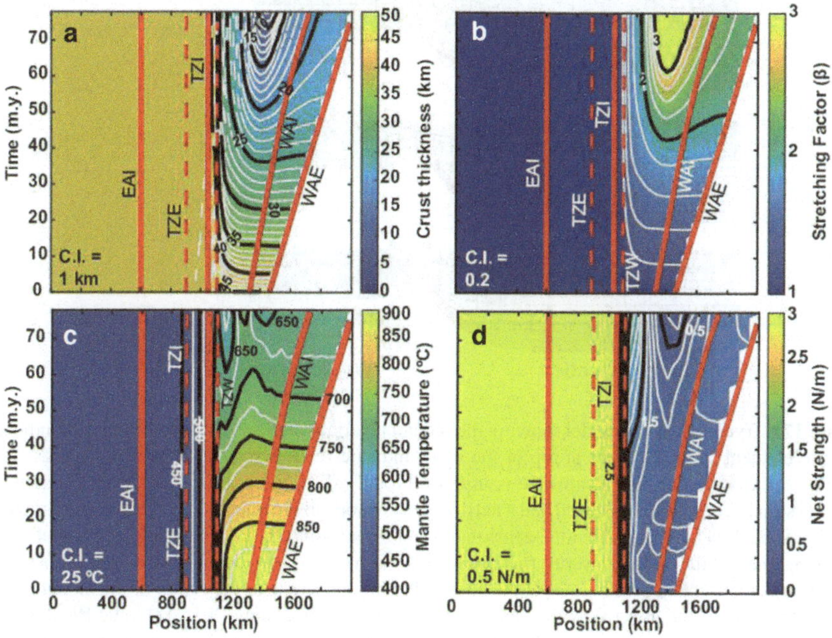

Fig. 16c Typical Class 2 model thermal and mechanical evolution. Model is shown in Fig. 9. **a** Thickness of the crust. **b** Stretching factor β. **c** The temperature at the top of the mantle. **d** Net strength of the lithosphere. Solid red lines show positions through a time of reference points located at the West Antarctica edge of the model (PWAE), within the interior of West Antarctica (PWAI), in the middle of the region where the pre-rift crust and lithosphere thicknesses are transitional between East and West Antarctica (PTZI), and within the interior of East Antarctica (PEAI). Dashed red lines mark the boundaries of the region with the transitional pre-rift lithosphere (PTZW on the eastern, or West Antarctica side, and PTZE on the western, or East Antarctica side). C.I.—contour interval(after Harry et al. 2018)

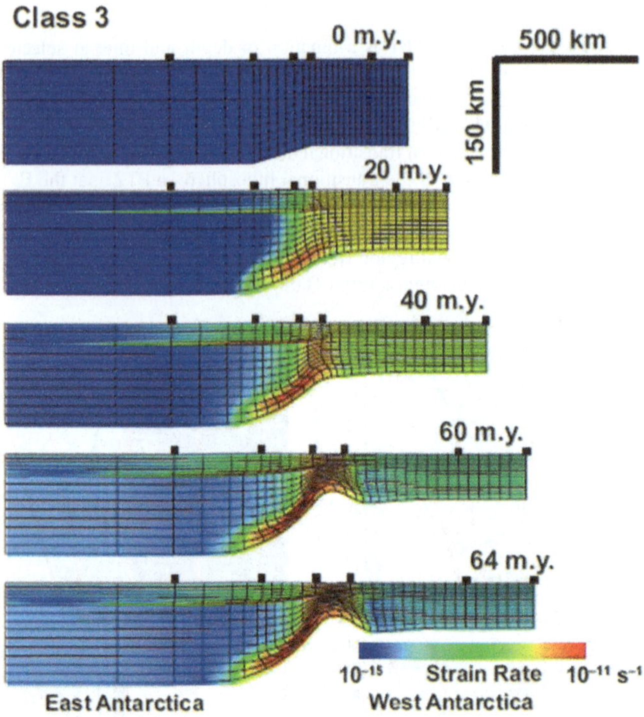

Fig. 17a Typical Class 3 model showing the second invariant of the strain rate. Black lines indicate finite element mesh. Squares at top of the mesh indicate reference positions shown in Figs. 15b and 15c. Moderately high strain rate is initially broadly distributed throughout West Antarctica. Extension becomes more focused and a narrow rift develops in the transition zone between East and West Antarctica by 60 m.y. Minor extensional strain extends into East Antarctica in this model, with a detachment at the base of the crust (narrow layer of high strain rates) decoupling deformation in the crust and mantle. High strain rates at the base of the lithosphere near the center of the model result from the flow of the lowermost lithosphere mantle into the extending West Antarctic Rift System region and away from the region of transitional lithosphere separating East and West Antarctica. The simulation was terminated at 64 m.y. when the crust in the necking region had thinned to less than 5 km (after Harry et al. 2018)

The near absence of earthquakes in the West Antarctic rift system probably results from a combination of primarily sparse seismograph coverage and, secondarily, suppression of earthquakes by the ice sheet as described by (Johnston 1987), showing very high seismicity shortly after deglaciation in the Ross Embayment followed by abnormally low seismicity at present (Muir Wood 1989). The evidence of high temperatures at shallow depth beneath the Ross Sea continental shelf and the adjacent Transantarctic Mountains is supportive of thermal uplift of the mountains associated with lateral heat conduction from the rift and can also explain the volcanism, rifting, and high elevation of the entire rift shoulder to the Ellsworth- Horlick- Whitmore Mountains. Previous researchers (Johnston 1987; Muir Wood 1989; Behrendt et al.

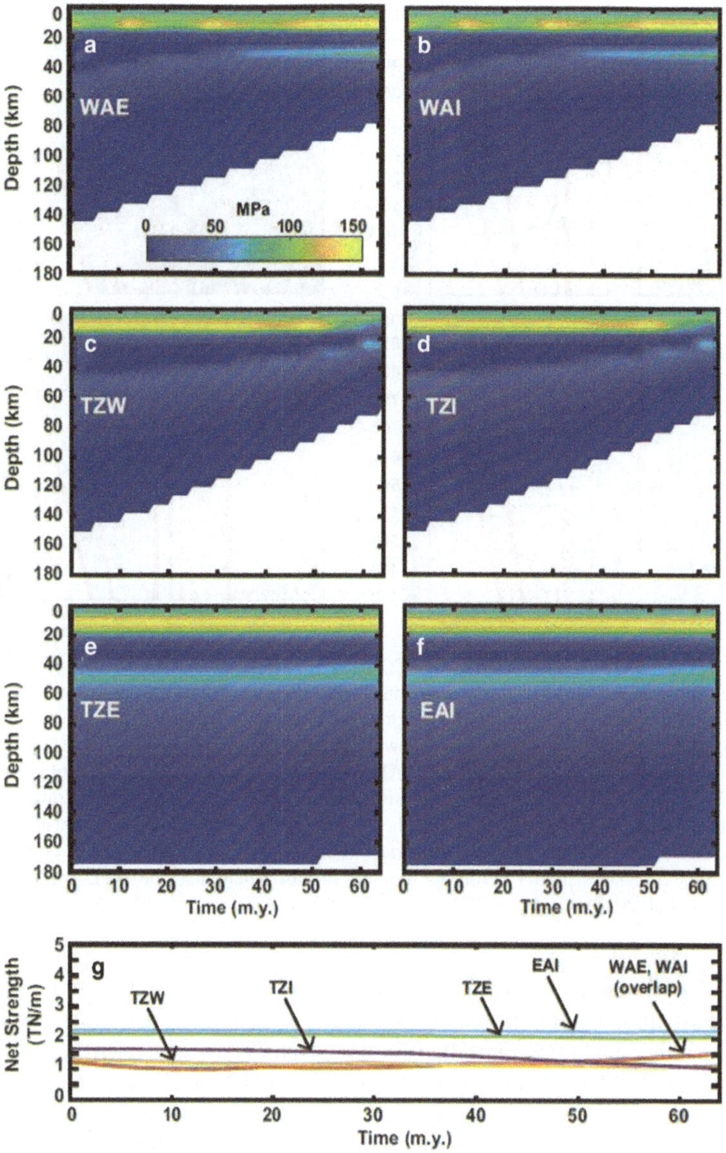

Fig. 17b Yield stress for Class 3 model as a function of depth and time at selected horizontal positions. The model is shown in Fig. 15a. Yield stress is computed using the model temperature and rheological structure and a constant strain rate of 10–15 s–1. **a** Position PWAE, at the West Antarctica edge of the model. **b** PWAI, in the interior of the West Antarctic Rift System. **c** PTZW, at the West Antarctica edge of the region of transitional lithosphere between East and West Antarctica. **d** PTZI, in the interior of the region of the transitional lithosphere. **e** PTZE, at the East Antarctica edge of the region of the transitional lithosphere. **f** PEAI, in the interior of East Antarctica. **g** Net strength of the lithosphere for each position (the vertical integral of a–f) (after Harry et al. 2018)

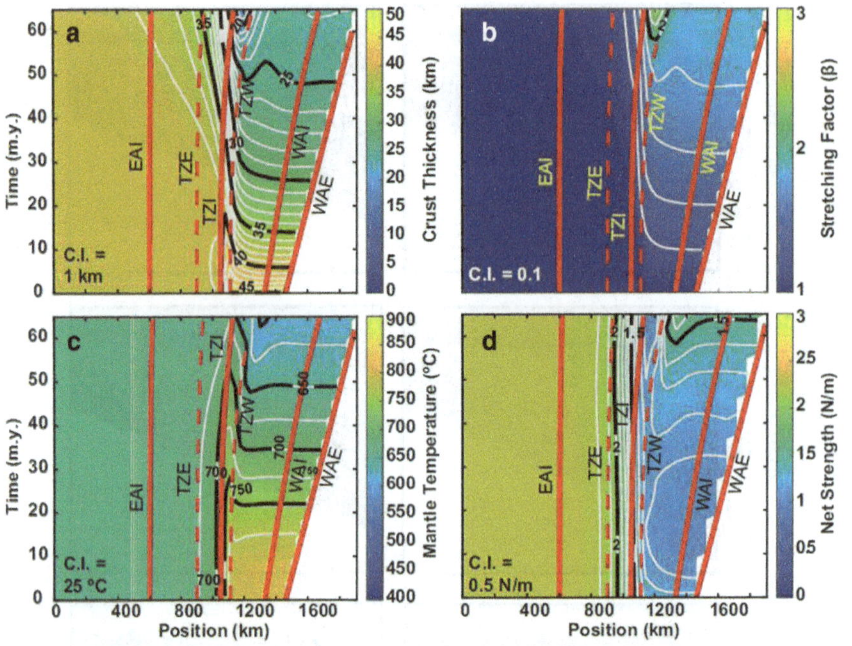

Fig. 17c Typical Class 3 model thermal and mechanical evolution. The model is shown in Fig. 15a. **a** The thickness of the crust. **b** Stretching factor β. **c** The temperature at the top of the mantle. **D** Net strength of the lithosphere. Solid red lines show positions through a time of reference points located at the West Antarctica edge of the model (PWAE), within the interior of West Antarctica (PWAI), in the middle of the region where the pre-rift crust and lithosphere thicknesses are transitional between East and West Antarctica (PTZI), and within the interior of East Antarctica (PEAI). Dashed red lines mark the boundaries of the region with the transitional pre-rift lithosphere (PTZW on the eastern or West Antarctica side, and PTZE on the western, or East Antarctica side). C.I.–contour interval (after Harry et al. 2018)

1991) inferred that the Gondwana breakup and the West Antarctic rift are part of a continuously propagating rift that started in the Jurassic when Africa separated from East Antarctica, including the failed Jurassic Transantarctic rift. Rifting proceeded clockwise around East Antarctica to the separation of New Zealand and the Campbell Plateau about 85–95 Ma and has continued (with a spreading center jump) to its present location in the Ross Embayment and West Antarctica. The Cenozoic activity of the West Antarctic rift system appears to be continuous in time with rifting in the same area that began only in the late Mesozoic. Although the mechanism for rifting is not completely explained and the study suggested that a combination of the flexural rigidity model based on the previous study of Stern and ten Brink (1989) explains the plausible mechanism for the Ross Embayment and the thermal plume or hot spot concepts. They explained that propagating rift may have been "captured" by the thermal plume.

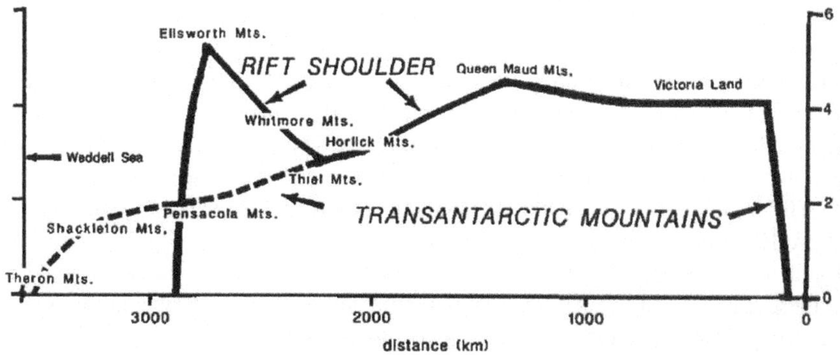

Fig. 18a Generalized topographic profile along the crest of the highest peaks parallel to and along the Cenozoic West Antarctic rift shoulder (Figs. 1 and 2) (solid line) from the Noah coast of Victoria land at the fight to the Ellsworth Mountains at left compared with lower topography (dashed line) of highest peaks in the Transantarctic Mountains from the Horlick Mountains to the Weddell Sea. Low areas within the mountains probably are the result of glacial and fluvial erosion, differential uplift, and transversed own faulting and are not shown in these profiles (after Behrendt et al. 1991)

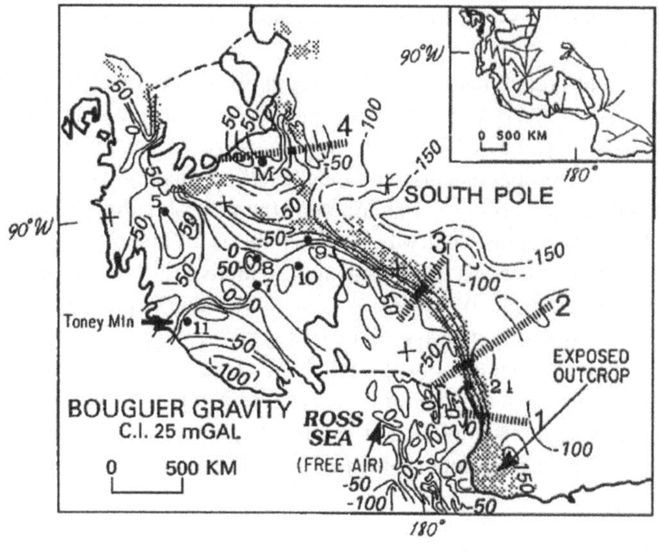

Fig. 18b Bouguer anomaly contour map for West Antarctica compiled from dam collated at seismic reflection stations (about 30–40 km speed) where ice thickness measurements were made by overs now traverse parties led from 1956 to 1964 by Behrendt et al. (1991). The inset map shows the locations of these over snow traverses. Additional data: Ross Ice shelf from the map of Robertson et al. (1982). The free air anomaly map over the Ross Sea shelf is from Davey and Cooper (1987). The numbered bands indicate areas where data density allowed reasonably accurate calculation of gradients: 1, Duerbaum et al. (1989) 2, Robinson (1964); Smithson (1972); Robinson and Splettstoesser (1984); 3, Robinson (1964); Robinson and Splettstoesser (1984); 4, Behrendt et al. (1974). Seismic velocity columns for different structures are indicated by Bentley and Clough (1972) are indicated (after Behrendt et al. 1991)

New magnetic, gravity and sub-glacial topography data allowed us to undertake the region's first comprehensive geological interpretation. Lithospheric domains and their bounding faults, including the suture between Indo-Antarctica and Australo-Antarctica have also been studied by Aitken et al. (2014) as shown in Fig. 17a–c Sub-glacial sedimentary basins, including the Aurora and Knox Sub-glacial Basins and the previously unknown Sabrina Subglacial Basin, have also been imaged using magnetic data. The commonality of structure in magnetic, gravity, and topography data suggest that pre-East Antarctic Ice Sheet (EAIS) tectonic features are a primary control on sub-glacial topography. The preservation of this relationship after glaciation suggests that these tectonic features provide topographic and basal boundary conditions that have strongly influenced the structure and evolution of the EAIS.

The magnetic intensity estimates are found to have a very good correlation with corresponding gravity data for the Antarctic region as shown in Fig. 18a–b. In this context aeromagnetic surveys resulted in the Antarctic region by Behrendt et al. (1991) successfully delineated late Cenozoic volcanic rocks in the Antarctic region and they showed that high-amplitude short-wavelength magnetic anomalies with outcrops of late Cenozoic volcanic rocks in the Marie Byrd Land area have long been discussed in the scientific community. The prominent (>1000nT) linear magnetic "inuous Ridge" anomaly was interpreted as caused by volcanic rocks of unknown age.

It has been interpreted that most anomalies in the area covered within the closed "2" contour of as shown in Fig. 19a–c over the ice-covered area of Marie Byrd Land (Byrd Sub-glacial Basinal) where depths to sources are < 1 km below the base of the ice whilst some are probably caused by older shallow sources. The character of the magnetic field changes abruptly, as marked by the north-trending part of the " > 2" contour in Fig. 19a–c, about halfway between the outcropping volcanoes of Made Byrd Land and the Ellsworth Mountains representative profile). This "break" first described by Behrendt and Wold (1963), is quite apparent in other profiles.

Total magnetic intensity map assimilated using airborne/ground magnetic data and isostatic gravity map from the global satellite data (Fig. 20a–b) unravelled the intricate structural heterogeneities beneath the region having corroboration with short-wavelengh high magnetic amplitude of varying magnetic susceptibility greater than 100nT/100 km (Fig. 21) anomalies mapped by Behrendt et al. (1991).

8 Conclusions

Conduction of integrated seismo-geophysical studies in the Antarctic region is very much warranted to explore the region for its better geo-scientific understanding to explore tectonic intricacies and geodynamical implications in the light of seismogenesis of the Antarctic region. Applications of seismo-geophysical tools of different types have their constraint because of hostile climatic conditions for which special design of instrument-protectors under thick ice sheets in the Antarctic region. Integrated seismo-geophysical studies in corroboration with potential geophysical

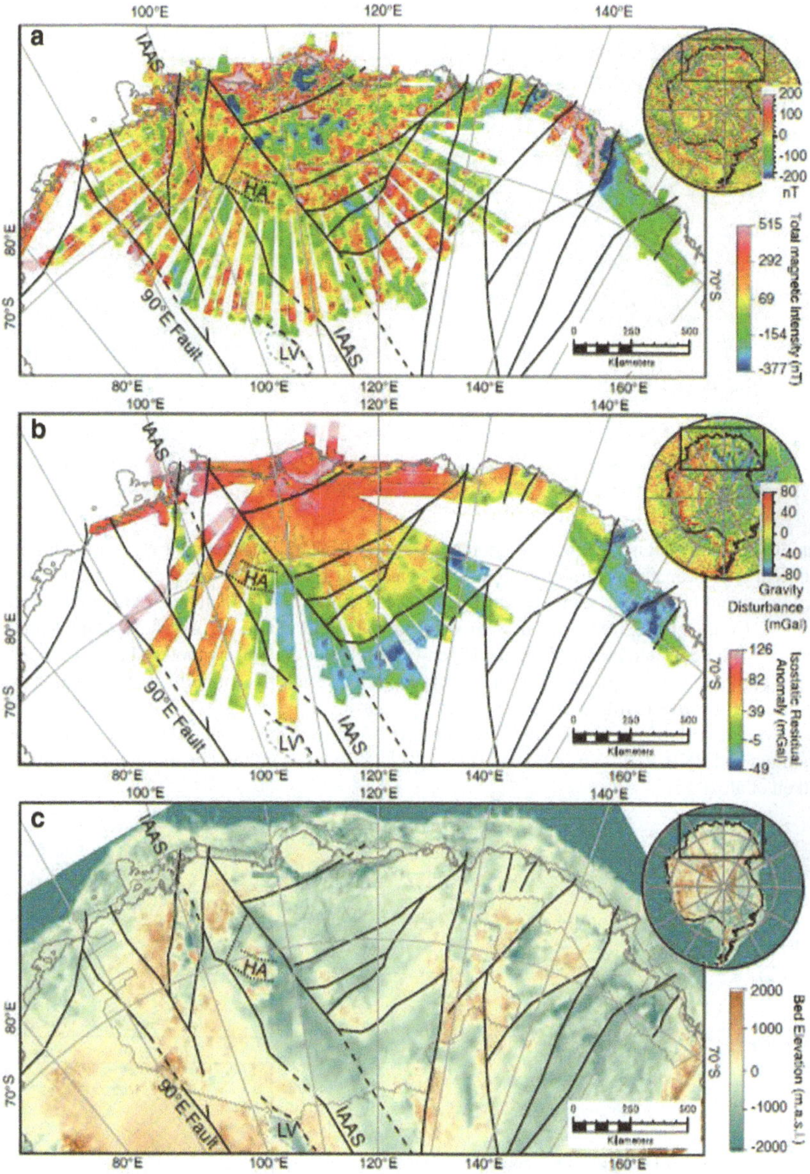

Fig. 19 **a** Total magnetic intensity anomaly (TMI) image and **b** isostatic residual gravity anomaly image. Insets show TMI from the World Digital Magnetic Anomaly Map (WDMAM) (Korhonen et al. 2007) and gravity disturbance from Eigen-6 s (Förste et al. 2011). **c** Subglacial topography from ICECAP data (within grey polygon) and Bedmap-2 (Fretwell et al. 2013). Solid lines indicate observed faults, long dashed lines indicate fault segments overprinted by younger features, and short dashed lines indicate the fault-related fjords of Highland A. IAAS–Indo-Australia-Antarctic Suture; LV—Lake Vostok; and HA—Highland A (after Aitken et al. 2014)

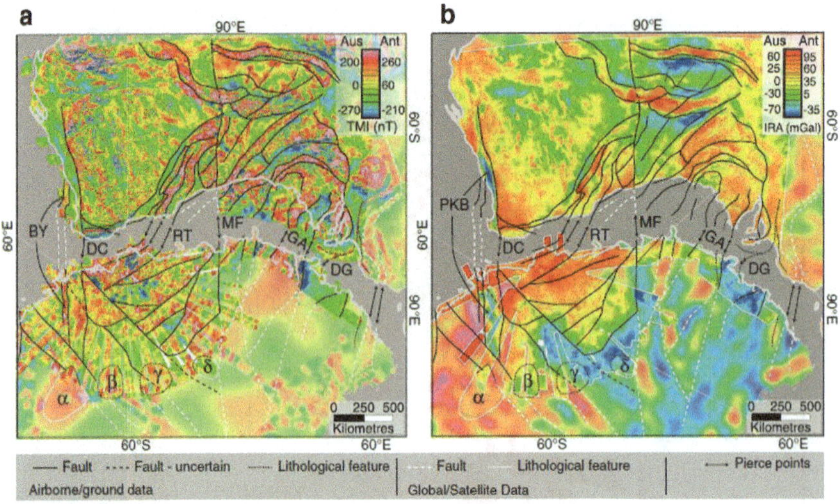

Fig. 20 **a** Total magnetic intensity anomalies and **b** isostatic residual gravity anomalies in the Leeuwin Gondwana reconstruction at 160 Ma. The color stretch is centered on the mean value for each continent but has the same dynamic range. Interpretation also included TMI values from the EMAG2 (Maus et al. 2009) (not shown) and World Digital Magnetic Anomaly Map (Korhonen et al. 2007) models and gravity from the Eigen-6 s satellite-only gravity model (Förste et al. 2011). Pierce points indicate regions where intercontinental correlations can be made. The major examples are BY—Bunger Hills-Yallingup Shelf; DC—Darling-Conger Fault; RT–Rodona-Totten Fault; MF—Mundrabilla-Frost Fault; GA—Gawler-Terre Adelie (Fitzsimons 2003); DG—Delamerian Granites; and PKB-Perth-Knox Basin. The α, β, γ, and δ indicate the late granite batholiths (after Aitken et al. 2014)

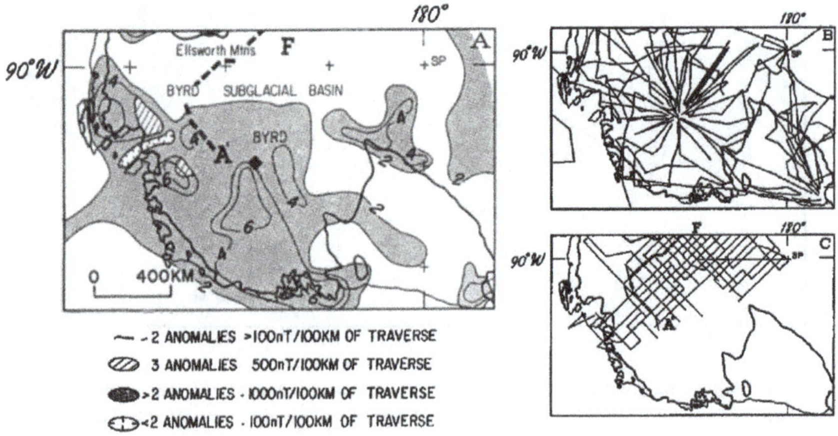

Fig. 21 A map showing the frequency of occurrence of short-wavelength high amplitude anomalies greater than 100nT/100 km of flight line average of 1° of latitude grid (111 km × 111 km) [after Behrendt et al. 1991]

(gravity and magnetic) methods can shed important light on intricate geodynamic processes involved with intriguing landscape evolution of the Antarctic plate since the breakup of Gondwana, West Antarctic Rift System (WARS), different patterns of exhumation events that occurred between the Early Cretaceous and Cenozoic. Several causative factors associated with natural and anthropogenic are found still enigmatic in sense to unravel the fact how the genesis of earthquakes are related to the glacial-dynamics and glacial mass change-induced earthquakes (GMCIE) due to melting of ice can be extensively studied using huge high-quality seismic data recorded by different seismic networks of global, regional, and local category. The Antarctic is evolving and our collaborative seismo-geophysical studies would provide deep insight into geodynamical processes in more detail.

References

Adams RD (1982) Source properties of the oates land earthquake. In: Craddock C (ed) Antarctic geoscience. The University of Wisconsin Press, Wisconsin, pp 955-958

Adams RD, Hughes AA, Zhang BM (1985) A confirmed earthquake in continental Antarctica. Geophys J Int 81(2):489-492

Adie, J. Raymond (1962). The Geology, of Antarctica, Book Series: Geophysical Monograph Series, https://doi.org/10.1029/GM007p0026.

Aitken ARA, Young DA, Ferraccioli F, Betts PG, Greenbaum JS, Richter TG, Roberts JL, Blankenship DD, Siegert MJ (2014) The subglacial geology of Wilkes Land, East Antarctica. Geophys Res Lett 41:2390-2400. https://doi.org/10.1002/2014GL059405

Anandakrishnan S, Alley RB (1997) Tidal Forcing of Basal Seismicity of Ice Stream C, West Antarctica, Observed Far Inland. J Geophys Res 102(B7):15183-15196. https://doi.org/10.1029/97JB01073

Anthony Robert E, Aster RC, Wiens D, Nyblade A, Anandakrishnan S, Huerta A, Paul Winberry J, Wilson T, Rowe C (2014) The Seismic Noise Environ Antarctica, Seismol Res Lett 86(1)

Ardhuin F, Stutzmann E, Schimmel M, Mangeney A (2011) Ocean wave sources of seismic noise. J Geophys Res 116:C05002. https://doi.org/10.1029/2011JC006952

Armadillo E, Bonaccorso A, Bozzo E, Caneva G, Capra A, Falzone G, Ferraccioli F, Gandolfi S, Mancini F, Privitera E, Vittuari L (2002) Geophysical features of the Mt. Melbourne area, antarctica, and preliminary results from the integrated network for monitoring the volcano. In: Gamble JA et al.(ed) Antarctica at the close of a millennium, vol 35. The Royal Society of New Zealand Bulletin, Wellington, pp 571-577

Aster RC, McNamara DE, Bromirski PD (2008) Multidecadal climate-induced variability in microseisms. Seismol Res Lett 79(2):194-202

Aster RC, McNamara DE, Bromirski PD (2010) Global trends in extremal microseism intensity. Geophys Res Lett 37(14):L14303. https://doi.org/10.1029/2010GL043472

Aster R, McIntosh W, Kyle P, Esser R, Bartel B, Dunbar N, Johns B, Johnson J, Karstens R, Kurnik C, McGowan M, McNamara S, Meertens C, Pauly B, Ri- chmond M, Ruiz M (2004) Real-time data received from Mount Erebus. EOS Trans 85(10), 99. https://doi.org/10.1029/2004EO100001

Banaccorso A, Gambino S, Privitera E (1997) A Geophysical to the Dynamics of Mt. Melbourne (Northern Victoria Land, Antarctica). In: Ricci CA (ed) The antarctic region: geological evolution and processes. Terra Antarctica Publication, Siena, pp 531-538

Bannister S, Yu J, Leitner B, Kennett BLN (2003) Variations in crustal structure across the transition from West to East Antarctica. Southern Victoria Land. Geophys J Int 155(3):870-884

Barrett PJ (1981) Late cenozoic glaciomarine sediments of the Ross Sea, Antarctica. In: M.J. Hambrey, W.B. Harland (eds) Earth's Pre-Pleistocene Glacial Record. Cambridge University Press, Cambridge, pp 208–211

Barrett PJ, Henrys SA, Bartek LR, Brancolini G, Busetti M, Davey FJ, Hannah MJ, Pyne AR(1995) Geology of the margin of the Victoria Land Basin off Cape Roberts, southwest Ross Sea. In: Cooper AK, Barker PF, Brancolini G (eds) Geology and seismic stratigraphy of the Antarctic margin. Antarctic Research Series, vol 68. Washington D.C., American Geophysical Union, pp 183-207

Behrendt JC, Richard JW (1963) Aeromagnetic survey in West Antarctica 1963. The University of Wisconsin Geophysical & Polar Research Center, Department of geology, Research report Series https://minds.wisconsin.edu/bitstream/handle/1793/64933/pr63-1.pdf?sequence=1%26isAllowed=y

Behrendt JC (1964) Crustal geology of Ellsworth Land and the Southern Antarctic Peninsula from gravity and magnetic anomalies. J Geophys Res 69:2047-2063

Behrendt JC, Henderson JR, Meister LJ, Rambo W (1974) Geophysical investigations of the Pensacola Mountains and adjacent glacierized area of Antarctica, U.S. Geol. Sun: Prof Pap 844, p 27, 2 plates

Behrendt JC, LeMasurier WE, Cooper AK, Tessensohn F, Trehu A, Damaske D (1991) Geophysical studies of the west antarctic rift system. Tectonics 10(6):1257-1273

Behrendt JC, LeMasurier WE, Cooper F Tessensohn AK, Tr6hu A, Damaske D (1991) The West Antarctic rift system-A review of geophysical investigations, in Contributions to Antarctic Research II, Antarct. Res. Ser., AGU, Washington, D.C., in press

Bentley CR, Crary AP, Thiel E, Ostenso NA (1960) Structure of west antarctica. Science 131:131-136

Bentley CR, Clough JW (1972) Seismic refraction shooting in Ellsworth and Dronning Naud Land, in Antarctic Geology and Geophysics. lnt Univ Geol Sci Ser B 1:683-691, Adie RJ (ed). Universitetsforlaget, Oslo

Bentley CR (1973) Crustal structure of Antarctica, Crustal structure based on seismic data, IUCM Proc-A Symposium. Tectonophysics 20:229-240

Bhattacharyya J, Shearer P, Masters G (1993) Inner core attenuation from short-period PKP (BC) versus PKP (DF) waveforms. Geophys J Int 114(1):1-11

Biswas, N.N. (1971). The upper mantle structure of the United States from the dispersion of Surface waves Ph.D dissertation, University of California, Los Angeles

Biswas NN, Knopoff L (1974) The structure of the upper mantle under the United States from the dispersion of Rayleigh waves. Geophys J R Astron Soc 36(3):515-539

Block S, Hales AL, Landisman M (1969) Velocities in the crust and upper mantle of Southern Africa from multi-mode surface wave dispersion. Seismol Soc Am Bull 59:1599-1629

Bolt BA, Niazi M (1964) Dispersion of Rayleigh waves across Australia. Geophys J Roy Astron Soc 9:21-35

Bromirski PD, Duennebier FK, Stephen RA (2005) Mid-ocean microseisms. Geochem Geophy Geosyst 6(4):Q04009. https://doi.org/10.1029/2004GC000768

Butler R, Lay T, Creager K, Earl P, Fischer K, Gaherty J, Laske B, Park J, Ritzwoller M, Wen L(2004) The global seismographic network surpasses its design goal. Eos Trans. AGU 85(23):225-229

Cande SC, Mutter JC (1982) A revised identification of the oldest sea-floor spreading anomalies between Australia and Antarctica. Earth Planet Sci Lett 58:151-160

Cande S, Stock J, Mueller D, Ishihara T, Tikku A (1998) Seafloor spreading constraints on Cenozoic motion between East and West Antarctica. EOS Trans Am Geophys Union 79:F906

Center NGD (2006) 2-minute Gridded Global Relief Data (ETOPO2) v2. National Geophysical Data Center, National Oceanic and Atmospheric Administration. https://doi.org/10.7289/V5J1012Q

Cox KG (1988) The Karoo Province. In: MacDougall JD (ed) Continental flood basalts. Kluwer Academic Publishers, Dordrecht, pp 239-271

Craddock C et al (1969) Geologic Maps of Antarctica. In: Bushnell V (ed) Antarct. Map Folio Ser. Am. Geogr. Soc. New York

Crotwell HP, Owens TJ, Ritsema J (1999) The TauP Toolkit: Flexible seismic travel-time and ray-path utilities. Seismol Res Lett 70(2):154-160. https://doi.org/10.1785/gssrl.70.2.154
Dalziel IWD, Elliot DH (1982) West Antarctica: problem child of Gondwana. Tectonics 1:3-19
Damaske D, Behrendt J, McCafferty A, Saltus R, Meyer U (1994) Transfer faults in the western Ross Sea: new evidence from the McMurdo Sound/Ross Ice Shelf aeromagnetic survey (GANOVEX VI). Antarct Sci 6:359-364
Davey FJ, Cooper AK (1987) Gravity studies of the Victoria land basin and Iselin bank. In: The Antarctic Continental Margin Geology and Geophysics of the Western Ross Sea, Earth Sci. Ser., 5B, edited by A.K. Cooper and F.J. Davey, pp. 119-138, Circum-Pacific Council for Energy and Natural Resources, Houston, Tex.
Davey FJ, Brancolini G (1995)The Late Mesozoic and Cenozoic structural setting of the Ross Sea region. In: Geology and stratigraphy of the Antarctic margin. Antarctic Research Series Vol. 68. Washington D.C., American Geophysical Union. pp 167-182
De Wit M, Bergh H, Nicholatsen L (1988) Geological map of sectors of Gondwana. American Association of Petroleum Geologists and University of Witwatersrand
Dimitrova L, Georgieva G, Raykova R, Gurev V, Georgiev I (2015) Experimental seismological and GNSS equipment in extreme conditions in preparation for complex studies in the area of the Bulgarian Antarctic Base. Geophys Res Abstracts 17:EGU2015-11069. EGU General Assembly
Dirscherl Mariel, Andreas J. Dietz, Stefan Dech, Claudia Kuenzer (2020). Remote sensing of ice motion in Antarctica - A review, Remote Sensing of Environment, 237 (2020) 111595.
Divenere V, Kent DV, Dalziel IWD (1996) Early Cretaceous paleomagnetic results from Marie Byrd Land, West Antarctica: implication for the Weddellia collage of crustal blocks. J Geophys Res 100:8133-8152
Drewry DJ (1983) Antarctica: Glaciological and Geophysical Folio, 9 Sheets. University of Cambridge, Cambridge, Scottish Polar Research Institute
Duerbaum HJ, Druivenga G, Geipel H, Merkel G (1989) Gravity Measurements along a traverse from Mount Melbourne to the Polar Plateau in North Victoria Land Antarctica. Geol. Jahr. Reihe E 38:231-243
Dziak RP, Parlk M, Lee WS, Matsumoto H, Bohnenstiehl DR Haxel JH (2009) Tectono-magmatic activity and ice dynamics in the bransfield strait back-arc basin, antarctica. In: The 16th international symposium on polar science. Incheon, pp 59-68
Ekstrom G, Nettles M, Tsai VC (2006) Seasonality and Increasing Frequency of Greenland Glacial Earthquakes. Science 311(5768):1756-1758. https://doi.org/10.1126/science.1122112
Findlay RH, Skinner DNB, Craw D (1984) Lithostratigraphy and structure of the Koettlitz Group, McMurdo Sound, Antarctica. NZ J Geol Geophys 27:513-536
Fitzgerald PG (1992) The Transantarctic Mountains of southern Victoria Land: the application of apatite fission track analysis to a rift shoulder uplift. Tectonics 11:634-662
Fitzgerald P (2002) Tectonics and landscape evolution of the Antarctic plate since the breakup of Gondwana, with an emphasis on the West Antarctic rift system and the transantarctic mountains. Royal Soc New Zealand Bulletin 35:453–469
Fitzsimons ICW (2003) Proterozoic basement provinces of southern and southwestern Australia, and their correlation with Antarctica. Geol Soc Spec Publ 206:93-130
Förste C et al (2011) EIGEN-6: A new combined global gravity field model including GOCE data from the collaboration of GFZ-Potsdam and GRGS-Toulouse, EGU General Assembly 2011
Fouda AA (1973) The upper mantle structure under the stable regions. PhDdissertation, University of California, Los Angeles
Fretwell P, Pritchard HD, Vaughan DG, Bamber JL, Barrand NE, Bell R, Bianchi C, Bingham RG, Blankenship DD, Casassa G et al (2013) BEDMAP2: Improved ice bed, surface and thickness datasets for Antarctica. Cryosphere 7:375-393
Gabriel VG, Kuo JT (1966) High Rayleigh wave phase velocities for the New Delhi, India, Lahore, Pakistan profile. Seismological Society of America, Bulletin 56:1137-1146
Galperin EI, Nersesov IL, Galperina RM (1986) Borehole seismology and the study of the seismic regime of large industrial Centres. Reidel, Dordrecht, Netherlands, p 315

Garcia R, Tkalčić H, Chevrot S (2006) A new global PKP data set to study Earth's core and deep mantle. Phys Earth Planet Inter 159(1-2):15-31. https://doi.org/10.1016/j.pepi.2006.05.003
Garrett SW, Harrod DB, Mantripp DR (1987) Crustal structure of the area around Haag Nunataks, West Antarctica: new aeromagnetic and bedrock elevation data. In: McKenzie GD (ed) Gondwana S.: Structure, Tectonics, and Geophysics, vol 40. Geophys. Mottogr. Ser., AGU, Washington, D.C., pp 109-115
Gleadow AJW, Fitzgerald PG (1987) Uplift history and structure of the Transantarctic Mountains: new evidence from fission track dating of basement apatites in the Dry Valleys area, southern Victoria Land. Earth Planet Sci Lett 82:1-14
Gonzales-Ferran O (1982) The Antarctic Cenozoic volcanic provinces and their implications in plate tectonic processes. In: Craddock C (ed) Antarctic Geoscience. University of Wisconsin Press, Madison, pp 687-694
Grob M, Maggi A, Stutzmann E (2011) Observations of the seasonality of the Antarctic microseismic signal, and its association to sea ice variability. Geophys Res Lett 38(11):L11302. https://doi.org/10.1029/2011GL047525
Gunn BM, Warren G (1962) Geology of Victoria Land between the Mawson and Mulock Glaciers. Antarctica New Zealand Geol Surv Bull 71:157
Harry DL, Anoka JL (2018) Sumant Jha; Geodynamic models of the West Antarctic Rift System: Implications for the mantle thermal state. Geosphere 14(6):407-2429. https://doi.org/10.1130/GES01594.1
Hansen SE, Graw JH, Kenyon LM, Nyblade AA, Wiens DA, Aster RC, Huerta AD, Anandakrishnan S, Wilson TJ (2014) Imaging the Antarctic mantle using adaptively parameterized P-wave tomography: evidence for heterogeneous structure beneath West Antarctica. Earth Planet Sci Lett 408:66–78
Hansen Samantha E, Reusch AM, Parker T, Bloomquist DK, Carpenter P, Graw JH, Brenn GR (2015) The Transantarctic Mountains Northern Network (TAMNNET): Deployment and Performance of a Seismic Array in Antarctica. Seismol Res Lett 86(6)
Hansen SE, Julia J, Nyblade AA, Pyle ML, Wiens DA, Anandakrishnan S (2009) Using S wave receiver functions to estimate crustal structure beneath ice sheets: An application to the Transantarctic Mountains and East Antarctic craton. Geochem Geophys Geosys 10(8). https://doi.org/10.1029/2009GC002576
Harrington HJ (1958) Nomenclature of rock units in the Ross Sea region. Antarctica, Nature 182:290
Harry Dennis, L, Jourdan L. Anoka, and Sumant Jha (2018). Geodynamic models of the West Antarctic Rift System: Implications for the mantle thermal state, Geodynamic models of the West Antarctic Rift, Geosphere, v. 14, no. 6, p. 2407–2429. https://doi.org/10.1130/GES01594.1
Hasselmann K (1963) A statistical analysis of the generation of microseisms. Rev Geophys 1(2):177-210
Helffrich G, Kaneshima S, Kendall JM (2002) A local, crossing-path study of attenuation and anisotropy of the inner core. Geophys Res Lett 29(12):1568. https://doi.org/10.1029/2001GL014059
Hill GJ (2020) On the Use of Electromagnetics for Earth Imaging of the Polar Regions. Surv Geophys 41:5-45. https://doi.org/10.1007/s10712-019-09570-8
Himeno T, Kanao M, Ogata Y (2011) Statistical Analysis of Seismicity in a Wide Region around the 1998 Mw 8.1 Balleny Islands Earthquake in the Antarctic Plate. Polar Sci 5(4):421-431. https://doi.org/10.1016/j.polar.2011.08.002
Iritani R, Takeuchi N, Kawakatsu H (2010) Seismic attenuation structure of the top half of the inner core beneath the northeastern Pacific. Geophys Res Lett 37(19):L19303. https://doi.org/10.1029/2010JB007942
Irving JCE, Deuss A (2011) Hemispherical structure in inner core velocity anisotropy. J Geophys Res 116:B04307. https://doi.org/10.1029/2010JB007942
Isse T, Nakanishi I (2001) Inner-core anisotropy beneath Australia and differential rotation. Geophys J Int 151:255–263

Ivan M, Marza V, de Farias Caixeta D, de Melo Arraes T (2006) Uppermost inner core attenuation from PKP data observed at some South American seismological stations. Geophys J Int 164(2):441-448. https://doi.org/10.1111/j.1365-246X.2006.02847.x
Jankowski, E.J., D.J. Drewry, and J.C. Behrendt, Magnetic studies of upper crustal structure in West Antarctica and the boundary with East Antarctica, in Antarctic Earth Science, edited by R.L. Oliver, J.B. James, and J.B. Jago, pp. 197-203, Australian Academy of Science, Canberra, 1983.
Jin YK, Lee DK, Nam SH, Kim Y, Kim KJ (1998) Seismic Observation at King Sejong Station. Antarctic Peninsula, Terra Antarctica 5:729-736
Johnston AC (1987) Suppression of Earthquakes by Large Continental Ice Sheets. Nature 330(1987):467-469. https://doi.org/10.1038/330467a0
Kaminuma K, Ishida M (1971) Earthquake Activity in Antarctica, Antarctica. Record 42:53-60
Kaminuma K, Dibble RR (1990) Seismic Activity of Mount Erebus 1981-1988. Polar Geoscience 4:142-148
Kaminuma K (1995) Seismicity around the Antarctic Peninsula. Polar Geoscience 8:35-42
Kaminuma K (2000) A Revaluation of the Seismicity in the Antarctic. Polar Geoscience 13:145-157
Kaminuma, K., A Possibility of Earthquake Swarms around ORCA Sea Mount in the Bransfield Strait, the Antarctic, In: Y. Kim and B. K. Khim, Eds., Proceedings of the Joint International Seminar: Recent Interests on Antarctic Earth Sciences of Korea and Japan, 2001, pp. 23-34.
Kanao M, Ishikawa M, Yamashita M, Kaminuma K, Brown LD (2004) Structure and evolution of the East Antarctic lithosphere: Tectonic implications for the development and dispersal of Gondwana. Gondwana Res 7:31-41
Kanao, M. and Kaminuma, K., Seismic Activity Associated with Surface Environmental Changes of the Earth System, around Syowa Station, East Antarctica, In: D. K. Futterer, et al., Eds., Antarctica: Contributions to Global Earth Sciences, Springer-Verlag, Berlin, Heidelberg, New York, 2006, pp. 361-368. http://dx.doi.org/https://doi.org/10.1007/3-540-32934-X_45.
Kanao M, Storchak D, Dando B (2012) Evaluation of long-period detectability of teleseismic events at Syowa Station. Antarctica. International Journal of Geosciences 3:809-821
Kanao M, Maggi A, Ishihara Y Yamamoto M-Y, Nawa K, Yamada A, Wilson T, Himeno T, Toyokuni G, Tsuboi S, Tono Y, Anderson K (2012) Interaction on seismic waves between atmosphere-ocean-cryosphere and geosphere in polar region. In: Kanao M et al (ed) Seismic Waves-Research and Analysis, InTech. Publisher, Rijeka, 2012, pp. 1-20. http://dx.doi.org/https://doi.org/10.5772/1400.
Kanao M (2018b) Seismological studies on the deep interiors of the earth viewed from the Polar region. Polar Seismol - Adv Impact. https://doi.org/10.5772/intechopen.78552
Kuge K, Fukao Y (2005) High-velocity lid of East Antarctica: evidence of a depleted continental lithosphere. J Geophys Res110:B06309. https://doi.org/10.1029/2004JB003382
Masaki K (2014) Seismicity in the Antarctic Continent and Surrounding Ocean. J Earthquake Res 2014(3):5-14
Kellogg KS, Rowley PD (1989) Structural geology and tectonics of the Orville Coast region, Southern Antarctica Peninsula, Antarctica. U.S.G.S., p 25
Knopoff L (1969) The upper mantle of the earth. Science 163:1277-1287
Knopoff L, Vane G (1973) Antarctic Seismological Studies. Antarctic J 256-257:1973
Kordy M, Wannamaker P, Maris V, Cherkaev E, Hill G (2016) Three-dimensional magnetotelluric inversion using deformed hexahedral edge finite elements and direct solvers parallelized on SMP computers, part I: forward problem and parameter jacobians. Geophys J Int 204:74-93
Kordy M, Wannamaker P, Maris V, Cherkaev E, Hill G (2016) Three-dimensional magnetotelluric inversion using deformed hexahedral edge finite elements and direct solvers parallelized on SMP computers, part II: direct data-space inverse solution. Geophys J Int 204:94-110
Korhonen JV et al (2007) Magnetic Anomaly Map of the World; Map published by Commission for Geological Map of the World, supported by UNESCO, 1st ed., GTK, Helsinki, Finland

Lawrence JF, Wiens DA, Nyblade AA, Anandakrishnan S, Shore PJ, Voigt D (2006) Rayleigh wave phase velocity analysis of the Ross Sea, Transantarctic Mountains, and East Antarctica from a temporary seismograph array. J Geophys Res 111(B6). https://doi.org/10.1029/2005JB003812
Lawver LA, Royer J-Y, Sandwell DA, Scotese CT (1991) Evolution of the Antarctic continental margins. In: Thomson MRA, Crame JA, Thomson JW (eds) Geological evolution of Antarctica. Cambridge University Press, Cambridge, pp 533-539
Lawver LA, Gahagan LM, Coffin MF (1992) The Development of paleo seaways around Antarctica. In: The Antarctic paleoenvironment: a perspective on global change. Antarctic Research Series vol. 56. Washington D.C., American Geophysical Union, pp 7-30
Lawver LA, Gahagan LM, Dalziel IWD (1998) A tight fit-Early Mesozoic Gondwana: a plate tectonic perspective. In: Origin and evolution of continents. Memoirs of the National Institute of Polar Research, Special Issue. Tokyo, National Institute of Polar Research, pp 214-229
LeMasurier WE, Thomson JW (eds) Volcanoes of the Antarctic Plate and Southern Oceans. Antarctic Research Series, vol 48. AGU, Washington, D.C., p. 489
Lisker F, Läufer AL (2011) Thermochronological research in northern Victoria Land (Antarctica): a key to the final pre-disintegration paleogeography of Panthalassian Gondwana. Polarforschung, Bremerhaven, Alfred Wegener Institute for Polar and Marine Research & German Society of Polar Research 80(2):100-110
MacAyeal D, Okal E, Aster R, Bassis J (2009) Seismic Observations of Glaciogenic Ocean Waves (Micro-Tsunamis) on Icebergs and Ice Shelves. J Glaciol 55(190):193-206. https://doi.org/10.3189/002214309788608679
Maus S et al (2009) EMAG2: A 2-arc min resolution Earth Magnetic Anomaly Grid compiled from satellite, airborne, and marine magnetic measurements. Geosyst. Geochem. Geophys. 10:Q08005. https://doi.org/10.1029/2009GC002471
McGinnis LD, Bowen RH, Erickson JM, Aldred BJ, Kreamer J (1985) L, East-West Antarctic boundary in McMurdo Sound. Tectonophysics 14:341-356
McKelvey BC, Webb PN (1962) Geological investigations in southern Victoria Land, Antarctica. Part 3: Geology of Wright Valley. NZ J Geol Geophys 5:143-162
McLeod IR (1960) An outline of the geology of the western portion of Australian Antarctic Territory, Union Geodesique et Geophysique Internationale, Monographie 5, p 75 (Antarctic Symposium, Buenos Aires, November 17-25, 1959)
McNamara DE, Buland RP (2004) Ambient noise levels in the continental United States. Bull Seismol Soc Am 94:1517-1527
Mishra OP, Neloy K, Sweta BD, Vikas K, Jagvir S, Singh VP, Ghatak M, Shashank S, Anurag T, Sasi KG, Ravikant M, Poorti G (2021) Glacial mass change induced earthquakes in the Himalayan region of South Asia and its bearing to understand Arctic glaciers dynamics: proxy of climate change, In: Understanding Present and Past Arctic Environments, pp 433–455. https://doi.org/10.1016/B978-0-12-822869-2.00025-6
Mishra OP, Zhao D (2003) Crack density, saturation rate and porosity at the 2001 Bhuj, India, earthquake hypocenter: a fluid-driven earthquake? Earth Planet Sci Lett 212(3–4):393–405. https://doi.org/10.1016/S0012-821X(03)00285-1
Mishra OP, Zhao D (2004) Seismic evidence for dehydration embrittlement of the subducting Pacific slab. Geo Res Lett 31(9). https://doi.org/10.1029/2004gl019489
Muller C, Eckstaller A (2003) Local Seismicity Detected by the Neumayer Seismological Network, Dronning Maud Land, Antarctica: Tectonic Earthquakes and Ice-Related Seismic Phenomena, IX International Symposium on Antarctic Earth Science Programme and Abstracts, Potsudam, p 236
Muir Wood R (1989) Extraordinary deglaciation reverse faulting in Northern Fennoscandia in, Earthquakes at North-Atlantic Passive Margins. In: Gregersen S, Bashamp PW (eds) Neotectonics Postglacial Rebound. Kluwer Acad. Pub., Dordrecht, Netherlands, pp 141-173
Murayama T, Masaki K, Masa-Yuki Y, Yoshiaki I, Takeshi M, Yoshihiro K (2015) Infrasound array observations in the Lutzow-Holm Bay region, East Antarctica. Polar Sci 9:35–50
Nanometrics (2005) Trillium 120P Seismometer User Guide. Nanometrics Inc., Kanata, Ontario

Nanometrics (2015) SQLX: Quality Assessment Tool. Nanometrics Inc., Kanata, Ontario

Nettles M, Ekstrom?m G. Glacial Earthquakes in Green- land, and Antarctica. Ann Rev Earth Planetary Sci 38:467-491. https://doi.org/10.1146/annurev-earth-040809-152414

Nettles M, Wallace TC, Beck SL (1999) The March 25, 1998, Antarctic Plate Earthquake. Geophys Res Lett 26(14):2097-2100. https://doi.org/10.1029/1999GL900387

Niazi M, Johnson LR (1992) Q in the inner core. Phys Earth Planet Inter 74(1-2):55-62. https://doi.org/10.1016/0031-9201(92)90067-6

Olaf E, Hofstede C, Diez A, Kristoffersen Y, Lambrecht A, Mayer C, Blenkner R, Hilmarsson S (2015) On-ice vibroseis and snow streamer systems for geoscientific research. Polar Sci 9(2015):51-65

Ohtaki T, Kaneshima S, Kanjo K (2012) Seismic structure near the inner core boundary in the south polar region. J Geophys Res Solid Earth 117(B3). https://doi.org/10.1029/2011jb008717

Ortiz R, Garcia A, Aparicio A, Branco I, Felpeto A, Del Rey R, Villegas MT, Ibanez JM, Morales J, Pez- zo Del E, Olmedillas JC, Astiz M, Vila J, Ramos M, Viramonte JG, Risso C, Caselli A (1986-1995) Monitoring of the Volcanic Activity of Deception Island, South Shetland Islands, Antarctica. In: Ricci CA (ed) The Antarctic Region: Geological Evolution and Processes. Terra Antarctica Publication, Siena, pp 1071-1076

Paul F (2002) Tectonics and landscape evolution of the Antarctic plate since the breakup of Gondwana, with an emphasis on the West Antarctic Rift System and the Transantarctic Mountains. Royal Society of New Zealand Bulletin 35(2002):453-469

Peterson J (1993) Observation and modeling of seismic background noise, U.S. Geol. Surv. Tech. Rept. 93-322:1-95

Rao SVR, Ramachandra L, Prem Kishore T, Chaitanya A, Akilan GS, Srinivas GB, Navinchander EC, Malaimani, Ravikumar N (2007) Twenty-First Indian Expedition to Antarctica, Scientific Report, 2007 Ministry of Earth Sciences, Technical Publication No. 19, pp 197-210 Studies on Seismotectonics and Geodynamical Processes between Antarctica and India

Reading AM (2002) Antarctic Seismicity and Neotectonics, In: Gamble JA et al (ed) Antarctica at the close of a millennium, vol 35. The Royal Society of New Zealand Bulletin, Wellington, pp 479-484.

Richter CF (1958) Elementary seismology. In: Freeman WH (ed). San Francisco, p 768

Robertson JD, Bentley CR, Clough JW, Greischer LL (1982) Sea-bottom topography and crustal structure below the Ross Ice Shelf, Antarctica. In: Craddock C (ed) Antarctic Geoscience. University of Wisconsin Press, Madison, pp 1083-1090

Robertson SD, Wiens DA, Shore PJ, Smith GP, Vera E (2002) Seismicity and Tectonics of the South Shetland Islands and Bransfield Strait from the SEPA Broadband Seismograph Deployment, In: Gamble JA et al. (ed) Antarctica at the close of a Millennium, vol 35. The Royal Society of New Zealand Bulletin, Wellington, pp 549-554

Robinson ES (1964) Geological structure of the Transantarctic Mountains and adjacent ice-covered areas, Antarctica, Ph.D. dissertation, Univ. of Wis., Madison, p 291

Robinson ES, Splettstoesser JF (1984) Structure of the Transantarctic Mountains determined from geophysical surveys. In: Turner MD, Splettstoesser JF (eds) Geology of the Central Transantarctic Mountains, Antarctic Res. Ser. vol 36. AGU, Washington, D.C., pp 119-162

Shimizu H, Hiramatsu Y, Kawasaki I (2015) Search for latitudinal variation of spectral peak frequencies of low-frequency eigenmodes excited by great earthquakes. Polar Sci 9(1):17-25. ISSN 1873-9652. https://doi.org/10.1016/j.polar.2014.07.002

Smithson SB (1972) Gravity interpretations in the Transantarctic Mountains near McMurdo Sound, Antarctica. Geol Soc Am Bull 83(3437-3442):1972

Soloviev DS (1959) The Lower Paleozoic schists of Oates Coast, East Antarctica, Union Geodesiqueet Geophysique Internationale, Monographie 5, pp 78-79, 1960(Antarctic Symposium, Buenos Aires, November 17-25, 1959)

Song X, Richards PG (1996) Seismological evidence for differential rotation of the Earth's inner core. Nat 382:221–224

Souriau A, Roudil P (1995) Attenuation in the uppermost inner core from broad-band GEOSCOPE PKP data. Geophys J Int 123(2):572-587. https://doi.org/10.1111/j.1365-246X.1995.tb06872.x
Stagg HMJ, Willcox JB (1992) A case for Australia-Antarctica separation in the Neocomian (ca. 125 Ma). Tectonophysics 210: 21-32
Stern TA, Brink UST (1989) Flexural uplift of the transantarctic mountains. J Geophys Res 94(10):315–330
Storchak DA, Kanao M, Delahaye E, Harris J (2015) Long-term accumulation and improvements in seismic event data for the polar regions by the International Seismological Centre. Polar Sci 9:5-16. https://doi.org/10.1016/j.polar.2014.08.002
Storey BC, Dalziel IWD, Garrett SW, Grunow AM, Pankhurst RJ, Vennum WR (1998) West Antarctica in Gondwanaland: crustal blocks, reconstruction, and breakup processes. Tectonophysics 155(381-390):1998
Storey BC (1996) Microplates and mantle plumes in Antarctica. Terra Antarct 3(91-102):1996
Storey BC, Vaughan APM, Millar IL (1996) Geodynamic evolution of the Antarctic Peninsula during Mesozoic times and it's bearing on Weddell sea history. In: Storey BC, King EC, Livermore RA (eds) Weddell Sea Tectonics and Gondwana breakup. Geological Society, Special Publication, London, pp 87-103
Stutzmann E, Schimmel M, Patau G, Maggi A (2009) Global climate imprint on seismic noise. Geochem, Geophys, Geosyst 10:Q11004. https://doi.org/10.1029/2009GC002619
Takaki I, Kanao M (2015) A quantitative evaluation of the annual variation in teleseismic detection capability at Syowa Station. Antarctica, Polar Science 9(2015):26-34
Takahiko M, Kanao M, Yamamoto M-Y, Ishihara Y, Matsushima T, Kakinami Y (2015) Infrasound array observations in the Lutzow-Holm Bay region. East Antarctica, Polar Sci 9(2015):35-50
Tanaka S, Hamaguchi H (1997) Degree one heterogeneity and hemispherical variation of anisotropy in the inner core from PKP(BC)-PKP(DF) times. J Geophys Res 102:2925-2938. https://doi.org/10.1029/96JB03187
Tom C (1995) The Exploitation of Antarctica's Natural Resources and the Evolution of the Antarctic Treaty System: An Overview Tom Cioppa, file:///C:/Users/SHRA/Downloads/bsb3-3_cioppa.pdf.
Tanimoto T (2007) Excitation of microseisms. Geophys Res Lett 34:L05308. https://doi.org/10.1029/2006GL029046
Torsvik TH, Gaina C, Refield TF (2008) Antarctica and global paleogeography: From Rodinia, through Gondwanaland and Pangea, to the birth of the Southern Ocean and the opening of gateways. In: Cooper AK, Barrett PJ, Stagg H, Storey B, Stump E, Wise W, Anderson J, Barron J, Bart P, Blankenship D, Davey F, Diggles M, Finn C, Fitzgerald P, Florindo F, Francis J, Futterer D, Gamble J, Goodge J, Hammer W, Helton P, Ivins E, Kyle P, LeMasurier W, Mayewski P, Naish T, Passchier S, Pekar S, Raymond C, Ricci CA, Studinger M, Sugden D, Thorn V, Wilson T (eds) Antarctica: A Keystone in a Changing World: Washington, D.C., USA, The National Academies Press Proceedings of the 10th International Symposium on Antarctic Earth Sciences, pp 125-140
Traer J, Gerstoft P, Bromirski PD, Shearer PM (2012) Microseisms and hum from ocean surface gravity waves. J Geophys Res 117:B11307. https://doi.org/10.1029/2012JB009550
Tsai VC, McNamara DE (2011) Quantifying the influence of sea ice on ocean microseism using observations from the Bering Sea. Alaska. Geophys. Res. Lett. 38:L22502. https://doi.org/10.1029/2011GL049791
Tseng TL, Huang BS, Chin BH (2001) Depth-dependent attenuation in the uppermost inner core from the Taiwan short period seismic array PKP data. Geophys Res Lett 28(3):459-462. https://doi.org/10.1029/2000GL012118
Tsuboi S, Kikuchi M, Yamanaka Y, Kanao M, March T, 25, (1998) Antarctic Earthquake: Great Earthquake Caused by Postglacial Rebound. Earth Planets Space 52(2000):133-136
Usui Y, Hiramatsu Y, Furumoto M, Kanao M, (2005) Thick and anisotropic D" layer beneath Antarctic Ocean. Geophys Res Lett 32:L13311. https://doi.org/10.1029/2005GL022622

Usui Y, Hiramatsu Y, Furumoto M, Kanao M (2008) Evidence of seismic anisotropy and a lower temperature condition in the D" layer beneath Pacific Antarctic ridge in the Antarctic Ocean. Phys Earth Planet Interior 167:205–216. https://doi.org/10.1016/j.pepi.2008.04.006

Veevers JJ, Powell CM, Rotts SR (1990) Review of seafloor spreading around Australia. I, Synthesis of the patterns of spreading. Australian J Earth Sci 38:373-389

Vila J, Ortiz R, Correig AM, Garcia A (1992) Seismic Activity on Deception Island. In: Yoshida Y et al (eds) Recent Progress in Antarctic Earth Science. Terra Science Publication, Tokyo, pp 449-456

Wannamaker P, Hill G, Stodt J, Maris V, Ogawa Y, Selway K, Boren G, Bertrand E, Uhlmann D, Ayling B, Green AM, Fucht D (2017) Uplift of the central Transantarctic Mountains. Nat Commun. https://doi.org/10.1038/s41467-017-01577-2

Warren G (1969) Sheet 14-Terra Nova Bay-McMurdo Sound area. Geologic map of Antarctica. Antarctic Map Folio Series, Folio 12-Geology. American Geographical Society

Watson T, Nyblade A, Wiens DA, Anandakrishnan S, Benoit M, Shore PJ, Voigt D, VanDecar J (2006)P and S velocity structure of the upper mantle beneath the Transantarctic Mountains, East Antarctic craton, and the Ross Sea from travel time tomography. Geochem Geophys Geosyst 7(7). https://doi.org/10.1029/2005GC001238

Weaver SD, Adams CJ, Pankhurst RJ, Gibson IL (1992) Granites of Edward VII Peninsula, Marie Byrd Land: orogenic magmatism related to Antarctic-New Zealand drifting. Geol Soc Am Spec Pap 272:281-290

White RS, MacKenzie D (1989) Magmatism at rift zones: the generation of volcanic continental margins and flood basalts. J Geophys Res 94:7685-7729

Wilson T, Bell R (2011) Earth structure and geodynamics at the poles. Understanding Earth's Polar Challenges: International Polar Year 2007-2008:273-292

Wilson TJ (1995) Cenozoic transtension along the Transantarctic Mountains-West Antarctic rift boundary, southern Victoria Land, Antarctica. Tectonics 14:531-545

Wilson TJ (1999) Cenozoic structural segmentation of the Transantarctic Mountains rift flank in southern Victoria Land. Global Planet Change 23:105-127

Yamamoto M-Y, Ishihara Y, Kanao M (2013) Infrasonic waves in Antarctica: a new proxy for monitoring polar environment Inter. J Geosci 4:797-802. https://doi.org/10.4236/ijg.2013.280 0498

Zhou C, WangZemina ED, SunJiabinga (2008) Remote sensing application In Antarctic Inland areas. In: The International Archives of the Photogrammetry, Remote Sensing and Spatial Information Sciences, vol. XXXVII, part B8. Beijing

Revealing the Contemporary Kinematics of Antarctic Plate Using GPS and GRACE Data

P. S. Sunil, Ajish P. Saji, K. Vijay Kumar, M. Ponraj, S. Amirtharaj, and Ajay Dhar

Abstract Revised horizontal and vertical plate velocities of the Antarctic continent in ITRF2008 and the impact of elastic and viscoelastic deformations over the continent due to Antarctic Ice Sheet (AIS) variations are simultaneously estimated using GPS and GRACE data for the period 2005–2015. The improved GPS time series and resulting horizontal and vertical velocities indicate that East Antarctica is subsiding significantly, whereas West Antarctica is experiencing uplift with transitional subsidence along the Trans-Antarctic Mountain ranges. According to the ongoing elastic deformation and AIS mass variations from GRACE data, the East Antarctic area is subsiding at a rate of 1 mm/yr. The elastically corrected or GRACE corrected vertical deformation also exposes the deformation patterns associated with the viscoelastic vertical deformation in terms of East Antarctica subsidence and West Antarctica upliftment. The GIA model values also agree well with elastically corrected vertical motions when validated with elastically uncorrected and corrected GPS vertical velocities. Hence we reveal that the outcome of the elastically corrected vertical deformation in the Antarctic region is very well connected to the long-term viscoelastic changes akin to AIS mass variations.

1 Introduction

Antarctica is almost completely encircled by divergent or conservative plate margins, occupying the main plate's unique structural setting (Hayes 1991). However, the lithospheric intra-plate movements taking place in Antarctica can be instigated by various factors. Long-term mass changes in the Antarctic Ice Sheet (AIS) since the Last Glacial Maximum (LGM) cause viscoelastic behavior due to Glacial Isostatic Adjustment (GIA) at first (Farrell 1972). Furthermore, the Earth's crust reacts more

P. S. Sunil (✉)
Department of Marine Geology and Geophysics, CUSAT-NCPOR Centre for Polar Sciences, School of Marine Sciences, Cochin University of Science and Technology, Kochi, India

P. S. Sunil · A. P. Saji · K. V. Kumar · M. Ponraj · S. Amirtharaj · A. Dhar
Indian Institute of Geomagnetism, Mumbai, India

N. Khare (ed.), *Assessing the Antarctic Environment from a Climate Change Perspective*, Earth and Environmental Sciences Library,
https://doi.org/10.1007/978-3-030-87078-2_18

elastically in response to the current mass changes of the AIS. Climate variables such as temperature and humidity control these behaviors associated with snow accumulation and ablation rates on the surface (Huybrechts 1994). The unloading of ice mass can root for slight convergence around the main discharge center in terms of horizontal motions. In contrast, in the case of viscosity, which is an elastic tendency that includes both the lithosphere and the underlying mantle, the divergence obtained is different (Bouin and Vigny 2000).

While conferring the intraplate deformation of the main Antarctic plate, West Antarctica moves independently of East Antarctica during the geological period. The evolution of the western Antarctic and its relationship to the eastern Antarctic have profound implications for the reconstruction of the Gondwana continent, as well as global-scale plate interactions, paleoclimate, and paleo-biogeography (Dalziel and Elliot 1982). Paleomagnetic data suggests that West Antarctica experienced a clockwise rotation of about 175–155 Ma and a counterclockwise rotation of about 155–130 Ma in comparison to East Antarctica (Grunow 1993). As a result, Antarctica can be divided into two structural domains: East Antarctica, which has a stable Precambrian shield, and West Antarctica, which has a more complicated assemblage of accreted terrain (Morelli and Danesi 2004). These two domains are separated by the Transantarctic Mountains (TAM), a 3500 km long-range with elevations up to 4500 m (Brink et al. 1997).

To study the lithospheric dynamics of the Antarctic plate, both, horizontal and vertical displacements of the West and East Antarctic plates must be taken into account. Over the last few decades, geodetic research has focused on calculating the current rate of relative plate motion and comparing it to the rate for millions of years (DeMets et al. 1990; Argus and Heflin 1995). Even though the Antarctic Plate moves/rotates very slowly (Denton et al. 1991; James and Ivins 1998), advances in space geodesy allow for extremely precise and accurate spatial observations in the Polar Regions, particularly in Antarctica, which aid in the study of crustal dynamics, post-glacial rebound, ice mass balance, and other topics (Argus and Peltier 2010; King et al. 2016). The displacement rates of geodetic monuments established over any part of the Earth's surface can be measured with a precision of less than 1 mm/yr using modern space geodetic techniques like very-long baseline interferometry (VLBI), Global Positioning System (GPS), Synthetic Aperture Radar Interferometry (InSAR) etc. and subsequent observations can be compared with predicted values from lithospheric deformation models of the polar regions (Dietrih et al. 2004; Ohzono et al. 2006). In addition, the expansion of Gravity Recovery and Climate Experiment (GRACE) satellite observations conjunction with GPS measurements, which can analyze temporal changes in the Earth's gravitational field and estimates changes in mass near the Earth's surface with incomparable precision, provide the current elastic response of the lithospheric plate with high resolution and accuracy (King et al. 2005; Bevis et al. 2009; Williams et al. 2014).

Contemporary deformation and mass loss of Antarctica is measured using geodetic observations such as the GPS and GRACE measurements, which provide constraints for the GIA model to meet (Dietrih et al. 2004; Ohzono et al. 2006;

Thomas et al. 2011; Williams et al. 2014). Over the Antarctic plate, several continually observing GPS stations are available, especially along the Antarctic coast to assess the elastic crustal deformation in response to the ongoing present-day AIS mass changes, long-term viscoelastic reaction due to GIA, and likely tectonic motion. However, the lack of a network of ground-based GPS observations still limits the vital tasks in the investigation of kinematics and deformation of the Antarctic plate. To address the spatial variation of the Antarctic plate's contemporary kinematics associated with the present-day ice mass variations, in the present study we use continuous GPS data from 27 permanent GPS stations distributed across the Antarctic continent established by different countries, including the 1 continuous observing GPS stations by India at Maitri (Schirmacher Oasis), and 12 International GNSS Service (IGS) stations (Fig. 1).

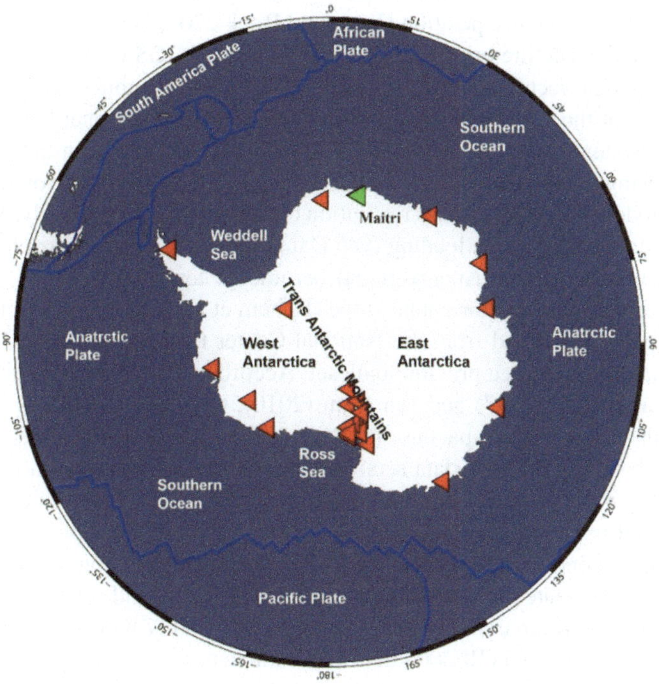

Fig. 1 Map shows the study region Antarctic continent in white color surrounded by the blue color Southern Ocean. The blue lines indicate the plate boundaries of the South Polar Region. The red triangles represent the GPS data used in the present study. The green triangle indicates the location of the Indian Antarctic research base station Maitri

2 Data and Analysis

2.1 GPS Network and Data

We have used continuous GPS data observed from a site (Maitri, established by Indian Institute of Geomagnetism) at the Indian Antarctic research base Maitri at Schirmacher Oasis, East Antarctica, 12 sites from the International GNSS Service (IGS) network, and 14 sites from other national networks during 2005–2015 (Fig. 1). In this study, the data from GPS stations that had been in operation for more than three years were analyzed to achieve accurate results (Blewitt and Lavalle 2002). The GAMIT/GLOBK 10.5 software from (Massachusetts Institute of Technology, USA) is used to obtain the time series of all GPS stations in ITRF2008 (Altamimi et al. 2011). The 24-h GPS dual-frequency and two-phase code observations are used to calculate the daily position (King and Bock 2005; Herring 2005). The first step is to use the satellite orbit parameters published by IGS to eliminate the ionospheric coupling effects through the ionosphere-free liner combinations. In addition, the influence of the Earth tide was corrected using the International Earth Rotation Reference System (IERS2010), and the influence of tidal loads on the ocean was corrected using the finite element solution model (FES2012). The use of sea-bottom pressure models with a time'space resolution of 6 h and 0.5°, respectively, eliminates the effect of non-tidal ocean loading (van Dam et al. 2012). The tropospheric delay correction was employed using a global barometric temperature (GPT) model to estimate zenith delay and horizontal slope (Bohem et al. 2007). The effect of atmospheric loading is derived from the National Center for Environmental Protection (NCEP) reanalysis surface pressure data set. According to the method described in Van Dam and Wahr (1987) and van Dam (2010), this dataset contains daily files of 4 periods with a degree spacing of 2.5° × 2.5°. Daily solutions were created by averaging the obtained GPS data sets over 6 h of atmospheric pressure to enable compatibility with the daily solution. The remaining GPS variant time series was then retrieved using two sets of daily solutions (GPS and atmospheric load solutions). In the second step, the computed loose constrained solutions are then passed to GLOBK to estimate the station position, velocity, orbit, and rotation parameters of the Earth using Kalman filtering. We used a set of quality IGS sites to implement the coordinate system in ITRF2008. It is important to note that when processing GPS data using GAMIT/GLOBK, we did not take into account the effects of hydrological or AIS loads; therefore, the surface deformation caused by this effect persists in the remaining GPS time series.

2.2 Common Mode Errors

Common Mode Errors (CME) exist in GPS time series and are one of the major causes of influencing the accuracy and reliability of location coordinates and the secular variation of GPS locations (Wdowinski et al. 1997; Nikolaidis 2002). The seasonal variations and antenna offsets are the prominent CMEs in the GPS signal. In general, the seasonal signal is modeled by the first and second harmonics of the sine functions, and the model is subtracted from the observed time series. Nikolaidis (2002), Herring (2003), and Tian (2011) proposed the following analytical function to obtain the secular trend, which represents the strain accumulation due to interseismic plate motion:

$$d(t) = c + vt + X\sin(\omega t + \varphi_1) + Y\sin(2\omega t + \varphi_2) \\ + a\left[\ln\left(1 + t/\tau_{\log}\right) \text{ or/and } \left(1 - e^{-(t/\tau \exp)}\right)\right] \quad (1)$$

where $d(t)$ is the displacement as a function of time t, v the linear velocity, c the coseismic offset, X the amplitude of the annual cycle, φ_1 the phase offset of the annual cycle, Y the amplitude of the semi-annual cycle, φ_2 the phase offset of the semi-annual cycle, $\omega = 2\pi/T$, the amplitude of postseismic decay, $\tau_{\log}$ the decay time corresponding to logarithmic decay, and $\tau_{\exp}$ is the decay time corresponding to exponential decay.

Equation (1) was used to achieve the improved time series and resulting horizontal and vertical velocities containing secular components of tectonic motion. Figure 2 and Fig. 3 show the resultant horizontal and vertical linear/secular motions at all GPS sites in ITRF2008 as calculated using the least-squares method suggested by Tian (2011). The sigmas of the coordinate estimates in the time series are used to calculate the parameter estimates and uncertainties with either white or flicker noise assumptions (Herring 2003). To separate the elastic deformation associated with the contemporary AIS mass changes in the Antarctic region, the GRACE-derived deformation associated with mass fluctuations has been removed from the GPS vertical components.

2.3 GRACE Data

The GRACE satellite's gravity mission was launched in March 2002 to detect mass changes caused by disturbances in the Earth's gravitational field caused by continental hydrology and ice melting (Tapley et al. 2004; Wahr et al. 1998). The errors tracked by GRACE include the total contributions of groundwater, soil water, surface water, snow, ice, and biomass. Based on GPS signals (time series), we used the monthly solution of the field harmonic product GRACE RL03 (Lemoine et al. 2013) of the French organization Groupe de Recherche in Space Geodesy (GRGS). According to Swenson et al. (2008), the center motion coefficient of 1 degree is determined by the Stokes coefficient. To generate outliers, we used the coefficients of the spherical harmonic model and converted the spherical harmonic model to a quality change according to EWH. The spherical harmonics are the 80th and 80th orders, corresponding to a spatial resolution (half-wave) of approximately 250 km

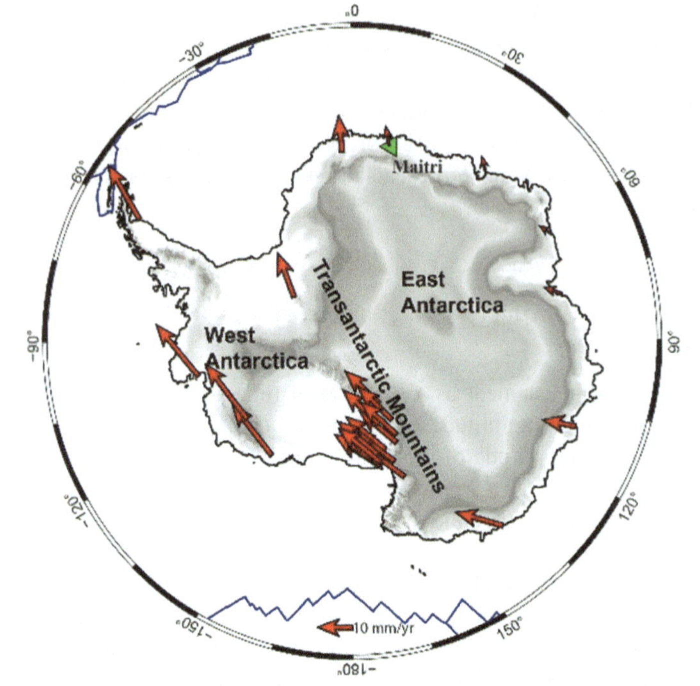

Fig. 2 Map portrays the GPS-derived Antarctic plate horizontal velocity in ITRF2008 after correcting the common-mode errors

(Wahr et al 1998).The inversion of spherical harmonic coefficients of gravity signals yielded vertical and horizontal surface deformations or displacements due to mass variations (Wahr et al., 1998). The horizontal and vertical elastic deformation was described by Farrell (1972) as follows:

$$\Delta H(\theta, \lambda, t) = R\sum_{l=1}^{\infty}\sum_{m=0}^{l} P_{lm}(cos\theta).[\Delta C_{lm}(t)\cos m\lambda + \Delta S_{lm}(t)\sin m\lambda].\frac{l_l}{1+k_l} \tag{2}$$

Davis et al. (2004) expressed the vertical elastic deformations as:

$$\Delta V(\theta, \lambda, t) = R\sum_{l=1}^{\infty}\sum_{m=0}^{l} P_{lm}(cos\theta).[\Delta C_{lm}(t)\cos m\lambda + \Delta S_{lm}(t)\sin m\lambda].\frac{h_l}{1+k_l} \tag{3}$$

where θ and λ are the latitude and longitude of the observed point at time *t*, *R* is the Earth Radius, *l* and *m* are the degree and order of the spherical harmonic model,

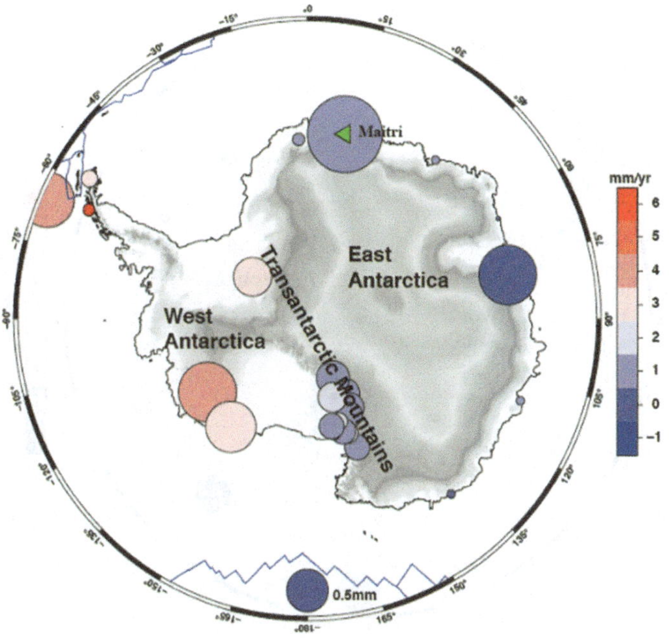

Fig. 3 Map illustrates the GPS-derived Antarctic vertical velocity in ITRF2008 after correcting the common-mode errors. The red shades indicate the uplifting region and blue shades represent the subsidence region. The size of the circles indicates the uncertainty of the GPS-derived vertical rate

P_{lm} is fully normalized Legendre functions, $\Delta C_{lm}(t)$ and $\Delta S_{lm}(t)$ are the spherical harmonic coefficient anomalies of GRACE data and h_l and k_l are elastic Load Love numbers (Pagiatakis 1990). The elastic deformation associated with present-day snow-mass variations was estimated for each GPS location in terms of the Earth-mass centre, as per Eq. (3). We used the Gaussian smoothing on a regional average with a radius of 250 km window and global forward modeling to eliminate correlated GRACE error and bias from GRACE mass changing (Chen et al. 2015). The vertical trend of elastic deformation estimated from the spherical harmonic coefficients of GRACE data over the Antarctic from January 2005 to January 2015 is represented in Fig. 4. The GRACE-derived mass variation-induced displacements in vertical directions are indicated as circles.

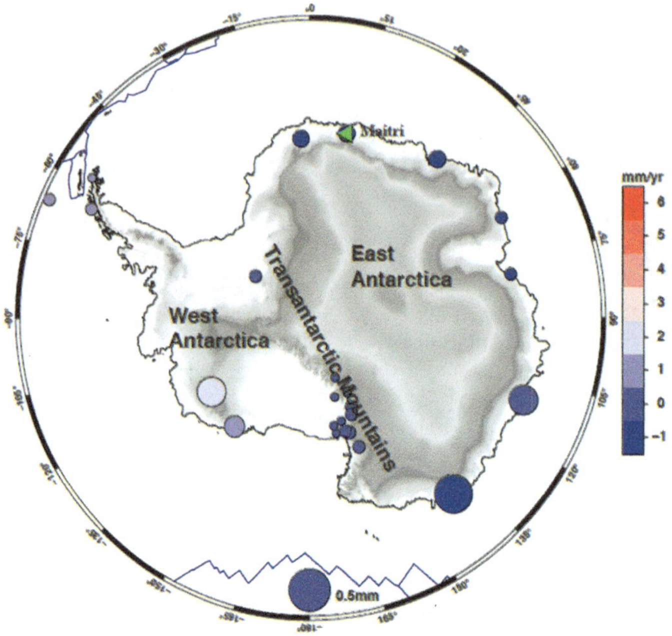

Fig. 4 Map shows the GRACE-derived vertical elastic deformation rate of Antarctica. The red shades indicate the uplifting region and blue shades represent the subsidence region. The size of the circles indicates the uncertainty of the GPS-derived vertical rate

3 Results and Discussion

3.1 GPS-Derived Crustal Motions

Deformations occur in Antarctica due to a combination of factors, including rigid platform rotations, internal tectonic movement, the viscoelastic glacial isostatic adjustment in response to previous ice mass changes, and elastic movement due to present ice mass fluctuations (Farrell 1972; Huybrechts 1994). GPS determined crustal deformation gives a unique proxy record of AIS mass changes due to the Earth's elastic and viscoelastic response to ice mass accumulations and ablation. The improved time series and resulting horizontal and vertical velocities containing secular components of tectonic motion were obtained using Eq. (1). Preliminary assessment of horizontal and vertical time series GPS and GRACE show that the seasonal variations are present in both horizontal and vertical components, however, the signal is more prominent in vertical displacement. Table 1, Figs. 2 and 3 represent the GPS-derived, both horizontal and vertical deformation pattern of Antarctica.

A tectonic plate motion, viscoelastic, and a local component are included in the measured GPS velocities. The local component is often substantially smaller than

Table 1 Station code, coordinates, improved velocity rates in ITRF2008

Sl. No	Sataion Code	Latitude	Longitude	North Velocity			East Velocity		
				Rate (mm/yr)	RMS_{raw} (mm)	RMS_{fit} (mm)	Rate (mm/yr)	RMS_{raw} (mm)	RMS_{fit} (mm)
1	BRIP	–75.79	158.46	–12.39	2.2	0.3	08.03	2.2	0.8
2	BURI	–79.14	155.89	–12.30	1.9	0.3	06.13	2.1	0.7
3	CAS1	–66.28	110.51	–09.64	2.9	0.2	01.6	3.0	0.5
4	COTE	–77.80	161.99	–12.06	2.1	0.4	08.19	1.9	0.5
5	CRAR	–77.84	166.66	–10.87	1.9	0.2	09.27	2.4	1.2
6	CRDI	–82.86	–053.19	11.11	7.8	6.4	07.12	3.0	0.9
7	DAVR	–68.57	077.97	–04.71	2.3	0.4	–02.87	2.1	0.4
8	DEVI	–81.47	161.97	–12.03	2.1	0.3	06.71	2.3	0.6
9	DUM1	–66.66	140.01	–11.70	3.8	0.4	08.14	4.7	0.8
10	E1G2	–77.53	167.14	–10.13	1.9	0.3	10.14	1.8	0.3
11	FIE0	–76.14	168.42	–11.30	2.4	0.4	10.16	2.2	0.4
12	FLM5	–77.53	160.27	–12.23	1.8	0.2	08.08	1.9	0.7
13	FTP4	–78.92	162.56	–11.98	1.8	0.3	08.03	2.0	0.7
14	HOOZ	–77.53	166.93	–11.40	2.4	0.3	10.80	3.1	0.6
15	IGGY	–83.30	156.25	–12.17	3.6	0.9	05.17	4.5	2.9
16	LWN0	–81.34	152.73	–12.43	2.0	0.6	04.47	2.0	0.6
17	MACZ	–77.53	167.24	–10.38	1.8	0.2	06.20	2.0	0.9
18	MAIT	–70.72	011.50	05.10	2.4	1.1	–02.00	2.7	1.3
19	MAW1	–67.60	062.87	–02.15	2.5	0.3	–03.69	2.4	0.3
20	MCAR	–76.32	–144.30	–04.92	4.7	2.1	17.46	6.2	3.5

(continued)

Table 1 (continued)

Sl. No	Sataion Code	Latitude	Longitude	North Velocity			East Velocity		
				Rate (mm/yr)	RMS_{raw} (mm)	RMS_{fit} (mm)	Rate (mm/yr)	RMS_{raw} (mm)	RMS_{fit} (mm)
21	MIN0	–78.65	167.16	–11.49	3.5	1.0	09.03	3.6	0.3
22	SCTB	–77.84	166.75	–11.81	2.2	0.3	09.29	2.0	0.4
23	SDLY	–77.13	–125.97	00.63	3.2	0.4	19.35	2.3	0.4
24	SYOG	–69.00	039.58	02.8	2.7	0.4	–03.87	2.7	0.3
25	TOMO	–75.80	–114.66	–11.42	4.3	1.8	05.18	6.5	1.8
26	VESL	–71.67	–002.84	10.25	2.1	0.2	–00.43	2.1	0.4
27	WHN0	–66.01	–060.55	–12.56	2.1	0.2	05.21	2.2	0.4

the plate rotation at intraplate station sites. Thus, in ice-covered regions like Antarctica, the signal emerging from viscoelastic GIA overrules the elastic local signal of deformation. The spatial view of the horizontal velocity map gives a clear picture of the clockwise rotation of the Antarctic continent (Fig. 2). As per the paleomagnetic record, this rotation of the Antarctic plate has been in motion since 175–155 Ma. However, during 155–130 Ma a counter-clockwise rotation was also been recorded (Grunow 1993). Consequences of these past clockwise and counter-clockwise rotations portioned Antarctica into two structural domains, the West and East Antarctica, and these two domains are separated by TAM (Brink et al. 1997; Morelli and Danesi 2004). It may be noted that the horizontal velocity map illustrates that West Antarctica rotates faster than East Antarctica, where the magnitude of the velocity vectors almost doubles compared to East Antarctica. Ghavri et al. (2017) suggested that the Antarctica plate is surrounded by both, convergent and divergence plate margins, hence the plate motions are large in West Antarctica. However, Zanutta et al. (2018) reported that the thickness of the Earth's crust differs between East and West Antarctica, and the intraplate relative velocities calculated from GNSS data demonstrate that movements between the two regions are insignificant.

In terms of GPS-derived vertical deformation (Fig. 3), East Antarctica exhibits significant subsidence while West Antarctica exhibits nearly equal levels of uplift. Nonetheless, the vertical deformation of GPS sites along the TAM indicates a level of subsidence intermediate to that of East Antarctica. It should be noted that the spatial pattern of these velocities does not match any GIA model (Bevis et al. 2009). As a result, the vertical rate estimated from the GPS time series alone cannot be explained by an elastic response to current ice loss. The study, however, supports the existence of discrepancies between the GIA-modeled and observed uplift rates, which could be attributed to deep-seated, regional-scale structures (Zanutta et al. 2018).

3.2 Elastic Deformation

For the precise estimation of the viscoelastic behavior due to GIA, the signal separation associated with elastic deformation is critical (King et al. 2012). Because the vertical displacement velocity obtained via GPS analysis is the sum of the GIA-induced viscoelastic and the elastic components. Thus proper correction of elastic deformation is required to extract the GIA-induced crustal deformation with accuracy. Figure 3 shows the GRACE-derived short-term elastic deformation caused by the current AIS mass variation at each GNSS site in this study using Eq. (3). However the horizontal elastic deformation derived using Eq. (2) was negligible, hence here we discuss the results in terms of vertical elastic deformation.

Estimation of the vertical rate of the Antarctic region has been carried out by different investigators. Thomas et al. (2011) and Argus et al. (2014) reported that following the breakaway of the Larsen B Ice Shelf in February 2002, elastic vertical crustal deformation rates in the Northern Antarctic Peninsula dramatically accelerated in response to the abrupt ice mass loss. Lutzow-Holm Bay, part of Drowning

Maud Land, has recently seen a significant increase in surface mass as measured by GRACE (Velicogna et al. 2020). Snowfall has increased in the region recently, as indicated by this mass increase. However, an assessment of GIA solutions based on GPS velocity filed by Li et al. (2019) suggested that GIA and AIS loading have little impact in East Antarctica, where vertical motion is negligible.

In the present study, we calculated the elastic deformation due to the current AIS mass variation at each GPS site using GRACE data. GRACE has recently observed a widespread gain in surface mass over the Antarctica region. The AIS-induced displacements determined from GRACE using Eq. 2 are shown as circles in Fig. 4. According to the estimated deformation, the East Antarctica region experiencing subsidence at a rate of ~1 mm/yr due to AIS associated elastic deformation.

3.3 Elastically-Corrected Deformation

Seasonal hydrological loading and unloading over continents elastically deform the Earth's surface (Farrell 1972). Seasonal signal contributions from AIS or hydrological mass variations affect the location time series and velocities of GPS stations, as recorded in Antarctica and elsewhere respectively (Blewitt and Lavallée 2002). In order to derive the viscoelastic deformation, we have subtracted the GRACE-derived vertical deformation rates from the GPS-derived counterparts assuming that the remaining vertical deformation in the Antarctic region is associated with the long-term viscoelastic deformation.

In terms of AIS adjusted GPS velocity vectors, Fig. 5 represents the secular rates of the vertical deformation of Antarctica. The GRACE corrected vertical deformation, exposes two unique deformation patterns of eastern side subsidence and western side upliftment. It may be noted that the maximum upliftment is taking place in West Antarctica and especially in the Antarctic Peninsula region. As reported by (Hattori et al. 2021), here, we believe that these corrections are the vertical motion of GIA due to changes in the viscoelastic AIS mass variations. Thus the results highlight the GIA-induced vertical trend of the Antarctic continent. It is also worth mentioning that the GIA's ice-sheet margin forecasts are heavily reliant on ice-sheet models. To further analyze the disparities in the GPS-derived and modeled GIA signals, additional GIA model estimates using alternative ice models are necessary.

3.4 Validation with GIA Model

A numerical simulation of global glacial isostatic adjustment is referred to as a GIA model. ICE-6G_C (VM5a) is a novel model (Argus et al. 2014) of the Late Quaternary ice age's last deglaciation episode (Fig. 6). Compared to the previous GIA models, ICE-6G_C (VM5a) model has been explicitly refined by applying all available GPS

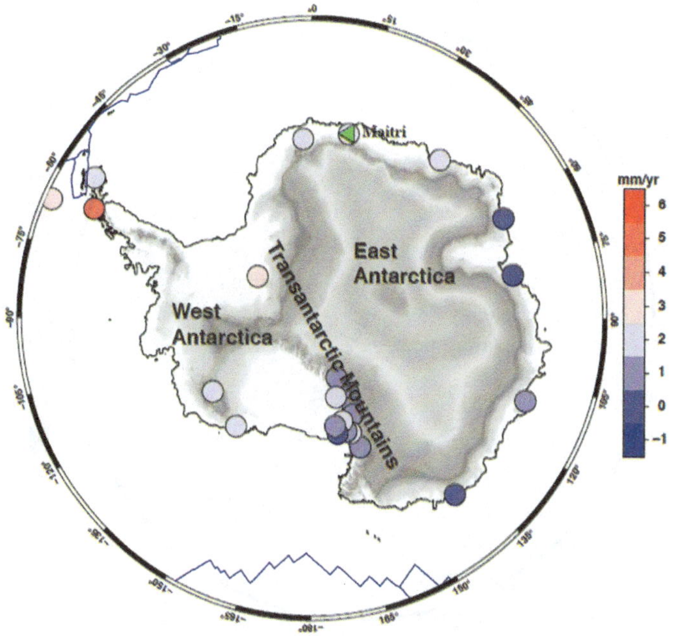

Fig. 5 Map represents the elastically corrected (i.e. GPS minus GRACE) vertical deformation rate of Antarctica. The red shades indicate the uplifting region and blue shades represent the subsidence region

measurements of vertical crustal motion that can be used to constrain the thickness of local ice cover and its removal timing.

In order to evaluate the correlation between the vertical velocities of the elastically uncorrected and corrected GPS values and the prediction of GIA model (ICE-6G_C [VM5a]), we present a linear fit between the vertical velocity of the GPS station and the predicted value of each GIA model in Fig. 7. It may be noted that the linear fit of the annual vertical amplitude of the GPS-GRACE (i.e. elastically corrected) values better fit with the GIA model (slope = 0.86; correlation coefficient = 0.83) compared to the fit of elastically uncorrected GPS velocities (slope = 0.78; correlation coefficient = 0.75). I.e. higher the correlation coefficient, the larger the degree of coincidence between the vertical velocities measured at GPS stations and the GIA model's predictions. This indicates that the elastically corrected (GPS–GRACE) site velocities are in best agreement with the GIA model velocities. The differences in the observations, and model can be ascribed to ice-load histories, different computation methodologies, and earth model parameters (Groh et al. 2012).

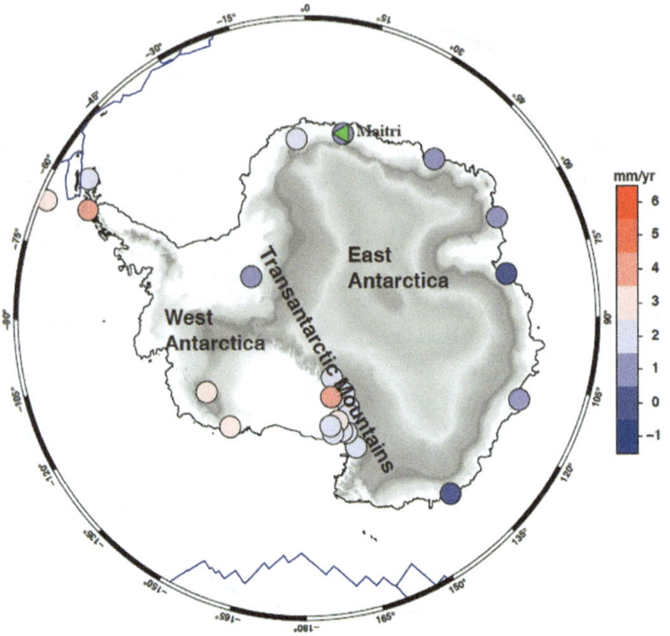

Fig. 6 Map illustrates the GIA model (ICE-6G_C [VM5a]) estimates of the vertical crustal motion values at the GPS locations used in the present study. The red shades indicate the uplifting region and blue shades represent the subsidence region

4 Conclusions

In regions like Antarctica, the short-term elastic response from current mass changes, the long-term viscoelastic impact from ice-sheet mass variations, and sometimes ongoing tectonic processes all work together to create horizontal and vertical crustal displacements. These short and long-term deformations can be captured very precisely with continuous GPS monitoring. The separation of these signals, on the other hand, is critical. In terms of Antarctic tectonic plate motion, the improved GPS time series and resulting horizontal and vertical velocities demonstrate that East Antarctica is subsiding significantly, whereas West Antarctica is experiencing almost equal uplift with transitional subsidence along the TAM regions. However, according to ongoing elastic deformation and current AIS mass fluctuations from GRACE data, the East Antarctic area is subsiding at a rate of 1 mm/yr. To derive long-term viscoelastic deformation GRACE-derived vertical elastic deformation rates were subtracted from GPS-derived equivalents. Further, the elastically adjusted or GRACE corrected vertical deformation highlights the deformation patterns in terms of East Antarctica subsidence and West Antarctica upliftment associated with the long-term viscoelastic vertical deformation. The validation of the GIA model with elastically

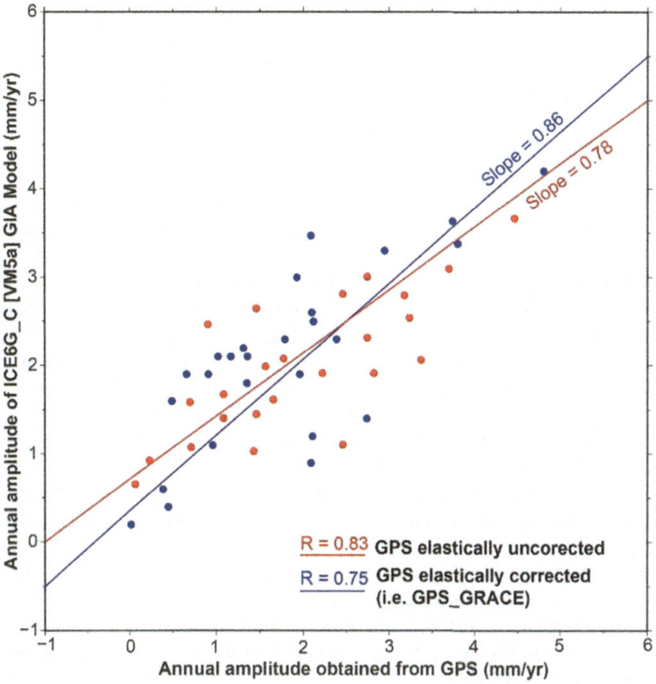

Fig. 7 Correlation plot between the annual amplitude estimated from the seasonal signals observed in GPS and elastically corrected GPS signals with the annual amplitude of the GIA model ICE6G_C [VM5a]. Red and blue lines indicate the estimated best fit line to the GPS elastically uncorrected and elastically corrected linear fit with slope and correlation coefficient (R) values

corrected and uncorrected GPS vertical velocities show that the GIA model values agree well with elastically corrected (GPS—GRACE) site velocities.

References

Altamimi Z, Collilieux X, Metivier L (2011) ITRF2008: An improved solution of the international terrestrial reference frame. J Geod 85(8):457–473. https://doi.org/10.1007/s00190-011-0444-4

Argus DF, Heftin MB (1995) Plate motion and crustal deformation estimated with geodetic data from the global positioning system. Geophys Res Lett 22:1973–1976

Argus DF, Peltier WR (2010) Constraining models of postglacial rebound using space geodesy: a detailed assessment of model ICE-5G (VM2) and its relatives. Geophys J Int 181:697–723

Argus DF, Peltier WR, Drummond R, Moore AW (2014) The Antarctica component of post-glacial rebound model ICE-6G_C (VM5a) based on GPS positioning, exposure age dating of ice thicknesses, and relative sea level histories. Geophys J Int 198:537–563

Behrendt J (1999) Crustal and lithospheric structure of the West Antarctic rift system from geophysical investigations: a review. Global Planet Change 23(1–4):25–44

Bevis M, Kendrick E, Smalley R Jr, Ian D et al (2009) Geodetic measurements of vertical crustal velocity in West Antarctica and the implications for ice mass balance. Geochem Geophys Geosys 10:Q10005. https://doi.org/10.1029/2009GC002642

Bohem J, Werl B, Schuh H (2006) Troposphere mapping functions for GPS and very long baseline interferometry from European Centre for Medium-Range weather forecasts operational analysis data. J Geophys Res - Solid Earth 111(B2):B2406

Blewitt,G, Lavallée D (2002) Effect of annual signals on geodetic velocity. J Geophys Res 107(B7:ETG 9–1–ETG 9–11). https://doi.org/10.1029/2001JB000570

Bouin M-N, Vigny C (2020) New constrains on Antarctic plate motion and deformation from GPS data. J Geophys Res 105-B12:28279–28293

DeMets C, Gordon RG, Argus D, Stein S (1990) Current plate motions. Geophys J Int 101:425-478

Denton G, Prentice ML, Burckle LH (1991) Cainozoic history of the Antarctic ice-sheet. In: Tingey RJ (ed) Geology of Antarctica. Oxford Univ Press, New York, pp 365–433

Dietrih R, Rülke A, Ihde J et al (2004) Plate kinematics and deformation status of the Antarctic Peninsula based on GPS. Glob Planet Ch 42:313–321

Farrell WE (1972) Deformation of the Earth by surface loads. Rev Geophys 10:761–797

Gharavi S, Catherine JK, Ambikapathy A, Kumar A, Gahalaut VK (2017) Antarctica Plate Motion. Proc Indian Natn Sci Acad 83(2):437–440

Groh A et al (2012) An investigation of glacial isostatic adjustment over the Amundsen Sea Sector, West Antarctica. Global Planet Change 98:45–53. https://doi.org/10.1016/j.gloplacha.2012.08.001

Grunow AM (1993) New paleomagnetic data from the Antarctic Peninsula and their tectonic implications. J Geophys Res. https://doi.org/10.1029/93JB01089

Hattori A, Aoyama Y, Okuno J, Doi K (2021) GNSS Observations of GIA-Induced Crustal Deformation in Lützow-Holm Bay, East Antarctica. Geophys Ress Lett 48:e2021GL093479. https://doi.org/10.1029/2021GL093479

Hayes DE (1991) Tectonics and age of the oceanic crust: circumAntarctic to 30^0S. In: Hayes DE (ed) Marine geological and geophysical atlas of the circum-Antarctic to 30^0S. American Geophysical Union, Washington, D.e, pp 47–56

Herring TA (2003) MATLAB Tools for viewing GPS velocities and time series. GPS Solutions 7(3):194–199. https://doi.org/10.1007/s10291-003-0068-0

Herring TA 2005) GLOBK, Global Kalman filter VLBI and GPS analysis program, Version 10.2, Report, Department of Earth, Atmospheric and Planetary Sciences, Massachussetts Institute of Technology.

Huybrechts P (1994) Formation and disintegration of the Antarctic ice sheet. Ann. Giaciol. 20:336–340

King MA, Penna NT, Clarke PJ (2005) Validation of ocean tide models around Antarctica using onshore GPS and gravity data. J Gophys Res 110:B08401. https://doi.org/10.1029/2004JB003390

King MA, Bingham RJ, Moore P, Whitehouse PL, Bentley MJ, Milne GA (2012) Lower satellite-gravimetry estimates of Antarctic sea-level contribution. Nature 491(7425):586

King MA, Whitehouse PL, van der Wal W (2016) Incomplete separability of Antarctic plate rotation from glacial isostatic adjustment deformation within geodetic observations. Geophy J Int 204:324–330

King RW, Bock Y (2005) Documentation of the GAMIT GPS Analysis Software, Massachusetts Institute of Technology.

Li W, Li F, Zhang S, et al. (2019) An assessment of GIA solutions based on high-precision GNSS velocity field for Antarctica. Solid Earth Discuss [preprint]. https://doi.org/10.5194/se-2019-101

Lemoine J-M, Bruinsma S, Gégout P, Biancale R, Bourgogle S (2013) Release 3 of the GRACE gravity solutions from CNES/CRGS. Geophys Res Abstracts. 15(EGU2013-11123):2013

Morelli A, Danesi S (2004) Seismological imaging of the Antarctic continental lithosphere: a review. Global and Plan Ch 42(1–4):155–165

Nikolaidis R (2002) Observation of geodetic and seismic deformation with the Global Positioning System. Ph.D. thesis Univ of Calif, San Diego San Diego

Ohzono M, Tabei T, Doi K, Shibuya K, Sagiya T (2006) Crustal movement of Antarctica and Syowa based on GPS measurements. Earth Planet Space 58:795–804

Swenson S, Chambers D, Wahr J (2008) Estimating geocenter variations from a combination of GRACE and ocean model output. J Geophys Res - Solid Earth 113:B8

Tapley BD, Bettadpur S, Ries JC, Thompson PF, Watkins MM (2004) GRACE measurements of mass variability in the Earth System. Science 305(5683):503–505. https://doi.org/10.1126/science.1099192

Ten Brink US, Hackney RI, Bannister S et al (1997) Uplift of the transantarctic mountains and the bedrock beneath the East Antarctic ice sheet. J Geophys Res 102(B12):27603–27621

Thomas ID, King MA, Bentley MJ, Whitehouse PL, Penna NT, Williams SDP, et al. (2011) Widespread low rates of Antarctic glacial isostatic adjustment revealed by GPS observations. Geophys Res Lett 38(22). L22302. https://doi.org/10.1029/2011GL049277

Thomas ID, King MA, Bentley MJ, Whitehouse PL et al (2011) Widespread low rates of Antarctic glacial isostatic adjustment revealed by GPS observations. Geophys Res Lett 38:L22302

Tian Y (2011) iGPS: IDL tool package for GPS position time series analysis. GPS Sol 15(3): 299–303. https://doi.org/10.1007/s10291-011-0219-7

Tregoning P, Ramillien G, McQueen H, Zwartz D (2009) Gacial isostatic adjustment and non stationary signals observed by GRACE. J Geophys Res 114:B06406. https://doi.org/10.1029/2008JB006161

van Dam T (2010) NCEP derived 6 hourly, global surface displacements at 2.5 × 2.5 degree spacing. [Available at http://geophy.uni.lu/ncep-loading.html]

van Dam T, Collilieux X, Wuite J, Altamimi Z, Ray J (2012) Nontidal ocean loading effects in GPS height time series. J Geodyn https://doi.org/10.1007/s00190-012-0564-5

van Dam TM, Wahr JM (1987) Displacements of the Earth's surface due to atmospheric loading: Effects on gravity and baseline measurements. J Geophys Res 92(B2):1281–1286. https://doi.org/10.1029/JB092iB02p01281

Velicogna I, Mohajerani Y, Landerer F, Mouginot J, Noel B, Rignot E, et al. (2020). Continuity of ice sheet mass loss in Greenland and Antarctica from the GRACE and GRACE follow-on missions. Geophys Res Lett *47*(8):e2020GL87291. https://doi.org/10.1029/2020GL087291

Wahr J, Molenaar M, Bryan F (1998) Time variability of the Earth's gravity field: Hydrological and oceanic effects and their possible detection using GRACE. J Geophys Res 103(B12):30205–30229. https://doi.org/10.1029/98JB02844

Wdowinski S, Bock Y, Zhang J, Fang P, Genrich J (1997) Southern California permanent GPS geodetic array: Spatial filtering of daily positions for estimating coseismic and postseismic displacements induced by the 1992 Landers earthquake. J Geophys Res 102(B8):18057–18070. https://doi.org/10.1029/97JB01378

Williams SDP, Moore P, King MA, Whitehouse PL (2014) Revisiting GRACE Antarctic ice mass trends and accelerations considering autocorrelation. Earth Planet Sci Lett 385:12–21

Zanutta A, Negusini M, Vittuari L et al (2018) New geodetic and gravimetric maps to infer geodynamics of antarctica with insights on Victoria Land. Remote Sens 10:1608. https://doi.org/10.3390/rs10101608

The manufacturer's authorised representative in the EU is Springer Nature Customer Service Centre GmbH, Europaplatz 3, 69115 Heidelberg, Germany. If you have any concerns regarding our products, please contact ProductSafety@springernature.com

Printed and bound by CPI Group (UK) Ltd, Croydon, CR0 4YY
30/09/2026
02225291-0013